D0148448

GEOLOGY AND ENGINEERING

GEOLOGY AND ENGINEERING

THIRD EDITION

Robert F. Legget
Consultant,
Ottawa, Canada

Allen W. Hatheway
Professor of Geological Engineering,
University of Missouri-Rolla,
Rolla, Missouri

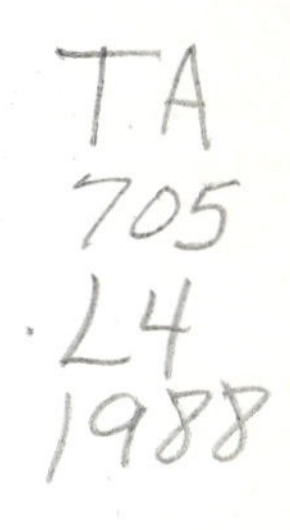

McGRAW-HILL BOOK COMPANY

New York St. Louis San Francisco Auckland Bogotá Caracas Colorado Springs
Hamburg Lisbon London Madrid Mexico Milan Montreal New Delhi Oklahoma City
Panama Paris San Juan São Paulo Singapore Sydney Tokyo Toronto

This book was set in Times Roman by the College Composition Unit
in cooperation with Waldman Graphics, Inc. The editor was B. J. Clark; the cover was designed by Joan O'Connor; the production supervisor was Denise L. Puryear. Project supervision was done by The Total Book. Arcata Graphics/Halliday was printer and binder.

Cover Credit

Geology and Engineering: their intimate association is shown here at the left abutment and emergency spillway of Williams Fork Dam, Grande County Colorado.

The insert shows the engineering geologic map of the migmatite coast rock, requiring engineering evaluation of its ability to anchor the dam and to pass the PMF overflow. Courtesy Denver Water Department.

GEOLOGY AND ENGINEERING

1 2 3 4 5 6 7 8 9 0 HAL HAL 8 9 3 2 1 0 9 8

ISBN 0-07-037063-X

Library of Congress Cataloging-in-Publication Data

Legget, Robert Ferguson.
Geology and engineering.

Includes index.
1. Engineering geology. I. Hatheway, Allen W.
II. Title.
TA705.L4 1988 624.1'5 87-15181
ISBN 0-07-037063-X (Text)
ISBN 0-07-037064-8 (Solutions Manual)

6-1-88

ABOUT THE AUTHORS

ROBERT F. LEGGET has over 50 years' experience in fieldwork, teaching, and research in civil engineering. After graduating from the University of Liverpool, he worked as a civil engineer in Scotland and Canada and taught at Queen's University (Kingston) and the University of Toronto. In 1947 he established the Division of Building Research for the National Research Council of Canada, serving as its director until his retirement in 1969. Mr. Legget is a former president of The Geological Society of America.

ALLEN W. HATHEWAY is a seventh generation Californian trained in geology (University of California at Los Angeles) and in geological and civil engineering (University of Arizona). Prior to joining the University of Missouri-Rolla faculty, he was with west coast and east coast geotechnical firms and taught (adjunct) at the University of Southern California and Boston University. He is past chairman of the Engineering Geology Division of The Geological Society of America and past president of the Association of Engineering Geologists. He is registered as a geologist, an engineering geologist, and as a civil and geological engineer.

CONTENTS

PREFACE

This volume is intended to provide engineering students with an introduction to the practice of civil engineering and, concurrently, to the geologic input that is essential for all civil engineering work. *Every* site utilized for civil engineering structures or works is unique; there will be differences between even adjoining sites, due to the underlying geologic structure and stratigraphy. Accordingly, planning and design of every civil engineering project must start with a study of site and local geology.

Throughout the book, a main and continuing theme is the vital importance of *observation,* during site investigations, throughout the entire construction process, and in maintenance and regular inspection of completed facilities. *Observation* is far more than just seeing. "You see" said Sherlock Holmes to his friend Dr. Watson, "but you do not observe." Many of the Holmes adventures, by Conan Doyle, could well serve as supplementary reading to this book.

In consequence, treatment of the subject is entirely descriptive. Mathematics is an essential tool in the design process, a procedure in civil engineering that is entirely dependent upon assumptions made as to ground conditions at the design site. This book deals with the accuracy, or otherwise, of those assumptions upon which the success and safety of all structures depend. Computers now occupy a unique place in data management and structural design, but all that they make possible so effectively is ultimately dependent upon the *actual* geological conditions at the site. It is with these site conditions that this book is concerned.

The volume is at once a third edition of the first author's book of the same name (first edition 1939, second in 1962) and a major abridgement of the *Handbook of Geology in Civil Engineering* (by Robert F. Legget and Paul F. Karrow, McGraw-Hill, 1983) made with the agreement of the Publishers and Dr. Karrow. Treatment of the subject is therefore by means of carefully selected and usefully illustrative case histories. Essential references to more detailed accounts of the more important histories discussed are given at the end of each chapter. For convenience in use, these reference lists have been kept very short and, in general, include only periodicals (such as *Engineering News-Record*), which are available in all University engineering libraries. Since all the case histories (and many more) are to be found in the *Handbook* noted, readers are directed to the much more detailed reference lists which it contains for those not given herein.

Metric units are used throughout the book (but always with Imperial units in parenthesis), in keeping with international practice. The authors, with great respect, must dissociate themselves from the spelling of *metre* as *meter* since the International Organisation for Standardisation (I.S.O.) has agreed that *metre* is the accepted spelling, and this is now in universal use except in the United States of America. Answers and questions for all chapters are contained in an accompanying booklet.

The authors are indebted to I. Noffke and M. Jacques for the expert typing of the final version of the text, and to all who so kindly assisted with the provision of illustrations for the *Handbook,* a selection from which now provides the illustrations in this text.

The authors venture to hope that as a by-product of the use of this book, some readers may come to appreciate that geology is indeed "the People's Science" (as was said over a century ago), a reminder that it can be studied by all and that even an elementary knowledge of the science can aid greatly in the appreciation of all scenery and especially of the beauties of the earth.

The proper use of geology as the starting point of all civil engineering achievement will become ever more important with the passing years and the gradual utilization of more favorable sites. Despite all the pressures that will come to bear on them through advancing technology and mounting demands, civil engineers should always remember the words of Francis Bacon, written as the modern world began to emerge:

NATURE, TO BE COMMANDED, MUST BE OBEYED.

Allen W. Hatheway

Robert F. Legget

ACKNOWLEDGMENTS

McGraw-Hill would like to thank the following reviewers for their useful comments: Jerry D. Higgins, Colorado School of Mines; Carol Simpson, The Johns Hopkins University; and James R. Sims, Rice University.

THE CIVIL ENGINEER AND GEOLOGY

Every branch of civil engineering deals in some way with the surface of the earth, since the works designed by the civil engineer are supported by or located in some part of the earth's crust. The practice of civil engineering includes the design of these works and the control and direction of their construction. *Geology* is the name given to that wide sphere of scientific inquiry which studies the composition and arrangement of the earth's crust. This book is concerned with the application of the results of this scientific study to the art and practice of the civil engineer.

At the start of the nineteenth century, before engineering had become the highly specialized practice it is today, many civil engineers were also active geologists. William Smith is the outstanding example of these pioneers. Robert Stephenson combined geological study with his early work in railway construction, and other well-known figures in the annals of engineering history were also distinguished in geology.

Today there is widespread recognition of the vital importance of the science of geology to those who practice the art of civil engineering. Geology is commonly included as a basic subject in courses of training for civil engineering (it should be included in *every* civil engineering course); civil engineering papers contain frequent references to the geological features of the sites of works described; and soil mechanics, the generally accepted scientific approach to soil studies, provides a common meeting ground for civil engineer and geologist and a means of fostering their cooperation.

In the past, geological considerations frequently have been featured prominently in the study and discussion of failures of civil engineering works. In fact, to some engineers, geology may still be thought of as merely a scientific aid to the correct determination of the reasons for some of the major troubles that develop during or subsequent to construction operations. Although the assistance rendered by geologists and by the study of geological features in such "postmortem" considerations is valuable, the very fact that geological features may have had something to do with these failures suggests with abundant clarity that the best time to consult a geologist or study geological features is *before* design and construction begin. In this way, the science can serve the art in a constructive rather than merely a pathological manner. It will later be seen, as applications of geology are considered in some detail, that this constructive service of the science can not only prevent possible future troubles, but it can also suggest new solutions to engineering problems and can often reveal information of utility and economic value, even in preliminary work.

The more obvious effects of geological features on major civil engineering works may be seen in the underground railway services in London and New York. In London, because the city is built on a great basin of unconsolidated material (including the well-known London clay), tube railways, located far below ground level and built in clay that was easily and economically excavated, have provided an admirable solution to one part of the city's transportation problems. In New York, on the other hand, the surface of Manhattan Island on which the city is located is underlain to a considerable extent by Manhattan schist. Underground railways had to be constructed in carefully excavated rock cuts just below surface level, as innocent visitors to that great city learn if they happen to stand on a ventilation grating when a train passes in the subway below. Many similar instances of the profound effect of local geological characteristics upon major civil engineering works could be cited, but all would serve to emphasize the same point: how closely the science and the art are related and how dependent civil engineering work generally must be upon geology.

The science of geology stands in relation to the art of the civil engineer in just the same way as do physics, chemistry, and mathematics. The importance of the latter sciences to the civil engineer is never questioned; they are always considered the necessary and inevitable background to civil engineering training. It would be inconceivable for any engineer worthy of the name to be unfamiliar with the chemistry of simple materials. Ignorance of the nature of the materials on which or in which civil engineers are to construct their work should be equally inconceivable. There is, however, an important point of difference to be noted in the relation of geology and parallel sciences to civil engineering. These parallel sciences are utilized directly by the engineer, since mathematical and physical methods are important in many branches of design work. Geology, however, renders a service less obviously direct, in that the findings of the pure science are applied to the specific problems of the engineer. In the case of construction problems, for example, it is the task of the engineering geologist to state the probable difficulties, and the task of the engineer to overcome them. In the case of materials of construction, it is the geologist who says where and how they may be found, and the engineer who obtains them and puts them to use.

This is a significant point of difference, and it leads to another matter of great importance. The geologist, who has the whole of the earth's surface for a laboratory, encounters in every locality purely local problems that will not be duplicated elsewhere. Thus, every application of the science or its methods to engineering work will be in some respect unique. In this sense, too, there is a difference between the relationships of civil engineering to geology and to associated sciences. Although local characteristics will vary, the fundamental geological principles applying to them do not. And these guiding principles constitute the basis of geological study, study that can be seen to be an essential part of the training of every civil engineer.

There is today increasing concern over the conservation of the natural environment, especially in relation to construction projects. Environmental impact statements are now often mandatory before construction can be started. The preparation of these important documents must start with a consideration of geology, since geological factors very largely determine the natural environment. Construction operations will inevitably disturb local geological formations and so affect the environment. With sound design, environmental conditions can be restored when construction is complete. In some cases, they are even improved from what they were before work started, yet this restoration must be assured before the necessary permits are granted. One of the many great contributions that geology can make to the practice of civil engineering is in providing a firm and understandable basis for formulating valid environmental impact statements.

TRAINING IN GEOLOGY

There is no special branch of geology that will provide all that is necessary for civil engineering students. Training in geology and not in just a single specialized course is most desirable, considering the way in which geologic principles have to be applied in civil engineering. Such emphasis would take the place of other engineering specialist studies. Some university curricula reflect this appreciation of geology.

FIGURE 1.1
The Roman bridge at Lavertezzo, Switzerland. (*Courtesy Swiss National Tourist Office, Toronto.*)

Geologic training for civil engineers must obviously be general; it must provide the student with a good grasp of principles and of the interrelation of various branches of geology. Attention has to be concentrated on the branches of geology which are of special importance in civil engineering practice—physical geology, structural geology, and petrology. Study of geologic maps and sections and examination of thin sections of rocks under the petrological microscope give special emphasis to lectures.

Field experience is of fundamental importance in all training. No course in geology for civil engineers can be regarded as complete without a reasonable period of time spent on geologic observation and survey. Local conditions will dictate how fieldwork will be arranged, but a continuous period of one or two weeks spent in a suitable locality will usually be more effective than any number of shorter periods fitted into a regular schedule. Although usually impossible to realize, a geologic and topographic survey camp would be an ideal combination.

The second part of geologic training for civil engineers should be a study of the geologic lessons learned in actual engineering practice. The average student will have had little experience with engineered construction. If she or he can start this after an introductory study of the applications of geology, a double purpose will be served to great advantage. Courses on foundation engineering and construction methods present excellent opportunities for geologic lessons.

Classroom instruction is only an introduction to what becomes a lifelong study for the majority of civil engineers; application of geology to their work should become an essential part of that basic experience that supports *engineering judgment,* the most prized possession of all members of the profession. This experience can be gained much more efficiently if the young engineer possesses a fundamental knowledge of geology. Fortunately, geology is coming to be widely recognized as an important aid to the civil engineer. Perhaps the best indication of this recognition is to be found in the almost invariable mention of site geology in published papers on civil engineering work, especially in such periodicals as *Engineering News-Record.* This is encouraging, since the presentation of any description of an engineering structure without some reference to foundation-bed conditions or to other corresponding geological data is equivalent to presenting a paper on a bridge without referring to the loading used in design or the materials used in construction.

PRACTICAL EXPERIENCE

Knowledge derived from wide experience is not so common as some people would suggest. It is almost intuitive by nature; certainly it is much more than merely factual knowledge derived from long observation. Some could not acquire such intuitive judgment even after a lifetime of outside experience; others acquire it easily and early in their lives. All so-called "practical experience" may not lead to sound intuitive judgment. Sometimes the "practical person" is the one who practices the theories of 30 years ago. In the past there have been many engineers, and there probably still are some, who have never troubled to find out what geology is, yet they know instinctively many of the absolutely critical geologic principles. The dividing line between what may truly be called *practical experience* and a cultivated appreciation of geological features is quite indeterminate.

Whether the civil engineer gains an appreciation of geology by training or by intuition, that appreciation can be of vital service in the field. Knowing, at least to some degree, what lies hidden beneath the surface of the ground, the engineer will be able to direct exploratory work more accurately. And in construction, every step taken in connection with excavation and foundation work will have a new and added significance for the resident engineer who is aware of all that geology can mean in the supervision of such work.

The important work of Dr. C. P. Berkey and his co-workers in connection with the Catskill Aqueduct for the water supply of New York City is a telling example of the constructive use of geology. The geologic investigations associated with the tunnels that now exist under the Mersey River at Liverpool, England, and particularly the work of Prof. P. G. H. Boswell in connection with the great vehicular tunnel, offer similar evidence. Hoover Dam on the Colorado River in the United States and many other successful dams testify silently to the value of the assistance that civil engineers have obtained from geological specialists, as also do bridges, large and small, around the world. The foundations of the great bridges of New York and San Francisco are especially notable testimonials. Comments on these few examples of the geologist's contribution to engineering could be extended to fill many pages, but other examples of cooperative work quoted throughout this book will be even more impressive.

Usually, engineering geologists are called upon to assist in engineering work in the capacity of consultants; the practice of having a geologist as a member of the official board of consultants is a fine North American tradition. During the years immediately prior to World War II, a few major engineering organizations added geological units to their permanent staffs. Experience during the war years, in both civilian and military engineering work, acted as the catalyst so that today one finds that almost all large civil engineering organizations have their own engineering geological staffs. Notable examples in North America are the U.S. Army Corps of Engineers, the Tennessee Valley Authority, the U.S. Bureau of Reclamation, the U.S. Geological Survey (one of the pioneers), and (today) most large engineering firms. On smaller works, numerous consulting geologists give similar useful service. Geological surveys in all countries are willing to assist when staff members are available and conditions appropriate. Geology departments of universities are often willing and able to assist through the service of specialist staff members. This gives reciprocal benefits, since such work provides experience of great value in teaching.

GEOLOGISTS AND CIVIL ENGINEERING WORK

This book is not intended primarily for the use of geologists, but it may be useful for engineers to note that most geologists enthusiastically welcome the opportunity to cooperate on civil engineering work; their only regret is that the opportunity does not occur more frequently. This suggestion is confirmed by the contributions made by geologists to the discussion of engineering papers, contributions such as the statement made many years ago by Dr. T. Robinson of the Geological Survey of Great Britain during discussion of a paper at the Institution of Civil Engineers, London:

> The records of the Geological Survey showed conclusively that closer cooperation between the geologist and the engineer would be greatly to the advantage of both, and it was a pity that there was no very direct way in which geologists could be kept informed of the progress of important excavations.

Practically the only way geologists can learn of new exposures made by civil engineering operations is through the engineer in charge of the work. This is yet another reason why every civil engineer must have a basic training in geology—not to be an expert in the field but to appreciate the vital significance of geology in all civil engineering operations, and to know when expert geological advice is needed.

When geologists are called in to advise in civil engineering work, they will act in conjunction with the engineers responsible for the work. Thus, the need

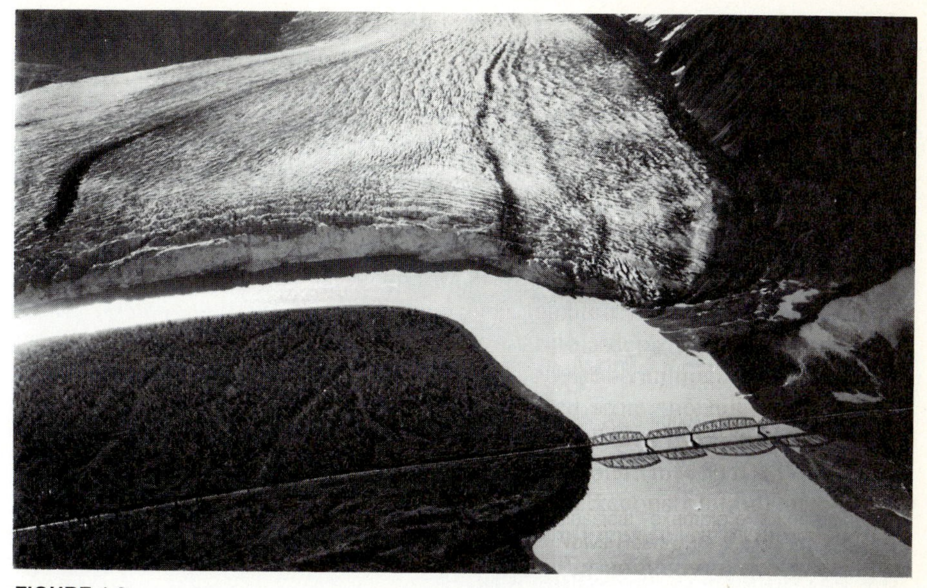

FIGURE 1.2
The Copper River Bridge, Alaska Railroad. (*Courtesy Boston Museum of Natural Science; photo by Bradford Washburn.*)

for cooperation arises between the civil engineer and the engineering geologists, the practical builder and the scientist. Their cooperation may lead to a valuable partnership, and it often is the source of considerable personal pleasure. Their partnership is, in some ways, a union of opposites, for the approach of the two to the same problem is psychologically different. The geologist analyzes conditions as they are; the engineer considers how existing conditions can be changed so that they will suit a specific plan. The geologist draws on his or her analysis to cite problems that exist and suggests troubles that may arise; the engineer has to solve the problems and overcome the troubles. The final responsibility for decisions concerning the project must always rest with the engineer; but in coming to such decisions, the engineer will be guided by and will probably rely upon the factual information provided by the geologist. It is not without reason that it has been suggested that, although mining geologists should be optimists, all engineering geologists should be confirmed pessimists.

This joint work calls for a fine degree of real cooperation. The engineering geologist has to remember that what the engineer wants is a clear picture, presented as concisely as possible, of the geological conditions related to the work, with a view to their practical utilization. On the other hand, the engineer must remember that the geologist is a geologist, not an engineer, and cannot be expected to deliver the kind of report that would come from another engineer. In many cases, the most effective results can be achieved if the engineer is able to give to the geologist at the outset a list of specific questions to be answered. The questions may relate to geological conditions, to the necessity for and the location of further exploratory works, such as boreholes and exploratory pits, or similar matters. The engineer should be willing to give the geologist the liberty of pursuing within reasonable limits any aspects of purely scientific interest that may develop in the course of the main task. The geologist, on the other hand, should be willing to explain to the engineer what eventual benefits may accrue from such attention.

Limited numbers of engineers, especially in North America, have been trained in recent years also as geologists. These *geological engineers,* graduating from universities in both the United States and Canada, can assess geologic data and define the limits of geological conditions at construction sites, determine engineering properties and characteristics of site materials for design, and furnish geological designs, especially in rock and in dealing with groundwater.

THE PATTERN OF CIVIL ENGINEERING

When a new project comes up for consideration, the civil engineer will initially require some firsthand knowledge of the site area. Site investigations then become the basis for complete contract drawings and specifications. When financial arrangements are made, tenders (bids) will be invited, a contract awarded, and construction begun. When the project is complete, it must be periodically inspected for safety and function and maintained in good order. Brief comments on the application of geology to each of these main divisions of civil engineering procedure may be helpful. The civil engineer generally visits the site area, even if only for a hurried tour of inspection, and studies descriptive literature on the district. If geologic reports are included in the literature, and if the topography of the area is studied with due regard to the significance of the local geology, then the engineer will get a more valid appreciation of design and construction conditions than if the geology were neglected. To a civil engineer trained in geology, elementary study of site-area geology will demonstrate the general structure of the ground being considered and the origin of leading features of the local topography that will be of importance in the work.

Preliminary studies may be made in more detail if civil engineering work is to cover a large area. General reconnaissance surveys will probably have to be made, and general topographic maps either prepared or checked. Simultaneously with this work, geological reconnaissance can always be carried out with advantage. The local geology can be studied in more detail and correlated, although still in a general way, with engineering requirements. It is not often that civil engineers will be called upon to undertake extensive work of this nature. When the need does arise, special organizations are usually recruited to undertake the work. An hour's overflight with a local air charter service should always be considered.

Having obtained a general idea of the district in which the work is to be done, the civil engineer will next proceed with preliminary plans and estimates. Gradually these will be evaluated and discussed, until finally an accepted scheme or design is evolved which can then be prepared in detail. All this work can be carried out only if the engineer possesses an adequate and detailed knowledge of the ground in which the work is to be located and of the natural materials available at or near the site. This essential information will be obtained by means of detailed geologic fieldwork and by exploratory investigations through such means as boreholes and exploratory pits.

The final design of the civil engineer will usually be incorporated into a set of contract plans and specifications. On the basis of these plans bids for performing the work involved will be called for from contractors. In those projects which are carried out by direct administration (on the "force account" of the owner's own employees) instead of by contract, a complete set of drawings equivalent to a set of contract plans will still be necessary, and the equivalent of a contract specification will be needed for the guidance of the engineers in charge of construction operations.

The preparation of these documents marks a definite change in the engineer's work and responsibilities. When issued to a successful bidder and made the basis of a formal contract, they become legal documents entitling the contractor to certain rights, and taking control of construction operations, to some extent, out of the hands of the owner and the owner's representative, the resident engineer. If, therefore, the application of geology can in any way assist in making the preparation of contract documents more effective and less open to question, the science will be rendering one of its most valuable services.

The chief aim of site exploratory work and associated geologic studies is to provide accurate information about subsurface conditions at the site of the proposed work and about the availability of suitable construction materials. Subsurface conditions affect, and often control, the engineer's design, and also the construction methods adopted by the contractor. Availability of materials often influences design, especially from the economic angle, and will have an appreciable effect on construction planning. The engineer must therefore include in the contract documents as much information as possible regarding the site and the available materials.

On contract plans the engineer should give full details of the logs of boreholes, exploratory pits, and other subsurface explorations. These should be given not only in section but also in a general *plan of explorations* that shows their location with respect to the outline of the work to be constructed. Only with these records on hand can the contractor, if so minded, take advantage of all the information that the engineer has available relative to the work. In many cases, and when possible, it will be advisable to show the nature of the geologic structure adjacent to foundations, instead of the usual diagrammatic representation of rock or unconsolidated material. In certain special cases such as tunnel work it will be desirable to show on the profile drawings the geological formations and major geologic structural elements anticipated along the main axes of the work to be undertaken.

The same criteria should pertain in the preparation of civil engineering specifications. As a general rule, the opportunity to use geologic information in specifications will arise in one or more of four ways:

1 In the provision for possible alterations in design due to variations encountered in subsurface conditions
2 In the provision of information relating to available materials of construction
3 In the clauses relating to choice of methods of construction
4 In reference to the measurement and payment for excavation

The first of these is closely related to site investigation work. The second calls for the clearest and fullest explanation of all the facts known about the properties, characteristics, and location of important matter that is generally and advisedly left to the selection of the contractor, usually with a qualification clause such as: "The contractors are to submit to the engineers a statement with drawings showing how they propose to carry out the works, but any approval of the engineers is not to relieve the contractors of any liability that devolves upon them under this contract."

This provision, as indeed the whole essence of a civil engineering contract, makes the engineer morally responsible for giving the contractor *all* available information since this may assist the contractor in construction planning and methods. At the same time, it leaves contractors free to apply their own accumulated experience and special skills most efficiently in carrying out the contract works. Geological information thus included in the specification, in addition to what is shown on the contract drawings, will further assist the contractor and enable the engineer to fulfill his or her obligations.

The fourth instance, the relating of geologic information to the excavation effort, is of such importance that it will be considered in some detail. Contracts for civil engineering projects are formulated in such a way that (1) the work being carried out will be to the satisfaction of the owner, (2) will fulfill the requirements of the engineer, and (3) will provide appropriate safeguards for the contractor. *General conditions* define the scope of the contract; *specifications* and *contract drawings* detail the design of the engineer; and, in the usual contract, *quantities* and *unit prices* define the extent of the contractor's operation. Great care should be exercised in preparing these contract documents in order to avoid disputes, but success is not always achieved, as the record of court cases regarding engineering contracts makes clear. It is probably safe to say that no one feature of civil engineering contracts has been responsible for more disputes than the engineer's classification of material to be excavated as "earth" or "rock" and the consequent cost of this part of the

FIGURE 1.3
The Quintette Tunnels on the (former) Kettle Valley Railroad, British Columbia. (*Courtesy Public Archives of Canada, No. PA 501,046.*)

work performed. As unit prices for rock excavation may be ten or twelve times as much as the corresponding unit prices for earth excavation, the possibility of disputes arising from questionable classification of material involved will be obvious.

The excavation of solid rock which must be drilled and blasted is rarely questioned in this connection, nor is the removal of loose sand and gravel or soft clay. But in between these two extremes there may be materials that cannot easily be classified as either "earth" or "rock" unless a reference basis is adopted *before* contract operations begin. Such doubtful material is often termed *hardpan*. This term should *always* be avoided by engineers, if they wish to avoid trouble. It is essentially a popular term and is sometimes applied to special local gravel deposits whose unusual hardness has been caused by partial cementing of the rock fragments. Hardpan has been the subject of many lawsuits, mainly because the best definition one can give to it is that it is material that proved harder to excavate than the contractor had anticipated! It is a name not generally included in geologic nomenclature, and there are no satisfactory definitions of it in engineering literature. Glacial *basal till* (*lodgement till* or *boulder clay*) is often described as hardpan, but in this case as in practically all others that material can be more accurately described as compact gravel, sand, and clay with boulders. If material of this kind has to be removed during construction operations, possible methods to be used for its excavation should be indicated and accompany its description in contract documents.

The designations used for describing material to be excavated should be as few and as simple as possible, e.g.:

Hard-rock excavation should designate excavation of crystalline rock (granite or hard sedimentary rock such as limestone) which has to be drilled and blasted.

Loose-rock or weak-rock excavation should designate excavation of blocky limestone which does not require blasting but which cannot be removed by hand shovels or picks.

Soft-rock and earth excavation should designate excavation of disintegrated crystalline rock, weathered rock (i.e., saprolite), clay, sand, and gravel material which can be handled with shovels and similar tools.

It may finally be noted that throughout the rest of this book, the term *soil* will refer to unconsolidated natural materials, *weak rock* will describe rock of low compressive strength, and *rock* will refer to solid bedrock, either in place or excavated.

CONSTRUCTION OPERATIONS

Every cubic meter* removed during excavations, every unusual loading applied to a natural foundation bed, every pile driven into the ground—in fact, every construction operation in which the existing condition of the earth's crust is affected—is associated with geological features of some kind. Preliminary investigations of the relevant geology should therefore be of considerable value not only to the resident engineer on construction work but also to the contractor who is undertaking the work. Throughout this book, most of the examples mentioned will confirm in one way or another the vital importance to contractors of advance geological information.

The geologic information available at the beginning of a job can be fully effective, however, only if it is constantly compared with actual geologic conditions as they are revealed during construction. It is essential, therefore, in all civil engineering construction work that a regular and constant watch be kept on geological conditions as they are revealed, and that, in addition to the usual construction progress records, an adequate and complete record of leading geological features be kept. This can easily be done if at least one of the engineers on the job has been trained in geological fieldwork. In some cases, geological training has been made a prerequisite for appointment to a resident engineer's staff.

*The authors, with great respect, must dissociate themselves from this spelling; the International Organization for Standardization (ISO) has agreed on *metre, and this spelling is now in general use except in the United States of America.*

Geologic information obtained as construction progresses has a threefold practical value. First, such information acts as a check on the assumptions made with regard to geologic conditions, so that, in the preparation of the final design for the work to be constructed, any variation from these assumed conditions can be incorporated into the design before it is too late. Second, the revelation of the actual geology of the working site enables the contractor to check the suitability of construction plans and equipment. Third, if the geologic record is kept in a satisfactory manner, it may prove of inestimable value at some future time if further work has to be carried out at the same location, or if there are contract litigations.

A significant geologic confirmation was made in connection with cement grouting work carried out in parts of the famous Severn Tunnel in the southwest of England. This tunnel, which carries a main line of what used to be the Great Western Railway (now the Western Region of British Railways) under the river Severn, was started in 1873 and completed only against tremendous difficulties. The contractor, T. A. Walker, kept a complete geologic record throughout the construction of the tunnel, which pierces the Trias and Coal Measure strata. The rocks encountered varied from marl and shale to limestone and sandstone. The existence of this valuable and complete record assisted the engineering authorities of the railway company almost 60 years later, as a basis for an extensive grouting program.

On the other hand, far too many examples could be given of the trouble and expense caused by lack of comparable records for other works. A well-known tunnel in the United States, for example, had to be resurveyed within seven years of its completion in connection with the installation of a concrete lining, since no records were then available of the timber sets used, the final tunnel cross sections, or of the geologic strata encountered.

MAINTENANCE

All civil engineering works have to be regularly inspected and maintained in good condition. This careful but routine work must include not only man-made structures but also the adjacent ground. Inspections of bridge piers, to check against scouring, and of dams, to check against erosion of foundation-bed material caused by leaks, are two of the more important aspects of maintenance work in which geologic features may be of special significance; reference will be made to them in later chapters. All engineered structures and works must be regularly inspected, with careful attention given to geologic features.

CONCLUSION

Geology has a vital part to play in all engineering operations which interact in any way with the ground, and thus the science is critical to the entire field of civil engineering. Most of the chapters in this volume are arranged to illustrate this interconnection—bridges, dams, and water-supply and marine works to

FIGURE 1.4
The Royal Gorge suspension bridge over the Arkansas River, Colorado. (*Courtesy The Royal Gorge Company, Cañon City, Colorado.*)

name a few. For ease of reference each chapter has been made as self-contained as possible. Appendixes provide a short glossary of geologic terms and guides to sources of further information.

It will be noted, possibly with some surprise, that the treatment of the subject is descriptive; no mathematics is introduced into the discussion. This is a natural course to follow and is one deliberately chosen. In their scientific work the authors are guided by Lord Kelvin's famous and oft-quoted dictum to the effect that "when you can measure what you are speaking about, and express it in numbers, you know something about it." But civil engineering is essentially an art, certainly in all its practical aspects, and its successful practice depends in large measure upon the exercise of sound judgment. Sound judgment in turn comes from long experience based on acute observations. Of nothing is this more true than in the study of the natural ground conditions upon which, or in which, civil engineering works have to be constructed. It is here that the judgment of the engineer, aided immeasurably by the skilled observations and studies of the engineering geologist, can make such great contributions. Such judgment is called for in solving difficult construction problems that must be solved in the field and without any delay; in selecting a final site from among various alternatives, with great economic issues dependent upon the correct decisions; or in determining when excavation of a deep foundation may safely be stopped despite possible imperfections in the material to be used as a foundation bed for a large structure. Mathematics is of no direct assistance on occasions such as these, vital as its background role in relation to designs and computations may be. Here there is no substitute for good judgment and sound observation.

Field geology is based entirely upon such powers of observation, coupled with the ability to deduce the presence and orientation of three-dimensional geologic structures from surface features. The civil engineer is trained to think in terms of three dimensions in relation to structural design; equally important, if not more so, is the corresponding ability to visualize the spatial character of subsurface geologic conditions. The geologist visualizes these conditions automatically, frequently relying on *block diagrams* as a means of portrayal. It may not be easy, but it is possible for the civil engineer to gain such ability quickly. It is an ability that is of unusual service in field engineering work. Study of simple block diagrams is a first step. Books with colored block diagrams can be of even greater assistance. But it is in the field that the art of three-dimensional appreciation of ground structure can most effectively be gained.

Equally important is a full appreciation of the importance of water in rock and soil. Most engineers will have occasion to observe the remarkably diverse effects of water on many solid materials and especially on soils. Dry soils cause little trouble; wet soils can cause havoc on a job. So it is in many geologic formations. Therefore, the civil engineer must not only learn to visualize subsurface conditions in spatial terms but must always realize the added complications that the presence of varying amounts of groundwater may cause. That groundwater may vary in just the same way as surface water should become a commonplace in an engineer's thinking. And the relation of both to the weather must always be remembered, especially if field studies are carried out in fine, dry weather. One of the basic tools of the engineer in the study of any construction site should be a simple summary of local weather conditions.

These concepts are basic to a full appreciation of the role that geology can and must play in the design and construction of civil engineering works. The concepts are relatively straightforward and simple, and their application calls for no unusual skills such as may be possessed by only the gifted few. The ability to "look below the surface," to realize that surface conditions may be vastly different from those only a few feet below ground level, to keep in mind the ever-present possibility of groundwater movement, to visualize the significance of unusual topographic features, and to be able to relate these to geologic processes of the past—all these are characteristics of observation in which every civil engineer can be trained, even to a large extent self-trained.

Such observational skills can be exercised not only on the construction job but in most trips to add greatly to the pleasure of travel and to the enjoyment of scenery. In this book, however, they are shown specifically as being critical to the successful completion of civil engineering work in all its variety and with all its continuing challenge. The works to be described are carried out by human beings for the benefit of their fellows. It may well be remembered, therefore, that behind all impersonal discussions of scientific applications and engineering endeavors lie relationships between people and the cooperation of many of them working toward a common goal in rendering useful service to their community. This book would not be complete without at least a passing reference to this human background.

FURTHER READING

It is hoped that all users of this book will examine, when occasion permits, one of the most remarkable publications ever issued by the United States Geological Survey, its *Professional Paper,* No. 950. First published in 1979 and subsequently reprinted, the *Paper* is entitled "Nature to be Commanded...(Earth Science Maps Applied to Land and Water Management)," by G. D. Robinson and A. M. Spieker. It is unusually large in page size (40.5 by 30.5 cm, or 16 by 12 in) to accommodate the brilliantly colored maps and the 150 photographs illustrating the six case histories which are presented, all testimony to the main thesis of this volume.

CHAPTER 2

ROCKS AND SOILS

Earth was the English word used to translate the one Greek word used by Aristotle to describe all the material making up the crust of the earth. The Romans differentiated between that part of the crust which was solid—*petra* (rock)—and that which was fragmented—*solum* (soil or ground). With the development of modern scientific studies, a clear distinction was made by John

Evelyn in 1675 between rock and soil. This distinction was in universal use in the English language and recognized by geologists and engineers alike until the early years of the twentieth century. (*Topsoil* was an additional term, used to denote the upper few inches of soil in which organic growth takes place.) William Smith, for example, used both words in his descriptions in his famous *Map of the Strata* as early as 1815. At the end of the century, G. P. Merrill, in his famous *Rocks, Rock Weathering and Soils* published in 1897 uses the two important words in the accepted manner, both in his title and in his text.

Some geologists have come to adopt the restrictive pedological use of the word *soil* so that today there is, most unfortunately, resulting confusion in geological literature. On the other hand, engineers have continued to use the word *soil* in its original sense, while naturally making clear distinction between soil types such as clays, silts, sands, and gravels. Table 2.1 illustrates the variations in this simple and most useful word. Throughout this volume, the word is used in its geological and engineering sense; correspondingly, the word *rock* is used to denote all bedrock.

The study of *rock masses (solid rocks)* dominates geological studies. Only a relatively few geologists are interested in detailed soil studied. This geologic emphasis on rock is readily understood when it is remembered that most soil encountered in North America, for example, have been formed within the last two million years, whereas most of the other geological phenomena studied cover a period of well over four billion years. In normal practice, the civil engineer will be concerned far more with soils than with bedrock. It is with soils that the most difficult problems arise. When problems do arise in work with rock, as in tunneling, the engineer will usually have recourse to the advice of

TABLE 2.1
VARIATIONS IN USAGE OF THE WORD *SOIL*

CLASSICAL GEOLOGY	CIVIL ENGINEERING	HORTICULTURE	PEDOLOGY (Soil Profile)	GLACIAL GEOLOGY	Weathering Profile
Topsoil	Topsoil	Topsoil	A_1 Horizon (SOLUM)	Zone 1	Humus
			A_2 Horizon (SOLUM)	Zone 2	Stable Silicates
Soil	Soil	Subsoil	B Horizon (SOLUM)	Zone 3	Clay and Oxides Enriched
				Zone 4	Carbonates leached
			C Horizon		
				Zone 5	Oxidised
				Zone 6 (unconsolidated rock)	Unaltered
				(consolidated rock)	Bedrock

specialists. Despite this, however, all civil engineers must be familiar with the principal rock types and with the elements of geological structure and stratigraphy.

ROCK AS AN ENGINEERING MATERIAL

Minerals

Minerals are the basic building blocks of the earth's crust, and aggregations of minerals comprise rocks and soils. Minerals are classified mainly by their chemical composition. The most important rock-forming mineral groups are the silicates, oxides, and carbonates. Each mineral species has physical and chemical characteristics which allow identification; the most common minerals are readily identified in a hand specimen. Crystal form (commonly not well developed), cleavage and fracture, hardness, specific gravity, reaction with dilute acid, and color are among the most significant properties easily observed in the field. Some familiarity with common minerals is useful in identifying various rock types. A few minerals and their characteristics are listed in Table 2.2.

Igneous Rocks

Igneous rocks are of two main classes: *extrusive* (those poured out at the earth's surface), and *intrusive* (large rock masses which have not cooled in contact with the atmosphere). Initially, the rocks of both classes were molten magma. Their present state results from the manner in which they solidified.

If a violent volcanic eruption takes place, some materials will be emitted with gaseous extrusions into the atmosphere, where they will cool quickly and eventually fall to the earth's surface as volcanic ash and dust. The main product of volcanic action is a lava flow emitted from within the earth as a molten stream which flows over the surface of the existing ground until it solidifies. Extrusive rocks are generally distinguished by their usual fine-grained texture.

Intrusive rocks, which cool and solidify under pressure and at great depths, are usually wholly crystalline in texture, since the conditions of cooling are conducive to crystal formation. Such rocks occur in masses of great extent, often going to unknown depths. Although originally formed deep underground, intrusive rocks are now widely exposed because of earth movement and erosion processes.

Hypabyssal rocks are intermediate in position between extrusive and major intrusive rocks. They occur in many forms, the main types of which are indicated in Fig. 2.1. *Dikes* are large wall-like fillings cutting across normal bedding planes in the earth's crust. *Sills* are large conformable sheets intruded into other formations parallel to their structure.

Chemical analyses of igneous rocks show that they are essentially composed of nine elements: silicon, aluminum, iron, calcium, magnesium, sodium,

TABLE 2.2
SOME COMMON MINERALS

Mineral	Specific gravity	Hardness	Crystal	Cleavage or fracture	Color
Quartz	2.65	7	Hexagonal	Curved like glass	Clear, milky gray
Feldspar	2.5–2.8	6	Tabular	2-planar at 90°	White, gray, pink
Mica	2.9–3.1	2.5–3	Hexagonal	One plane	Black, brown, gray
Hornblende	3.1–3.5	5–6	Tabular	2-planar at 60°	Black
Pyroxene	3.2–3.7	5–6	Tabular	2-planar at 90°	Dark green
Calcite*	2.7	3	Rhombic	3-planar at 60°	White, pink, gray
Dolomite†	2.9	3.5–4	Rhombic	3-planar at 60°	White, gray, brown

* Reacts strongly with acid
†Reacts weakly with acid

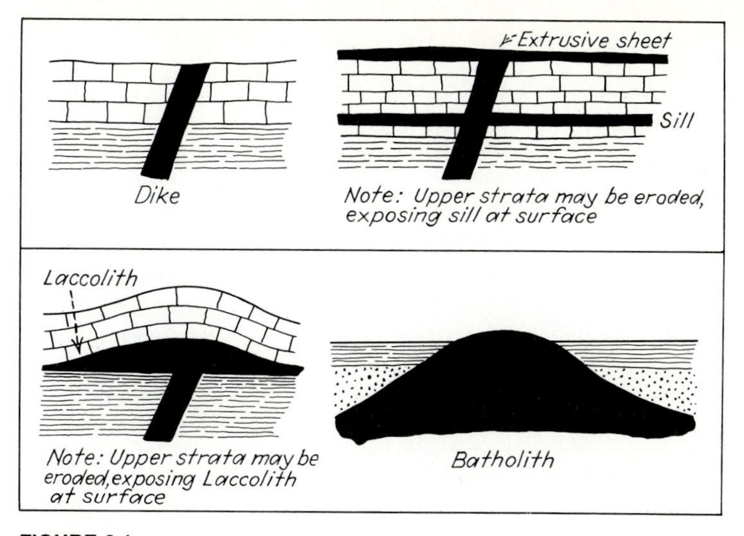

FIGURE 2.1
Some forms taken by igneous rocks.

potassium, hydrogen, and oxygen. These occur naturally in combination, generally as silicates, oxides, and hydroxides. Although the proportion of oxides varies considerably, the chemical composition of rocks considered as a whole varies within quite narrow limits. Within these limits, despite the steady gradation from one composition to another and the varying and quite distinct mineral constitution of different rocks, certain groups can be distinguished. Chemical and mineral composition, in association with the mode of origin of the rock, has therefore been adopted as a basis for a general classification of crystalline igneous rocks. Silicon dioxide (silica), often crystallized as quartz, is one of the main mineral constituents of igneous rocks, and by the silica content a broad dividing line is fixed. Table 2.3 lists the main types of igneous rocks and suggests broad lines of classification.

Sedimentary Rocks

This great group of rocks may properly be regarded as *secondary rocks,* because they generally result from the weathering and disintegration of existing rock masses. These rocks are somewhat loosely named, since *sedimentary* is truly descriptive of but part of the group. Water is not a depositing agent in the case of all these secondary rocks, nor are they always found stratified, another title which has been suggested. In view of the common use of the word *sedimentary,* that term has been retained here and elsewhere in this book to denote this entire group of secondary rocks. Fossils are found almost exclu-

TABLE 2.3
CLASSIFICATION OF IGNEOUS ROCKS

	Acidic (quartz)	Intermediate (little or no quartz)*		Basic (no quartz)
Most common minerals	Orthoclase	Orthoclase	Plagioclase	Plagioclase
	Oligoclase	Biotite		Olivine
	Mica	Hornblende	Biotite	Hypersthene
	Hornblende	Augite	Hornblende	Augite
	Augite		Augite	
Plutonic	Granite	Syenite	Diorite	Gabbro
Hypabyssal	Quartz porphyry	Granite porphyry	Monzonite	Diabase (dolerite)
Volcanic	Rhyolite	Rhyolite	Andesite	Basalt

*Reference should be made to the text for comment on the indefinite divisions noted.

sively in sedimentary rocks; nevertheless, sedimentary rocks are often lacking in visible fossil remains.

The distribution of sedimentary rocks over a wide area throughout the world is a result of the great land movements that have taken place in past geological eras. These movements are often vividly demonstrated by the existence of marine deposits, including fossil seashell remains, in places now elevated considerably above the nearest lake or seacoast. Marine deposits are found in the upper regions of the Himalayas and in many other parts of the world thousands of feet above sea level. The top of Mount Everest, the highest point of land in the world, is limestone, a sedimentary rock.

Sedimentary rocks are generally found in quite definitely arranged beds, or *strata,* which were horizontal at one time, but are today sometimes displayed through angles up to 90°. This bedding, or stratification, is a direct result of the method of formation; the material was deposited evenly over a lake or sea bottom or in some tropical swamp. Similarly, sedimentary rocks are being formed today by mud being washed down rivers into lakes and seas and by such marine organisms as coral in tropical seas. Measured by human standards of time, the building-up process is almost infinitesimally slow; geological time periods must be used as a basis for comparison.

Sedimentary rocks can conveniently be classified in three general groups. Briefly, the divisions include those rocks which are (1) mechanically formed, (2) chemically formed (evaporites), and (3) organically formed. The so-called "mechanical processes" leading to rock formation are the action of wind, frost, rain, snow, and daily temperature changes. All these can be classed as weathering influences which lead to the formation of surface soil, rock talus (or scree breccia), and fine deposits of rainwashed material and blown dust such as brick earth and loess, as well as some special types of clay. Both physical and chemical weathering processes, acting together but with differing relative strength under differing circumstances, act to break down rock into fragments and solutions. These fragments form the clastic sediments—gravel, sand, silt, and clay—while the chemical solutions are carried away invisibly in surface or subsurface water. Water is the most effective agent for eroding, transporting, and depositing sediments. Finally, glacial action has been and still is a potent factor in rock formation, having given rise to extensive glacial deposits over wide areas of the world.

All these different types of deposits, when subjected to great pressure, and possibly to the action of chemicals and heat, will be transformed into rocks such as sandstone. Evaporites, their mode of origin indicated by their name, are chemically formed rocks; typical examples include rock salt, gypsum, anhydrite, and the various types of potash rocks found in some areas. Limestone is perhaps the best known of the organically formed rocks, being generally the accumulation of the remains of marine organisms; it occurs in many different forms. Coal and phosphate rock are other rocks that are obviously of organic origin. Table 2.4 shows some of these interrelationships.

Metamorphic Rocks

Many natural processes have helped to change sedimentary and igneous rocks into metamorphic rock types. The principal ones are the intense stresses and strains set up in rocks by severe earth movements and by excessive heat from the cooling of nearby intrusive rocks or from permeating vapors and liquids. The action of permeating liquids appears to be particularly important. The results of these actions are varied. The metamorphosed rocks so produced may display features varying from complete and distinct foliation of a crystalline structure to a fine, fragmentary, partially crystalline state caused by direct compressive stress, including also the cementation of sediment particles by siliceous matter.

Foliation is characteristic of the main group of metamorphic rocks; the word means that the minerals of which the rock is formed are arranged in felted fashion.

TABLE 2.4
SOME EQUIVALENT FORMS OF SEDIMENTARY AND METAMORPHIC ROCKS

Sediment	Consolidated rock	Metamorphic product
Till	Tillite	Metatillite
Gravel	Conglomerate	Metaconglomerate
Sand	Sandstone	Quartzite
Silt	Siltstone	Argillite or slate
Clay	Shale or mudstone	Argillite or phyllite
Lime mud	Limestone	Marble
Peat	Bituminous coal	Anthracite coal

Each layer is lenticular and composed of one or more minerals, but the various layers are not always readily separated from one another. It will be appreciated that these characteristics are different from the flow structure of lava and also from the deposition bedding which occurs in unaltered sedimentary rocks. *Schist* is the name commonly applied to such a foliated rock, and the various types of schistose rock are among the best-known metamorphic rocks.

The nature of the original rocks from which metamorphic rocks were formed has been and still is a matter of keen discussion. Briefly, it may be said that some metamorphic rocks are definitely sedimentary in origin, some were originally igneous rocks, and some are of indeterminate origin. The presence of fossil remains in certain crystalline metamorphic rocks is proof enough of their sedimentary origin. On the other hand, uninterrupted gradation from granite and other igneous rock masses to well-defined schistose rocks is equal proof of the igneous origin of some metamorphic rock types. Of the former types that are sedimentary in origin, *marble* (altered limestone) is a notable example; its appearance frequently shows the organic remains from which it was originally formed. Quartzite can often be seen to have been formed from sand grains, and the great variety of slates are all obviously of clay or mudstone origin. The rocks classed generally as schists are of varying composition. Mica schist is a crystalline aggregate of mica and quartz and occasionally other minor minerals. *Gneiss* is a term somewhat loosely applied. It is generally used to distinguish a group of rocks similar to the schists but coarsely grained and with alternate bands of minerals of different composition.

Distinguishing Rock Types

The civil engineer has a natural interest in the study of rocks because they are used not only as materials of construction but also as foundation beds and host ground for many structures. The engineer should therefore be able to distinguish at least the main rock types seen in the field. For all practical purposes of the civil engineer, field tests and simple microscopic examinations will suffice for identification of most of the common rock types. More detailed petrographical investigation will be necessary for projects involving rock excavation, underground construction, or selection of riprap.

Simple equipment is all that is needed for the field investigation of rocks—a geologist's hammer, a pocket magnifying lens, a steel pocketknife for hardness tests, and a small dropper bottle of hydrochloric acid for the determination of mineral carbonates. A small magnet is often useful, since the mineral magnetite can be separated from other associated minerals (when crushed) by running a magnet through the mixture. The hardness test utilizes a scale based on the relative hardness of selected minerals (Table 2.5).

Investigation of rocks in the field necessitates close examination of a hand specimen which may be obtained from an outcrop by a sharp blow from the hammer. A clean, fresh rock surface must be used for all examinations. The word *fresh* is used to indicate that the rock surface must have been freshly

TABLE 2.5
RELATIVE HARDNESS OF MINERALS

Hardness*	Mineral	Test characteristic
1	Talc	Can be scratched with a fingernail.
2	Gypsum	
3	Calcite	Can be cut very easily with a penknife.
4	Fluorspar	Can be scratched easily with a penknife.
5	Apatite	
6	Feldspar	Can be scratched with a penknife but with difficulty.
7	Quartz	
8	Topaz	Cannot be scratched with any ordinary implement. Quartz will scratch glass; topaz will scratch quartz; corundum, topaz; and a diamond, corundum.
9	Corundum	
10	Diamond	

* The numbers given are used as relative hardness numbers, relative only since the actual hardness value of talc is about 0.02, whereas that of a diamond runs into the thousands.

broken, since most rocks weather on their exposed surfaces to some extent, and thus do not show their true character. From an examination of such a fresh surface, one can usually see whether the rock is crystalline or not.

Of the noncrystalline rocks, shales, which are consolidated fine sediments, are usually hardened clay or mud and have a characteristic fracture. Generally dull in appearance, shale can be scratched with a fingernail. If it breaks into irregular laminae, the shale is *argillaceous;* if gritty, *arenaceous;* if black, it may be *bituminous;* and if it effervesces on the application of acids, it is *calcareous*. Slate can easily be recognized by its characteristic fracture or cleavage and its fine, uniform grain; in color it may vary from black to purple or even green. All these rocks demonstrate their argillaceous nature by emitting a peculiar earthy smell when breathed upon.

Limestone is one of the most widely known sedimentary rocks; it can often be distinguished by the presence of fossils, but a surer mark of distinction is that it effervesces briskly when dilute hydrochloric acid is applied to it. Marble is a crystalline (metamorphic) form of limestone, generally distinguished by its crystalline texture but always effervescent when treated with dilute acid. Dolomitic limestone is generally dark in color. It effervesces slowly when treated with cold hydrochloric acid, but more quickly when the acid is warm. Flint and chert, compact siliceous rocks of uncertain chemical or organic sedimentary origin, occur often as nodules in limestone beds.

Conglomerates are, as their name implies, masses of waterborne gravel and sand, as denoted by rounded shapes, cemented together in one of several ways into a hard and compact mass. *Sandstone* is the general term used to describe such sedimentary cementation of sand alone. *Quartzite* is a metamorphosed type of sandstone in which the grains of rock have been cemented together with silica so strongly that fracture takes place through the grains and not merely around them. *Grit* is a term sometimes used to denote a coarse-grained hard sandstone containing angular fragments (Fig. 2.2).

The identification of some igneous rocks and metamorphic rocks not yet mentioned is not quite so straightforward as the determinations so far described. These rocks are usually crystalline, but the crystals may vary in size from those of coarse-grained granite to those so minute that they must be examined under a microscope. Of the remaining metamorphic rocks, serpentine, a rock composed wholly of the mineral of the same name, is generally green to black, fairly soft, and greasy or talclike to the touch; the color may not be uniform. Serpentine is important to the civil engineer, since it is often a cause of instability in rock excavation. Gneiss may be recognized by its rough cleavage and typical banded structure, which shows quartz, feldspar, and mica with a coarse structure. Schists may be distinguished from gneiss by their essentially fissile character. In all schists, there is at least one mineral that crystallizes in platy forms (mica, talc, or chlorite) or in long, oblong blades of fibers, giving the rock a cleavage parallel to the flat surface.

Granite is a typical example of an igneous rock; it is widely distributed and constitutes an important igneous rock type. Granite is composed of quartz

FIGURE 2.2
Cross-bedded Wingate Sandstone, Johnson Canyon, Kansas County, Utah. (*Courtesy U.S. Geological Survey; photo J. Gilluly.*)

(clear), orthoclase feldspar (white or pink), some mica, and possibly hornblende. All the crystals are about the same size or equigranular, and the quartz (the last mineral to separate) occupies the angular spaces between the other crystals. This latter characteristic marks a granitic structure. An average composition is 60 percent feldspar, 30 percent quartz, and 10 percent dark, minor minerals.

The texture of igneous rocks varies from a coarse, equigranular structure to an *aphanitic* (without visible crystals) structure in which crystallization cannot be seen with the unaided eye. In a *porphyritic* structure, the constituent minerals occur as much larger crystals than the remainder (the large crystals are called *phenocrysts*). The other minerals may appear as a crystalline groundmass or, alternatively, they may be aphanitic. As a general rule, the acid igneous rocks tend to be lighter in color than the basic rocks; granite, therefore, is pale because of the predominance of feldspar, a light-colored mineral. Diorite has a texture similar to that of granite, but it contains no free quartz. In gabbro—the corresponding basic rock—feldspar is subordinated but is still an important constituent; hornblende, pyroxene, and olivine are dark minerals which make the rock dark and give it a high specific gravity. Diabase is a similar basic rock with a smaller grain size; basalt is the corresponding aphanitic rock.

Granite porphyry and diorite porphyry are similar to granite and diorite in composition, but they have a porphyritic structure, feldspar being the most usual phenocryst. The porphyries are a common group of rocks. They are hypabyssal and occur both in lava flows and as sills, dikes, and laccoliths. Rhyolite, an extrusive rock, corresponds in composition to granite, an intrusive rock. Andesite bears a similar relationship to diorite; it contains no quartz. The aphanitic type of both rhyolites and andesites is known as felsite, a name that includes most of the large group of light-colored aphanitic igneous rocks; the corresponding dark rocks are classed as basalts. Glasslike rocks are found only in the vicinity of cooled lava flows; obsidian, a lustrous, dark rock, is the most common variety. Pumice is simply a frothed type of glassy rock.

The foregoing list includes the main common varieties of igneous rock. It must again be emphasized that there is no hard-and-fast dividing line between a given variety and the adjacent one in scale. It cannot be too strongly urged that study of rock structures in situ is the only truly reliable method of becoming familiar with the rock types described. Indoor examination of hand specimens will help the beginner, but there is really no substitute for examination of fresh rock specimens in the field.

GEOLOGIC STRUCTURE

Structural features encountered in geology are regularly met with in normal civil engineering work. Two widely used terms, *strike* and *dip,* are used to describe the present position of strata of rock with reference to the existing ground surface (Fig. 2.3). The *strike* of a rock layer is the compass direction of a line considered to be drawn along an exposed bedding plane of the rock so

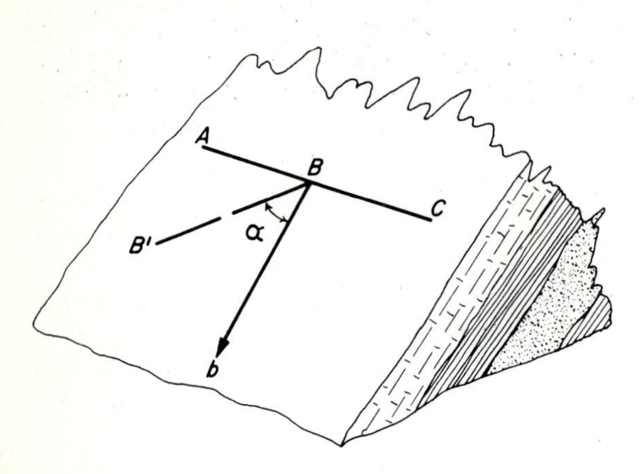

FIGURE 2.3
Diagram illustrating the dip and strike of rocks.

that it is horizontal; obviously, there will be only one such direction for any particular rock layer. The *dip* of the bed is the angle between a horizontal plane and the plane of the bedding, measured at right angles to the strike; it is thus a measure of the inclination of the bed to the horizontal plane. These two particulars, together with the usual topographical information, definitely locate a rock surface in three-dimensional space. As such, they are as invaluable to the civil engineer as they are to the geologist.

It is necessary to have at least a general geological survey made of the area in which the works are to be situated, even when the nature of the rock or rocks to be considered is known. Assumptions as to the fabric of the rock being studied must be made before theoretical studies can begin. These assumptions may be correct, but they also may not be. Structural geology has become a specialty in itself and the civil engineer should be familiar with the main approaches taken to describe complex geologic structure.

Discontinuities

Geological history is more important when the continuity of rock strata is considered. Only rarely will one encounter bedrock which is continuous within the area of study and which exhibits no discontinuities.

Most uses of rock for engineering purposes require careful recognition of *discontinuous surfaces* (discontinuities). These natural fracture surfaces of rock masses come in a variety of types, depending on their origin. The most common are the *bedding planes* of sedimentary rocks. When found in metasedimentary rocks, their metamorphic equivalents are *foliation planes*. All rock types contain the most common of discontinuities, *joints*, which were formed in cooling of crystalline and extrusive rocks and in regional tectonic compressive and tensile stress fields associated with plate tectonic movement. Those joints that have undergone limited shear displacement are known as *shear planes;* those with more than just a few centimeters of displacement are termed *faults*. Both shear-displacement discontinuities are also found in bands of parallel and semiparallel fractures and are then termed *shear zones* and *fault zones*. These zones are commonly a few centimeters to several decimeters in width. Major fault zones, however, such as the San Andreas of California, exceed more than a kilometer in width in some instances.

Where present, discontinuities generally take control of the engineering characteristics of a rock mass. Hence, *rock mass characterization* techniques have been developed to account for the spacing, orientation, lengths, and surface characteristics of discontinuities in assessing particular sites for rock excavation and underground construction. Rock mass characterizations must take special care to recognize the existence of spatial rock volumes of different lithologic and structural character (*structural domains*).

Rock mechanics takes discontinuities fully into consideration by its use of clastic mechanics. Even here, however, certain assumptions must be made as to the regularity of discontinuities and the uniformity of the resulting blocks. It is therefore essential that geological studies demonstrate the accuracy or the inaccuracy of the assumptions that must be made. Again, geological history is the first guide that must be utilized. The basic character of the rock (sedimentary, metamorphic, or igneous) will be the starting point, since that will indicate the origin of the rock and its major components. Knowledge of tectonic history since the rock was first formed will be the second guide. If there have been no significant earth movements, then jointing as a result of temperature changes will be the first possibility to be explored, once bedding planes (where they exist) have been determined.

If, however, there have been significant orogenic disturbances, then a careful geological survey of an appreciable area surrounding the site of the works in question will be essential. The overall geological structure for this area must be determined in order to see what folding has taken place since rock formation and whether there has been any faulting. This is a study that cannot be done merely by studying the geology at the site in question; a regional approach must be followed.

From the simple diagrams then presented, it can be seen that whenever rock strata have been subjected to folding, those near fold axes (such as on the upper part of an anticline or the lower part of a syncline) will have been subject to tensile forces. Jointing is therefore to be expected, the joints being an expression of tensile failures. In the case of gentle folds, the joints may not be obvious to the eye, but they must be anticipated and special field investigations conducted to determine whether they are present or not.

The existence of faults will be indicated by a detailed geological survey of the region involved. They may not be evident at the surface if there is soil cover. Not only must the presence of faults be determined but also their location (as intersected by inclined drill holes) and throw. Stresses caused by major faulting may cause ruptures in rock a considerable distance from the fault itself, so that in all cases of faulted rock the most detailed reconnaissance must be carried out in advance of any decisions about rock excavation. Even then it may not be possible to locate all failure surfaces, but their possible presence can be anticipated.

Bedding Planes

The method of formation of sedimentary rocks leads in many cases to an uneven disposition of material and to an uneven distribution of pressure on deposits. Thus, a distinct variation in the physical qualities of a sedimentary bed at different levels, as well as changes in the thickness of a bed, may result. Sedimentary strata often thin out completely, which can cause confusion in geological mapping. These variations are of minor importance, but another subsidiary feature is worthy of special note. It is what is called *cross,* or *false, bedding,* caused by a special process of deposition, i.e., by currents from varying directions.

Joints in a Typical Rock Mass

It will soon be seen that, in addition to bedding planes that may be visible (in a sedimentary rock), fractures can also occur in other planes roughly at right angles to bedding planes. These fractures give rise to a blocklike structure, though the blocks may not be separated from each other. Such fractures are generally known as *joints,* or *joint planes,* and result from internal stresses either during the cooling of the rock or during tectonic displacement. Joints are sometimes filled with mineral that has crystallized out from solution, e.g., quartz and calcite. Some remarkably intricate formations of this type may be found in which the filled joints are so small that they are barely visible. In sedimentary rocks, jointing is generally regular; in granite, it is often irregular. In basalt, it leads to the peculiar polygonal column formation that is a striking feature of such locations as the island of Staffa off the west coast of Scotland and the eastern Washington state area where the Columbia plateau basalts occur (Fig. 2.4). Joints are of great importance to the civil engineer, and numerous references to them will be found in the main part of this book.

Folding

Folding is perhaps the simplest structural feature. Simple regular folds, as shown in Figure 2.5, are termed *anticlines* and *synclines* respectively, according to the type of bend. *Anticlines* are upfolds; *synclines* are downfolds, convex

FIGURE 2.4
Nature's sculpture, the result of jointing in basalt, Grand Manan, New Brunswick. (*Photo: R. F. Legget.*)

and concave upward respectively. This essentially simple type is not always found in practice. The first, and most general, variation is the inclination of the axis of the fold with the result shown. In certain special cases, folding may not be confined to one direction. If a rock mass is subjected to tectonic bending stresses all around, a domelike structure may occur. If the simplest basic type is extended, many variations can be obtained, such as double folds, reversed folds, and even the recumbent and fanlike structures indicated. Several outstanding examples of this type of structure could be given, including the European mountain group of which Mont Blanc is a leading member. It will be realized that if the angle through which rock is folded is at all appreciable, unequal stresses will be set up in the rock mass, a matter of consequence to the civil engineer. If such distorted material is underground, civil engineering operations may meet with unexpected results when the previously restricted stresses are released.

Faults

When subjected to great pressure, the earth's crust may have to withstand shear forces in addition to direct compression. If the shear stresses so induced become excessive, failure will result. Movement will take place along the plane of failure until the unbalanced forces are equalized, and a *fault* will be the

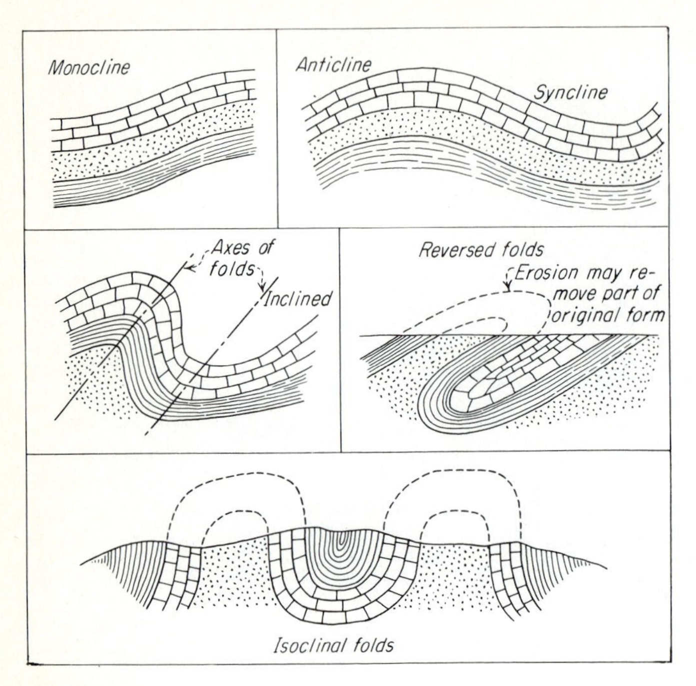

FIGURE 2.5
Diagrammatic representation of some types of folds in rock strata.

result. Figure 2.6 shows in simple diagrammatic form the types of fault most generally encountered, with the relative displacement of the various strata clearly shown. In the simple fault, the terms most commonly used are indicated. The first fault shown is a normal fault in which the *hade* (or inclination with the vertical) is always in the direction of the downthrow. The *throw* of the fault is the vertical displacement; the *heave,* the lateral displacement. In the diagram, the fault is shown as a plane surface. In practice, the rock on one or both sides of the fault is frequently badly shattered into what is termed *fault breccia*. Faults are characterized by their general direction as *dip faults* and *strike faults*. *Step faults* and *trough faults* are terms that will readily be understood by reference to the diagram. *Reverse faults* are the opposite of normal faults, having the hade away from the direction of the downthrow. As the diagrams show, faults are almost always inclined to the vertical. The upper face of inclined faults is known as the *hanging wall,* the lower face as the *footwall*.

Simple diagrams suggest the complicated structures that can and do result from faulting, especially where the fault plane cuts across several rather thin strata. All these will be displaced, relative to one another, at the fault. After denudation has taken place, their surface outcrops will be confusing (Fig. 2.7). The civil engineer's usually clear appreciation of three-dimensional drawings

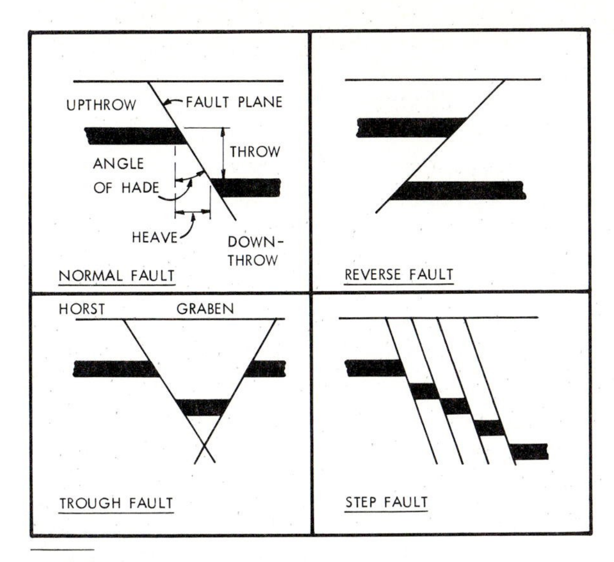

FIGURE 2.6
Diagrammatic representation of some types of faults in rock strata.

FIGURE 2.7
A normal fault on beds of sandstone and fakes, exposed at Mossend, Scotland. (*Courtesy of the Geological Survey of Great Britain; Crown Copyright.*)

will be of great service in the quantitative (as distinct from the qualitative) investigation of faulting, together with all allied calculations.

Faults may vary greatly in size. In length, they may range from a few feet to hundreds of miles. In throw, they may include movements of a few inches or less (even pebbles have been found with perfect fault sections in them) to thousands of feet, as in the case of the faults in the plateau region of Arizona and Utah, some of which cross the Grand Canyon. Such dislocations may cause no unusual disturbance of topographical detail, although in some cases they may be the main factor affecting the physical character of large areas. In Canada there are many notable conspicuous faults. For example, Lake Timiskaming, which separates Ontario from Quebec in the north, lies in a fault depression.

Denudation

Erosion of rock and soil masses by wind, water, and ice is known collectively as *denudation*. Nearly all landforms have been markedly affected by such processes. The effects of erosion are most pronounced in inclined sedimentary strata, which may be worn down unevenly, according to the overall hardness of the strata encountered. Other notable results of erosion are important in deciphering the history of river systems, where erosion may have been accelerated in valley bottoms along faults, or in which ancient erosion has led to buried river valleys.

Unconformity

Occasionally a sequence of sedimentary beds will be encountered where, beyond a particular surface (*unconformity*), the strike and dip of bedding changes, more or less notably. The surface of unconformity juxtaposes two distinct bedding attitudes and represents a time interval (hiatus) during which the upper surface of the older rocks was eroded and the landscape was tilted so that only the younger surface was then horizontal. Where the surface indicates only a period of erosion without tilting, the appropriate term is *disconformity*, a special case of unconformity.

ROCK CHARACTERISTICS

Every application of rock mechanics is unique, since no two geological exposures are ever identical. Rock characteristics are of two types, those of the *intact* rock (such as in unfractured laboratory specimens) and of the *rock mass*. Every specimen of rock that is tested gives results that can be applied only to the location from which it came and to no other. As the volume of test information for different kinds of rock has steadily accumulated, it has been only natural that similar information on the same rock type should be collected and assessed. This information in turn suggests the range of test values that might be expected for other incidences of the same rock type. Such generalities, of course, are made less valuable if the states of weathering or alteration are different. These general engineering properties of various rock types apply mainly to resistance to excavation, to tunneling machine design, and as a general indicator of ground support conditions in underground construction, and must be used with discretion.

Rock Properties

The term *intact rock* is sometimes used to indicate rock which is free from joints, bedding planes, breaks, or shear planes. This rock is so intact that it can be tested by means of samples in a laboratory where its mechanical properties can be accurately determined. It is the task of the engineering geologist to show to what degree the rocks actually in place at the site being studied approach the ideal condition that has been assumed. This procedure is a sound one, provided the limitations of the theoretical solutions are not forgotten. The elegance of the mathematical solutions to some slope-stability problems, especially if assisted (as is now possible) by computer graphics, tends to obscure the fact that some geological feature of the actual rock face in the field may invalidate significant theoretical assumptions made initially. Mathematical simplification is a common feature of engineering design, although the uncertainty of geological structure and of rock characteristics in the field makes the tempering of theory with practical considerations of rather more importance here than is usual. Accordingly, the remainder of this short section will discuss the practical considerations that must always be associated with the results of rock mechanics theory so that the latter may be used judiciously and with confidence. Full treatments of mechanics will be found in the excellent texts on rock mechanics now available.

To a degree, rock strength can be studied by means of carefully procured samples tested in a laboratory. Sampling of rocks for this purpose is an easier matter than the sampling of soils, but care must be taken to protect samples from damage to their integrity. Although shear stresses are those that are usually critical in rock masses, a first test will usually be of the compressive strength of the rock. When this test is properly carried out it will also be the means of determining the value of Young's modulus.

Deere has prepared a valuable series of charts relating compressive strength with modulus of elasticity. Peck has combined a number of these, as well as other useful concepts of Deere, in one chart which is reproduced as Figure 2.8. The value of this useful summation will be obvious at once. The general agreement in the strengths shown for such a wide variety of rocks, when related to the corresponding values for Young's moduli, is striking and significant. The great range in the compressive strengths, however, is equally impressive, especially when the use of the logarithmic scales is kept in mind. The groups of rocks shown are those most commonly encountered on engineering works and

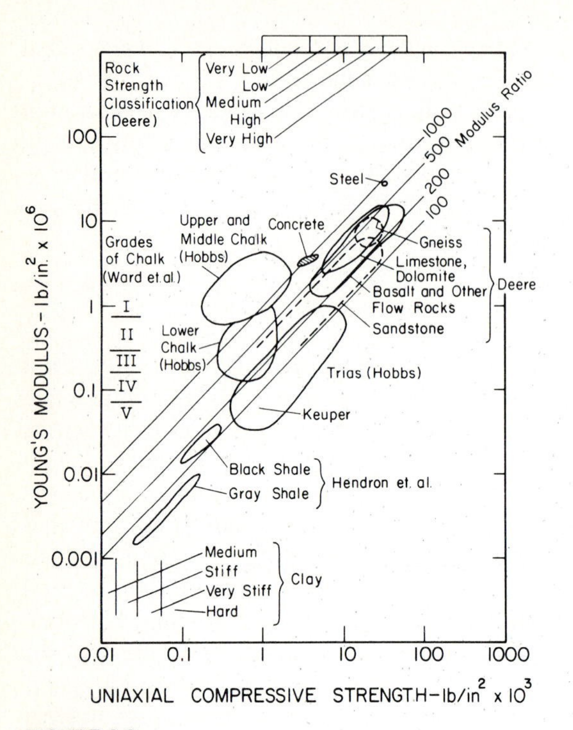

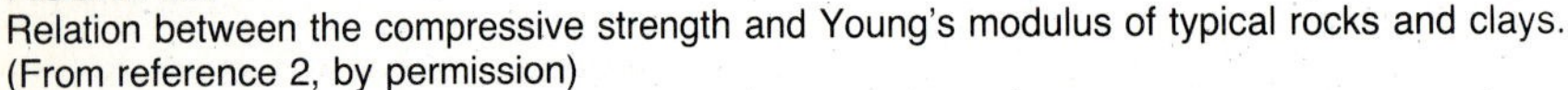

FIGURE 2.8
Relation between the compressive strength and Young's modulus of typical rocks and clays. (From reference 2, by permission)

so the chart has a double value, not only illustrating the need for judicious use of the concept of "rock substance" in rock mechanics studies, but also showing the wide range of qualities involved in the use of what is so loosely called "bedrock."

Mechanical properties thus obtained will provide the necessary values for insertion in equations of stability. But it will have been obvious to the reader long before this that there are some major qualifications. If all rock substance were homogeneous and isotropic, all would be well, but this is seldom the case. Sedimentary rocks provide the most obvious and probably the most widespread exception, especially if bedding planes are well delineated, as they are in many limestones and shales. Even in sandstones the mode of deposition of the original sand is often obvious, though bedding planes may not be clearly evident. In all such rocks, the relation of the major axis of test loads with the bedding planes must be accurately determined, since there will usually be considerable difference between results from tests performed at right angles to bedding planes and those performed parallel to them. Differences of the same type may be found during testing of igneous rocks displaying flow structures and foliated metamorphic rocks such as schists. The bonding of individual grains may similarly affect test results, since some bonding materials will lose strength when wet. All tests, therefore, should be carried out both on dry and wet samples. Despite the cost, it is desirable to obtain a number of comparable test results on any major rock type involved in rock mechanics studies so that appropriate statistical correlations can be made.

All that has been said so far related to rock that is reasonably uniform, if not homogeneous, typical of the sort of rock exposure that is found in temperate regions. But in warmer areas where rock weathering is a normal feature of the terrain, or in areas where geological processes have themselves created differentiation between varying horizons of the same rock, the determination of rock strengths that may be used in the application of rock mechanics theories becomes more complex. In a useful paper Matula cites the case of a Triassic quartzite from the Carpathians, formed originally from beach sands cemented during their diagenetic evolution by silica to form a solid, hard, and brittle rock. The resulting sound rock has a strength of 2,840 kg/cm^2 (18 tons/in^2) and Young's modulus of 600,000 kg/cm^2 (3,800 tons/in^2). The region from which this rock comes was subjected to folding in Cretaceous times, with large-scale faulting in some areas. Rock that has been subject to these disturbances has been shown to lose as much as two-thirds of its strength and modulus. Further Tertiary faulting and hydrothermal alteration have reduced the same rock in places to an almost cohesionless sand upon which tests could not even be conducted.

GROUNDWATER

Discontinuities affect not only the application of rock mechanics theory but they will, in most cases, lead to problems with water. Some discontinuities show tight contacts between their surfaces, but most surfaces will be open to some degree. They will thus permit entry of rainwater at the surface after heavy rains, assuming that only light precipitation will be retained in the soil cover. Once within the fractured rock mass, most of this *cleft water* will move with a remarkable rapidity, but essentially only through the discontinuities.

Problems associated with the presence of water and its movement through rock masses relate to weakening of the rock mass, alteration of the properties of tunnel muck, and with removal of water as a nuisance to construction. Any estimate of quantity that can be made prior to the start of work will be of much assistance in planning construction methods and schedules. When excavation is complete, drainage facilities will have to be provided for any flowing water that enters the excavated area. It may even be necessary to calculate flow nets for the finished cross section in order to determine the probable nature of groundwater movement in the finished works.

Rock stability calculations for the finished excavation must allow for hydrostatic pressure induced in the discontinuities, no matter how small. Such pressures will have a profound influence on the effective stresses in the rock mass, a concept now well understood and appreciated because of the pioneer work of Karl Terzaghi. The concept is one of great importance and must be rigorously examined for all cases in which groundwater may be present. If, for example, stability of rock slopes within a reservoir is being investigated, then the effective pressures that will exist after the water level in the reservoir has been raised must be carefully evaluated. These pressures may even have the effect of moving two sides of a steeply sloping reservoir inward after water is stored, rather than outward as might be thought, at first sight, to be the case.

When groundwater is going to play a significant role in rock stability, it will be necessary as a permanent precaution to install measuring devices (piezometers) to record its position or pressure. Design work may be aided if these installations are made as soon as design work begins, so that assumptions made about groundwater based on overall preliminary studies may be checked and verified prior to their use in design calculations. The dynamic nature of groundwater (which will be stressed in the next chapter) as well as its annual variation (already mentioned) must always be kept in mind.

IN SITU STRESSES

Field measurements of stresses in rock, and other factors involved in the behavior of rock masses, might be thought to be a subject of limited interest. In April 1977, however, there was held in Zurich, Switzerland, a full-scale international symposium devoted entirely to the subject of field measurements in rock mechanics, which became the title of the valuable two-volume proceedings that resulted from the meeting. This is yet another way in which rock mechanics studies have so rapidly advanced.

Good instruments are available today for measuring the stresses in rock in place. One of the pioneers in this field (one of the "lone workers" mentioned earlier) was Hast in Sweden. With the instruments he developed, Hast observed that there were significant in situ stresses in the Precambrian rocks of his country. Since his time, and especially in quite recent years, there have been an increasing number of similar observations in other countries and areas, one of the most notable being that part of northeastern North America centering around Niagara Falls. Movements of rock masses in response to in situ stresses have been observed at Lockport (on the New York State Barge Canal) for more than 50 years and in wheel pits of one of the Canadian power stations at Niagara Falls since soon after the start of the century. "Pop-ups" in quarry floors are further evidence of these stresses.

In later chapters of this text more will be said about the serious problems caused in some engineering works by in situ stresses, problems which may be geologically significant. The matter is mentioned here to indicate yet another of the complications that can influence theoretical analyses of rock stability. If, however, the existence of in situ stresses is determined in advance of design studies, due allowance can be made for them.

FIELD TESTS AND FIELD OBSERVATIONS

The discipline of rock mechanics today embraces a sound body of theoretical knowledge, a comprehensive background of laboratory testing and analysis, and an increasing volume of case histories of "rock mechanics in action." But the discipline has also made a significant contribution to the development of testing in the field. When allied with necessary geological information, the results of careful field testing of rock can make important contributions to the solution of rock-stability problems.

Drilling is common to all site investigations. In specific studies of rock properties, the results of carefully supervised diamond drill work can give much valuable information. With good core recovery, some indication will be given of the amount of jointing present; some faults will naturally be revealed by drill-core records. Much good work has been done in photographing the sides of drill holes with specially designed borehole cameras and even with portable TV cameras which give useful images for immediate examination at the surface. By the use of ingenious "packers" in drill holes, sections of a hole can be isolated and then subjected to water pressure so that the permeability of different horizons and probable losses of water through joint systems can be determined. Deere has proposed the term *rock-quality designation (RQD)* to denote, by reference to percentage core recovery, the state of rock in situ.

Much useful information can, therefore, be obtained from a carefully laid-out program of test drilling, the choice of location of holes being assisted by an appreciation of the local geological structure. So useful (and often interesting) is this drill-hole information that it is all too easy to forget that it relates only to the rock in the immediate vicinity of the test hole, and so to only a minute proportion of the rock mass that is involved in any rock excavation of slope study. Geophysical methods, complementary to test drilling, can extend knowledge of the properties of the rock mass. The seismic method has generally been found to be most useful in investigations of the continuity and relative uniformity of rock masses. Especially valuable has been the use of suitably designed equipment for cross-borehole testing.

Strength tests in the field hold a special place in rock mechanics work; these tests give valuable information, but only at considerable cost. A worthwhile preliminary to any such testing is the use of the Schmidt hammer in a simple but effective type of "nondestructive testing." The standard hammer is allowed to drop onto a clean and suitable rock surface from a fixed height and its rebound is measured, giving figures which can usefully be applied in comparing rock types and in establishing the need for more elaborate testing.

Full-scale tests of rock in situ can be seen to be a desirable means of eliminating some of the uncertainties arising from the use of only theoretical and laboratory studies in major rock-stability investigations. These tests are, however, costly to conduct, even if they can be carried out while construction is proceeding and heavy equipment and skilled workers are available. In 1950, the strength of bentonite seams in the Niobrara chalk on which the Fort Randall Dam is founded was in doubt. It was finally decided to carry out shear tests on blocks of the rocks in question, blocks left in place after rock around had been carefully excavated.

The Carillon Dam on the Ottawa River, midway between Montreal and Ottawa, is founded on horizontal or gently dipping strata of dolomitic and calcareous limestones with minor shale horizons, the strata being part of the Beekmantown Formation of Paleozoic age. The strength of the shale strata was in question. Five test blocks, 0.76 m (30 in) square in plan, were prepared by the use of wire saws operating from 0.91-m- (36-in)-diameter Calyx drill holes. When fully instrumented, the blocks were vertically loaded with concrete blocks and then subjected to shear forces by means of horizontal jacks. Consistent results were obtained from four of the five blocks, leading to confidence in design assumptions.

Field Observations

Since geology is so clearly a vital counterpart of rock mechanics, how best can the necessary geological information be obtained and presented? Normal geological maps, if available, will provide a sound starting point; otherwise, geological survey work will have to be carried out and the geological map of the requisite area prepared. These regional maps supplemented by geological sections should give a good general picture of the geological structure of the area embracing the site under study and of the main rock types that will be encountered. These rock types must be described concisely in geological terms, and their lithology must be stated. A typical description might be "granite, gray, coarse-grained, uniform," with additional comments such as "high strength" if test results are available.

In almost all cases, a more detailed geological map will be necessary, a map based upon intensive survey work immediately around the site in question. Again, with the aid of good sections and possibly models, the designer can gain a good three-dimensional picture of the geological structure to be dealt with. Faults would naturally be shown on such detailed maps, which would also give an indication of the frequency and nature of any joints present. The dip and strike of all strata would be clearly delineated as a result of this detailed study. If steep slopes are involved (as, for example, in the case of an open mine pit), aids such as plane-table surveying and photographic surveying may have to be utilized.

Joints are of such importance in rock-stability studies that special attention must be given to them and every effort be made to gain an accurate assessment of their frequency. This is tedious but important work. Deere has suggested a convenient subdivision of joint frequency:

Very close	Less than 5 cm apart
Close	From 5 to 30 cm apart
Moderately close	From 30 cm to 1 m apart

Very close	Less than 5 cm apart
Wide	From 1 to 3 m apart
Very wide	More than 3 m apart

The nature of joints must also be observed and recorded with care, i.e., whether they are open or tightly closed. In the case of faults that can be examined, the nature of the fracture surface and the character of the gouge filling must be determined. A number of conventions and recording methods have been developed in rock mechanics work, some directed toward placing the information in a computer system. More details will be found in volumes dealing with methods in rock mechanics.

SOIL AS AN ENGINEERING MATERIAL

Soils represent a special category of engineering materials. Unlike most other engineering materials, soils originate from geologic processes, are deposited through geologic processes, and are changed considerably through time by geologic processes. Nearly all engineered works come in close proximity to soils. Problems and difficulties encountered in design and construction are many times more frequent when structures are to be founded on soil rather than on bedrock. Traditionally, soil has been viewed as being of peripheral importance to engineers, and of little economic significance. However, increasing emphasis is being placed on engineering geologic evaluation of soils, because of their importance in foundations in groundwater surveys, evironmental impact assessments, and waste management projects. It is now common for published geologic reports to contain sections dealing with Pleistocene geology, the great epoch of formation for most North American soils.

Soil mechanics is the universally accepted geotechnical term for the engineering study of soil. What is *soil* and what is *rock* remains a difficult matter of determination in many cases. Some of the softer shales, for example, grade imperceptibly into stiffer clays. Most of the transitional material is now becoming known as *weak rock,* though this book will consider only the end member materials. Soils, however, represent a distinct design-related challenge for the civil engineer. This is the material presenting inherent difficulties in terms of shear strength and volumetric shrinkage and swelling, with an all-too-frequent unwillingness to yield its pore water to drainage efforts. The engineer should therefore know something of the origin of soils and how transported soils have reached their present position. This knowledge will lead to an improved understanding of geotechnical soil characteristics and is an essential prerequisite to a full appreciation of the role of soil mechanics in civil engineering practice.

ROCK WEATHERING

Rock surfaces that have been exposed to the elements will show evidence of disintegration of a surficial layer. This *weathering* is a complex physical and chemical process. A sister process, alteration, also produces disintegration, but through the passage of at least slightly thermal groundwater, and at a depth below the ground surface. Alteration produces generally weakened (*bad ground*) tunneling conditions.

Rock weathering is always complex. It naturally varies by locale, by elevation, with the seasons, by the time of geologic exposure, and with different rock types. Civil engineers should be concerned with identifying the *agencies* of weathering and of learning to separate these from the *erosion* processes that transport the weathering products from origin to deposition.

Agencies of Weathering

Atmospheric gases are the catalyst of weathering, which itself is wholly dependent on the presence of water and on variations of temperature. Impurities in water lead to oxidation of rock minerals, notably the brown staining of rock in the presence of hydrated iron oxides, carbonates, and sulfates. Such reactions result in increases in volume and the subsequent pressure-wedging that disintegrates rock. Rainwater and groundwater containing small quantities of carbonic acid are further solvents of many rock types, especially the carbonates (such as limestone and dolomites).

Cold-temperature extremes are effective forces in expansion of water in natural rock fractures (joints), in which rock blocks are wedged apart, giving more and more surface area for attack by the elements. Variations of temperature lead to weathering attributable to flaking of surface layers of rock masses.

Products of Weathering

The degree to which rock weathers is controlled primarily by the weathering resistance of individual rock-forming minerals. These minerals respond differentially to weathering processes, meaning that each instance of rock weathering is a case of individual mineral breakdown; some minerals respond faster than others. Harder and more stable minerals (led by quartz) are the last to degrade, hence many soils have an appreciable content of quartz grains remaining from the attack of either or both weathering and alteration. Even the most stable minerals are continually subjected to degradation by rainwater, especially when it is charged with humic acid gained from contact with topsoil organic debris. Chemical disintegration of rock minerals becomes less active as depth increases. Eventually, after passing downward through several successive horizons of these *residual soils,* one will detect fresh, unweathered rock.

All soils, therefore, are the products of rock weathering. To appreciate soil properties fully requires therefore a general appreciation of the original rock from which the soil has been formed.

RESIDUAL SOILS

Soils formed by the direct in situ weathering of bedrock are common in many areas in the warmer parts of the world where bedrock is far below the surface. The observable soils are usually of great age and their origin is not immediately recognizable. Soil formation under tropical conditions is still only imperfectly understood and a matter of debate. Laterites have accumulations of iron and aluminum oxides at the surface, with silica content generally leached out to a lower horizon. Black cotton soils constitute another major and important group in tropical areas, but they are well recognized, and their properties are being gradually codified.

Residual soils have several interesting characteristics. *Residual soils* are the sharp-edged fragments of original rock, fragments that have not suffered rounding by transport. They may retain the structure and basic appearance of the original rock or a completely weathered rock may become a clay soil and bear no evidence at all of the parent material.

The physical and mechanical properties of residual soils naturally vary greatly, but they are determined in exactly the same way as those of glacial soils. Simple soil tests will sometimes give clear indication of varying origins for residual soils that may have the same appearance. Results of tests upon two soils encountered during field studies carried out for the Konkoure development in French Guiana are typical. Some residual soils, including the black cotton soils so common in India, Africa, and the Far East, have rather unusual properties when considered for engineering use. The most difficult of all these soils to deal with are those derived from the weathering of volcanic rocks such as lava flows and volcanic ash.

Less common clayey residual soils are those that are sandy. All over the world, the *laterites* are so common and so important as to deserve special mention. Laterites are found throughout the tropical regions. Because of an absence of organic material in the topsoil and a consequent lack of a humic-acid leaching agent, such soils have a high alumina and iron oxide content. The clay content is naturally soft, but hardens appreciably on exposure to the atmosphere.

Residual soils are found only in regions that have not been geologically disturbed over at least Pleistocene time (the last two million years). Except in isolated cases of deep pockets, glaciated terrane is generally devoid of residual soil; coastal areas, subject to sea-level fluctuations, seldom contain residual soils.

TRANSPORTED SOILS

The forces of erosion are one of the great modifying influences at work today on the surface of the earth. Throughout geologic time, erosion has led to the formation of most of the sedimentary rocks that cover three-quarters of the earth's land surface. Broadly, transported soils are classified as *colluvial* (gravitational), *aeolian* (wind-blown), *fluvial* (river-borne), and *marine* (related to the seas).

Erosion

As surely as soil is usually found transported from its original site of weathering, *erosion* is the beginning mechanism of transport. Forces of erosion are those that strip the degrading rock particles from the *outcrop* and place them in the motion of the transport. Wind, running water (rivers and coastal wave action), and moving (glacial) ice are the prime agents of erosion. While many eroded terranes are smooth and hardly noticeable as such, others, such as the Iranian badlands of the Persian Gulf, are striking examples of the action of wind and torrential rains.

Aeolian Deposits

Not only does wind as an agent of weathering erode rock, it further transports individual mineral fragments over appreciable distances, forming *aeolian deposits*. Sand dunes and *loess* are the most familiar of aeolian deposits. Sand dunes are typically made up of poorly stratified, low-density deposits of fresh, well-rounded, uniformly sized grains. The engineer need be concerned mainly with the low-density nature of such soils. Loess, on the other hand, is wind-transported glacial rock flour. It is found in a wide range of fine-grained compositions, and with a particularly irksome solid structure. These yellow-tinged soils are without stratification and occur in massive beds up to dozens of meters thick. Microscopically, loess is composed mainly of sharp-edged silt particles set against each other so precariously that addition of pore water precipitates a rapid and profound volumetric collapse. Loess, to be used as a foundation material, must be recognized as such and presaturated to induce volumetric collapse prior to placement of foundation loads. By the same token, construction cuts in loess will slough rapidly to the near-vertical condition noted in Figure 2.9.

A major distinction should be made between residual and transported aeolian soils. The peculiar characteristics of some types (such as loess), owing to their open structure, will affect soil sampling and result in findings of unusual soil properties, especially with regard to the action of water. In situ soil testing is generally called for. A well-accepted practice today is to saturate loess before placing loads upon it and then to consolidate the wetter material by external means. This increases its density, and when consolidation has proceeded far enough, shear strength will increase. Even lawn watering has resulted in settlement of adjacent houses in such soil. Geological recognition of the presence of loess, therefore, can be a vital contribution to engineering soil studies.

Fluvial (Alluvial) Soils

Alluvial soils, transported down rivers and streams, are usually readily recognizable in present watercourses, but they may also be found on the banks

FIGURE 2.9
Loess as exposed in a road cut adjacent to the railway station at Timaru, New Zealand. (*Photo: R. F. Legget.*)

of ancient rivers. Such conditions may not be at all obvious to the untrained eye, particularly if the area has been developed in any way for special uses. Knowledge of local geology will be a helpful guide to the existence of buried watercourses. Once their presence is recognized, then the characteristically variable and unconsolidated alluvial soils can be anticipated in soil sampling and testing. Incessant down-gradient movement of surface runoff is a magnificent transport mechanism, moving the largest group of transported soils. All manner of creeks, streams, and rivers transport large amounts of suspended and bed-load soil particles.

Fluvial sedimentation begins first in stream-gradient changes at the mouths of mountain canyons, resulting in *alluvial fan* deposition. These fans show crude stratification, with a wide variance in particle size distribution, both within and between individual beds; each relates to a major rainfall and floor-deposition event. As would be expected, boulders are common in the upper reaches of the fans; silts and sands are common at the toes of the fans. Where adjacent valleys are deep, alluvial fans are found in depths of hundreds of meters. Deposition also occurs wherever stream flow velocities drop significantly, such as at lakeshores and on floodplains. The main divisions are those of *marine* and *continental* soils (loosely called *sediments*). All of the processes of sedimentation can be observed in today's river systems, resulting in the most complex of all soil deposits, consisting of channels, bars, and pockets of soils ranging from clay through gravel. The engineer who attempts to design and construct on fluvial deposits without careful geologic assessment of the nature, grain size, and positional location of such materials may encounter real difficulties.

As rivers work their way to floodplains, at the low-gradient, *aggrading* (highly depositional) ends of their networks, river beds continue to build up, resulting in formation of *natural levees, floodplains,* and *alluvial terraces.* Older terraces, formed by ancient flood deposition and subsequent erosional planing of their surfaces, are often excellent places in which to prospect for construction borrow sites for gravel. The more arid the region, the greater the likelihood of encountering both layered and channel-fill deposits of cobbles and boulders, deposited during flash floods.

Gravitational Deposits

The forces of gravitation act right at the site of weathering. Individual particles of weathered rock fragments travel down the hill slopes by gravity, accumulating at the foots of slopes as *talus* or *colluvium* (Fig. 2.10). Where not removed by agents of transportation, such fragments remain and continue to weather further into soils. If surficial masses are further covered by a layer of topsoil and go unrecognized in site investigations, they may be incorrectly utilized for foundation purposes, with probably disastrous results, since low-density soils consolidate or deform considerably under structural loading.

FIGURE 2.10
Rock talus slopes on basalt cliffs on the west coast of Grand Manan Island, New Brunswick. (*Photo: R. F. Legget.*)

Glacial Soils

About one-third of the land area of the world was once covered by ice—nearly all of Canada, most of northern Europe, some parts of Asia, and part of South America. The hallmark of glacial deposits is their *heterogeneity*. Rock fragments and boulders are excavated and carried literally hundreds of kilometers from source areas. The particle size gradation of glacial deposits is as wide as is found in any soil and in-place density varies from *loose* to as *hard* as weak concrete. Unexpected cut-and-fill channels and pockets of sand and silt in otherwise hard lodgement till are to be expected. All in all, the geological complexities of glaciated topography are as great as in any other geologic terrane.

Glacial Erosion

Site exploration in glaciated terrane must be conducted with an appreciation of the nature of *glacial* erosion. Masses of ice several kilometers thick overrode the countryside slowly (decimeters per year), plucking and pushing loose all manner of jointed rock masses in their path. The ice masses then crushed and abraded these fragments to ever smaller boulders, cobbles, pebbles, and sand, finally producing *rock flour*. Much of the debris was picked up and incorporated into the ice mass, being later discharged in the melting of retreating ice sheets.

Glaciers are found in two types: *continental glaciers* of broad expanse and great thickness that tend to excavate and level the countryside, and *intermontane glaciers,* which are thinner, sinuous masses of ice that cannot overcome their mountainside constraints, and which move slowly down and out of valleys. Most engineered construction in glaciated areas will be strongly affected by the results of continental glaciation. The high degree of erosive smoothing of even the hard metamorphic rock of Manitoba may be appreciated by viewing Figure 2.11.

Just as ice sheets were able to excavate hill masses of hard crystalline rock, so were they able to further gouge and remove masses of lowland soils. Unexpected *buried river valleys* or *bedrock valleys* were also produced by complex preglacial or interglacial river erosion, and by subglacial streams operating below the ice sheets.

Types of Glacial Deposits

Glacial deposits are all considered as soil. These soils are found in an amazing variety of rock and mineral content, heterogeneity, and density. Unfortunately, in many geological and engineering publications, glacial deposits are not sufficiently identified for engineering purposes. It is common for glacial debris, in general, to be referred to as *drift* in North America (an undesirable term for engineering purposes). From an engineering standpoint, three basic types of

FIGURE 2.11
Effects of ice action over metamorphic rocks of the Cross Lake group, Manitoba; striae, polishing, and gouging can clearly be seen. (*Courtesy Geological Survey of Canada, photograph No. 117,969.*)

glacial deposits exist: *tills,* and *glaciofluvial* and *glaciolacustrine* deposits. *Tills* are the basic product of continental glaciation. Excavated rock and soil debris of the preglacial countryside, scooped up and carried along with the ice sheet, was finally deposited in irregular ridges and mounds (*lateral moraines* and *terminal moraines*) across the countryside, as loose, highly heterogeneous, granular *ablation till.* In contrast, the fine-grained, clay-rich material caught beneath the ice sheet was transported a far lesser distance, but ultimately consolidated under such extreme pressures to form *lodgement till* (also *basal till* or *boulder clay*). This till is a dense and very hard heterogeneous, clay-rich material, often found with compressive strengths equivalent to that of weak concrete (Fig. 2.12). Lodgement till also underlies the curious oval-shaped, streamlined hills known as *drumlins,* aligned with long axes parallel to the direction of glacial ice movement and resting together in fields of other such hills. Unexpected occurrences of either type of till can greatly affect site excavation and construction conditions. The engineer must always strive to detect the presence of till and take care in specifying which of the two varieties is present.

One of the most important contributions that geology can make to soil studies is at the same time one of the simplest—an indication that the site has

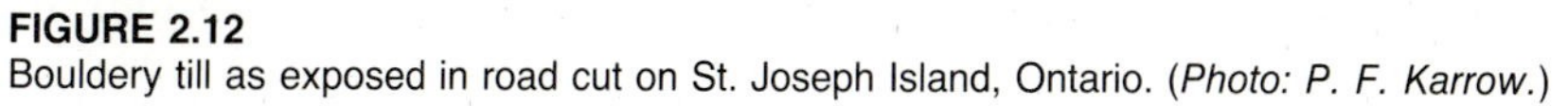

FIGURE 2.12
Bouldery till as exposed in road cut on St. Joseph Island, Ontario. (*Photo: P. F. Karrow.*)

been subjected to glacial action. This will be the case for almost all the northern section of the northern hemisphere and for very small corresponding areas in the southern hemisphere. The civil engineer must then be on the alert for the great variation in glacial soils that may be found in relatively short distances. The distinct possibility exists of encountering buried preglacial river valleys that were subsequently filled with glacial deposits that may themselves differ from the overlying soil. The civil engineer must also take into account the possible effects of ice pressure upon soils.

Glacial till, then, is a soil type to be expected. If a mixture of dense clay, sand, and gravel is found in boreholes, that mixture should not be characterized just as "hardpan" but should be carefully studied, if necessary by means of block samples, to determine its behavior under project conditions. If the direction of ice movement can be deduced, then high till densities and consequent toughness of the material, especially when dry, may be expected. If the material has to be used as a foundation bed, this feature will be desirable. If the material has to be excavated, however, this feature may be ruinous if not recognized before contract arrangements are made. If the till is the *ablation* variety, deposited in morainal form, then the reverse may be expected, and relatively low densities may be found with corresponding open structure. Ablation till was encountered in northern Quebec. Its perfect particle-size grading made it almost impermeable, but the addition of water filled its open voids. After heavy rains it quickly acquired the consistency of the justly famous local *soupe aux pois,* with results that can be imagined.

Glaciofluvial soils result from the meltwater discharge of retreating glaciers. Under an infinite variety of topographic and glacial melting conditions, these soils run the full gamut of particle-size gradation from cobble to clay. Since water played a major part in their transport and deposition, glaciofluvial soils tend to be more homogeneous than till. As these *outwash* flow channels broadened with distance from the receding glacier, stream velocity decreases brought about successive deposition of ever-finer materials. *Outwash plains* and their associated, roughly conical *kames* are fine sources of sands and gravels. Where outwash channels flowed below and away from the ice sheet, highly desirable, snakelike gravel ridges were often deposited in the form of *eskers*. Most urban eskers will have been destroyed by removal of their highly prized content by the late twentieth century. *Ice-contact deposits* exhibit characteristics of both till and outwash, and are generally found to cover only relatively small areas. Such soils originated from melt-front deposition of very wet ablation till.

With increasing distance from the melt front, suspended and bed-load materials of meltwater are reduced to the clay and silt fractions, resulting in glaciolacustrine deposits. Lake beds such as these covered thousands of kilometers in late Pleistocene time, leaving broad expanses of compressible, varved silts and clays throughout the lower margins of glaciated terrane.

It might be thought that there could be no such variations with sands, but many engineers have discovered otherwise, sometimes to their cost. The density of sand can vary appreciably, depending on size gradation and how the individual sand grains are packed together. It is a normal condition for sands to be close to their maximum density, but for geologic reasons not yet fully understood, sands of low relative density are sometimes encountered. The relative density of sands, therefore, is an important soil characteristic that can best be measured indirectly in the field by means of penetration tests; it may also be tested by unusually careful sampling. Geology cannot here be a certain guide, but if sand of low relative density is encountered at a site, geological comparisons between its location and other local sand deposits can usually be helpful in further exploration. When sand of low relative density has been found, it can be consolidated to a greater and more desirable density by a variety of engineering methods, such as pile driving, falling weights, the use of explosives, and the combined use of water and vibration.

Since glacial silts grade into glacial clays, they may be considered together. At one extreme, a coarse silt may consist of particles of fresh minerals that fully justify the name *rock flour* (essentially a very finely ground sand). When wet, such materials may appear at first glance to be identical with clays. Their true character may easily be detected when they are dry, but the use of simple soil tests will readily display the difference, even in the wet condition. The Atterberg tests, for example, will give low values for the plasticity indices of silts.

Once recognized, the rather difficult properties of glacial silts can be anticipated, and design and construction measures may be accordingly adapted. In some glaciated areas (along the north shore of Lake Superior, for example), it is possible to trace a gradual change in glacial deposits, all of which look alike, from silts (as described) to true glacial clays. The soil particles gradually change from fresh minerals to clay minerals, and the properties change correspondingly. Here is a case in which the simple tests of soil mechanics not only give the engineer the information needed, but, at the same time, reveal information of considerable geological significance.

For all fine-grained glacial sediments, a check upon the natural moisture content and its relation to the lower liquid limit is a first requirement in soil testing. In many areas, especially on and near the Precambrian Shield in Canada, the natural moisture content may be appreciably higher than the lower liquid limit. When this phenomenon is first detected by those who have not encountered it previously, they immediately suspect that the test results are wrong. But the results may be right. This unusual condition is again explained by the process of deposition. During the settling of the fine particles in the fresh water of glacial lakes, the particles came together in such a way that interparticle attraction led to what can best be described as an artificially loose "honeycomb structure" of the solid particles. Excess water is therefore held between the particles and gives the resulting soil an artificially high moisture content. When the soil remains in its natural position, this unusual condition is of no moment and may be .unsuspected. When such soils are disturbed, however, as they can be by engineering operations, the excess water may be released and quickly convert what had previously been a solid-looking material into a viscous liquid that will flow readily on low slopes until the excess moisture is lost and the soil "solidifies," having a new, lower moisture content.

Geological identification of these unstable glacial clays can clearly be a most useful guide; this points the way to the necessity for careful soil testing. The phenomenon of *varving* (the deposition of such soils in thin, alternating light and dark layers probably representing annual sedimentation cycles), is of great geological interest, even though the varving will not usually have any significant effect upon engineering uses of the soil (see Fig. 2.13).

Other glacial clays may have been deposited in seawater rather than in fresh water. Clays so deposited are unusually sensitive; they too have an abnormally high moisture content. Accordingly, they are quite unstable when disturbed and have been the cause of many disastrous landslides along the St. Lawrence Seaway and in Norway. Fortunately, the areas in which this has taken place are now fairly well recognized. The most extensive are in southern Scandinavia and in the valleys of the St. Lawrence and Ottawa rivers, with small extensions into connecting valleys.

The Atterberg limits give critical indications of the character of fine-grained soils. This can be illustrated in a general way by the graphical relationship of the lower liquid limit and the plasticity index for a group of typical glacial soils. Mention has already been made of another soil test that has considerable geological significance. This is the test upon enclosed soil samples, a test that relates their consolidation under load with time. The graphical records of this interrelation has a well-defined pattern which clearly reveals, for some soils, that the soil has been subjected to a previous long-term load that has already partially consolidated it. This value is known as *preconsolidation load*. The load may have been caused by overlying soil that was removed for some reason long before the sample was obtained. Commonly, the load will be an indication

FIGURE 2.13
Distortion in varved clay at center of a drained lake, Steep Rock Lake, Ontario. (*Courtesy: Steep Rock Iron Mines Ltd.; photo, N. Vincennes.*)

of the weight of ice that has at some time in the past stood upon the soil in question. This preconsolidation load is an essential factor in the determination of probable settlements when the soil under test is actually loaded by a structure, but the geological significance of the test will be at once apparent.

Erratics

In his monumental nineteenth-century studies of glaciation, the Swiss naturalist Louis Agassiz reported on large boulders found so far from their natural bedrock that he thought their transport must have occurred in a body of glacial ice. Large *erratic blocks* such as these, as well as smaller ones, should be routinely expected. Erratics in excess of 1,000 tons have been found at elevations *above,* and at great distances beyond, their parent outcrops. Their unexpected effect on the costs and manner of site-preparation excavation can be substantial. At the site of the cutoff trench for the Silent Valley Dam near Belfast, Northern Ireland, an erratic was encountered "as big as a cottage." It is prudent to obtain a diamond drill core to a depth of at least 3 m (10 ft) whenever "rock" is struck in exploration in glaciated terrane. Furthermore, an engineering geologist should seek clues by examining the core to determine if it represents site area bedrock or an unusual rock type.

Marine Deposits

Due to the gradually sloping nature of most coastal shorelines, marine deposits are generally found as (1) *shore zone;* (2) *shelf,* or *shallow-water zone;* and (3) *deep-sea zone.* Except for the shore zone, fine-grained materials dominate. Shore-zone deposits are characteristically heterogeneous in size and rock content, reflecting mainly the proximity of fluvial discharge points for continental sedimentary debris, as well as the content from wave erosion of shoreline cliffs.

Shelf or shallow-water materials (extending outward to the 200-m, 100-fathom, or 600-ft depth) are of engineering importance when there is need to construct wastewater treatment outfalls, incoming petroleum transport pipelines, or harbor works. There is much continuous wave-action sorting, in which finer material is gradually carried down into deeper water. Of particular interest will be the clays, silty muds, and organic oozes, all of which present extremely poor foundation characteristics. Crustal subsidence of the Pleistocene epoch has depressed many bedrock units into the shelf zone; the engineer may thus be warned that not all shelf areas will be directly floored with soil. Shallow-water deposits of late Tertiary and Pleistocene age may also appear on land, elevated by crustal rebound or sea-level reduction. Many of these soils are nearly horizontally bedded.

SOIL CHARACTERISTICS

Geologists and civil engineers generally maintain different levels of appreciation of soils: The geologist has an ingrained interest in origins, and the civil engineer is preoccupied with solid types and their particular engineering properties. Those who must use soils as an engineering material, however, soon learn that there is a universal need to classify soils correctly, as *engineering geologic units,* each unit having a reasonably predictable set of physical characteristics and geologic history. The expected engineering properties and behavior of such a unit will be narrowed down sufficiently and can be treated as a design-related engineering material. Thus, the common meeting ground between geologist and engineer is the mutual need to consider the particular soil being dealt with.

One of the first places that engineering geologists and geotechnical engineers look for information on soils are the agricultural *soil surveys* common throughout the world. These surveys commonly classify soil to a depth of 200 cm (6 ft) as a means of assessing land for agricultural productivity. Such surveys are produced by *soil scientists* or *pedologists* and contain several work products useful for geotechnical purposes—soil maps, soil profiles, soil classification, and some physical properties. Many soil surveys contain interpretations as to the nature of parent bedrock material underlying the upper soil horizons. An example of such a profile is shown in Figure 2.14.

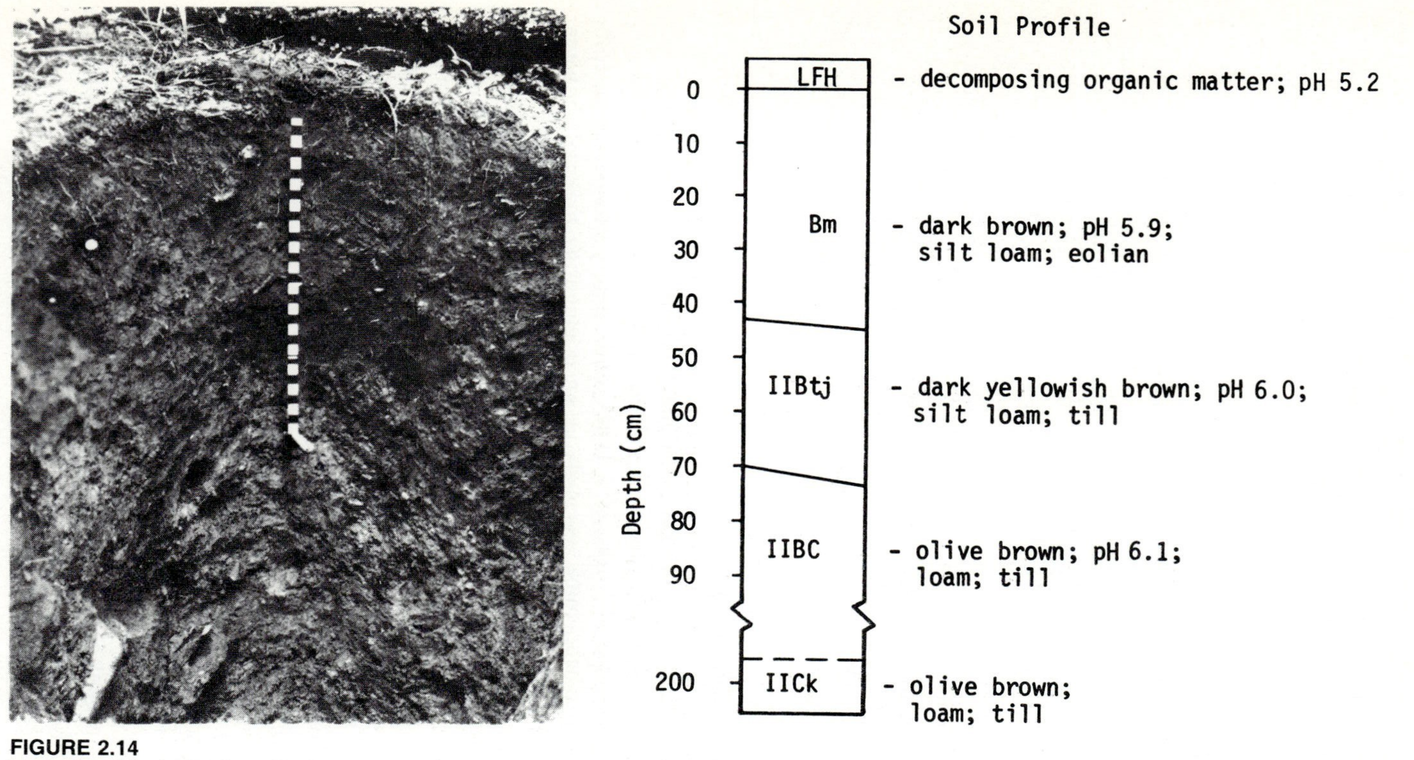

FIGURE 2.14
Typical pedological soil profile (showing schematic "symbolic" representation), from Pend d'Oreille Valley, British Columbia. (*Courtesy T. Vold, B.C. Ministry of the Environment, and J. Day, Ottawa.*)

In general, each particular set of origin and depositional conditions will produce a characteristic soil. In a geotechnical sense, this classification can be made on the basis of soil particle size and mode of deposition. Of the two groups of characteristics, particle size is the more important. This is shown dramatically by the fact that the most widely used means of classification engineering of soils today, the Unified Soil Classification System (USCS) developed by Dr. Arthur Casagrande in 1942 for the U.S. Army Corps of Engineers, relies on particle size as an entry point.

Each of the main groups of soil classification—*gravel, sand, silt,* and *clay*—is based on the *size* of its individual particles, without regard to the nature of the mineralogical content. Compounded descriptive names relate to physical properties, e.g., *loose, fine sand* and *stiff clay.* When in the field, choices of names used to describe soils are still a matter of individual judgment, in which the observer must detect changes between layers of soils encountered in exploratory borings or in exploratory pits. The practicality of the USCS method (or of other accepted schemes) is that, once carefully classified, a range of expected engineering properties is transmitted, in understanding, between engineers.

Gravels

Accumulations of naturally fragmented, unconsolidated rock fragments at least 2 mm in diameter are considered as *gravels*. The rock fragments become known as *pebbles* in the range of 4 to 64 mm; as *cobbles* at 64 to 25 mm (about 10 in); and as *boulders* at a greater diameter. Gravels are rarely found without some proportion of sand, and possibly silt. They are characteristic of shallow-water or river deposits. They may also be beach-formed, and some gravels are found concentrated in certain resource-valuable glacial landforms. The relative degree of roundness, or of angularity of gravel fragments and mineralogical freshness, depends on geologic history. Rounded fragments represent considerable transportation of the more resistant rock types. Angular fragments, on the other hand, usually represent relatively little transport and may be found in association with fragments of relatively soft rock, such as shale.

Sands

Sands are the fine granular materials derived from natural weathering or from artificial crushing of rock; sand ranges from 0.053 to 2.0 mm in diameter. A further subdivision between *fine sand* and *coarse sand* is made at 0.42 mm. A sandy soil will frequently contain some clay and silt mixed with sand-sized grains. Sand may contain up to 20 percent silt and clay, in which case it would be classified as a silty, sandy clay or a sandy, silty clay, depending on which percentage of the secondary constituent is greater.

The origin and occurrence of sands are similar to those of gravel and the two are often found together or as individual layers in one engineering geological unit. Sands formed from similar geologic processes commonly possess particular, similar characteristics. Dune sands, of extremely uniform size gradation, are often found far inland from the sea; river sand often contains relatively large amounts of gravel, silt, and clay. Glacial deposits may contain sand beds. Residual soils often contain horizons of sands, formed in place by selective weathering.

Since sands are often formed from disintegration of gravel, it follows that they will generally be composed of the harder and more stable minerals. Although some sands consist mainly of quartz, most will contain a small percentage of other minerals. In locations where the source materials are other than the usual rock types, sands of nearly pure calcite (oceanic islands) or fluffy ash (volcanic belts) can be expected. Glacial sands, in particular, due to their young age may contain minerals that are especially prone to weathering on exposure.

Sand particle shape varies from completely rounded grains to angular fragments. The former are rare and are found generally in arid regions. Sand particle shape depends principally on parent rock type and on the geologic history of transportation. Long travel distances in water tend to round the grains. As a rule, river sands are more angular than those found in lacustrine and marine deposits.

Silts

Particles in the range of 0.053 and 0.002 mm are known as *silts*. Silts are found in both *inorganic* (mineral fragment) and *organic* (decaying plant matter) varieties. Most inorganic silts have little or no plasticity, and many glacial silts are made up of fresh-mineral *rock flour*. Plasticity is generally due to the presence of plate-shaped clay minerals. Silts are often mistaken for clays on the basis of their typically gray color and apparent consistency when wet. Experienced observers can spot the difference by simple hand tests, such as shaking a moist pat of the soil; silt will lose water and demonstrate a glossy surface.

Clays

Deposits of particles smaller than 0.002 mm displaying a characteristic wet plasticity are known as *clays*. Such a term is size-related and should not be confused with the *clay minerals* that make up much of the content of most clay soils, in addition to such common soil minerals as quartz and feldspar, the main constituents of most rocks. Clays are made up of a combined clay and silt content of at least 50 percent. The flat, platelike particle shapes of clay minerals control the peculiar engineering properties of clay soils.

Clay soils are formed by many processes of rock weathering. They may be *residual* or *transported*. Transported varieties may be found on floodplains, and are of variable thickness and silt/sand content. They may be lake deposits, now elevated as the earth's crust rises, or as sea level falls; or they may be of estuarine or marine origin. The last are a particularly important group of soils since they are widespread and many exhibit characteristics of instability. *Lodgement tills (boulder clays)* are typically clayey soils.

Clays are susceptible to various geologic pressures and are found in conditions ranging from quite soft to extremely hard as their pore water is drained. Natural pressures such as extreme thickness of overlying marine sedimentary layers or of glacial ice sheets have drained pore water in some clay soils to such a degree that the soils, once exposed by excavation at constructed sites, may swell in stress-relief expansion.

Clay Minerals

Possibly the most important basic characteristic of fine-grained soils is the dominant type of mineral present. Clay soils consist primarily of clay minerals. Modern mineralogical methods, including differential thermal analysis, the electron microscope, and X-ray diffraction techniques, have made possible detailed analyses of individual clays and the study of different types of clay minerals.

Prior to the introduction of X-ray diffraction in 1923, it was known that clays consisted of aluminum, silicon, water, and often iron. Such chemical analyses, although interesting, were of little assistance in understanding the properties of different kinds of clay. Clay minerals are silicates made up of very thin sheets of aluminum and/or iron and magnesium. Some contain alkaline materials as essential components. Some argillaceous material may be amorphous, but this is not a significant component of normal clays. The characteristic properties of these crystalline forms go far in determining the physical properties of a clay. The main clay minerals can be grouped together as kaolinite, halloysite, illite, montmorillonite, and chlorite; a few others are occasionally encountered. Of special significance is the fact that there is little bonding force between the successive layers in montmorillonite; therefore, water can readily enter between the individual sheets and cause swelling, a condition which can lead to serious foundation trouble.

The formation of clay minerals is the direct result of weathering. Weathering is an extremely complex process and is still not fully understood. Climate plays an important role. If a given kind of climate persists in one area for a very long time, the same products of weathering may result despite differences in the parent, or fresh, material. Vegetation may also have a profound effect. The pH value of the water that percolates down from the surface (thus acting as an important element in breaking down fresh minerals) will be affected by the character of vegetation in combination with local weather, especially rainfall. The variety of clay-forming processes prior to, during, and since transportation from their place of origin naturally helps to explain otherwise strange variations in materials.

The well-accepted limiting size for clay-sized particles, 2 microns (2 μ), was not just a fortuitous choice, but a useful limited point below which the surface properties of particles begin to dominate the chemistry of the material. Below this size, the electric charges on individual particles tend to increase with decreasing size. It may seem that this discussion is getting too far away from soil mechanics and from geology, but the very important soil property of cation-exchange capacity can only be fully appreciated against this background. Under certain suitable conditions, some of the ions in clay will exchange; the most common exchangeable ions are calcium, magnesium, potassium, and sodium. As the ions exchange, properties may also change. If a change can be predicted and controlled, a useful method of altering clay soils is available.

One of the earliest applications of this method was utilized at the San Francisco World's Fair in 1939. A 3-hectare (7-acre) lagoon was a prominent landscape feature, lined with a 25-cm (10-in) layer of calcium clay of loamy texture. Tests with freshwater showed high leakage losses almost immediately. Because of the intended purpose of the lagoon and the time schedule, this was a serious defect. Laboratory studies showed that base exchange would result from contact between the clay and seawater, with its high sodium content, increasing the watertightness of the clay. The lagoon was therefore flooded with seawater for 45 days, after which freshwater was readmitted. Remarkably little leakage then took place, and the lagoon performed satisfactorily for the duration of the fair.[1]

The geological origin of clays is of considerable but indirect significance in the engineering study of clay soils. The process of weathering is of equal importance, but again indirectly, through the influence it has played in producing the clay minerals actually present in the samples being used. Differences in clay-mineral content will readily explain differences in soil properties, even those demonstrated so simply by the Atterberg limit tests. In most cases, such indirect evidence will naturally suffice for engineering purposes. Knowledge of the significance that clay minerals may have, however, is invaluable to the student of soils, for if unusual characteristics are encountered or suspected, a check on the clay minerals present—such as montmorillonite in relation to swelling clays—may prove of great utility.

ORGANIC SOILS

Although such soils are not usually encountered in the course of normal civil engineering projects, organic soils can cause real problems. These are masses of dead vegetation, often supporting a surface layer of living vegetal matter, but deteriorated to such a degree as to be able to hold up to 1,000 percent by weight of groundwater. Such topographic terms as *marsh, peat, bogs,* and *swamps* in the temperate zone and *muskeg* is Canada and Alaska are a clear indication of the existence of organic soils.

Organic soils are mainly Pleistocene in age, the modern successors to coal. They are seldom found in thicknesses of more than about 10 m (33 ft). Once found in a particular site region, their existence can be related to geologic history. Carbon-14 age determination is helpful in working out the pattern of distribution, since many of the deposits are less than the current age limit of that technique (50,000 years). Organic soils occur primarily in the temperate and subarctic zones. Tropical organic disintegration is so intense as to remove the possibility of accumulation of organic soil deposits.

Spongelike temperate-zone organic soil deposits characteristically load-consolidate in excess of 35 to 50 percent by volume, leading to the greatest of all potential settlements of engineered structures founded on natural materials. In some cases, such as in the Foundation Valley area of Los Angeles, organic soil pockets are not visible, being largely covered by later fluvial deposits and altered by previous site grading and agricultural activity. Engineering geologists here explore for these foundation-hazardous soil pockets by seeking evidence of alignment along former estuarine stream courses and by employing seismic refraction surveys.

PERMAFROST

Permafrost terrane, a condition of permanently frozen soil, covers about one-half of Canada, most of Alaska, and about one-third of the USSR. So delicate is the balance of water contained as ice in solid soil, that even the slightest variations from ambient ground temperature can induce sufficient thawing and subsequent differential or total settlement of buildings, as well as portions of roads and airfields. The culprit soils are silts and clays which, upon slight increases in soil temperature, will see conversion of ice to water (and then drainage) with corresponding consolidation. A proper solution for design on permafrost is to recognize the importance of the frozen ground condition, to utilize geologic principles to search for better-drained sands and gravels in the site area, and to minimize temperature fluctuations associated with construction and operation of each facility.

SOIL MECHANICS AND GEOLOGY

The unconsolidated materials, described as soil, found in the earth's crust constitute so large a part of the actual surface of the earth that few civil engineering operations apart from rock tunneling can be conducted without an encounter with soil of some type. Since foundations cannot always be carried to solid rock, the founding of structures on unconsolidated material is probably the most important part of foundation engineering. Troubles met during the construction of civil engineering works and after their completion are frequently due to failure of soils. Despite their significance in all phases of civil engineering work, soils were not studied and investigated by civil engineers in any general way until relatively recent years. The contrast between this neglect and the past century and a half of progress in the study and investigation of practically all other materials used by civil engineers is so marked as to be indeed a paradox. Nobody sells soils, however. It may be that the absence of commercial incentive is linked in some way with this past neglect. The development of testing and research laboratories under public control for public benefit was probably necessary in order to initiate the surprising progress in scientific soil study of the last five decades.

Soil Mechanics Today

One of the main features of modern soil mechanics is the impressive body of theory that has been developed dealing with all aspects of the states of stress in soils and with the deformation resulting from such stress conditions. Methods of calculation are now available for theoretically determining the stress which an imposed load will cause at any point below and around a foundation slab or the factor of safety against failure of a slope excavated in soil or constructed of soil. One can calculate the total settlement to be anticipated and the rate at which settlement will occur for any clay soil under load at its surface.

There are many theoretical approaches now available for solving structural problems involving the use of soil. All depend upon certain basic assumptions with regard to the properties of the soil being used. The investigation of soils in the laboratory and the values required for calculations have led to the development of special equipment for soil testing in the last three decades. In any modern soils laboratory one will find *consolidometers,* rather simple devices for determining the consolidation characteristics of small soil samples under increasing increments of load; *permeameters,* for the determination of soil permeability; and, of greatest importance, two types of *shear-testing machines*.

For simple tests and for shear determinations upon certain special types of soil, direct shear boxes are used. These are relatively simple machines in which the soil sample is placed in a split box so that it can be loaded vertically. While under load, the sample can be sheared along the break in the box by a horizontal force that can be measured. Of far more importance, however, are the triaxial compression machines that are now widely used. In these a cylindrical soil sample, sealed in a flexible membrane, is fitted into a larger transparent cylinder and held between top and bottom supports. The large cylinder is filled with an appropriate liquid to which pressure can be applied. With the restraining liquid under pressure and with loads applied vertically to the specimen through the top and bottom supports, the cylinder of soil can be subjected to a combination of three-dimensional stress, and loading can continue until the sample fails, usually on a definite shear plane.

Soil testing might appear to be a simple business. In practice, however, it requires much skill and delicacy of operation. Just as the accuracy of theoretical calculations depends on the accuracy of the soil properties assumed or determined, so will the accuracy of the results of the soil tests themselves depend upon the accuracy of soil sampling. The procurement of good soil samples, their accurate description, and the necessary determination of general soil conditions at the site are necessary to demonstrate the validity of samples obtained. *Undisturbed sampling* connotes no disturbance whatsoever of the sample itself, and the retention in the sample of the exact moisture content that the soil has when in place. This last result is achieved by the immediate waxing of soil samples upon removal from a borehole or by the use of special sampling tools, or sleeves, in which samples can be left, waxed at the open ends, until they reach the laboratory.

Soil Testing

The actual techniques of soil sampling and soil testing in the field are matters that do not require detailed treatment here; some notes by way of introduction to current methods will be found in Chapter 3. Suffice it to say that soil-sampling methods have been developed, even for cohesionless soils such as sands, that will give the laboratory worker a specimen of soil as close to the natural condition of the soil in the ground as it is humanly possible to obtain. The first tests will usually be so-called "indicator tests"—those which determine generally the type of soil. These are used as a basis for consideration of the more elaborate mechanical tests and for purposes of accurate description. Usually the first such test made will be to determine the natural moisture content. This is done by drying a small sample in an oven under controlled and standard conditions. Moisture content is always expressed as the ratio of the weight of the water contained to the weight of dry solid-soil matter, expressed as a percentage. Another introductory test will be made to determine the distribution of soil particles of difference sizes, the mechanical analysis of the soil. This test is performed by sieving down as low as the standard no. 200 mesh, and determining smaller particles (with some degree of overlap) by means of a sedimentation method, standardized, an application of Stokes' law. The results are plotted to a semilogarithmic scale; Figure 2.15 shows a few particle-size distribution curves for typical soils. It should be noted that, although it is especially useful for coarser soil mixtures such as glacial till, the record of mechanical analysis is not an infallible guide to solid properties for fine-grained soils. The subdivisions shown in Figure 2.15 for particles of sand, silt, and clay size are those most generally used in North America.

Soil samples will next be subjected to so-called "plasticity tests" to determine their Atterberg limits. These are limits of consistency named after the Swedish soil scientist who first suggested them. These limits are expressed as percentages of the weight of water at specific behavior points, compared with the weight of dry, solid soil in the sample. If water is slowly added to a perfectly dry sample of fine-grained soil and uniformly mixed with it, the soil will gradually assume some cohesion, probably first forming lumps. It will eventually reach a plastic stage at which it can be rolled out in a long, unbroken thread upon a solid surface. A simple test has been standardized to indicate the limit of water content at which this plastic stage is reached; this limit is called the *plastic limit*. If more water is added and the mixing continued, the soil will gradually achieve the state of a viscous liquid.

Another simple test, based on the flowing together of two parts of a pat of

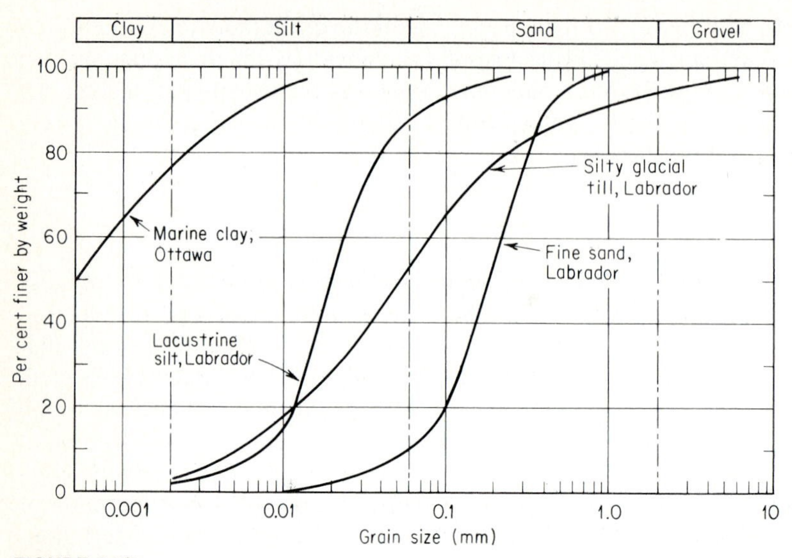

FIGURE 2.15
Standard chart showing the mechanical analysis of particle-size distribution for four typical soils.

the soil-water mixture in a standard cup under standard tapping, has been accepted for general use to indicate this liquid limit. The difference between the liquid and the plastic limits is a measure of the range of plasticity of the soil.

The third "limit" is a much lower value—that percentage of moisture content at which the soil sample stops shrinking as it dries. When this point is passed, the sample will usually grow lighter in color as the drying process continues. This is known as the *shrinkage limit,* but since its physical significance is not so great as that of the other two limits, it is not referred to so frequently.

While the liquid and plastic limits are simple in concept and relatively easy to determine, they are valuable indicators of soil characteristics. The *liquidity index* is obtained by dividing the difference between the natural moisture content and the plastic limit by the plasticity index. These terms appear almost universally in engineering reports as indicators of the general characteristics of a soil. When considered in combination with the appearance of a soil and its mechanical analysis, they assist in the determination of an accurate description of fine-grained soils such as "silty clay," "sandy silt," or "highly plastic clay."

Two comments must precede consideration of the place of geology in this overall picture of soil mechanics. First, it will be noted that no symbols or formulas have been used to describe the simple tests now standard in soil testing. Symbols and formulas are, after all, merely a kind of scientific shorthand, with no intrinsic meaning. The concepts described are so basic and simple that their significance can actually be shrouded by the veil of mathematical symbolism. Second, all the simple tests relate to the quantity of water present in a soil sample at different stages in its wetting or drying. Most of the subsurface problems that have to be faced by civil engineers in the conduct of their construction operations are caused by water and not by solid material as such, whether the material be soil or rock.

Links with Geology

Soil and rock mechanics have become vital parts of the scientific core of civil engineering. At the same time, the scientific study of rock and soils must have contact with geology; this, also, is natural and logical. Soil mechanics closely approaches a geological study when field investigations are involved. Soil mechanics gradually merges into geology at this one extreme and into structural engineering at the other.

When studies of soil in the field are considered from the engineering point of view, however, one may well ask how they can possibly be conducted without due reference to the local geology. One would imagine that it would be impossible to neglect geology, even if not calling it by that name. All too often, however, soil studies have been conducted in the field without benefit from contact with geology in any form, recognized or unrecognized. Sometimes no harm has resulted, but this has been through good luck rather than through good management. Poor results have often been obtained through patent neglect of geological features. In all too many cases, money has been uselessly expended because subsurface explorations were not coordinated with the local geology, and occasionally the neglect of geology has had disastrous results. Some of these will be recorded later in this book, always with constructive intent, since the proper study of failures can assist so greatly in preventing similar troubles on future works performed under similar circumstances. It can therefore be stated without qualification that soil studies in the field, no matter how carefully conducted, are incomplete without some consideration of the appropriate local geology. Herein lies perhaps the chief contact between geology and soil mechanics.

It is not without significance that the name *Geotechnique* has been adopted as the title of one of the leading English-language journals in the field of soil mechanics. The term *geotechnical engineering* indicates the steadily developing liaison between the scientific approach to the geology of rock and soils and the engineering investigation of their properties. It is inevitable that this interrelation should progress to the mutual benefit of both geology and engineering and to the continued advance of human understanding of the most common of all solid materials. In a paper published in 1955, Dr. Terzaghi emphasized the benefits of the relationship, noting:

> The geological origin of a deposit determines the physical properties of its constituents.... Therefore, the knowledge of the relation between physical properties and geological history is of outstanding practical importance.

CONCLUSION: DEALING WITH ROCK AND SOIL

Today's civil engineer faces siting, design, and construction of projects in a bewildering array of earth material types which are far greater than the usual number of manufactured materials used to frame and support the structures themselves. The term *bedrock* itself seems to connote a condition of sturdiness and incompressibility. Yet the range of rock strengths extends from the weak rocks bordering on soil to those that exceed good concrete in compressive strength. For even the strongest rock masses, pervasive discontinuities provide the basis for deformation brought about by stress concentrations related not only to structural loads, but from the very geometry of the portions of the facility that are cut into or lie within the rock mass.

Once the geology of a rock construction site has been defined, rock mechanics can be used as a powerful analytic tool in determining rock strength to meet the stresses that accumulate in and around engineered facilities. There will be great activity in rock construction in the closing years of the twentieth century—tunneling to provide underground space for a variety of purposes, and in general construction in sites previously deemed too difficult to use.

In the case of soil, one can hardly presume that the outstanding advances in soil mechanics of the past 50 years have solved all of the inherent problems related to construction in and on soils. Although a considerable fund of knowledge has been built up in explanation of general soil properties, these properties vary infinitely as a function of the formative processes of weathering, erosion, and deposition. Even the massive sandstones and shales may be seen as evidence of the same processes being present on earth, certainly as long as the planet has had the benefit of free water in the form of rain, snow, fog, and dew, in oceans, lakes, rivers, and streams, and as groundwater. Metamorphic rocks such as slates and some schists and gneisses were unquestionably once sedimentary rocks, originating from soil-forming processes. Truly, so much of the earth's sedimentary and metamorphic rocks are only recycled soils composed of the debris of weathering, alteration, and erosion.

For engineering purposes in North America, however, it is the vast bulk of Pleistocene geologic materials that is of such great importance to civil engineers. Many otherwise strange and perplexing variations in soil conditions can be explained and sometimes foreseen only by engineering geologists who are well versed in the Pleistocene history of the site region. Soil scientists and geomorphologists also deal exclusively with modern landforms and their soil veneer. These scientists are producing an ever-growing body of literature, much of it directed toward actual geographic locations of interest to civil engineers. Charlesworth's remarkable two-volume summary, *The Quaternary Era,* is an excellent starting point from which to appreciate more fully the engineering value of a better acquaintance with the Pleistocene epoch, its soils, and its landforms.

REFERENCES

1 C. H. Lee, "Sealing the Lagoon Lining at Treasure Island with Salt," *Transaction of the American Society of Civil Engineers, 106,* pp. 577–607 (1941).

FURTHER READING

There are available many excellent guides to the study of geology. Especially convenient and fitting into purse or pocket are the Golden Science Guides of the Western Publishing Company. *Geology* by F. H. T. Rhodes (1972), *Landforms* by G. F. Adams and J. Wyckoff (1971), and *Fossils* (1962) and the original *Rocks and Minerals* by H. S. Zim, P. R. Shaffer, and R. Perlman (1957) can all be warmly recommended.

The best of all introductions to geology in the field where it can best be learned is *Field Geology in Colour* by D. E. B. Bates and J. F. Kirkaldy, published in 1976 by the Blandford Press of Poole, Dorset, England. Pocket-sized and inexpensive, it contains 156 excellent color plates which well illustrate its guidance to field observations; it is an ideal guide for civil engineers. (Available from ARCO Publishing Co., 219 Park Avenue South, New York, N.Y.)

The wonderful store of information available from the U.S. Geological Survey can best by tapped by first consulting the Survey's *Circular No. 777,* by Clarke, Hodgson, and North, entitled "A Guide to Obtaining Information from the USGS, 1979."

The authors naturally hesitate to mention one book from the many now available on the science of geology, but, if only because of its incomparable illustrations, *Putnam's Geology,* now edited by E. E. Lawson and P. W. Birkeland, and published by the Oxford University Press, can be the exception. Its title refers to the author of the widely acclaimed original edition. It is well worth looking at during a library visit.

CHAPTER 3 GROUNDWATER AND CLIMATE

Seventy-five percent of all American cities derive their public water supplies from groundwater, a volume equivalent to about 20 percent of all water consumption in North America. Yet this is but a small fraction of the one million cubic miles of groundwater that is estimated to lie within half a mile of the ground surface. Though it is popularly assumed that the metropolis of

Greater London obtains its water from the rivers Thames and Lea, one-sixth of its total supply comes from deep wells, continuing a practice that goes back many centuries. And in India, at least 20 million acres (an area comparable with the total area of irrigated acres in North America) are irrigated with groundwater obtained from wells.

These few figures emphasize the importance of groundwater in today's world. This quantity of water relatively close to the earth's surface is believed to equal one-third of the total volume of water in the oceans. This vast reservoir of water is of great importance to those who have to work in the earth's crust. For the miner, groundwater may be a matter of life or death, and it may be all that swings a venture from success to disaster. Water below ground surface is also of vital importance to the civil engineer, not only as a source of water supply but also as the controlling factor in all drainage operations. Groundwater is a hazard encountered in tunnel driving and other underground operations; it almost always adds to the complexity and cost of foundation work.

Despite its significance, groundwater is often neglected in the practice of civil engineering. Water beneath the surface of the ground is all too often regarded merely as a nuisance during the course of construction projects, and is given due regard only when it leads to serious trouble. Important as is the study of soils for the civil engineer, its appreciation will not be fully effective unless there is an equal appreciation of the significance of groundwater and its characteristics.

CLIMATIC EFFECT

Rivers, streams, and springs are all reminders of a hydrologic cycle that seems to distinguish this earth as a planet. The cycle is but part of the vastly greater dynamic system of air and water around the entire globe, a system influenced by the earth's rotation and to a large extent controlled by movements of the "heavy air" in the lower 5 km (3 mi) of the atmosphere. The dynamics of this great system create what is known everywhere as *the weather,* the "weather machine" being a phenomenon of nature which is vast in extent and fascinating in complexity.

Climate is defined as the generalized weather at a particular site over a long period of time. The localized result of the general weather cycle is important to both engineers and geologists. Civil engineers must design all their structures in relation to all relevant climatic factors—snow loads on roofs, wind loads on all structures above ground, rainfall and its disposal, and the possibility of changing levels of the groundwater surface. Operations of the construction engineer are obviously also determined by climatic factors. Those with experience in construction know that any major variation from average climatic conditions at a building site can interfere with progress and change an anticipated profit into a serious loss. The geologist is interested in the long-term effects of climate since the weather is principally responsible for the geomor-

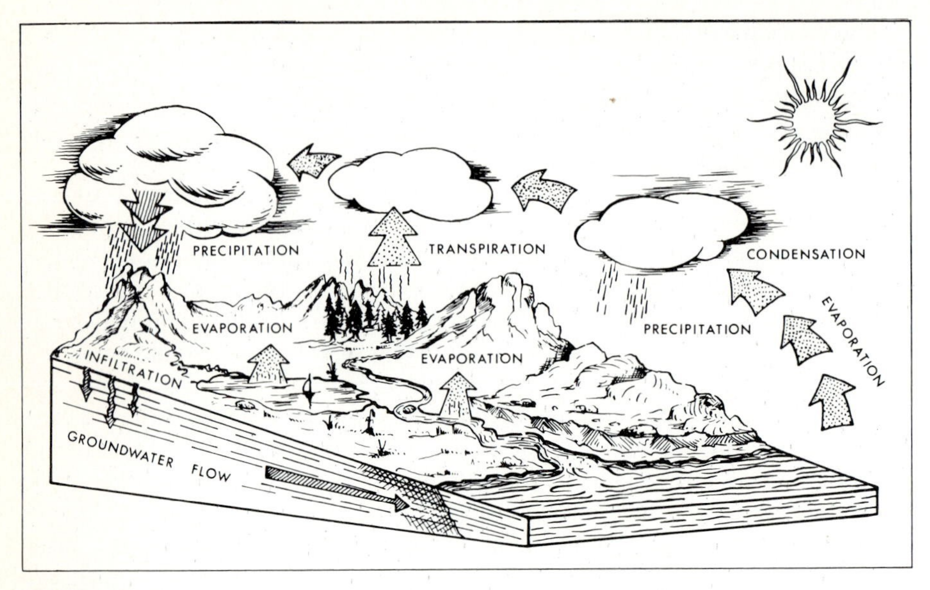

FIGURE 3.1
The hydrologic cycle. (*Courtesy Department of the Environment, Canada.*)

phic changes in the earth's landforms. Weathering of rocks and soils is clearly related to climate, slope stability is greatly affected by climate, and the devastating effects of floods and debris flows are the most dramatic evidence of all of the impact of climate on human affairs. There is a general neglect of climate as one of the natural features that must be studied in advance of design for civil engineering works.

THE HYDROLOGIC CYCLE

The hydrologic cycle provides the key to understanding the effect of climate. The driving force of the system is solar radiation; 47 percent of this radiation is transformed into heat. Temperature changes around the globe are the basic cause of air movements, and also of the movements of water. Seventy percent of the surface of the globe is covered with water, land surface being only 30 percent, so that evaporation from the sea, in particular, is on a scale that is difficult to visualize. It is conservatively estimated that there is 1.25 billion km^3 of water on the globe; of this, 97 percent is the saltwater which forms the oceans. Of the remaining 37.5 million km^3 of freshwater, about 87 percent is locked up, occurring as ice in polar regions and in glaciers, all but 10 percent of that in Antarctica. This leaves a mere 4.7 million km^3 as the freshwater with which all are familiar. Of this, about 90 percent constitutes groundwater. The remaining surface water, on which life depends—that held in lakes, rivers, and streams—amounts to a mere 0.01 percent of the overall total.

These figures provide a useful background against which to consider the dynamics of the whole system. Solar radiation drives water into the atmosphere from all surface waters. A small quantity of water will enter the atmosphere by transpiration from living organic matter, such as trees. A small part of the atmospheric water is held in clouds, which precipitate the water they hold as rain, hail, or snow. The precipitation that falls onto open water, by far the largest part, will repeat this first part of the cycle directly. It is that part which falls on the land that is of special significance in human affairs. Some will run off the surface directly, especially if it is bare rock or soil well protected by a natural cover of thick grass, and so find its way into watercourses and eventually back to the sea. That part which does not run off will percolate into the ground, either through cracks in the surface or directly into soil if it is at all pervious and if it is not already saturated. Figure 3.1 illustrates the cycle graphically.

"Soil water," illustrated in Figure 3.2, represents an average condition. If, however, a site is subjected to continuous and heavy rainfall, then the water level will rise to the maximum elevation it can attain, and any further precipitation will become runoff. If not anticipated, this situation can have serious results in the form of upwelling flow into the floor of excavations, loss of cutslope stability, and transformation of cohesive soil into unmanageable mud, and as a general hindrance to construction activities.

HISTORICAL NOTE

The existence of groundwater has probably been realized from the dawn of history. The coyote and other animals will dig down to accessible water. One of humanity's earliest documents, the twenty-sixth chapter of the Book of Genesis, discloses that biblical peoples had a thorough familiarity with groundwater conditions; other references can be traced throughout the Bible. The Romans were familiar with the use of wells; in England, they had one well 57 m (188 ft) deep, and they supplemented their well supplies by means of adits driven in the chalk. Throughout the Middle Ages, groundwater continued to be widely used, although its occurrence was not understood. Almost until the end of the seventeenth century, people generally conceded that hillside springwater could not possibly be derived from rain. Many explanations were advanced to explain its origin; one of the most interesting was that, owing to the curvature of the earth, the water in the middle of the ocean was actually at a higher altitude than the springs, and thus furnished the necessary head.

Not until the sixteenth century were these views questioned. Bernard Palissy (1510–1590) played an important role in developing the theory of the infiltration of groundwater. Pierre Perrault (1608–1680), one of the first to put hydrology on a quantitative basis, related the total rainfall and runoff for the basin of the river Seine in France. This work was roughly done, but it may be regarded as the starting point of modern hydrology, especially as it was just about this time that Edmund Halley (1656–1742) conducted the first known

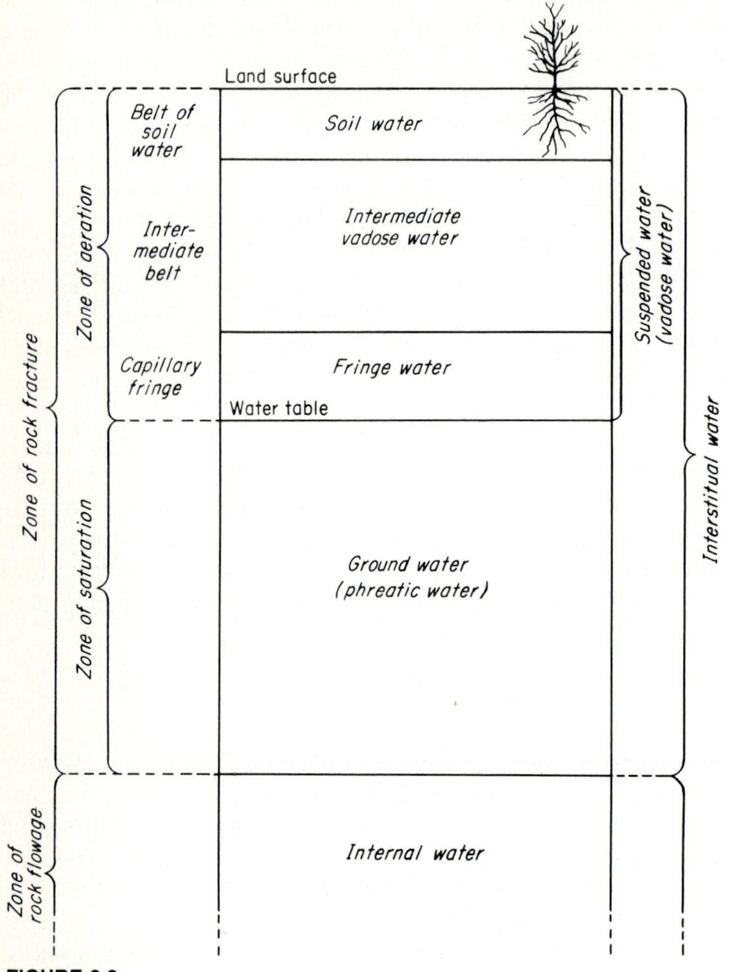

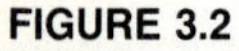

FIGURE 3.2
The main subdivisions of groundwater. (*Courtesy U.S. geological Survey, from W. S. Paper 494 by O. E. Meinzer.*)

quantitative experiments on evaporation and thus proved definitely the origin of rainfall. Thereafter, hydrological work developed steadily into the scientific study of today.

A scientific approach to groundwater problems was adopted and developed by several civil engineers early in the eighteenth century. One of these practitioners was William Smith, who was one of the first to relate the study of groundwater to the study of geology, applying the results to the solution of engineering problems. His works show a vivid conception of the main principles of groundwater location and movement. Two prime examples are the water-supply system for Scarborough and the agricultural drainage of ground at Combegrove, near Bath, in the south of England.

CHARACTERISTICS OF GROUNDWATER

That part of the rainfall which is absorbed into the ground (infiltration) is important in many ways. Absorption may account for almost the entire rainfall, as in the drier sections of Australia where there are many rivers that disappear and others that flow only for certain limited portions of the year. Even more striking is the elevated area of two million acres to the north of Casterton and Coleraine in the state of Victoria. This area acts as the "intake" for the Great Murray River Artesian Basin; it has an average rainfall of 62.5 cm (25 in) and no runoff. This is an extreme case. On the average, infiltration may be subdivided as follows:

1 Absorption directly by plants
2 Capillary return to the surface and there evaporated
3 Absorption in the molecular structure of minerals
4 Direct flows in the sea (in coastal districts), through springs and underground channels
5 Escape at the ground surface through springs or by feeding rivers
6 Retention in the earth

The first two divisions are important in agriculture, the last two in both civil and mining engineering; the fourth is sometimes important in engineering work.

Civil engineering considerations of groundwater refer almost wholly to that part which is retained. *Vadose water* describes that water which is still in the zone of aeration., i.e., in that part of the immediate subsurface which is not saturated with water. Vadose water affects the drainage design of shallow excavations and is naturally related to surface drainage and irrigation. Below the zone of saturation, what is termed *internal water* is located in bedrock (estimated by some to occur to depths of as much as 1 to 3 mi, terminating where rock pressure is so great that no interstices or fractures occur in which free water can exist. This deep water is primarily of interest only in deep mines and in crystalline rock repositories for high-level radioactive wastes. During the construction of the foundations for Harlem Hospital on the upper east side of New York City, a daily flow of several million gallons of groundwater was encountered, at a steady 20°C (68°F). It was effectively sealed off, but since there was no obvious source of the heat it contained, it may have been of plutonic origin.

On rare occasions, plutonic water is released from a primary origin at great depth in the form of hot springs. Those near Kivu in central Africa, at Karlovy Vary in Czechoslovakia, and at Bath in England are notable examples. The mines at Bendigo, Australia have springs at a depth of 1,370 m (4,500 ft). There are wells in the Great Australian Artesian Basin from which water is obtained so hot that it can be used for making tea or cooking.

Engineered works to be constructed in the vadose zone and in near proximity to the zone of saturation require a three-dimensional depiction of groundwater. As can be seen in Figure 3.2, the zone of saturation is topped by a *groundwater level* or *piezometric surface*. This surface is also known by some as the *water table*, but such terminology must be avoided on the basis of legal implications as the piezometric surface is rarely flat. Groundwater contour maps are the preferred method of portraying the elevation of the top of the zone of saturation.

If the earth's crust were of uniform composition, all that is essential for a summary of groundwater characteristics would have now been said. But underground conditions are far from uniform, and so the disposition of groundwater is one of the most complicated problems in geology. Essentially, the disposition of water below the saturation level depends on two main factors: (1) the texture of unconsolidated materials (soils) and bedrock, their composition, and relative porosity; and (2) their structural orientation and relationship to neighboring rocks. The variations possible in both these factors will illustrate the complexity of the distribution of groundwater.

Although conditions underground are variable, groundwater movement is in accordance with the law of gravity, apart only from the minor motion due to capillary action. The fact that groundwater *does* move should be emphasized. Cyclic vertical oscillations of the surface of groundwater affects many engineering projects. This movement is commonly observed in monitoring wells and in the intermittent flow of springs. Evidence of lateral movement along the gradient of sloping groundwater surfaces is perhaps not quite so clear.

It has been well recognized that underground water could travel for vast distances. For instance, it was known that in the artesian basins of Australia underground water traveled for hundreds of miles. There have been cases in the United States where it traveled 320 km (200 mi), with pressures which ran up to 150 m (500 ft) head at the surface and something like 600 m (2,000 ft) head at the bottom. In the Algerian Sahara it was known from borings that water came a distance of 480 km (300 mi) from the Atlas Mountains, passing along definite channels, some of which had actually been traced for a length of 112 km (70 mi).

These cases are extreme but they indicate how extensive underground travel can be. A more common example is the water supply of the city of Leipzig, Germany, obtained from an underground source that is almost a "stream," as it is 3.2 km (2 mi) wide and 12 m (40 ft) deep. A 550-m (1,798-ft) well at Grenelle, Paris, is known to draw water from an ultimate source near Champagne, 160 km (100 mi) away.

The usual means of determining the position of groundwater below the ground surface is to measure the water level in wells or boreholes; multiple measurements can be used to contour the groundwater surface. Sufficient time must always be allowed for equilibrium to be restored after completion of a borehole so that the observed level will not be transitory. Groundwater levels can generally be expected to vary throughout the year in response to local rainfall and water use. This annual variation is of great importance in excavation work, both surface and subsurface, and also in foundation design, road performance, frost action in soils, and in the siting of waste management facilities.

GROUNDWATER IN EARTH MATERIALS

Groundwater is found in all manner of earth materials. The distribution of groundwater in uniform coarse sand and in solid igneous rock are examples of extreme cases of the effect of the composition and texture of rock on groundwater distribution. The influence that the nature of earth materials can have on groundwater is explained by only a few terms. *Perviousness* and *permeability* are commonly used in this connection; they are defined as the capacity of a rock to allow water to pass through it. A pervious rock has communicating interstices of capillary or supracapillary size. Its degree of permeability cannot be correlated with any standard scale but is usually expressed as a volume of flow per unit of cross-sectional area per unit of time (such as liters per square meter per day). Porosity, on the other hand, is a measure of the interstices (pore spaces) contained in any particular volume of the rock. It is generally expressed as a percentage and indicates the aggregate volume of interstices to total volume.

Strange as it may appear at first sight, a high degree of porosity is no assurance of perviousness. Clay, for example, has a high porosity; examples have been found of newly deposited Mississippi clay with a porosity of between 80 and 90 percent. When it is saturated, however, it becomes impervious, the water it contains being held firmly to clay minerals by molecular attraction. Despite the fact that porosity and permeability are not always synonymous, the convenience of the property of porosity as an index to the water-bearing capacity of solid rocks and as one that can easily be investigated in the laboratory has resulted in much attention being devoted to it. Table 3.1 presents some typical values for common types of earth materials.

These figures are approximate only. For example, the porosity of the chalk found in southern England varies from 26 to 43 percent; that of the chalk found in Yorkshire is about 18 percent; and that of the chalk of Antrim, Ireland, is sometimes less than 10 percent. Seven sample cores of Bunter Sandstone (usually permeable and without discontinuities) taken from depths varying from 90 to 225 m (300 to 750 ft) gave porosity values from 13.2 to 30.2 percent and proved to have a permeability ranging from 0.02 to to 7.3 l/m^2 (0.06 to 20.9 gal/ft^2) per 24 hours. These figures serve to demonstrate the wide variation possible in an otherwise uniform type of rock. Despite this appreciable

TABLE 3.1
THE POROSITY OF EARTH MATERIALS

Type of rock	Maximum porosity* (Percent)
Soil and loam	Up to 60
Chalk	Up to 50
Sand and gravel	25 to 30
Sandstone	10 to 15
Oölitic limestone	10
Limestone and marble	5
Slate and shale	4
Granite	1.5
Crystalline rocks	Up to 0.5

*These figures are presented as illustrative only of the general trend of porosity in relation to various rock types. Tests on actual samples should be performed if a value for porosity is desired for design.

porosity, the permeability in the direction of bedding, i.e., perpendicular to the cores, was very small indeed and due only to flow taking place along discontinuities.

In chalk, water traverses along the bands of flints and rubble that frequently distinguish chalk strata. Another hydraulic feature of both sandstones and limestones is the significance of the material that often acts as a natural cement joining the individual grains to form the solid rock. An unusual but striking example of this is given by Monk's Park and Ketton stones. Both are British oölitic limestones in which the oölitic grains are microporous and similar in size and structure. In the Monk's Park stone, however, the intergranular spaces are filled with a nonporous matrix of crystalline calcite, whereas in the Ketton stone they form interconnected pores. The Ketton stone is therefore pervious and yields water freely, whereas the Monk's Park stone is impervious.

It is sometimes thought that porosity decreases with increase of depth as a result of the geostatic pressure of the overlying rock. The chalk of England, usually regarded as the most porous British rock, gives a very small yield because of the pressure of overlying rock on this relatively soft material; a good example is found under Crystal Palace Hill in South London.

In all but pervious rocks such as sandstone and some limestones, fissures are usually the leading factor in determining the water-bearing capacity. Below about 100 m (328 ft) of depth, the presence of open joints generally diminishes appreciably. Below this depth, a regular diminution of water supply from jointed rock may be expected, and therefore an economic pumping limit will soon be reached. In any case, pumping is not normally economic below depths of about 250 m (820 ft) and water cannot usually be regarded as "commercially" available unless under artesian pressure as considered later in this chapter.

Distribution rather than the mere location of groundwater is therefore the prime question in underground water surveys. For this reason, some general notes on the water-bearing properties of the more common rock groups follow.

Sand and Gravel These materials, both porous and pervious, are ideal water-bearing strata. Their wide use as artificial filtering and drainage media is well-established engineering practice. Extremely fine sands, however, are of little use in any of these categories.

Clays and Shales As a general rule, these will be useless as sources of groundwater. Although clays are often wet, the water present is not readily available. Hard shales may yield limited water from joints and bedding planes.

Sandstones Sandstones are variable in texture and composition. Some may be almost impermeable and others may be so pervious that water actually squirts out from them when under pressure, as from some of the Bunter Sandstone of England. Pervious sandstones form an admirable source of groundwater, providing not only a high yield, but acting as an effective filtering medium. For example, water from deep wells in New Red Sandstone is almost invariably clear, sparkling, and palatable.

Limestones This extensive group of rocks is second only to sandstones as a source of groundwater. Its importance is well indicated by the existence of underground dissolution channels, and associated karst terrane, as well as by the wide dependence on chalk strata as underground reservoirs. The water-bearing characteristics of chalk have already been noted. Furthermore, water in contact with limestones dissolves a small quantity of the rock, tending always to increase the pumped yield. It is calculated that with every million gallons of water pumped out of the chalk of the south of England, about 1,500 kg (3,300 lb) of chalk is also removed; thus pore spaces and any existing underground channels are correspondingly enlarged.

Crystalline Rock In general, these rocks are not classed as water-bearing, although few are absolutely dry when encountered in excavation. When decomposed as the result of weathering or some other cause, and also when fractured and fissured, they may yield appreciable quantities of good potable water. Fresh crystalline rocks in massive formation will not generally yield any useful quantity. Granites, when decomposed, may have a relatively fair yield; quartzites, slates, and marbles may yield a useful supply because of jointing; gneisses and schists, unless they have decomposed badly, cannot be relied upon for any appreciable quantity. In Cornwall, England, several important towns obtain their water from wells in granite, probably from infiltration into the blanket of disintegrated rock at the surface. In Maine there are many residential wells in igneous and metamorphic rock, but their yield is usually small (less than 40 lpm; 10 gpm) unless jointing is very marked.

GROUNDWATER QUALITY

The quality of groundwater is of vital importance, whether the water is to be used for industrial or for domestic purposes. In general, groundwater is free from bacteria, since the passage of water through the ground strata constitutes a natural filtering process. This does not remove, however, the vital necessity for routine bacteriological examination of all groundwater to be used for domestic purposes, especially when site geology would permit contamination of the groundwater from surface sources.

Groundwater will almost certainly contain dissolved solids and gases. Methane gas is usually formed by the decomposition of organic matter in the absence of free oxygen and may occasionally be encountered in water that has not traveled far from surface deposits or that has been in contact with strata of organic origin. It is dangerous, and if it is present in such quantities that it is liberated from the water on reaching atmospheric pressure, it can be a toxic or explosive hazard. Hydrogen bisulfide occasionally occurs and is easily detectable by odor if present in any appreciable quantity. Its origin may vary, but a possible source is the interaction of organic acids from surface deposits with underground sulfates. The gas is easily removed by aeration; its presence is often the chief feature of medicinal springwaters. Carbon dioxide is the most important gaseous impurity of groundwater; its origin is generally the atmosphere. It gives water a sparkle which is not unpleasant, and it is not therefore objectionable. But in water the gas makes a weak solution of hydrocarbonic acid, which acts as a solvent for several different rock constituents, and for this reason the presence of the gas is often significant.

Free carbon dioxide gas is occasionally encountered in civil engineering work in sufficient quantities to be troublesome. On the aqueduct that conveys water from Owens Valley to the Los Angeles area in California, severe corrosion was discovered in the steel pipelines. Tubercles formed very quickly, and constant cleaning and painting were necessary for adequate maintenance. Careful tests found that between the two ends of a 2,100-m (7,000-ft) tunnel the carbon dioxide content increased from a negligible amount to 4.2 parts per million (ppm). The tunnel was dewatered and examined; the gas could be heard hissing as it escaped through cracks in the concrete lining. The tunnel penetrates Soda Hill, a granitic body containing many highly sheared fault zones; limestone body strata are thought to exist nearby. Acidic water was the cause of the corrosion troubles. The remedy adopted was to lead the gas through special chases into a pipe beneath the tunnel invert which in turn was connected with a new adit, 600 m (2,000 ft) long, through which the accumulated gas could escape.[1]

Pure water will dissolve only 20 ppm of calcium carbonate and 28 ppm of magnesium carbonate, but water containing carbon dioxide will dissolve many hundreds of parts per million of these solids. These and similar solid impurities in groundwater may give rise to the main purification problems of the waterworks engineer. Water obtained from limestone strata is always suspect. The dissolved carbonates give the water what is termed *temporary hardness,* since the dissolved solids can be removed by simple chemical processes. If the dissolved solids are the corresponding sulfates, the resultant hardness of the water is termed *permanent,* since it cannot be removed by simple processes used for eliminating the carbonates. Chemical analyses are necessary to determine the degree of hardness of the water. If this exceeds about 200 ppm of calcium carbonate the water requires softening.

Trace elements are those minute quantities of the rare elements which are essential for healthful living. The absence of the necessary minute quantity of fluorine in water, for example, causes dental decay. Conversely, there are other elements, minute traces of which can be fatal. For example, several cases of arsenic poisoning occurred some years ago north of Kingston, Ontario. The presence of arsenic was finally detected in the farm well water used by the victims. This was traced to minute quantities of ferrous arsenate existing naturally in the limestone stratum through which the well had been excavated.

The presence of high concentrations of calcium and magnesium sulfates in groundwater can cause very serious trouble with concrete work in contact with the ground. Occurring chiefly in heavy clay soils, high-sulfate groundwater is common in western Canada. It was recognized as early as 1908, when problems were encountered with deterioration of concrete in Winnipeg. By 1918 these problems had become so serious that the Engineering Institute of Canada was requested to appoint a committee to look into the matter. A major study was undertaken, and the satisfactory results led to the use of sulfate-resisting cement.

The same problem is widespread in Great Britain. Sulfates of calcium, magnesium, and the alkali metals occur widely in the Mesozoic and Tertiary clays of Great Britain, including such widely known formations as the Keuper Marl, the Lias, and the Oxford, Kimmeridge, Weald, Gault, and London clays.

Sodium chloride may originate either in the sea or it may be of mineral origin. Rock-salt deposits are an obvious source of supply; some are "mined" by means of brine solutions. Although this system of mining is of great interest from the point of view of utilizing groundwater for mining purposes, it can have unusually serious consequences if the workings are so close to the surface that settlement can result.

To deal with the possible impurities in groundwater is to recognize the relation of such impurities to the geological conditions of the ground from which the water is being obtained. Preliminary study of the local geology will at least suggest the possible impurities to be expected in any prospective groundwater supply. Accurate knowledge of the relevant geology will demonstrate the course of any known impurities and indicate the probability of any increase or decrease of the amount of contamination. All civil engineers should be able to gain a general idea of the geology of the source area of the water supply of a town they may be visiting merely by washing their hands. If the water is hard, as from limestone, lathering with soap will be slow.

INFLUENCE OF GEOLOGIC STRUCTURE

Underground conditions affecting groundwater differ from the ideal case not only because of the wide variation of materials in contact with the water but also because of the way in which the various rock strata are arranged, in general and in relation to one another. The first variation from the ideal is the alternation of pervious and impervious beds. If the impervious stratum constitutes the surface layer, no water will normally penetrate to the potential underground reservoir provided by the pervious bed. A more usual case has the pervious bed at the surface. Rain will soak through until it reaches the main body of groundwater retained above the impervious stratum from which supplies can be drawn by means of boreholes or wells. This arrangement may be duplicated, and such a "sandwich" structure can be imagined. Such structures actually occur. The Middle Lias Formation in Northampton, England, only 28 m (93 ft) thick, contains no less than five water-bearing horizons separated by impervious clays, any four of which may be present at one location. The condition illustrated in Figure 3.3A will be interfered with only if the impervious stratum is pierced. This would seem hard to imagine, but cases have occurred in which wells have been drilled too deep. At Kessingland, near Lowestoft in England, the bottom of an old well penetrating 15 m (50 ft) of gravel was pierced, and the water tapped by the well ran down to the lower stratum. A similar instance happened in the drilling of a 21-m (70-ft) well at Nipigon in Ontario.

A further variation will be obtained when the strata are inclined instead of level, as indicated in Figure 3.3B. In this case, the inclined impervious stratum will constitute a barrier between the two pervious beds, so that the elevations of the respective groundwater surfaces need not be the same. A similar effect is noted when alternating layers constitute part of a fold. Many arrangements are naturally possible, but only one general case is illustrated. Figure 3.3C shows how water will collect in such a distorted stratum. The geologic structure underlying the city of London is of this type. Finally, Figure 3.3D demonstrates the effect that a fault may have on the distribution of groundwater in alternating strata. The variations possible in this simple case are dependent on the relative thickness of the strata, and the nature and throw of the fault. In the case illustrated, the barrier provided by the impervious stratum remains, but a slight increase in the throw of the fault would leave a gap in the barrier, with consequent alteration of the groundwater conditions. The shatter zone generally found in rock bordering a fault plane may also alter the water-bearing character of a rock by providing water passage along the planes of fracture.

Variations of the four simple cases already considered are illustrated in

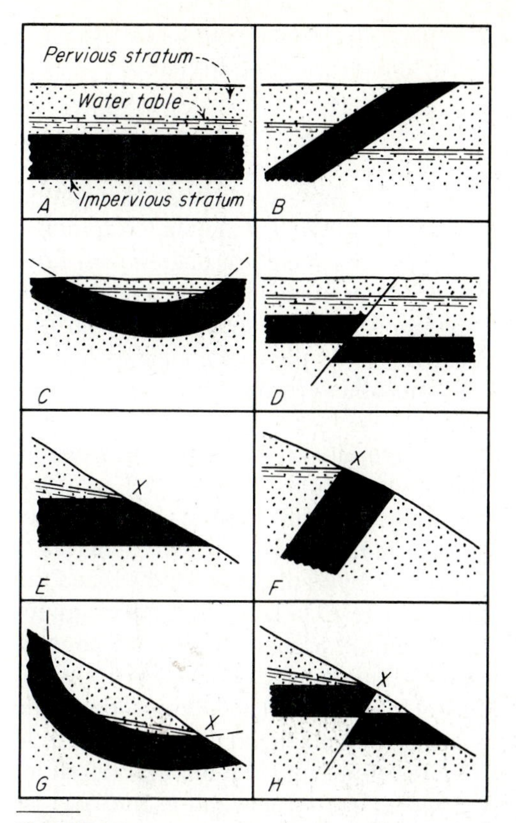

FIGURE 3.3
The effect of geological structure on groundwater distribution.

Figures 3.3E, F, G, and H, with the surface of the ground inclined; the altered groundwater conditions are clearly apparent. At the points marked *X*, bodies of groundwater will come into contact with the atmosphere. This means that the surface of the groundwater will not be level but will be inclined at the hydraulic gradient necessary for flow to take place through the material of the stratum. An important consequence is that, as a rule, the "groundwater surface follows the ground surface"—a broad statement, but useful as a guide.

Of particular significance is the variation in the yield of water obtained from the same rock at different positions along the range of a fold in which the rock has been distorted. If the folding has been anticlinal and is of relatively short radius, there will be a distinct tendency for fissures to develop. Similarly, if the folding has been synclinal, there will be a tendency for the rock to "tighten up" and have its usual perviousness reduced; this tendency is most marked in the softer rocks. This has been demonstrated in connection with the water supply of the municipality of Richmond, London, where the construction of a well and over 3,000 m (10,00 ft) of adits in the chalk failed to develop an appreciable supply because the location was on a syncline in the chalk formation.

GROUNDWATER MOVEMENT

The concept of groundwater movement is not so widely appreciated as might be imagined. In general, the movement follows D'Arcy's law, amended to allow for the particular hydraulic properties of pervious media through which flow takes place. Well established as is the study of groundwater hydrology, it can never be forgotten that the theory assumes certain specified subsurface conditions and that the practical application of the theory demands accurate information about subsurface conditions—in other words, an appreciation of local geology.

The water level in a producing well will naturally fall when pumping begins and will continue to do so until the hydraulic gradient in the material surrounding the well is such that flow into the well will occur at the rate equal to that at which water is being removed by pumping. For uniform material, a parabolic curve will approximate the depressed hydraulic gradient between the unaffected groundwater surface and the water level in the well, under equilibrium conditions. As this condition will obtain all around the well, a cone of depression will result (Fig. 3.4); the apex of the inverted conical form of the groundwater surface will be the water level in the well. The intersection of the cone of depression with the natural water level will, theoretically, be a circle which will mark the limit of the range of interference of the pumping operation. Any other well located within this circle will be interfered with to some degree, and its effective yield of water will be reduced. Such interference can be a critical matter in built-up areas where it is desired to have wells close together

FIGURE 3.4
Simplified diagram of the cones of depression formed when groundwater is pumped.

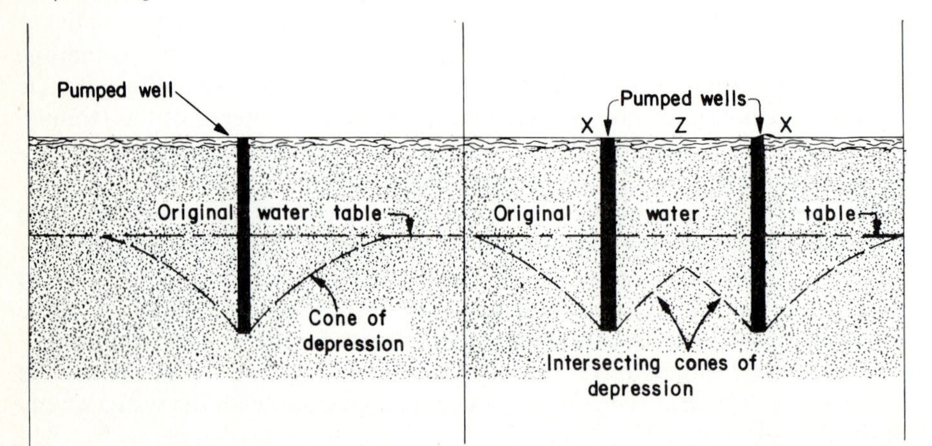

so that groundwater resources will be used to maximum extent. These ranges of influence are quite wide in very pervious strata. In Liverpool, England, for example, the effect of pumping has been noticed as far as 3.2 km (2 mi) away from a pumped well in pervious sandstone.

GROUNDWATER SURVEYS

Determination of groundwater conditions becomes a vital part of preliminary site investigations. Groundwater surveys generally call for trained and expert hydrological advice and a working knowledge of local geology. In addition, familiarity with the hydrological characteristics of the host rocks is essential before even general principles can be applied to the particular problem. The following is only an outline of what is involved in this survey work and is presented so that engineers engaged in groundwater problems may know how to approach them and may appreciate the advisability of expert assistance when the necessity arises.

The engineer's problem usually is to find water at such a depth, in such quantities, and of such quality that it can be utilized economically in the public service. Plants were the first indicators of groundwater to be used regularly, and vegetation still constitutes a fairly reliable guide. In the first century B.C., Vitruvius listed plants known to him to indicate the presence of groundwater near the surface of the ground and even endeavored to suggest when they were reliable guides.

O. E. Meinzer suggested the term *phreatophytes* for those plants that habitually grow in arid regions only when they can send their roots down to the groundwater surface.[2] His paper shows that the vegetation encountered will be some indication of the depth to the groundwater surface, and indicates that there is at least some basis for the idea that plants indicate the quality of the groundwater. In other parts of the world, information regarding plants which may serve as good guides can generally be obtained from observant local inhabitants.

Contour maps provide one means of estimating the quantity of groundwater available. Various methods may be used; their relative suitability is dependent on local conditions and the data available. One method is to attempt to prepare a water balance sheet for the area considered, measuring the total runoff and rain, estimating the losses due to evaporation and absorption by plants, and determining the remaining balance, which may normally be assumed to be the addition to the groundwater supply. Measurements may also be made regarding actual discharge of groundwater into watercourses from areas in which the groundwater surface is known to be reasonably steady. A time lag will always be found between the falling of rain and an increase in groundwater flow, but with this determined, estimates can be made of the percentage of the rainfall available as groundwater. The most reliable and most widely employed method is to determine and use the specific-yield factor for the rock in question. The one serious obstacle to the application of this method is the possible variation in the hydrologic nature of the local rocks, which may invalidate the calculations.

Specific yield describes the interstitial space that is emptied when the groundwater surface declines, expressed as a percentage of the total volume of material that is unwatered. Laboratory methods are convenient but the possible variations, not only of the materials tested but also of their structural arrangement in the area being studied, make these methods of doubtful value unless they are used as a guide to probable yield of groundwater before major operations are undertaken in the field. Effective field methods for estimating specific yield depend on the recording of water levels underground during pumping or recharging. Records of groundwater levels will always be the most useful guide to groundwater supplies, since inevitably they reflect all the subsurface variations that might be disregarded in the application of any experimental test results. Even in the use of recorded water levels, however, observations from adjoining boreholes and wells should be carefully correlated with one another and with what is known of geological conditions in the vicinity. In this way any underground irregularity may be detected from the data available, if such detection is possible. A record of one of William Smith's many engineering activities in England may be cited in view of its historic interest and its intrinsic illustrative value. In a letter dated February 24, 1808, from John Farey to Sir Joseph Banks, the following descriptive note has been found:

> We met the Reverend Mr. Le Mesurer, Rector of Newton-Longville near Fenny-Stratford, who related his having undertook to sink a well, at his parsonage house, within a mile or two of which no good and plentiful springs of water were known, but finding clay only at the depth of more than 100 feet, was about to abandon the design: Mr. Smith, on looking into his map of the strata, pointed out to us, that Newton-Longville stood upon some part of the clunch clay strata, and that the Bedford limestone appeared in the Ouse river below Buckingham, distant about eight miles in a northwest direction, and he assured Mr. L. that if he would but persevere, to which no serious obstacles would present themselves, because all his sinkings would be in dry clay, he would certainly reach this limestone, and have plenty of good water, rising very near to the surface; Mr. L. accordingly did persevere in sinking and bricking his well, and at 235 feet beneath the surface (the first 80 feet of which were in alluvial clay with chalk and flints, etc. similar exactly to what I have uniformly found on your estate at Revesby, and in the bottoms of many of your fen drains) the upper limestone rock (8 feet thick) was reached and found to be so closely enveloped in strong blue clay as to produce not more than 9 feet of water in the well in the course of a night; from hence, an augur hole was bored in blue clay, for some distance, to the second limestone rock which produced a plentiful jet of water, which filled and has ever since maintained the water, I believe almost up to the surface of the ground.[3]

SPRINGS AND ARTESIAN WATER

Springs sometimes provide a useful and pure water supply. They may give a continuous supply, but some springs can be affected by the pumping of groundwater in adjacent areas. Springs often accompany instability of ground and especially that of steep natural slopes, requiring that their origin and course must always be determined in advance of design. Figure 3.3E represents in simple fashion the geologic arrangement necessary for steadily flowing springs. One can readily see from this how spring flow may vary with time, but the relatively slow movement of groundwater will explain the lag between heavy rainfall and a variation in the flow of springs (Fig. 3.5).

Water is said to be *artesian* when the groundwater rises either up to or above ground level as soon as the water-bearing bed is pierced. Local geological structure, in which the water is stored in the bed under hydrostatic pressure, accounts for this phenomenon. The name *artesian* is one of the few in connection with subsurface conditions that is not self-descriptive. It originated from that of the province of Artois, France. Groundwater contained in chalk does not find its natural level until the overlying Tertiary clay is pierced.

Subartesian wells are those in which the groundwater is under hydrostatic pressure but not to such an extent that it is forced to ground level. As soon as it is freed it comes to rest at some point below ground level but above the top level of the water-bearing bed. The artesian pressure is due to the hydrostatic head created by groundwater confined by the same impervious strata but lying above the level at which it is tapped. This condition may arise in a syncline of alternating pervious and impervious beds, either in the trough of the fold or on the upper flank, as illustrated in Figure 3.6. Alternative structural arrangements and conditions can be imagined.

FIGURE 3.5
Springs coming out of limestone bedrock in the valley of the Saugeen River, Ontario. (*Courtesy Ministry of Natural Resources, Ontario.*)

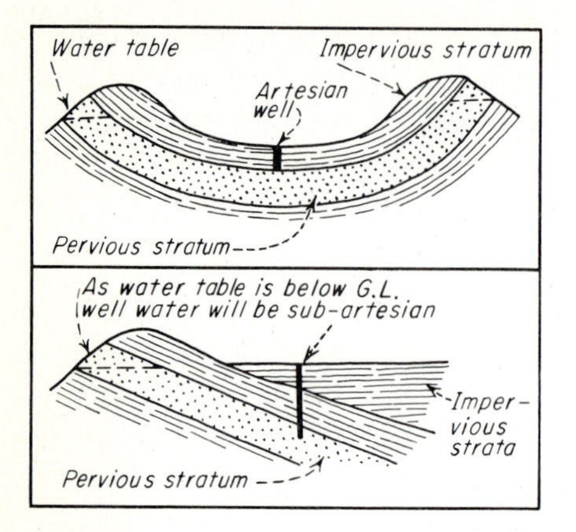

FIGURE 3.6
Simplified geological sections illustrating artesian groundwater conditions.

Artesian water may quite possibly exist beneath impermeable strata which form a river bed, with results such as the unusual occurrence shown in Figure 3.7. This photograph was taken after a borehole at the site of the Shand Dam in Ontario had penetrated lodgement till and entered the underlying Guelph Dolomite beneath the waters of the Grand River. The same condition was encountered in the 1977 construction of dry docks at the U.S. Naval Station at Bremerton, Washington. At Thebes, ancient Egyptians sunk access shafts and then bored 15- to 20-cm- (6- to 8-in)-diameter wells to intercept water-bearing sands, in many cases 120 m (400 ft) from the surface.

In basins, artesian water may be present at depths at which local conditions and demands will determine the relative economy of using such supplies, which explains the great variation in the depths to which artesian wells have been sunk. In Berlin, Germany, and in St. Louis, Missouri, it has been necessary to drill to depths of 1,000 m (3,280 ft); a well 476 m (1,585 ft) deep at Ottershaw appears to be about the maximum depth in England; at Boronga, New South Wales, 1,302 m (4,338 ft) deep. Along the Atlantic seaboard of the United States, artesian supplies at depths of a few hundred meters are relatively common. Among the most famous of these is the 30-cm (12-in) well at St. Augustine, Florida, which originally supplied 38 million liters (10 million gal) a day from a depth of 420 m (1,400 ft).

Most artesian water bodies are relatively small. There are some extensive regional artesian groundwater bodies such as the Great Artesian Basin in North and South Dakota, which has an area of about 39,000 km^2 (15,000 mi^2). It is widely used as a source of water supply. The water-bearing bed is Dakota Sandstone, and it is possible that some if not all of the artesian pressure may be due to the weight of the overlying superincumbent rock strata. Such a

FIGURE 3.7
Groundwater under artesian pressure flowing out of a drill hole casing pipe in the bed of the Grand River, Ontario, at the site of the Shand Dam. (*Photo: R. F. Legget.*)

phenomenon was discussed by Thales about 650 B.C. and later by Pliny as being the agent responsible for elevating seawater to the level of springs.

Australia is a continent in which artesian water is of major importance. The Great Australian Artesian Basin is certainly one of the natural wonders of the world (Fig. 3.8), covering an area of more than 1.5 million km^2 (590,000 mi^2) of the states of South Australia, Queensland, and New South Wales. Over 3,000 artesian wells draw a daily flow of over 2,250 million liters (600 million gal). The basin approximates the ideal complete artesian basin. The eastern rim is tilted up to a greater altitude than the western, and the main aquifer extends in a practically continuous body of very soft Jurassic sandstones. Artesian conditions extend over the greater part of the basin, becoming subartesian in the western section conditions. The water obtained is of good quality, but its temperature is high, extremely so in the parts which reach a depth of 1,500 to 1,800 m (5,000 to 6,000 ft). In the New South Wales portion, temperatures vary almost directly with depth—30°C (86°F) at 210 m (700 ft) to 50°C (122°F) at 900 m (3,000 ft).

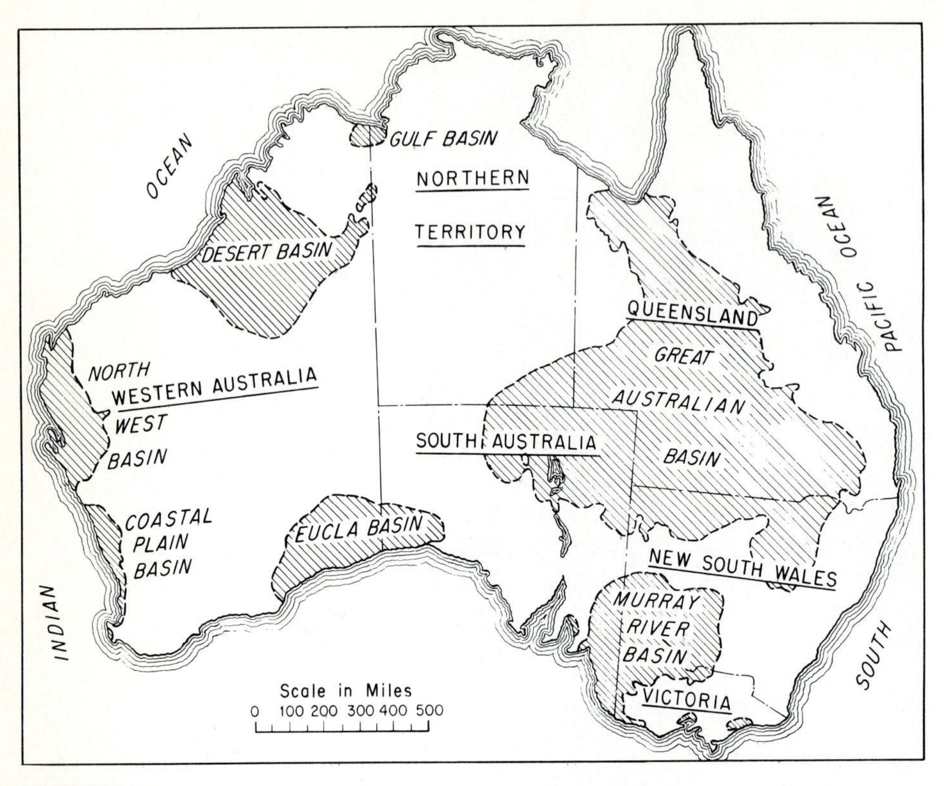

FIGURE 3.8
Sketch map of Australia, showing the extent of the principal artesian basins.

GROUNDWATER NEAR THE SEA

Sodium chloride is a common impurity in groundwater, developed through contact of groundwater with rocks having soluble content. Another source of sodium chloride and allied salts in groundwater is the serious problem of encroaching seawater in coastal areas. Basically, the matter is one of simple hydrostatics. Since the specific gravity of seawater is slightly greater than that of freshwater, a state of equilibrium will exist when the two are in contact in a porous medium, as illustrated in Figure 3.9. This superimposing of a lens of freshwater over seawater is called the Ghyben-Herzberg effect after the two scientists who investigated the matter in theory. Herzberg's original investigation of the interrelation of seawater and freshwater was carried out in Germany about 1900. The findings apply generally to all geological conditions in coastal areas and so should be kept in mind, not only with regard to pumping groundwater in coastal areas but also in connection with any civil engineering works which may interfere with natural groundwater conditions. The whole of the Florida peninsula and such islands as Oahu, on which Honolulu is situated, are most certainly completely underlain at depth by seawater. The state of

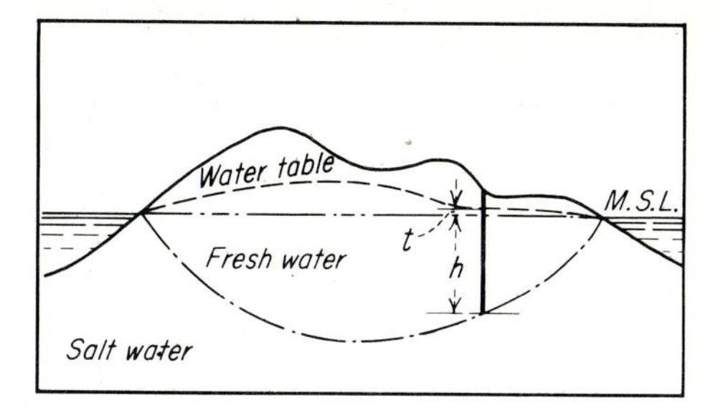

FIGURE 3.9
Diagram showing the Ghyben-Herzberg effect when freshwater overlies saltwater.

equilibrium is a delicate one. It can very easily be disturbed, and once disturbed it cannot readily be restored. Overpumping the freshwater will permit a largely irrevocable advance of the boundary between freshwater and salt.

This natural balance has been upset in many parts of the world. Wells along the shores of the Thames Estuary and in the Liverpool district of England, wells along much of the California coastline, and wells tapping the remarkable groundwater reserves under Long Island, New York, all have been made brackish when saltwater became mixed to varying degrees with freshwater because of excessive drawdown of the latter. The presence of brackish water in wells indicates the existence of water-bearing, i.e., pervious, strata. If the wells are adjacent to the sea, it will normally follow that at some time the freshwater will be in contact with the infiltrated seawater. The zone of diffusion between the two is relatively narrow, and thus any appreciable change from the normal position of the groundwater (such as can be caused by excessive pumping) will result in a corresponding change in the position of the saltwater. As the average salt content of seawater is about 35,000 ppm, and the normal maximum salt content for water for domestic use is about 3,000 ppm, it will be clear that it takes but a small contamination of freshwater by seawater to make the former unfit for human use. Rainfall onto the exposed land surface restores balance by replacing freshwater lost to evaporation and transpiration.

The case of Angaur Island, one of the Palau Islands of the Pacific, about 1,280 km (800 mi) southwest from Guam, is an illustration of the serious results that simple excavation can have under such circumstances. Angaur Island, roughly triangular in shape, has an area of only 8.3 km^2 (3.2 mi^2). Valuable phosphate deposits on the island have been mined steadily since 1908, although the commercially available deposits are now exhausted. In 1938, the Japanese introduced power-operated equipment to speed up the mining operations. With the availability of such improved methods, large excavations were soon completed. Some of these large pits went appreciably below sea level. Water from the freshwater lens overlying the main body of seawater began to form pools and induced flow of seawater. The previously potable groundwater supply on the island was contaminated. Remedial measures carried out by U.S.

forces after the war included sealing off artificial lakes and reducing their size, using coral rock in fill, with as much fine material included as possible. This is a small and isolated case, a vivid reminder of the delicate equilibrium that exists wherever seawater and freshwater are in contact beneath the ground, a situation that should never be forgotten when civil engineering works, especially pumping operations, have to be carried out near the seacoast.

RAINFALL

Rainfall patterns are influenced by topography, and topography is always an expression of the local geology. There is, therefore, an intimate relationship between rainfall intensity and local geology. Study of local rainfall records, therefore, must be an early part of all site investigations. Where no local recording stations exist, expert advice should be taken as to how the nearest available records can be best extrapolated for the site being studied. In such cases, for major projects, early installation of a recording rain gauge should be seriously considered. Close attention must be paid to the relation between the rainfall during the period of site study and the long-term average, since if any major deviation from the normal is occurring, then suitable allowance for the effect of this must be made. In particular, unstable slopes will often move during excessive rainfall. The reverse naturally holds true; slopes that appear to be quite stable in very dry weather may display instability when continuous heavy rain falls on the site. The Handlova landslide in Czechoslovakia is a telling reminder of the vital necessity of continuous vigilance and regular study of current rainfall records wherever unstable ground is known to exist.

Rainfall is a major contributor to flood flows in rivers and all smaller watercourses. If a site under investigation includes any watercourse, however small, every effort must be made to determine the extent of the floodplain and to correlate this with relevant rainfall records so that the necessary allowance can be made for possible future flooding. Floodplain mapping is now well developed in many urban and developing areas, but the danger of building on known floodplains remains.

Climatic records in many regions are so sparse that the engineer would do well to develop a sensitive eye to read the record contained in the landscape. No better example can be cited than the desert regions. These dry regions with low average rainfall seemingly contradict the weather records; the sensitive eye can see that the surface has been etched and eroded mainly by the effects of running water. While long intervals, perhaps years, may pass between storms, the rare, intense, localized storms provide spectacular evidence of the power of flowing water.

Drainage facilities will be dependent upon maximum runoff flows, and so upon intense rainfall. If possible, no major design should be completed and, correspondingly, no site investigation should be regarded as complete, until the site in question has been seen during the heaviest possible rainfall (Fig. 3.10).

FIGURE 3.10
What a difference the rain makes: A small-scale construction site after heavy rainfall. (*Photo: R. F. Legget.*)

TEMPERATURE AND WIND

Seasonal temperature variations are usually taken for granted and seldom thought about to any degree except by those with special interest in the weather. These variations are, however, of great importance in civil engineering works and they influence geological phenomena such as weathering. The average annual cycle of temperature should therefore be studied as a part of all site investigations. Maximum temperatures may affect concreting operations. Minimum temperatures will dictate the necessity of special precautions for winter operations, if below-freezing weather should be experienced. Alternations between freezing and thawing—freeze-thaw cycles—can be of importance in relation to the durability of exposed materials, such as weak rock, either in place or as used in construction. The duration of cold throughout a winter, when considered in association with the nature of the local soil, will affect the penetration of frost into the ground.

Temperatures are therefore an important factor in the local climate and are recorded at all weather stations, usually as the maximum and minimum for each day. When these are plotted on a chart which already contains a record of the long-term average temperatures, then daily variations from the normal can readily be seen, and judgments made accordingly. The duration of winter cold is measured as the total of degree-days for the season, a degree-day being the average daily temperature subtracted from 18°C (64°F).

The duration of winter has another effect of much importance in site studies. As underground construction for appropriate purposes steadily increases (Chap. 6), a knowledge of local ground temperatures becomes correspondingly important. Figure 3.11 illustrates a typical annual chart of ground temperature variation. In summer, the temperature of the ground close to the surface will be somewhat less than the local air temperature at midday, but still warm. Correspondingly, in winter the temperature close to ground surface will be just slightly higher than that of the adjacent air, unless there is a local cover of

FIGURE 3.11
The pattern of soil temperature variation, showing also the occurrence of permafrost.

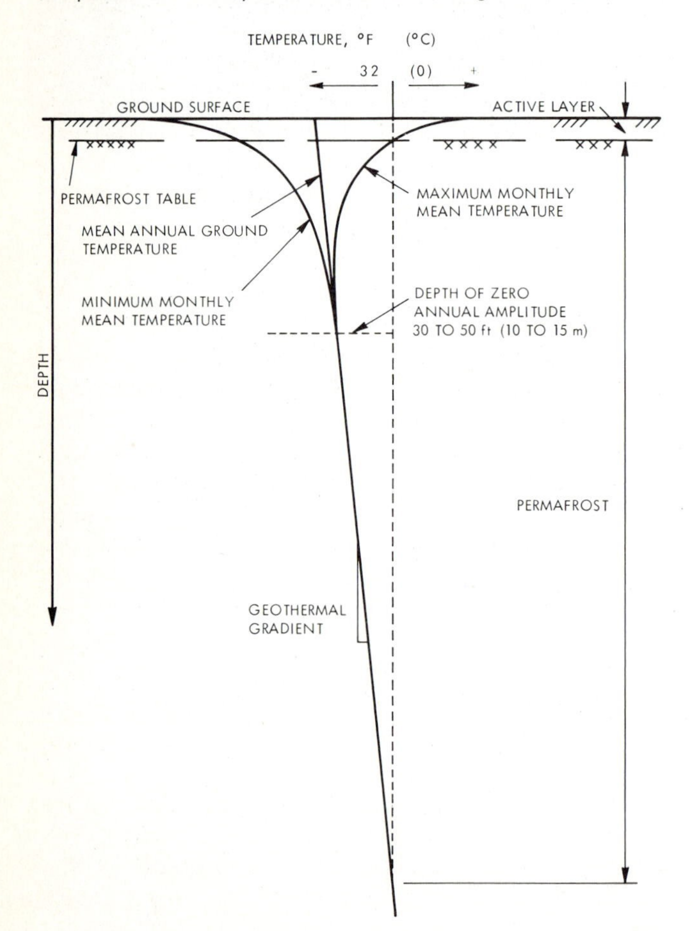

snow, in which case the difference will be more marked. Due to the thermal properties of the ground (whether soil or rock), the subsurface temperature will change rapidly with depth, decreasing in summer and increasing in winter. It is found that there is a hysteresis effect at a modest depth below the ground (3 m or 10 ft near the Canadian-U.S. border), due also to the insulating effect of the ground, and a corresponding time lag of up to six months. (This explains why, in northern regions, trouble with frozen pipes is experienced at its worst in the spring, not in the depth of winter).

Eventually, at a depth of not more than 10 m (33 ft) in temperate parts of the world, the annual variation of ground temperature becomes imperceptible. Below this, the temperature will increase slowly due to heat coming from the earth's core, a typical rise being 1°C (1.8°F) for every 90 m (295 ft). This general pattern is almost universal for temperate regions. A little thought will suggest that the mean ground temperature must be the mean annual air temperature for the locality, or very close to it. Where snow covers the ground for a part of each winter, there will always be a difference between the two figures of one or two degrees. Appreciation of the general pattern shown in Figure 3.11 will assist with many problems involving underground works, but the phenomenon of ground temperature variation has still further significance.

Since the mean ground temperature at any location is close to the average annual air temperature for that location, it follows that as the latter climatic factor decreases, so also will mean ground temperature. The average air temperature will steadily decrease the further one goes north. There will be some point on each meridian at which the average annual air temperature will be at freezing point, and beyond this it will fall increasingly below 0°C—so also will the mean ground temperature, with a slight lag. In all areas where the average ground temperature is below freezing, the ground is in the condition of *permafrost*. This word (although semantically inaccurate) is now universally used in the English language for this condition of the ground; it does not denote any special material.

Local air temperatures, therefore, are indeed significant in site studies, quite apart from their relevance to human comfort. Not many site studies will have to be carried out in permafrost areas, but it is worth noting that one such study in northern Canada did not take ground temperatures into consideration. The location was in the discontinuous permafrost zone, a fact undetected until construction was well advanced; necessary remedial work can best be left to the imagination. In all regions where frost is regularly experienced, and that includes a large part of North America and Europe, troubles may be experienced during construction or with the operation and maintenance of structures, as soil temperatures are not something of significance only in far northern parts.

CLIMATIC RECORDS

Every major country today has a national weather service responsible for recording all weather features throughout the country and for publishing these

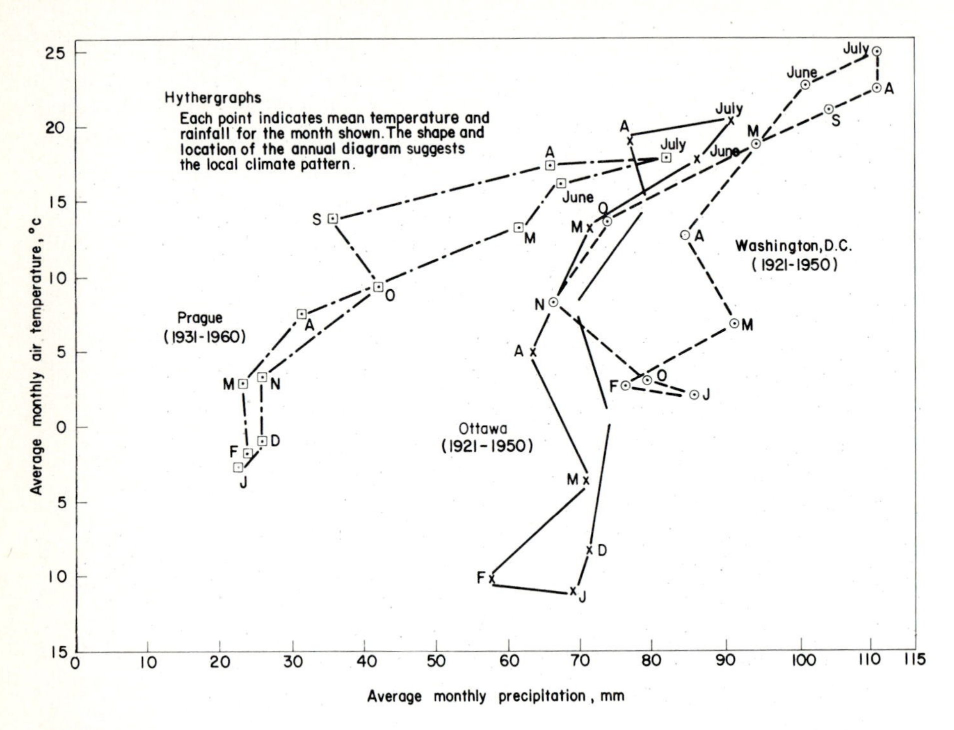

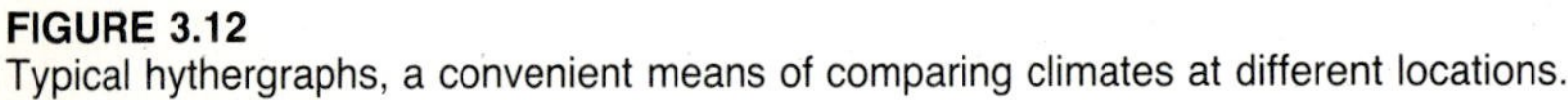

FIGURE 3.12
Typical hythergraphs, a convenient means of comparing climates at different locations.

in convenient form. Typical are the series of fine bulletins entitled *Climates of the States* prepared by the National Oceanic and Atmospheric Service. These are available for each state from the Superintendent of Documents in Washington, D.C. Information given in a typical bulletin includes "freeze data," long-term average air temperatures, and precipitation by month for locations throughout the state, as well as the extreme values so far recorded. Helpful maps with isolines for air temperatures and mean annual precipitation are also included. With every passing year, the duration of the available records increases and so all such publications are regularly issued in revised and updated versions. In Canada, the Atmospheric Environment Service of the Canadian Department of the Environment provides similar climatic information, as well as useful bibliographies and advice as to the availability of further and more detailed information.

On first looking at one of these climatic summary publications, the tables in which most of the records are presented appear rather foreboding. Those accustomed to reading tabular material will be able to select the information for the location in which they are interested without difficulty and to appreciate the general pattern that it presents. It takes an unusual grasp of figures, however, to visualize the "picture" of the annual variation in climate for one particular location. An elegant solution to this difficulty was proposed by Griffith Taylor in his geographical writings, but Taylor's solution, the *hythergraph,* has not yet achieved widespread use. In our opinion, it is a device that should be featured in every site report and in many other geological and engineering publications. It is illustrated in Figure 3.12, which presents several contrasting hythergraphs. By this simple system of plotting monthly average air temperatures and average precipitation, a 12-sided polygon unique to each location is obtained. As can be seen, the hythergraph provides a ready means of comparing at a glance the climates of any one location with that of another.

In individual hythergraphs it is possible to include a record of prevailing wind, also. Starting at each monthly temperature-precipitation point, a line is drawn to a length representing, to some scale, the intensity of wind for that month, an at an angle to the vertical corresponding to the wind compass direction; an arrow completes each vector. These have not been used in Figure 3.12 since the necessarily small scale would have made the resulting diagram too complex, but it can be done, thus giving all main climatic factors for one location in one simple diagram. So useful are hythergraphs as an aid to site study that we wish we could use them freely in later chapters of this book when describing cases in which climate was a decisive factor. This would be impractical, but we do commend the use of hythergraphs in all geotechnical reports upon site investigations.

CONCLUSION

Practicing civil engineers are becoming increasingly aware of the overall importance of groundwater in civil engineering work. The widespread comparative neglect of the subject more than likely is very often due to its inherent simplicity and the fact that water is so "common" a substance. Engineers are beginning to realize, however, that groundwater must be taken into consideration whether the works be large or small. Control of groundwater is often the key to the solution of stability and construction problems on all sizes of projects. In many cases, it can be pointed out that if there had been no groundwater present, there would be no problem. Often, it is the case that water, in association with soils that would have been quite stable if dry, is the cause of trouble, and once the water is removed or controlled, the troubles are quickly minimized.

The fact that this simple truth has so often proved to be the critical factor will serve to explain why this early chapter emphasizes the importance of groundwater in civil engineering work. Groundwater characteristics must be considered even before necessary preliminary site investigations are discussed, since adequate subsurface investigation must define local groundwater conditions. Such attention is essential to the study of local geology, which must now be thought of as more than the geology of rocks alone.

And as for climate, there is an old saying attributed to Mark Twain that "everyone talks about the weather but nobody does anything about it."

Although powerless to change the weather, the engineer has a responsibility to be aware of what it holds in store, both from the standpoint of structural design and problems during construction. "Forewarned is forearmed" is as appropriate for climatic factors as it is for geological factors. Both are essential parts of the environmental context or framework in which the engineer must work. The most careful assessment of the properties of geological materials under certain "normal" weather conditions may be defeated by major changes in those properties under "abnormal" conditions. Climate, therefore, must always be a vital part of all site investigations.

REFERENCES

1 "Corrosiveness of Aqueduct Reduced by Gas Drainage," *Engineering News-Record, 117,* p. 790 (1936); see also same journal, "Drains Remove CO^2 from Tunnel." *120,* p. 885 (1938).

2 O. E. Meinzer, *Plants as Indicators of Groundwater,* U.S. Geological Survey Water Supply Paper No. 577, Washington (1927).

3 T. Sheppard, *William Smith, His Maps and Memoirs,* Brown, Hull, England, p. 245 (1920).

CHAPTER 4

SITE INVESTIGATIONS

Before undertaking any design work for a project, a civil engineer must have full information about the earth material on which the structure is to be founded, constructed with, or in which the construction is to be carried out. This will necessitate a preliminary examination of the work site before designing is started and a thorough investigation of the site before detailed designs are prepared. Truisms? Admittedly, but they are truisms that have often been neglected in the past, sometimes with disastrous results and almost always with consequent monetary loss. This last statement might be characterized as too sweeping were it not for the many recurring examples from engineering practice. To the civil engineer with even a slight appreciation of geology, neglect of what is the obviously essential course of making thorough and complete site investigations before embarking on construction will appear to be utter folly. An engineer who failed to make such investigations might with propriety be compared to a surgeon who operated without making a diagnosis or to a lawyer who pleaded without having any prior discussion with a client.

To the engineer who has not studied geology, these analogies might appear farfetched, since in many cases the nature of the surface of the ground and experiences on similar adjacent works might seem enough evidence to warrant making certain design assumptions about subsurface conditions. Such reasoning has led to the occasional neglect of preliminary explorations in the past. Similarly, unquestioning reliance on only the information given by exploratory borings, considered without relation to the local geology, has often resulted in unforeseen difficulties in construction and may have led engineers to distrust the value of such preliminary work. All too often, however, it has been the difficulty of obtaining the necessary funds before the start of construction that has prevented engineers from carrying out the explorations that they knew were necessary.

Most civil engineers today recognize the supreme importance of this preliminary work. Readers who may entertain doubts about its value will not have to read far beyond this chapter before they find evidence that should set all their doubts at rest. But all engineers may have to deal with owners who have not actually seen the necessity of exploratory work. Subsurface predictions from geologic mapping and from image interpretation, good as they may be, and sufficient for very general reconnaissance purposes, are yet indirect. Such predictions must be checked by actual penetration of the subsurface, either by exploratory boring and sampling of the soil and rock encountered, or by excavation of small shafts or adits, which will also yield samples of materials for later testing in a laboratory. Exploratory excavations are primarily limited by cost and by the problems of deep penetration if more than shallow depths are to be studied. Exploratory drilling can now economically reach any depth likely to be needed in site investigations for civil engineering works. These techniques have changed little over the years, although they are being refined steadily with improved equipment. This is well demonstrated by the fact that the masterly post–World War II report on subsurface exploration by Hvorslev was reprinted in its entirety in 1968, being still regarded as probably the best guide to the subject in the English language.

One of the main problems in advance explorations of building sites is obtaining any money at all for this essential preliminary work. It must therefore be carried out efficiently and economically. This means that, if exploration is to be fully effective, it must be carried out in close association with geologic studies of the site and, as work proceeds, in the closest correlation with geological predictions. This, in turn, makes it quite essential that contractual arrangements for the conduct of test boring and drilling must always be flexible. This ensures that the program can be changed, expanded, or reduced, depending on how the gradual accumulation of facts from the borings confirms or disproves the underground conditions anticipated on the basis of surface geologic observations.

It is probably in the conduct of subsurface exploration that the engineering

TABLE 4.1
SUMMARY OF MAIN METHODS OF DIRECT SUBSURFACE INVESTIGATION

Method	Materials in which used	Method of work	Method of sampling	Value in foundation-bed investigations
Hand-auger boring	Cohesive soils and some granular soils above water table	Augers rotated and withdrawn carefully for removal of soil	Samples from augers always very disturbed	Satisfactory only for shallow investigations for roads, airports, and small buildings
Test pits	All soils above the water table	Hand excavation, necessary bracing, and due safety precautions	"Undisturbed" samples can be cut out as required	Most valuable; soil conditions can be studied in situ
Wash borings	All soils except the most compact	Washing inside a driven casing and retaining "sediment"	Material so obtained a rough guide only	Almost valueless; dangerous if limitations of method not fully appreciated
Dry-sample boring	All soils except the most compact	Washing inside a driven casing; sampling tool used at bottom of hole	Relatively "undisturbed" samples usually obtained	Most reliable of simple methods of boring; groundwater information lacking
Undisturbed sampling	All soils except the most compact	Forcing sampling tool into cased hole to get continuous core of soil	Best possible "undisturbed" samples	Best available method of soil sampling, with a wide variety of excellent sampling tools available
Core drilling	All solid rock and very compact soil	Rotating power-driven coring tools with diamond, shot, or steel cutters	Cores cut out and recovered from holes	Best method of studying bedrock and boulders
Caissons, trenching, and tunnelling	All types of ground	Regular construction methods used, with due safety precautions	"Undisturbed" samples obtained from working faces	Best of all methods, but expensive; essential for major works

geologist and the civil engineer come closest together, since the proper conduct of the work requires the closest attention of both. The actual work of excavation, test boring, or drilling is an engineering operation, the details of which lie outside the scope of this textbook. An outline of the methods will, however, be presented as a background for proper appreciation of subsurface exploration as an essential part, but only a part, of site investigation which is, in turn, so necessary for the proper execution of exploratory work. Table 4.1 summarizes the main methods.

PRELIMINARY INVESTIGATIONS

Economics of Preliminary Investigations

The *economics* of preliminary investigations should be considered in relation to the total cost of a project and of the savings that may be effected and the difficulties that it may avoid. Preliminary work includes geological surveys, geophysical surveys, exploratory pits, borings, and any further exploratory measures that may be called for in special cases. Examples of the percentage expenditures on such work are not easy to obtain because the totals are usually so small that they cannot suitably be featured in traditional cost summaries. It has been possible, however, to assemble some figures for interesting examples of tunnel, dam, bridge, and building construction; these are given in Table 4.2.

These records from differing works and from varied parts of the world show clearly that the cost of preliminary work relative to the total cost of a civil engineering project is small indeed. When compared with other percentages included in civil engineering estimates, the minimal economic significance of these costs will be appreciated. Special thought might be given to the comparison of this usual 1 or 2 percent for the total cost of site investigations with the item of "10 percent for contingencies," which is so frequently included in civil engineering estimates. For some types of work, especially marine work,

TABLE 4.2
PROPORTIONAL COST OF GEOTECHNICAL SITE INVESTIGATIONS FOR LEADING TYPES OF CIVIL ENGINEERING PROJECTS

Type of work	Range of cost of site investigation (as % of total cost)
Tunnels	0.3–2.0
Dams	0.3–1.6
Bridges	0.3–1.8
Roads	0.2–1.5
Buildings	0.2–0.5
Mean values	0.26–1.48
In round figures	0.25–1.50

Source: A summary prepared from many detailed costs in the records of the authors.

contingencies must be allowed for in any estimates of cost; but, in general, this broad item is often used to allow for just those uncertainties which thorough subsurface exploration tends to eliminate. Viewed in this light, the cost of preliminary work should not appear to be so high to an owner faced with spending money to determine the feasibility of the project being planned.

Consideration might first be given to cases in which adequate preliminary work was *not* carried out. For example, a contract for one tunnel included a large bonus payment for every day gained in completion before a certain fixed date. The engineers based their estimate of construction time on a calculated excavation rate through the quartzite exposed at the two tunnel portals. The tunnel line pierced an anticline, however, and it was found that the quartzite lay over easily excavated shale. The tunnel was driven very quickly and the contractor's bonus was so large that it led to a legal battle.

An even more serious case was that of a dam for water supply which was to be built in a valley as an earth-fill structure with a concrete core wall carried to rock at an assumed maximum depth of about 18 m (60 ft) below the valley floor. This figure was derived from exploratory borings which were stopped at what was believed to be solid rock but which later proved to be boulders of a glacial deposit. The core wall finally had to be carried to a depth of 59 m (196 ft), and the cost of the dam was increased correspondingly by a very large amount.

An example from bridge engineering practice is one in which a bridge was built upon a surface of "hardpan" in the river bottom. No examination of this surface was made, however, because of the difficulty of taking borings in a river current of 8 km/h (5 mph). When constructed, the bridge pier sank out of sight, causing the loss of the two adjacent spans of the bridge and a number of human lives. An examination was made later, and it was found that the "hardpan" was only a thin stratum with boulders overlying a deep layer of soft clay (Fig. 4.1).

These examples show that trouble can be encountered through neglect of preliminary investigations. What of the constructive contributions that such work can make to civil engineering practice? Many examples could be quoted, although the actual monetary savings effected are usually hard to assess. The successful completion of many notable tunnels, such as those under the river Mersey in England or those constituting the Catskill and Delaware aqueducts for the water supply of New York (Chap. 6) provide striking evidence of the value of preliminary geological studies and underground investigation. Evidence supplied by the Hoover Dam and a vast number of smaller dam structures founded on strata previously investigated is also striking. From the records of bridge construction there may be mentioned, to similar effect, the foundations for the San Francisco–Oakland Bay Bridge. Despite their unusual size and the record depths to which they were carried, their construction agreed closely with what was anticipated as a result of the preliminary exploration, on which the sum of $135,000 was expended.

A somewhat unusual example may be given to illustrate the value of underground exploration for building foundations. The Lochaber water power

FIGURE 4.1
Remains of the railway bridge across the St. Lawrence River near Cornwell-Massena shortly after the central pier had failed on September 6, 1898, dropping two spans into the river with serious loss of life. (*Courtesy Public Archives of Canada, Dominion Bridge Collection; photograph No. PA 108,929.*)

scheme in Scotland discharges tailrace waters partially in a rock tunnel, partially in open rock cut, and partially in open cut through glacial deposits. An extensive program of 312 m (1,040 ft) of shallow borings investigated all available powerhouse sites and corresponding routes for the tailrace. Results so obtained were carefully plotted, mainly to locate contours on the rock surface underlying the surface deposits; all borings were carried at least 1 m (3 ft) into the rock to make sure that boulders had not been encountered. Sixteen complete schemes for the powerhouse location were prepared and their overall costs estimated. The final decision was for the arrangement already described, for which the combined cost of tailrace, pipeline, and powerhouse excavation, on the average 12 m (40 ft) deep, was kept to a minimum. A substantial saving, which amounted to many times the cost of the exploratory work, was thus effected over the original plans.

Scope of Preliminary Investigations

When a project is first contemplated, the site should always be examined in a preliminary way, using such topographic maps as are available. As plans develop, more detailed and more accurate maps may be made. Upon the basis of these, economic studies can proceed and the feasibility of the project can be evaluated. Finally, if the work is to proceed, detailed designs are prepared, the necessary contract documents and drawings are assembled, tenders are called, a contract is awarded, and the job gets under way. This is a familiar pattern, followed with only minor variations for the vast majority of construction jobs, large and small. What is meant, then, by preliminary work supplementary to these well-recognized steps? An absolutely essentially complement of these preliminary studies of the topography of the construction site is an equally careful study of the geologic strata which will be encountered in excavation and utilized as foundation beds and construction materials. The procedure for this part of the preliminary investigation is straightforward, simple in its essentials, and easy to follow.

General site reconnaissance involves a cursory examination of those features that betray something of the local geology. The physiography of the country around will yield some idea of the geologic history, e.g., whether it is glaciated or not. But more accurate knowledge of the regional geology will usually be available through published geological reports. Real study of the site, therefore, may well begin in a library. The records of the local geological survey are a natural starting point. The extent of this procedure will vary from cases of wild, inaccessible country for which only the most general information is available to city building sites for which complete geological information is readily available in printed form.

Similar study of geologic records of adjacent works at urban sites may be rewarding, but for isolated sites there may be no such records. Assembly of all available information from borings put down in the area of the proposed work, if not already publicly available (as it is in some cities), can often be most rewarding. Examination of well records in the work area will often yield valuable information. When all such information has been obtained and studied, examination of the site itself is the next essential. In times past, this meant going to the site, but today a first approach may be made through the medium of aerial photographs. When all possible information has been obtained from aerial photographs, with alternative sites possibly greatly reduced in number and limited in area by this ingenious method, site study itself must be undertaken. A geological survey is the first site investigation called for on all but the most restricted sites and those in urban areas. The geological survey can often be done in remote locations concurrently with a site-area topographic survey.

Geologic fieldwork must observe rock, soil, and especially groundwater—its location, quality, and the impurities it contains. In materials suspected of having a high sulfate content, analysis of groundwater is a particularly important part of the field study. Soil and rock samples obtained during the course of fieldwork will frequently be subjected to laboratory tests. On large jobs, field testing of soil and rock may also be necessary, either by tests on existing ground surfaces or in specially excavated shafts, pits, or tunnels. Where material properties will exert a significant impact on construction, extraordinary testing such as trial excavations, placement of trial sections of fill, in situ permeability tests, or grouting trials may be necessary.

Actual practice will naturally vary from one job to the next, and the extent of preliminary work will depend on the size and location of the job. The overall pattern, however, remains the same—first, a search is conducted for all available written information on the regional geology. Second, a study is made of the site, first through aerial reconnaissance and then on the site itself, a study in which all the techniques of geological surveying and geophysical prospecting may be used. Third, an exploratory drilling program is developed, based on what has been determined of the local geology. This is checked continually as it progresses, against the gradually unfolding picture of the local geological structure so that maximum advantage can be obtained from every test hole sunk. Finally, any special tests and field investigations which may be indicated are carried out.

GEOLOGIC METHODS

Regional Geology

A general appreciation of the regional geology can be of direct assistance to the site investigation, and yet is often overlooked. General geologic reports on specific areas, such as quadrangles in the map framework of the United States, are a regular form of publication by national and state geological surveys. These maps and their descriptive memoirs can be of great value in indicating, for example, the possibility of two or three strata of lodgement till, since buried tills are often of unusual toughness due to the action of overriding ice. The direction of ice movements across the area will also be significant in indicating some features of the glacial deposits that may be encountered.

The origin of clay soils should also be outlined in general geological reports. In southern Scandinavia and in the valley of the St. Lawrence River and its tributaries, large areas are found in which the local clay has been deposited in saltwater, with consequent unusual characteristics. About 51,800 km^2 (20,000 mi^2) of the St. Lawrence Valley is covered with the Leda Clay (so named because of a common fossil found in it) but is more accurately described as the Champlain Clay. This clay soil has an open microstructure because of its mode of deposition. This often results in natural moisture contents that are higher than the corresponding liquid limits, indicating clearly that such soils are highly sensitive. Accordingly, they can lose strength rapidly if disturbed and, because of the breakdown of the internal structure, can lead to large surface settlements. Figure 4.2 shows the area in which these clays will almost certainly be encountered. Corresponding areas in southern Scandinavia are now equally well known.

Entirely different are the properties of glacial deposits found in and around New York City. Wisconsinan glaciation extended just beyond the tip of Manhattan, a terminal moraine being a distinctive feature across Staten Island

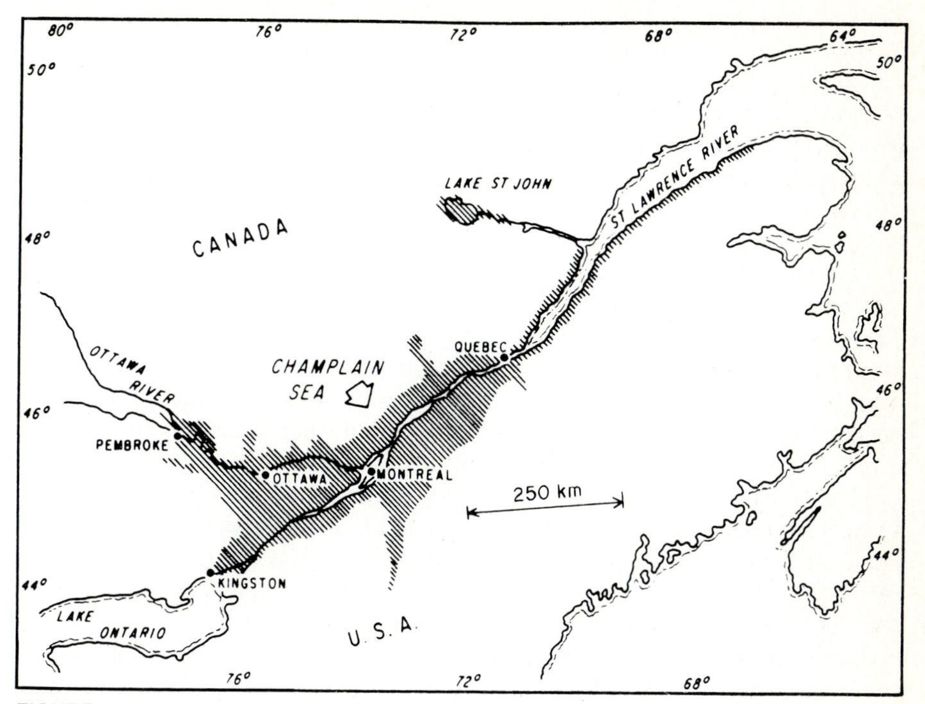

FIGURE 4.2
The full extent of the Champlain Sea in North America.

and Long Island. This moraine interfered with drainage of the melting ice sheet, the result being the formation of the glacial lakes Flushing, Hudson, and Hackensack. Soil was deposited in each of them, but the resulting sediments are far from identical. Lake Flushing occupied an area covering parts of the boroughs of Manhattan, Brooklyn, and Queens. Although it is popularly believed that New York lies on bedrock almost at the surface, these glacial deposits cover an extensive part of the great city. Those from Lake Flushing have been investigated in detail as a result of foundation studies for many of the large buildings built over them.

In earlier days, these buildings were founded on piles to bedrock, but more recently the glacial clay-silt-sand deposits have been used as bearing surfaces. This "bull's liver," as the deposits are colloquially known, is a variable succession of varved, silty, fine sand, silt, and clay. Laboratory tests have shown that these materials are preconsolidated, and that they must have been overridden by ice near the close of the glacial period, even though this is a conclusion contrary to some geological deductions. A fine summary of this experience has been written up by Parsons, whose paper is a prime reference for all who are engaged in engineering work in the area once covered by Lake Flushing. Figure 4.3 shows the regional geology deduced from this extensive subsurface exploratory work, one of the innumerable reciprocal contributions of engineering to geology.

Another different yet typical example of useful regional geological information from geological survey publications is the summary report by Scott and Brooker on the great area featured by swelling clay-shales of the Midwest. The widespread Upper Cretaceous Bearpaw Formation, first named in Montana, is widest in extent in southern Saskatchewan, being a problem rock that is therefore truly international in occurrence. Geotechnical testing laboratories show this shale to be overconsolidated and composed generally of silt- and clay-sized particles dominated by clay minerals of the montmorillonite type. The desiccation-induced overconsolidation is greater than that due to the present overburden, thus giving a clue to the geologic history of the area. The properties of the Bearpaw Formation naturally lead to many problems in engineering practice, some of which are directly dependent upon the local geology, since deposits vary throughout the extensive area of its occurrence.

British experience has provided useful examples of possible variations in subsurface conditions even though the bedrock at two adjacent sites may be identical. Oldham in Lancashire and Stoke-on-Trent in Staffordshire are less than 64 km (40 mi) apart, similar in topography, in history, and in development, and both are susceptible to problems due to coal-mining subsidence. Superficial judgment would conclude that the geological conditions underlying the two cities were similar, especially as both are underlain by Carboniferous strata. In fact, subsurface conditions are quite different because of fundamental differences in geological deposition. The Oldham coal beds are much more variable in vertical sequence and in horizontal continuity than those at Stoke-on-Trent. Depth to the bedrock at Oldham varies from nothing to 30 m (100 ft), and the bedrock is overlain by varying depths of lodgement till; whereas at Stoke-on-Trent a uniform glacial clay soil, always between 2 and 3 m (6 and 10 ft) thick, overlies the relative horizontal surface of the bedrock. Once these differences in the regional geology are appreciated, then it can be seen that site investigations in Oldham will be more difficult and more expensive than will corresponding studies at the other city.

Photo Interpretation

Concurrent with the remarkable development of aviation has gone a similarly significant development in the use of aerial photographs and other remote imagery for mapping and survey purposes, as well as for generalized geologic studies. It was only in 1919 that a real start was made in taking photographs from airplanes. Early photographs were made by adapting the rough-and-ready experience gained by a few intrepid flyers in World War I. The first topographic maps to be produced from aerial photographs were issued in 1923. The year 1931 saw the first use of aerial photographs for highway location. Not until 1935 does there seem to have been any wide-scale geologic use of aerial photogra-

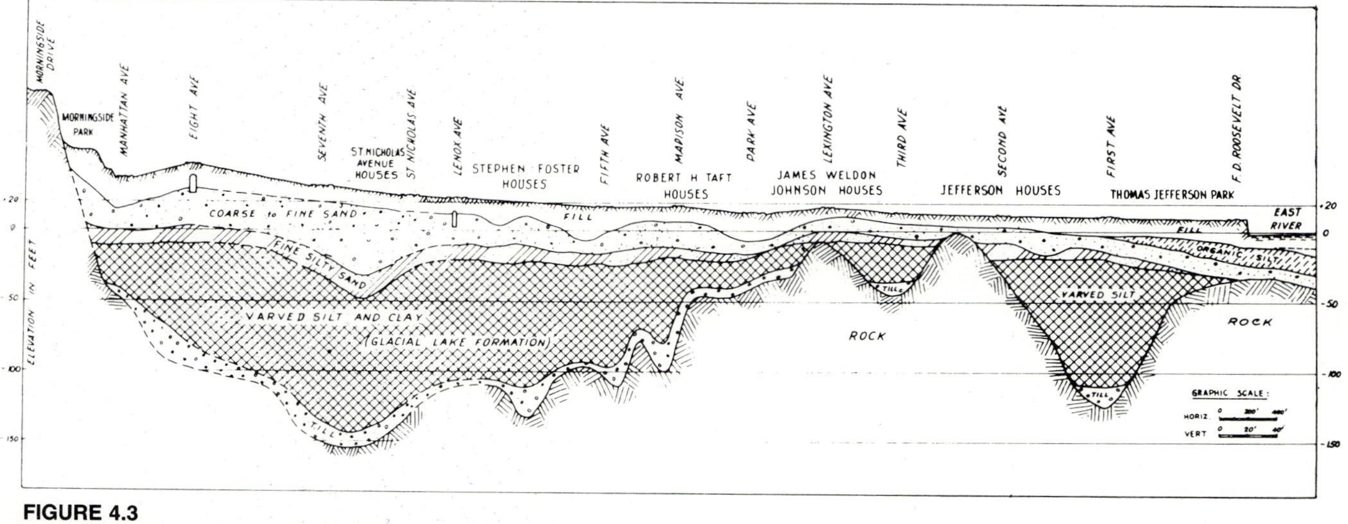

FIGURE 4.3
Geological section across upper Manhattan at 113th Street, New York. (*Courtesy American Society of Civil Engineers, from paper by J. D. Parsons in Proc. 102 (GT 6), p. 269, 1976.*)

phy. A significant application was made by the Dutch in that year in a reconnaissance study of New Guinea. Aerial photography is now used routinely as an aid to route location. Despite the fact that Canada is the second largest country in the world, photographic coverage is maintained on the entire country. Sequential air photo coverage, available for some locations for several decades, gives a vivid and even quantitative impression about geomorphologic processes such as shoreline erosion.

Geological interpretation of aerial photographs is now a well-accepted technique for which there are excellent printed guides. As with so many other modern technical developments, the services of the expert are advisable in order to derive maximum benefit from aerial photographs of areas to be studied. Aerial photographs are especially useful in betraying old earth movements such as landslides, which may be undetectable on the ground. Stereophotographs almost always form the basis for work of this sort. It is now possible to recognize many of the surface manifestations of permafrost from aerial photographs, and this is especially useful in the region of discontinuous permafrost (i.e., on the southern boundary of perennially frozen ground). Muskeg provides its own peculiar patterns when viewed from the air. Since these patterns are most revealing when seen in color, the use of strip-color aerial photographs has become a powerful technique in route location through terrain made difficult by extensive muskeg deposits (Fig. 4.4).

Location of sand and gravel or other borrow deposits is greatly assisted by aerial-photo interpretation of key physiographic features and local microrelief,

FIGURE 4.4
Muskeg country in Ontario as seen from a low-flying aircraft; scale is given by automobile on right. (*Photo: N. W. Radforth.*)

including gully shape and examination of telltale soil tones and vegetal patterns. Mollard and Dishaw have recorded the method of locating most of the available sand and gravel deposits in the plains of western Canada. Over 2,000 prospects were mapped in an area of 84,000 km^2 (32,000 mi^2). Experiences of this kind have suggested a 75 percent accuracy associated with this economical use of aerial photographs.

A similar approach is used for assessing the general geology of a construction site area, including study of photographic tone, ground texture, the pattern of physiographic features (especially drainage), orientation of bedrock discontinuities, and sedimentary rock bedding, and the interrelation of all these aspects with existing knowledge of typical geology for the area. This geologic information is recorded on maps that may themselves be prepared from aerial photographs; this is but one interlocking of photogrammetry and photo interpretation. The result can be a useful guide to further, more detailed site investigation. The larger the area to be investigated, the more useful will be the aerial photographs. As the area being studied gets smaller, so will the utility of aerial photographs decrease until they cease to be a necessary or useful aid at all, as in the case of urban building sites.

Even with optimum use of the best possible photographs and the most expert interpretation, aerial-photo interpretation is not a substitute for investigations on the ground. The efficient use of aerial photographs can save considerable time and effort in eliminating obviously unsuitable sites or routes. Their proper interpretation can point clearly to features calling for special study on the ground. But in the final analysis, accurate knowledge of the subsurface conditions must be obtained by careful subsurface investigation.

Ground truth is the term now generally used to indicate the essential field verification that must be made for even the best of aerial-photo interpretation. The term is found in the first paragraph of a paper describing one of the most extensive and urgent uses of photo interpretation in recent years, the survey of the proposed route for the Trans-Alaska Pipeline system. This 1,270-km (789-mi) route traverses varying terrain between the Arctic and the southern coast of Alaska, including 950 km (590 mi) of permafrost. Before the pipeline could be designed, accurate information on all ground materials, permafrost conditions, and the environment was obtained on a photomosaic base for a strip 2 mi wide, to a scale of 1:12,000, on which all borings were indicated (Fig. 4.5). A generalized physiographic map was prepared, using results from the borings, in association with ground truth determination. The physiographic provinces were further subdivided into 55 typical *landforms,* and on this basis the line was successfully designed and constructed. It was estimated that something on the order of 20,000 borings would have been necessary had ordinary exploration standards been followed.

Satellite images have now joined traditional air photos as part of the data base for many modern geologic studies. Taken at higher elevations and of larger areas, they have provided a broader view of the earth's surface and have led to the discovery of many new faults in its crust. They are of particular interest in

FIGURE 4.5
An aerial view of part of the Alaska pipeline; here the pipeline (center) is elevated on steel piles to prevent ground thawing. Middle Fork of the Koyukuk River on left, Yukon to Prudhoe Bay highway on right; location in central Brooks Range. (*Courtesy R. A. Kreig and Associates, Anchorage.*)

mineral exploration (including petroleum) and in the siting of nuclear power plants. One might assume that such small-scale imagery is of little use in engineering geology, but this is not entirely true.

Correlation with Geology

It cannot be too strongly emphasized that underground exploratory work must always be considered supplementary to and conditioned by previous studies of local geologic structure. There are countless examples of complex structures existing beneath relatively simple-looking ground surfaces. It is unusually important in previously glaciated areas where glacial debris now covers an original bedrock surface which may be entirely unrelated to present-day topography. Even without the existence of glacial conditions, however, disastrous results have been known to occur when reliance was placed on the results of boreholes that had not been correlated with the local geology.

Consider as an example (not taken from actual practice but typical of many a river valley) the conditions shown in Figure 4.6. Cursory surface

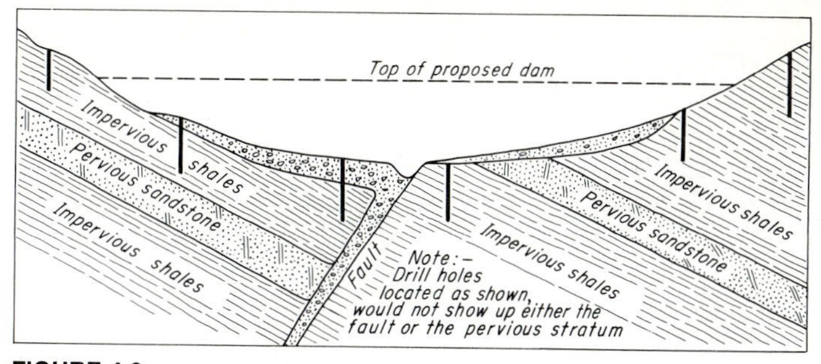

FIGURE 4.6
Geological section showing how a fault may go undetected by exploratory drilling.

examination of exposed rock in the immediate vicinity of the proposed dam would disclose outcrops of shale only. Borings put down to confirm what these rock outcrops appear to suggest might give an entirely false picture of the subsurface conditions across the valley, possibly with disastrous results. How could geology assist in such a case? Any geological survey made of this dam site would include at least a general reconnaissance of the neighborhood, and it is almost certain that, either by observation of the outcrops along the riverbed or through some peculiar features of local topography, the fault would be detected. Even if this were not done directly, a detailed examination of the outcrops of shale on the two sides of the valley would almost certainly reveal differences between the two deposits, possibly minute, possibly even of their fossil contents, but enough to show that they were not the same strata or formation, thus demonstrating some change in structure between the two sides of the valley. Experienced engineering geologists generally expect faults to occur, as part of the physical conditions that lead to formation of valleys.

Figure 4.7 shows a case in which casual surface examinations and the use of the information given by the boreholes shown would be misleading because of the existence of folds in the strata, the outcrops of which are covered by glacial debris. Figure 4.8 illustrates the necessity, where the strata involved dip steeply across the site being investigated, of correlating the strata encountered in one drill hole with those pierced by the adjacent hole. If this were not done

FIGURE 4.7
Geological section showing how a fold may go undetected by exploratory drilling.

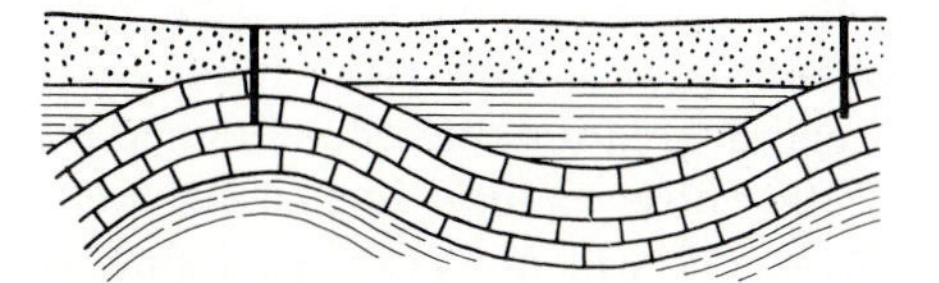

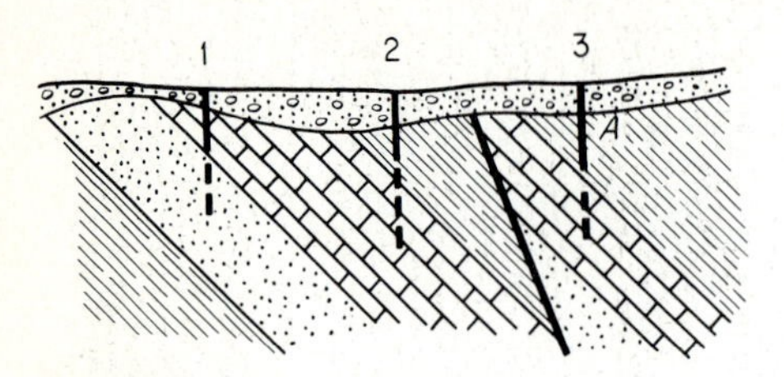

FIGURE 4.8
Geological section showing the desirability of correlating steeply sloping strata in exploratory drilling.

and hole 3, for example, had been put down only as far as the point marked *A*, the existence of the fault would have been undetected unless it were discovered through surface geologic investigations. In instances such as those illustrated by Figure 4.8, inclined borings, sometimes placed perpendicular to structural dip or bedding, most economically penetrate the subsurface.

Geologic Mapping

The initial simplicity of geological field observations is almost dangerous, certainly to a beginner. The mapping involves little more than accurate observations with the unaided eye and simple measurements with hand instruments, yet it calls for an unusual concentration of attention and an appreciable background of experience. After making a general reconnaissance of the area to be surveyed, the geologist must carefully examine all exposures (outcrops) of rock, noting and recording the nature of each exposure and its strike and dip. These direct observations are, however, but the start of investigation. Although in topographic mapping a record is made only of what can be seen, in geologic mapping the contacts of adjacent beds are the main goal of the survey, and these are quite often hidden from sight by surface deposits. Thus, the direct examination of rock exposures must be followed by a detailed examination of all topographic features that may give a clue to hidden geologic structure.

Escarpments, indicating a relatively hard stratum; unusual deviations of watercourses and waterfalls, also often indicating a hard-rock layer; the elevation and location of springs; these and many other topographical features come within the critical review of the geologist in the field. The natural slope of the ground must always be remembered, especially when consideration is being given to such surface features as are demonstrated by the soil; here the upper limit of all special effects is clearly the only one that can be used with certainty. Naturally, a contact of two adjacent beds can only rarely be traced, or even seen except in isolated places; its general location must be obtained by inference from the accumulated evidence of other observations. In a similar way, the various structural features noted in Chapter 2 will not often be found in anything but minor and infrequent locations. Occasionally, complete sections of rock folds are exposed, but generally, folding will be revealed only after the study of numerous dip observations. Faults are sometimes shown a little more clearly, especially if they are of comparatively recent origin and their effect on topographic detail is still unobscured. Glacial structures and effects are sometimes clearly indicated.

These notes are only an introduction to geologic field methods. Experience in the field under expert guidance is the only sure training, although much help can be obtained from the writings of experienced geologists and geological engineers. One excellent guide is listed at the end of Chapter 2.

The impression may have been given that geologic mapping can be carried out only in country exhibiting noticeable topographical variations and frequent outcrops of solid rock. This is not the case; special geologic maps have often been prepared to show the nature and structural arrangement of superficial deposits above solid rock. These maps began with the *maps of drift* of the Geological Survey of England in 1863, and since that time such maps have gradually come to be regarded on an equal footing with bedrock (solid) maps, especially in countries with extensive glacial deposits.

Mapping methods for soil are similar to those for rock, but there is added emphasis on the close examination of soil constituents. Geologists are often aided by the excavation work of burrowing animals, not only because of the spoil dumps which they accumulate outside their burrows but also because of the well-known habits of the main species—e.g., the habit which rabbits have of choosing dry sand deposits. Hand augers and backhoes provide a simple and ready means of obtaining actual sections through deposits with the expenditure of little energy or cost—in marked contrast to the use of exploratory drilling in rock strata.

Because soils are derived from parent material whose location can usually be readily determined from the overall geological picture of a region, detailed study of soil constituents is now being developed as an indirect method of "prospecting." Geochemical *prospecting* is already a term of real meaning. Vogt, a pioneer in this work in Norway, noticed many years ago that the health of grazing cattle was affected by changes in the chemical composition of the soils on which they grazed. These changes were due to variations in the underlying bedrock, which Vogt mapped. Study of vegetation is therefore also developing as a special aid to geological surveying, and *geobotanical prospecting* is another term that may be encountered and with which the civil engineer should be familiar. Naturally, such studies of soils are closely linked with the pedological soil-survey practice of soil scientists. The identical field methods are employed but they usually penetrate to a depth of less than 1.5 m. Correlation between surficial geologic units and the general pattern of agricultural soil distribution is an essential technique of the engineering geologist. Correspondingly, there is a growing volume of information relating to pedological soil subdivisions and engineering properties, especially in highway and airport engineering.

EXPLORATION TECHNIQUES

Exploratory Boring and Sampling

The purpose of underground investigations must be clearly appreciated at the outset. Only in special cases is the purpose merely to find out where the hidden surface of rock lies. As a rule, it is also necessary to identify the material penetrated before rock is reached, not only for purposes of construction but also for the preparation of designs, especially when foundations are to be constructed in the unconsolidated material. For important foundations and when rock has to be excavated, the nature of the rock penetrated by drilling must also be investigated. Usually, therefore, the aim of preliminary work is not only to provide information about underground boundaries between essentially different materials but also to procure samples of the materials for laboratory analysis and testing. In view of the attention now paid to the testing of all soils, it is almost always of the utmost importance to obtain relatively undisturbed samples.

The simplest type of exploration in unconsolidated material can be done by probing with a steel rod. In shallow ground, depths to rock may sometimes be explored in this way, but the method is extremely limited in its application. The next stage involves the use of a manually powered soil auger, a boring tool described by its name, which may be obtained equipped with many special devices for penetrating different kinds of material and for procuring samples that, although not undisturbed, are often satisfactory for preliminary testing in a laboratory. The auger bores the hole, and samples of cohesive material are obtained by withdrawing the tool and cleaning off the material adhering to it. Holes through sand have to be lined with pipes, but special drills and devices with which samples may be obtained are available for this type of work. A soil-auger set, complete with the usual extra fittings and capable of boring to 9 m (30 ft), weighs less than 45 kg (100 lb) when packed for carrying and is relatively inexpensive. It is strange that more use is not made of this equipment, at least for small-scale work.

The soil auger could be very usefully substituted for a type of boring, generally called *wash boring,* which was widely adopted by engineers in the past. In wash boring, a hole is excavated by washing inside a pipe casing which is forced into the ground as the hole is made. Material excavated is trapped in a washtub, where it is examined and identified (if possible). Wash borings may be useful in some cases to show where a hidden rock surface is located, but they are worse than useless as a means of exploring unconsolidated material since they can give very misleading information. They reveal the material penetrated only in a thoroughly disturbed state, generally with much of the finer material washed away. They give no clue to underground water conditions and no sample of any use at all for testing. A simple modification of wash boring provides a much more satisfactory method. A cored hole is sunk by washing, but the washing is stopped at intervals (determined by the nature of the ground) so that a sampling device may be lowered into the hole, forced into the undisturbed material at the bottom of the hole, and withdrawn with a sample enclosed. Washing out then continues until another sample is required. This method also has the disadvantage of not immediately disclosing groundwater conditions.

Limited in scope and efficiency though they may be, auger boring and wash boring should not be written off completely. If their limitations are kept in mind, especially in the character of the material washed up from a wash boring, then the two methods have a place in subsurface investigations—in isolated locations for preliminary site studies or for quick checks on conditions underground when no other equipment is available. Reliance upon results obtained only from wash borings has, in the past, led to many sad experiences and even to some disasters, the natural character of soils penetrated (especially lodgement till) being often completely destroyed by the action of washing. Wash borings, therefore, are to be avoided whenever possible and used only with the greatest discretion when no other course is open.

Soil coring, either by rotary drilling, by percussion sinking of holes, or by continuous sampling with well-designed sampling tools, is eminently desirable for all modern site studies. Soil-sampling tools can then be used to obtain undisturbed samples at every desired depth in the holes formed by drilling or driving, when continuous soil sampling is not required. The design of samplers for this purpose has progressed considerably since Alexandre Collin stressed in 1846 the need for samples of clay that were in their "natural state." A variety of excellent tools is now available for obtaining good samples from all the main types of soils. Piston samplers are a common type, but for friable soils the so-called "Dennison sampler" (developed by the U.S. Army Corps of Engineers at the former Dennison District, Texas) is useful, because loss of the sample is minimized by spring guards. Freezing, chemical injection, and impregnation with asphalt have all been employed for procuring undisturbed samples of cohesionless soils such as sands. The necessity of obtaining undisturbed samples arises from the vital importance that low-density soils may be subject to volume loss, in-place compaction by earthquake motion, or machine vibration.

A parallel development has brought about the *Shelby tube* device, used to force thin-walled steel tubing into the undisturbed ground, without the aid of wash boring. The steel tubing is so arranged that the lower section can be removed for sample retrieval. Soil sampling in this way has now become an accepted practice in soil mechanics. Measurement of the amount of standard force (as driving blows) required to penetrate successive soil strata can be used as a guide to soil properties. Emphasis must again be placed upon the importance of obtaining good, undisturbed soil samples. Pioneer work of this type was carried out in the exploratory work for the San Francisco–Oakland Bay Bridge. Undisturbed samples were obtained along the line of the bridge from depths up to a maximum of 82 m (273 ft) below low-water level. Thus, the unusually deep foundation for this great structure could then be designed with accuracy and confidence.

On the east coast of the United States, another pioneering test-boring program was carried out in 1935 and 1936 in connection with the Passamaquoddy tidal-power scheme then projected by the United States government. This work was all in exposed coastal waters and was hampered by severe winter climatic conditions, tidal currents in excess of 8 knots, and a tidal range of 8.84 m (28 ft)—a combination of adverse conditions that could hardly be equaled. Despite this, soil sampling and core drilling into the rock over a large area of the sea near the coast was carried out in conjunction with geologic investigations for foundation-bed conditions. The deepest hole put down during these interesting operations was in 36 m (120 ft) of water and through 50 m (164 ft) of unconsolidated material above the rock, which was then core-drilled.

Rivers also represent deep-water sampling challenges. Explorations for the Tappan Zee Bridge across the Hudson River required borings to depths of 88.5 m (295 ft), with undisturbed soil samples for laboratory testing. On small jobs, on land as over water, the same attention must be paid to exploratory boring operations in order to ensure accurate records of all soil strata penetrated and the best possible undisturbed soil samples for testing in the laboratory. In this way, preliminary geologic studies can be corrected, confirmed, or extended, in an accurate determination of the subsurface conditions that will be encountered when construction commences.

Exploratory Drilling in Rock

When rock surface is reached, either in a test pit or in test boring, a change in method is necessary if the rock is to be penetrated. Usually the rock is penetrated, if only to determine its nature and to make sure that it is not just a boulder. Various methods are available, but reference is made here only to core drilling, in which a cylindrical hole is drilled around a central core which is periodically broken off from the bedrock at its lower end and removed from the hole for examination. The rotary drilling machine used for this purpose is fitted with a special bit equipped with black diamonds (diamond drilling), chilled-steel shot (shot drilling), or removable steel cutting teeth. The choice of equipment will depend to some extent on the nature of the rock, while the size of hole will depend on the drilling tool available and on the anticipated depth of the hole. Core drilling is a highly specialized operation, necessitating skilled workmanship and experienced supervision. Percussion drilling is an older means of penetrating rock which is no longer widely used for exploratory work, as it does not deliver core for examination and testing.

The idea of diamond drilling originated in Switzerland in 1863. The first machine was hand-operated, but a steam-driven machine was used as early as 1864 in the Mont Cenis Tunnel between Italy and France. The bit speed was only 30 rpm, and the penetration was about 30 cm (12 in) per hour. The first United States patent for a steam-driven diamond drill was issued in 1867, and by 1870 a 250-m (750-ft) hole had been drilled with the first American machine in the search for coal near Pottsville, Pennsylvania. Steam continued to provide the motive power for diamond drilling until World War I, after which smaller gasoline engines took over. With higher speeds possible and with greater flexibility because of the use of lightweight materials, diamond drilling has steadily improved, so that today it is an efficient and reliable process in the hands of expert operators and may be carried to depths of more than 300 m (1,000 ft) with portable rigs or on barges over water.

In some types of soft or disintegrated rock, it will sometimes prove difficult to "recover" complete sections of core by drilling, and the engineering geologist at the rig must estimate, with the assistance of the operator's observations and experience, the exact nature of the material that is crushed and lost due to the drilling action. The amount of core recovered will vary with the type of unweathered rock penetrated; typical average values are quartzite, 90 percent; granite, 85 percent; sandstone, 70 percent; limestone, 60 percent; shale, 50 percent; slate, 40 percent. These figures are naturally only a guide to what may be expected.

Diamond bits leave an especially clean, smooth borehole. Consequently, it is possible to utilize cored holes for other purposes, such as in water pressure monitoring wells (as in a dam foundation), or in a hole which will be capped and filled with water under pressure to note if the hole "holds water" or not. Television inspection is now routinely carried out in drill holes of relatively large diameter (about 10 cm or 4 in and over) by the use of a periscopic video camera equipped with an electric light. The rock walls thus can be examined and video-recorded. Simple trigonometric relationships can be used to ascertain the strike and dip of joints and other discontinuities passing through the hole. Spatial orientation is provided by an internal compass and an accurate depth gage. One of the earliest of such instruments was the NX photographic camera developed by geologists of the U.S. Army Corps of Engineers to investigate faulted foundation rock encountered during the construction of Folsom Dam. At Folsom Dam, details of the critical fault zone were revealed, as well as delicate changes in rock coloring and fractures as small as 0.25 mm (0.01 in) in aperture (opening), and even the groundwater surface.

There are now commercially available stereoscopic cameras with which stereo photos have been taken at depths up to 3,000 m (10,000 ft). Borehole devices have now been sufficiently miniaturized to operate in 7.5-cm- (3-in)-diameter boreholes, giving a reasonable image on a 56- by 18-cm (21- by 7-in) television screen. A notable example of underwater closed-circuit television was the study of the Columbia River bed before designs and plans could be made for the closure of the Dalles Dam. Preliminary surveys by divers were unsatisfactory because of high river flow velocities. The camera, suspended by a steel cable, was operated from a strongly moored barge. The riverbed character, i.e., bare rock or gravel, could be distinguished, leading to development of plans for the closure in which over 2.3 million m^3 (3 million yd^3) of fill was placed.

It is, however, essential that civil engineers in particular remember that holes can be drilled at any desired angle, since inclined holes, horizontal holes, or even holes inclined upward may be necessary. A good example is the use of inclined holes in the investigation of rock conditions in the bed of the Hudson River, New York. Horizontal holes were used extensively in the investigation of rock conditions for the Pennsylvania Turnpike, with access through abandoned mine tunnels. About 1,050 m (3,500 ft) of horizontal holes were drilled to supplement the 3,000 m (10,000 ft) of vertical holes in the thorough study made of the geology of the new turnpike tunnels. One of the horizontal holes was successfully drilled to the almost unprecedented length of 435 m (1,450 ft) from the east heading of the Tuscarora Tunnel. At Ripple Rock, British Columbia, before its demolition, overhead drilling was carried out from an access tunnel extended out under the two rock mounds that were being geologically characterized for removal by blasting.

Large-Diameter Exploratory Holes

Large-diameter exploratory holes or shafts are especially valuable for in-place inspection. Such holes are either advanced by percussive chiseling and crushing of rock or drilled with machines of the "Calyx" type, with diameters varying from a minimum of 1.1 m (42 in) to 1.8 m (72 in) in diameter (Fig. 4.9). The method of drilling and removal of cores is similar to that followed for smaller holes, although breaking off the core from the bedrock is sometimes difficult and necessitates the use of special wedging devices or blasting. The holes are of use

FIGURE 4.9
Typical Calyx drill core (36-in diameter) laid out for inspection of core losses. (*Courtesy U.S. Bureau of Reclamation.*)

only if leakage of groundwater into them can be removed by a small pump. Special precautions must always be taken to keep a supply of fresh air at the bottom of the holes, particularly if blasting has been used for core removal.

When the holes are completed and cleaned out, geologists and engineers in charge may be lowered down in suitable cradles. With the aid of portable lights they can carefully inspect the surrounding rock exactly as it occurs in place. They can investigate boundaries (geologic contacts) between beds, study discontinuities, and make a thorough and complete exploration of the rock with certainty and convenience. If the holes are drilled after foundation grouting, the efficacy of the grouting operation can thus be checked against the inevitable uncertainty of grout penetration. The cost of holes of this size may be considerable but it is commensurate with the great advantages that they present for underground investigation.

The use of these large-diameter holes in civil engineering work appears to have developed initially in the United States, although they have now been used successfully in several other countries. Early applications were mostly in connection with dam-foundation work, as for the Grand Coulee and Norris dams. A particularly significant early application was at the site of the Prettyboy Dam, constructed to impound water for supply to the city of Baltimore, Maryland. The site is topographically a good one (the valley is quite deep), but geological conditions required unusual precautions in the excavation of the cutoff trench below the base of the main structure, a procedure which called for placing about 145,000 m^3 (190,000 yd^3 of concrete. The rock formation beneath the dam is mainly mica schist with some limestone, gneiss, and intruded quartz, and it has been twice subjected to major tectonism and metamorphism. Faulting was therefore to be expected. The exposed rock was weathered to a considerable degree and extreme care would have to be taken in blasting where required for foundation excavation. Calyx core shafts, 0.9 m (36 in) in diameter, were sunk so that the consulting geologist could be lowered into them to study the schist bedrock in place. Accurate geotechnical profiles were constructed to show the position and dips of all faults and large seams. Disintegration of the schist was discovered along the hanging-wall side of a major fault, and considerably increased the expected volume of excavation.[1]

Large-diameter drill holes of this type were widely used during the great program of dam building of the early years of the Tennessee Valley Authority. One particularly interesting application was utilized in exploring the unusually complex subsurface at Watts Bar Dam. Large holes were drilled within cofferdams at various locations across the river at the dam site. The rock was penetrated to depths up to 21 m (68 ft) and provided for a complete inspection of the foundation rock.

Exploratory Pits, Shafts and Tunnels

Another means of investigating superficial deposits is the exploratory pit, an excavation large enough for a person to observe and sample comfortably. In practically all cases of depths greater than 1.5 m (5 ft), pits must be shored with timber or removable aluminum supports, the design and placing of which must always be carefully checked. Clearly, there is a limiting depth to which test pits may reasonably be carried, which is usually determined by machine excavation capabilities and active soil pressure.

The main use of pits is for the investigation of relatively shallow depths of unconsolidated material and particularly for the study of surface deposits of gravel, sand, or clay for use in construction. Exploratory pits have the advantage that the deposits penetrated can be easily examined in place, and undisturbed samples can be obtained, when necessary, from the bottom of the pit. They will disclose underground water conditions in the material penetrated, but clearly they can be used only on dry land. When extended in length, pits become exploratory trenches; the main technique for identifying and assessing active earthquake-producing faults.

When exploratory pits are used to investigate surface deposits of construction material, it will be necessary to take regular samples of the materials encountered. Sampling is a matter that does not always receive the attention that it deserves in engineering work; it must not be the haphazard collection of small quantities of material that one might imagine it to be. Two objectives are important in sampling: (1) obtaining samples typical of average conditions for all material to be investigated and (2) obtaining samples representative of maximum and minimum characteristics of the material.

When investigating material to be used in construction, both objectives must be remembered, although the first will usually be the controlling factor unless the material is extremely variable in composition. Samples must always be taken at regular intervals in materials of the same appearance. From a large number of these, thoroughly mixed at the site, one or more average samples should be taken. These may then be used for testing, but they must always be accompanied by at least one unmixed sample, selected as typical for the stratum of material being studied, to check on the correctness of the average samples obtained. (Naturally, this procedure applies only to materials that can easily be mixed and not, for example, to stiff clay, from which a number of individual samples must be selected.) All samples must be immediately marked with full information about the location from which they have been taken.

Where windlasses and manual labor were formerly the means of excavation and sampling, backhoes are now routinely used to dig pits. Pits can usually be dug, logged, and filled at a rate of about eight to twelve a day. Occasionally, on major projects, subsurface conditions are so critical and so uncertain that nothing less than an exploratory shaft will suffice. One of the early Tennessee Valley Authority projects, the Gilbertsville Dam, provided such a case. The ground consisted of limestone bedrock overlain by about 15 m (50 ft) of water-bearing soil. It was decided that a shaft would be the best means of determining soil properties, the character of the rock surface, and the way in which steel sheet piling behaved when driven through the local overburden. In addition, a shaft would provide a means for inclined drilling into the bedrock. A ring of steel sheet piling was driven to rock, and excavation was carried down in the open, with bracing installed as required. A bad blow occurred, however, 4.2 m (14 ft) from bedrock, and it was then decided to freeze the soil around the shaft. This was accomplished with a brine-circulating system; thus, the shaft was unwatered and excavation was completed. Bedrock was found to be a dense, black siliceous limestone into which 25 of the 50 steel piles had penetrated as deep as 12.5 cm (5 in); six of the interlocks between piles had, however, failed. Steeply inclined boreholes were then carried out from the bottom of the shaft, as shown in Figure 4.10. A large volume of ground was thereby explored from a single shaft.[2]

Special exploratory or *pilot* tunnels are frequently constructed in advance of the excavation of major underground works or for unusual explorations such as those into the lower parts of massive landslides. Such was the 2,500-m-(8,200-ft)-long pilot tunnel beneath Loveland Pass, about 100 km (60 mi) west of Denver, prior to the construction of a major highway tunnel. (More will be said about this subject when tunnels are considered later in this textbook.) Equally remarkable has been the construction of a full-size 20-m-(65.6-ft)-long section of subway tunnel in Atlanta, Georgia, purely for experimental and exploration purposes prior to the building of Atlanta's new subway system. Prior to the construction of a new water-power plant on the Columbia River at Revelstoke, British Columbia, the ancient but major Downey landslide had to be evaluated, the foot of which would be inundated when the power dam was complete and closed. The slide had a volume of about 1.5 billion m^3 (1.95 billion yd^3), and its stability had to be assured. It is not surprising to find that one of the measures taken to study its condition in advance of final construction was

FIGURE 4.10
Cross section of the special exploratory shaft at Gilbertsville Dam, showing use of inclined boreholes for subsurface investigation. (*Courtesy Engineering News-Record;* from reference 2.)

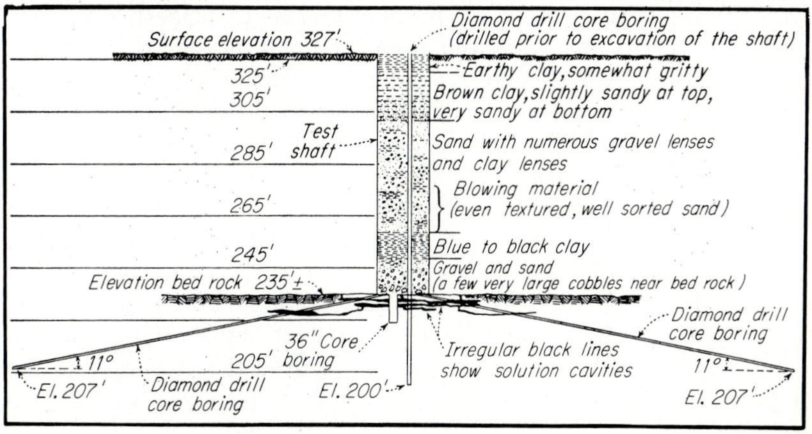

the driving of an adit 250 m (820 ft) into the slide near its base. Exploratory tunnels were used to similarly good effect in advance of final designs for parts of the Snowy Mountains project in Australia.

In Situ Tests

Most civil engineering projects require only relatively simple exploration techniques. The use of small-diameter borings and drill holes is obvious. These have been supplemented by in situ simple tests of soil and rock, using the original boreholes. One of these simple measurements is the resistance to penetration of casings or sampling tools as they are forced into the ground. Therefore, the *standard penetration test* has developed in which a specially designed soil sampling device is forced into the soil stratum exposed at the bottom of a bore hole, while rate of penetration and force are recorded.

Another type of borehole device is one equipped with small vertical vanes. If this is forced into the undisturbed soil at the bottom of a bore hole and then turned with a torque-measuring device, which equates to the shear resistance of the soil, correlations can be made with more accurate laboratory values. Such vane shear tests are not confined to small devices that can be lowered into bore holes. Convenient portable instruments have been developed for use at or near the ground surface or where handwork is possible, such as at the bottom of an exploratory pit. Once the idea of utilizing boreholes for carrying out field tests is accepted, the wide scope of this supplementary aspect of subsurface exploration can readily be appreciated. Field permeability tests can also readily be conducted both in soil and rock.

Such sophisticated tools and instruments are now available that it is easy to forget that sound and useful exploration can still be carried out with the aid of very simple tools. Modern methods and equipment should, naturally, be used whenever possible and economical, but a few examples from the early days of civil engineering still have sound lessons to suggest. One such case is the 1860s Intercolonial Railway from Montreal to Halifax, Canada's first main-line railway. Chief engineer Sandford Fleming had a number of major bridges to design, two of the most important being across the two branches of the Miramichi River in northern New Brunswick. Located near the junction of the two branches, and so only about 400 m apart, the two sites were assumed to be identical; preliminary simple test boring seemed to confirm this. Actual excavation showed that foundation beds were different. Fleming immediately had better augers prepared and carried out further borings into the riverbed. He also got his blacksmith to attach small plates to boring rods, had these lowered to the bottom of some of the borings, loaded them, recorded the settlement, and so deduced the character of the soil from penetration tests similar to those of today, but made well over a century ago.[3]

Supervising Exploration

The accuracy of underground exploratory work is of supreme importance. To achieve accuracy is to require exceptional skill and wide experience on the part of the person in charge of subsurface exploration. Employers should therefore give particular attention to an engineer's experience and reliability. Many large engineering organizations, such as highway and public works administrations, maintain special field exploration divisions. Consulting engineering firms are not able to maintain regular boring crews, and must engage boring contractors for this work. The necessary contract documents must be prepared to provide a wide degree of flexibility due to variations that may be revealed as the work proceeds. Under no circumstances whatsoever should a boring contract be awarded on a "lump-sum" basis. Unit prices must be secured, probably with a guaranteed minimum number of holes and total depth of drilling or boring, with modified prices for operations in excess of certain specified limits. Although the usual practice of calling for bids will probably have to be followed, contract award should not be made on the basis of price alone. Due regard should be paid to the experience and reliability of the respective contractors who bid and to the drilling equipment they have available, descriptions of which should be required with the tenders.

Exploratory work is of value only if a complete and accurate record of the results is prepared for the use of the engineer in charge and all advisers. This is so obvious that there would seem to be little need to emphasize the necessity of obtaining accurate records. Frequently, however, recording is left to the drilling supervisor, who has occasional visits by an engineer or geologist. This is often done even when a considerable amount of money is being spent on the exploratory work. Even experienced and conscientious foremen should never be left in charge of the records, not only because they will probably not be well versed in record work but also because it is important that the results of the exploratory work be checked by an independent, trained professional person.

Either a qualified engineer or engineering geologist should always be present throughout exploratory work to watch its progress and keep the necessary records. If a geologist can undertake this task, so much the better, but this will be possible only when a large geological field staff is available. A senior-level geologist or geological consultant should occasionally visit the site of the exploratory work while the work is in progress so that the results may be seen firsthand and discussed with the professional staff members in charge. This insistence on accuracy helps to assure the success of the entire construction project.

A precise record of the nature of all strata must be kept to achieve the second objective of subsurface investigations—study of the materials encountered. This *boring log* is supplemented by samples of each stratum and by the rock core. The importance of preserving soil samples in their undisturbed state has already been stressed and cannot be overemphasized. All soil-sample tubes must be immediately sealed (with paraffin wax or other suitable sealant), right at the drilling site so that the samples will lose no moisture. Additional samples should be procured for visual inspection, since the samples to be used for laboratory testing must not be handled on the job after they are sealed. If this point has to be emphasized to field engineering staff, a simple demonstration of the soil water evaporation rate may be made by exposing a small sample of moist soil on a delicate chemical balance.

Rock core represents an expensive investment, and so must also be carefully stored, in flat wooden boxes divided into narrow compartments, each wide enough to hold one core and, for convenience, of standard length. Standard (N through H) core is placed directly in these compartments as it is obtained exactly in the sequence of its location in the core holes. Inscribed blocks are placed at the ends of earth core run and at the gaps representing core losses.

Finally, a prime requirement of all records of exploratory work is that materials encountered should be accurately described. If it is possible to examine samples of all the materials after the hole has been made, it will not be imperative to keep accurate descriptive notes as the work proceeds, provided all samples and cores are properly correlated with the progress records. It will be a distinct advantage, however, if accurate terminology is used in the day-to-day records. The engineer who is keeping records should therefore be familiar at least with main rock groups and the distinctions between the various grades of unconsolidated material. The latter should always be described by the use of the appropriate geological term—gravel, sand, silt, or clay, or a combination of two or more of these—together with an accurate notation of the physical condition of the material, e.g., whether it is hard packed or very loose. Popular terms such as "hardpan," or local terms such as "pug," should be avoided. If an exact measure of the state of the material is desired, this must be obtained by using a penetration device of some type. Naturally, the more uniform such field descriptions are, the more useful they will prove to be. There is no widespread agreement, even in the English-speaking world, on terms to be used for the field description of soils, but a number of printed guides are available.

GEOPHYSICAL METHODS

Geophysical exploration methods can be a useful and economical supplement to geologic mapping, exploratory borings, and backhoe pits. A full appreciation of the value of the methods, however, demands an awareness of the contributions which physics is now making to the development of geological knowledge. Classical geology was essentially a descriptive science, based on careful observation in the field and in the laboratory, aided by careful reasoning in deducing general principles from individual observations. With the turn of the twentieth century, however, increasing attention was paid to the contributions that could be made by application of the principles of other sciences. Geochemistry, paleobotany, and statistical analysis are now accepted geologic techniques. The greatest transformation in geological study has arrived through the application of *geophysics*. A witty writer has described geophysics as "geology once removed." Like many a witticism, this saying carries more than a germ of truth. When one hears geophysics defined as "the science concerned with the constitution, age, and history of the earth, and the movements of the earth's crust," it is difficult to distinguish geophysics from geology, as defined by some workers. Geophysics can be thought of as providing a framework within which classical geological studies can helpfully be viewed. The general methods of geophysics when applied to detailed geological problems broaden the limits of human understanding of the world in a rewarding and challenging manner.

Geophysical methods constitute only another exploratory aid to geological surveying, so that, as applied to civil engineering, they must never be regarded as anything more than aids. They will not disclose more than will a good set of boreholes—and usually not so much—and they can never be used without specific and constant correlation with direct geological observation information. In fact, a preliminary geological survey is essential before geophysical methods can be applied with any certainty of success, since these methods require knowledge of certain general conditions, of local geology if interpretation is to be at all effective. The most favorable condition occurs when rock underlies a shallow superficial deposit and the physical characteristics of the two are markedly different. This condition, in connection with civil engineering work, occurs most frequently as a deposit of glacial debris overlying a solid rock body. It is in the study of the geology of such deposits, especially regarding the depth to bedrock, that geophysical methods have generally been utilized in civil engineering practice.

Seismic Methods

As early as 1846, Robert Mallet suggested that artificial earthquakes (and thus earthquake waves) could be created for experimental purposes by exploding gunpowder on land or on the sea floor. This idea is the basis for geophysical seismic methods. Since the vibrations set up by earthquakes, either real or artificial, do not travel at the same speed in different media, the existence of a change of medium may be detected. It was not until the early part of the twentieth century that satisfactory instruments were devised for measuring and recording seismic vibrations reaching points of observation.

Artificial ground motion is produced by blasting. High-strength gelatins and dynamites are most suitable, and the charges should be buried to obtain best results. The two types of body waves so set up in the ground—elastic earth waves, as they are sometimes known—are due, respectively, to longitudinal and to transverse vibrations. The latter occur in both the vertical and horizontal planes. Other minor types of surface waves may be generated, but they are secondary to the body waves. Of these, the longitudinal waves travel faster than the transverse waves and so will be the first to reach the point of observation. Both types of waves travel through different kinds of rock with different velocities, and they will be *refracted* as they pass from one medium to another. This fact is the basis of geophysical seismic methods of investigation. Records of observed vibrations are taken at different distances from the location of the explosion, and the results are correlated with known facts about wave travel in different media (Fig. 4.11).

Seismographs are used for recording the vibrations reaching the points of observation. The essential part of the seismograph is a heavy mass suitably mounted by a nonrigid suspension system in an equipment case. The inertia of the mass will tend to keep it in a state of rest, whereas the cover will tend to move in accordance with vibrations reaching it through the ground; the instrument thus measures the movement of the cover relative to that of the mass. Usually, a clockwork mechanism that produces a continuous photographic record of the vibrations is included. Time intervals and the instant of firing the explosive charge are also marked on the same photoelectric strip chart of cathode ray type.

In recent years, a simplification in technique has been introduced by the commercial production of an easily operated recorder, actuated by the waves created by the fall of a heavy sledgehammer upon a steel plate in contact with the ground. The method is simple to use and is now widely accepted in engineering work. Most recording units now include *signal enhancement,* whereby up to 12 or 24 separate geophones accept simultaneous wave forms, and each hammer blow strengthens the solid-earth response, while obviating stray signals from other human activities or natural noise in the vicinity.

Electrical Methods

Earth materials possess electrical properties of wide variation; the two most commonly used in geophysical work are *conductivity* and its reciprocal *resistivity*. The differences in conductivity of different rock types are so large that the range of variation is much greater than that of any of the other geophysical properties so far discussed. The usual methods of investigation depend on passing a current through a section of the earth's crust between two electrodes placed at a fixed distance apart, and exploring the nature of the ground adjacent to or between them by means of two or more (potential) electrodes inserted into the ground at specially selected points (Fig. 4.12).

Another technique of great utility in boreholes is the use of resistivity methods in wells or drill holes, usually known as "electrologging," or *borehole geophysical logging*. Three insulated conductors are housed in a downhole sonde with their ends at different depths; the deepest will be the current electrode, and the other two the potential electrodes. The recording instruments are chart-recorded at the surface and great speed can be achieved—up to 300 m (1,000 ft) of hole can be examined in an hour. This application is becoming increasingly important, its worth having been proved during numerous siting and licensing studies for nuclear power plants. In these instances, a relatively few site geological boreholes in soil and rock were continu-

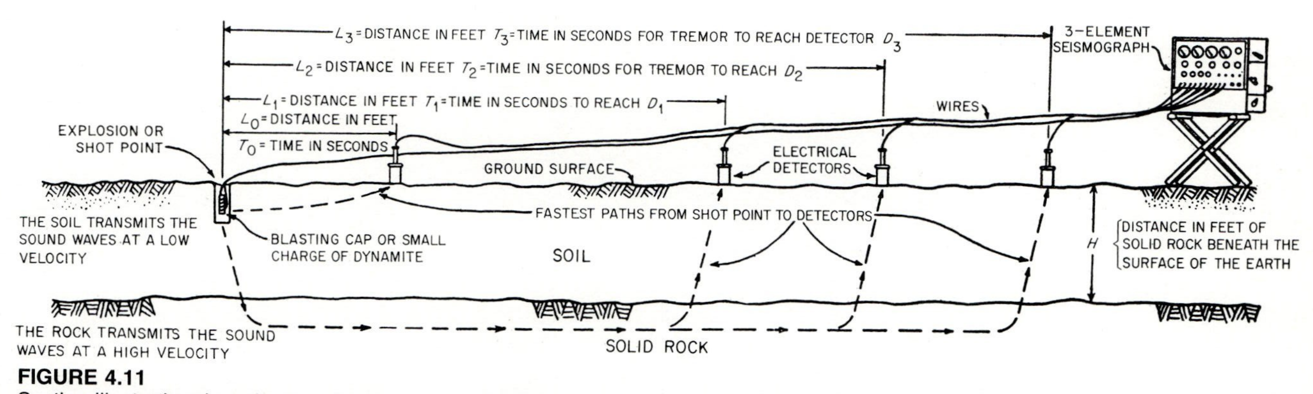

FIGURE 4.11
Section illustrating the principle of the seismic reflection method of subsurface investigation.

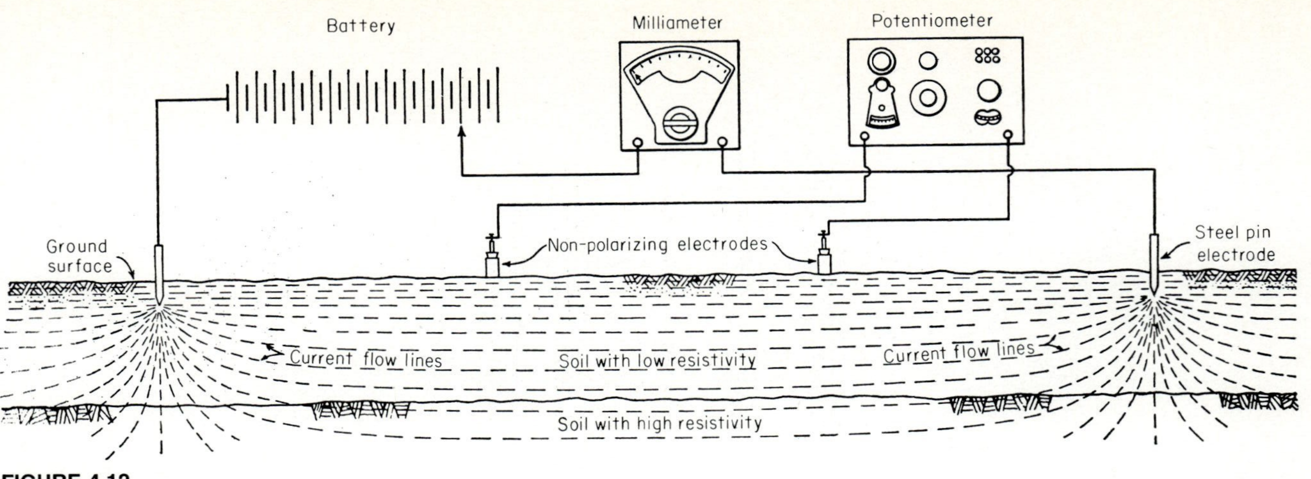

FIGURE 4.12
Section illustrating the electrical resistivity method of subsurface investigation.

ously sampled and usually logged, forming the basis for obtaining geophysical signature so that remaining borings could be drilled without extensive sampling.

Magnetic Methods

The phenomenon of terrestrial magnetism has been known for a long period, and it is not surprising that magnetic methods of investigation are the oldest of all geophysical methods for studying underground conditions. The earth's magnetic field is relatively constant in direction and strength and can be measured by a magnetic needle. When the needle is freely suspended on a vertical axis, it will come to rest in a direction known as the *magnetic meridian;* when it is suspended on a horizontal axis, it will come to rest at an angle with the horizontal called the *magnetic dip,* if the axis is perpendicular to the magnetic meridian. The earth's magnetic field is not absolutely constant; it varies slightly at any locality in a regular manner and has a slow seasonal change and a rapid daily variation; it may also be affected irregularly by magnetic storms. In addition, the field may vary in an unusual manner between one location and an adjacent point, owing to the presence underground of some material possessing the property of a permanent magnet. Minerals of iron most notably possess this property. Magnetite, oxide of iron, and pyrrhotite, a sulfide of iron, are the minerals most notable for this property. Some other minerals and rocks, however, also have weak magnetic effects. By employing suitably sensitive instruments it is possible to map the distribution of certain types of concealed rock and mineral formations by detecting these effects.

Instruments based essentially on the simple magnet have been developed for measuring these local magnetic anomalies. The more usual types of such instruments are known as *magnetic variometers,* and they may be of either the vertical or the horizontal type. In place of a magnet, a rotating coil may be used to measure the vertical or horizontal magnetic force; the induced current is measured by electrical circuitry. The modern magnetic variometer is a complex instrument, but despite this, magnetic methods of terrestrial or airborne investigation are capable of detecting magnetically susceptible strata, some fault zones, and buried caches of drummed hazardous waste.

Gravitational Methods

The fundamental law of gravitation was announced by Isaac Newton in 1687. It stated that the force of attraction between any two bodies is directly proportional to the product of their masses and inversely proportional to the square of the distance between them. Originally, the law was associated only with large bodies such as those of the solar system. Scientists soon turned their attention, however, to the attraction exerted by large mountain masses. Bouguer appears to have been the first to experiment in this direction; he carried out some experiments at the volcano Mount Chimborazo (in what is now Ecuador) between 1737 and 1740 in which he used a pendulum at a height of 2,700 m (9,000 ft) and again at sea level. Similar work was done in Scotland by Maskelyne in 1775 and in the following century by various experimenters.

Gravity measurements are generally made with *gravimeters*, or gravity meters. These instruments are sensitive to changes in the horizontal component in gravity with an accuracy of one million-millionth part of the total gravitational attraction. Modern models are very compact devices, little larger than a large vacuum flask; it is possible to survey up to 100 stations a day with such instruments.

Applications in Civil Engineering

Surficial Deposits Investigation of shallow surficial deposits is one of the most effective applications of geophysical methods, both seismic and electrical. Several state highway departments of the United States now include this work as a part of the regular exploration work for sand and gravel deposits and in the classification of highway excavation before the material is moved. The State of Minnesota pioneered the use of electrical prospecting methods for locating road material (sand and gravel); it is reported that over 100 prospective sites were tested for this purpose in a little more than a year and that 98 percent of the subsequent sampling pits verified the predictions. In Massachusetts, the U.S. Geological Survey some years ago had a steady program of seismic investigation of glacial deposits for highway construction. Good agreement was found between borings and refraction interpretations on well over 200 separate studies.

The 200 km (123 mi) of Massachusetts Turnpike (running westward from the vicinity of Boston to the border of New York State) provided many sites of deep rock cuts without adjacent exposure of bedrock. Seismic investigation was used in the selection of final cut locations in glaciated terrane. In general, results from the seismic work checked with depths to bedrock obtained by borings within 60 cm (2 ft). During winter months, frozen ground behaved as a separate geological stratum, but the work was continued successfully throughout the winter. Geophysical methods were also used with success in the planning and design of the Ohio Turnpike to which further reference is made later in this book.

Groundwater Can geophysical methods be used successfully to determine the existence and position of groundwater? This is often the one matter that cannot be foretold from geological investigations alone; boreholes are often necessary to verify only the groundwater surface. Geophysical methods by themselves will not "discover" groundwater, yet they can be of considerable assistance in determining subsurface conditions that are favorable to the occurrence of groundwater, although a drill hole must always be put down to make the actual confirmation. Resistivity of rock depends mainly on two factors, porosity of the rock and salinity of the solution held in the pores and discontinuities of the rock. In general, resistivities are high for dense, impervious rocks and low for porous, water-bearing rocks. Bruckshaw and Dixey give the following typical values for resistivity (i.e., the electrical resistance of a cubic centimeter of material when the current flows parallel to one edge).

Fresh crystalline rocks	20,000–1,000,000 ohm cm
Consolidated sedimentary rocks	1,000–50,000 ohm cm
Recent unconsolidated formations	50–5,000 ohm cm

Variation in resistivity values provides a means for direct determination of the groundwater surface. The upper surface of groundwater occurring in a uniform medium will appear as the resistivity boundary between two different media, the wet and the dry material respectively. This ideal case occurs rarely in practice, but its extension to three- and four-layer conditions presents possibilities for direct groundwater determination. In some instances, satisfactory results have been achieved by a careful correlation with detailed geologic knowledge, but on other trials results have been disappointing.

Construction Operations Geophysical exploration is mainly a twentieth-century development. Early examples achieved notable results in civil engineering applications. The Bridge River Tunnel was completed in 1930 for the British Columbia Electric Railway Company, Ltd. The 3,960-m (13,200-ft) tunnel passes through Mission Mountain at a depth of 720 m (2,400 ft) below the summit. It was constructed in connection with hydroelectric development and is located 175 km (100 mi) northeast of Vancouver. A geological survey of the surrounding area was made before bids were called; the structure disclosed is shown in Figure 4.13. Tunnel driving started at both ends, and work at the northern end proceeded without trouble. About 360 m (1,200 ft) in from the southern portal, however, bad ground was encountered as a very pronounced fault zone was met. The greenstone was found to be finely crushed and extensively altered to talc and serpentine, and the whole zone was highly water-bearing; so much pressure was exerted on the timber supports that a short length of the timbers was completely demolished.[4]

The probability of encountering such material had been anticipated and confirmed by exploratory borings, but it was necessary to find out how wide the zone was. This was done by the application of the resistivity method, then used by the Schlumberger Electrical Prospecting Methods firm. It was found that the high water content of the fault zone had a resistivity much different from that of the surrounding hard greenstone, and it therefore proved possible to determine the width of the zone on three trial locations from the tunnel center line. The second of these was adopted, and the tunnel was diverted by 45 m (150 ft) from its original location through this fault zone. As a result of the geophysical survey, it was estimated that the troublesome section of tunnel

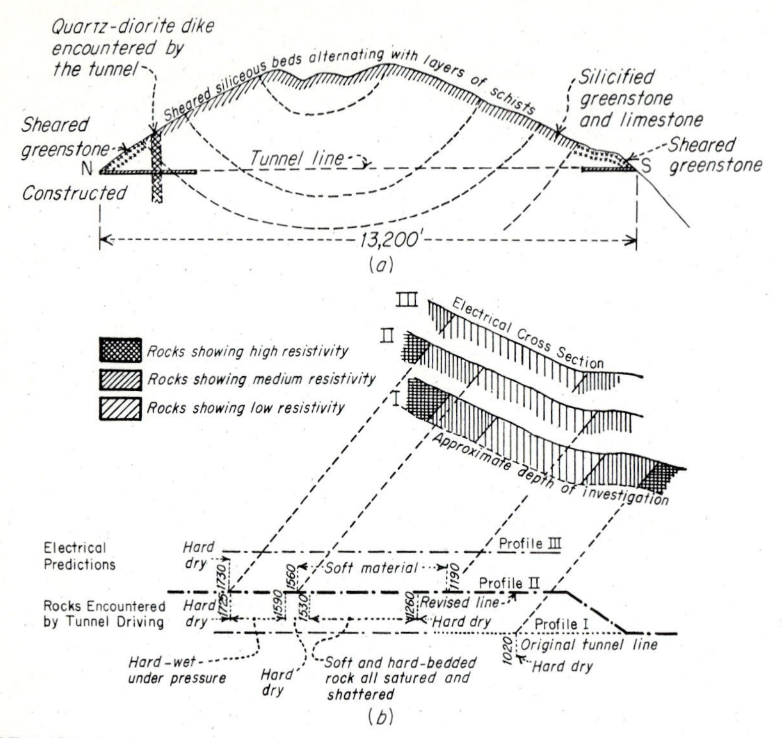

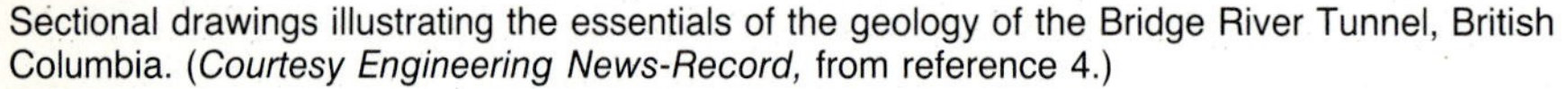

FIGURE 4.13
Sectional drawings illustrating the essentials of the geology of the Bridge River Tunnel, British Columbia. (*Courtesy Engineering News-Record,* from reference 4.)

would be 111 m (370 ft) wide. Excavation proved it to be about 81 m (270 ft) wide, with another short section of water-bearing rock some 18 m (60 ft) farther on. The engineers in charge of the tunnel work were then prepared to drive through this relatively short stretch of water-bearing material with the knowledge that satisfactory conditions could soon be expected.

Geophysical methods applied to dam-foundation problems have been principally concerned with determining the depth to the surface of solid rock through superficial deposits. What is believed to be the first application of geophysical methods to civil engineering work in North America was made in 1928 at the site of the Fifteen Mile Falls hydroelectric station of the New England Power Association on the Connecticut River near Littleton, New Hampshire. The local bedrock of Lower Paleozoic schist and occasional intrusions of granite was overlain by glacial deposits of lodgement till, outwash gravel, and sand. The presence of boulders made exploratory boring operations expensive, and so electrical resistivity was employed to investigate the two available sites. A buried valley was known to exist beneath the glacial deposits, and its presence was detected under an overburden of 90 m (300 ft), as confirmed by later drilling carried out as a check. Depths of the verification borings were planned on the basis of the previous electrical prospecting, and varied between 69 and 118 percent of the confirmed depths. Similar work was done in some of the early studies on the international section of the St. Lawrence River; drilling in this case revealed an average error of only 6.6 percent.[5]

Geophysical methods are now a worldwide standard component for preliminary investigations at potential dam sites and at the proposed locations of many other types of facilities. These investigations are almost always carried out in association with a drilling program, and the combination of the two methods has usually proved to be an economical and speedy means of subsurface investigation. Australian engineering practice has also made extensive use of geophysical investigations, especially at some of the dam sites in the great Snowy Mountains Hydro-Electric scheme. One of these lies on Spencer's Creek, a tributary of the upper Snowy River, at an elevation of about 1,698 m (5,660 ft); the valley is flat-bottomed because of glacial action and is underlain generally by slightly weathered granite. A potential dam site was presented by a natural stream break through a large terminal moraine. The site called for use of the moraine itself as the left abutment for the dam. Geological mapping, drilling, exploratory pits, and seismic surveys were all employed by engineering geologists in their thorough study of the site. Much difficulty was experienced because of the presence of granite boulders, which were usually considerably weathered and which complicated the permeability tests. The seismic results, however, validated depth to rock as established by borings and the dam has been constructed and is performing satisfactorily (see Fig. 4.14).[6]

Probably the most extensive use of geophysical methods in the practice of civil engineering has been in the United States. All the larger federal engineering organizations, notably the U.S. Army Corps of Engineers, the Bureau ofReclamation, and the Federal Highway Administration, employ geophysical methods regularly in their exploration work. Typical of these uses was the application of the seismic-refraction method for determining the bedrock profile at the site of the Englewood Dam on the Cimarron River in Oklahoma, a project of the Bureau of Reclamation. Only four boreholes were available as controls for the first geophysical work; 15 seismic horizons were then determined with good agreement with predictions from borehole logs and 12 additional drill holes along the projected axis of the dam. Eight final borehole correlations showed that the seismic interpretations of bedrock depth were accurate within an average error of 0.9 m (3 ft) at an average depth of 9.9 m (33 ft). The seismic method gives equally reliable results in sand and gravel and in clay overburdens. Direct comparison of the costs of these two complementary methods suggests a ratio of about 7 to 1 for the cost of exploratory drilling as compared with that for the geophysical work.

URBAN SITE EXPLORATION

Throughout the world, a large proportion of civil engineering work is carried out in urban areas, where most large buildings are located. Associated foundation

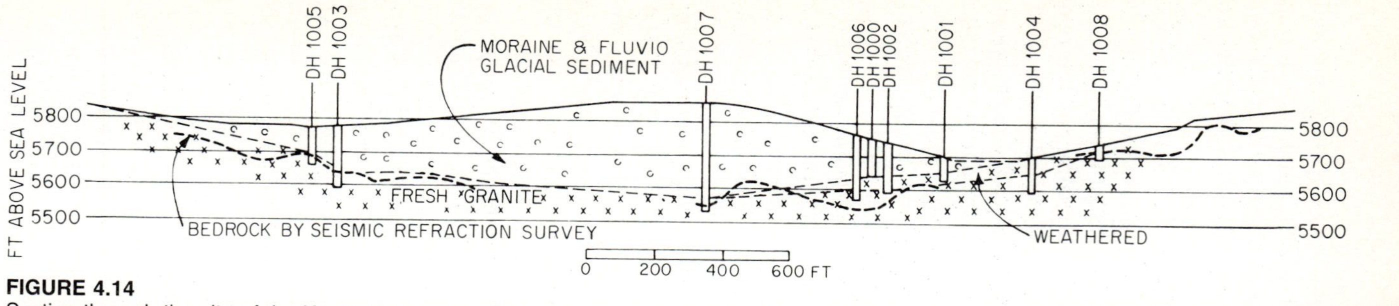

FIGURE 4.14
Section through the site of the Kosciusko Dam on Spencer's Creek, part of the Snowy Mountains project, Austrialia, showing results of the seismic survey. (*Courtesy Snowy Mountains Hydro Electric Authority, New South Wales, Australia.*)

investigations present the civil engineer with some of the most challenging problems in which all the disciplines involved (with the possible exception of aerial-photo interpretation) must be brought into play. There is, however, the common complication that all too often the site will be covered up by the roadways and sidewalks of city streets. Nothing about the underlying geology can be determined, in all but exceptional cases, from surface observations.

Civil engineers charged with the design and construction of urban public works are well aware that the geology beneath city streets will be just as complex, as interesting, and as vital to their work as is geology of sites far beyond city limits. Because of the usually restricted size of urban building sites, and the complications caused by the existence of underground works throughout all urban areas, they know also that the details of the subsurface for all sites within municipalities must be known with an unusual degree of accuracy. The presence of so many existing structures in every urban area would suggest that somewhere, most probably in the city hall or its equivalent, there will be a convenient collection of subsurface information assembled from earlier construction operations and ready for use as a guide for site investigations for new works. Would that this were so! In all but exceptional cases such information banks do not exist. Fortunately, however, there are some additional aids to help with urban site investigation, aids supplementary to those already considered for the general case. It is for this reason, and because of the unusual importance of geology beneath city streets, that this separate treatment is included as a subject of remarkable importance in civil engineering.

Site area searches for published records of urban geology should begin with the respective national or state (or provincial) geological survey. Some surveys have published this useful information in the form of maps, others as smaller-scale maps supplemented by descriptive memoirs. If there is a local engineering society with its own publications, it may prove to have information of value. Local natural history organizations, even though normally thought of as "amateur" bodies, will sometimes be found to have issued useful geological guides. Correspondingly, some of the more active local museums have interested themselves in the geology underlying their own cities, with resulting publications. And, quite naturally, local geological societies or local sections of national societies have likewise sometimes provided such guides for public information and enjoyment.

Boston, Massachusetts Boston must be given pride of place in any listing, since in the first published copy of the *Journal of the Boston Society of Civil Engineers* (1914) there appeared a paper by J. R. Worcester, entitled "Boston Foundations," in which an assembly of boring records was featured. As a subsequent development of the interest aroused by this paper, the society set up a Committee on Boston Subsoils, the purposes of which were stated as follows:

> The purpose of this Committee is to gather data regarding the character of the subsoils in Boston and adjacent areas, and to present it to the Society in such formas

to add to the general knowledge and to make it available for reference by any who may wish to get a clear idea of the geological construction under this City.

This committee made extensive reports in 1931 and 1934 which contained the locations of over 9,000 borings shown on 16 sectional maps. The borings are so well located that it was possible to prepare six geologic profiles through the city, to make maps of the rock and boulder-clay contours, and to locate the colonial-era shorelines of Boston and Cambridge. The rock and boulder-clay (lodgement till) maps were made by the geologist member of the committee who had started a study of the area as early as 1923. This important work continued through the years, still under the aegis of the Boston Society of Civil Engineers. Starting in 1949, the society's Foundations Committee published a series of papers giving, in concise form and accompanied by key maps, full details of the boring logs available for various parts of the Boston area. These reports were issued in 1949, 1950, 1951, 1967, 1969 and 1984. This invaluable collection of records for one city is now being used by the U.S. Geological Survey for the preparation of one of its very few urban geologic mapping projects. It has been well said that Boston has probably been more completely probed, drilled, cored, and investigated than any other American city.

New York, New York New York City can, however, give Boston a good run for its money. As early as 1902, the U.S. Geological Survey published the New York City Folio as no. 83 of its *Atlas;* this was followed in 1905 by the publication of Bulletin no. 270, *The Configuration of the Rock Floor of Greater New York,* based on a study of over 1,400 borings. The municipal engineers of the City of New York were active in this field since 1915 and published several notable papers in their journal. Planning the West Side Highway in 1933 provided a spur to what has proved to be probably the most extensive collection of subsurface information available for any city. In the 1930s a Works Progress Administration (WPA) project led to the preparation of "The Rock Map of Manhattan." More than 17,000 borings were plotted in this map when it was first prepared; the number of records available has greatly increased in the years since then. A similar project was started at the same time for that part of New York outside the borough of Manhattan; 27,000 boring records were used for this mapping of earth and rock borings project. Through the years these fine record maps and compilations of surface information have been extended and improved and are available for public consultation at the various borough offices.

San Francisco and Oakland, California One of the most significant of the early postwar efforts in this direction was the work of the U.S. Geological Survey in producing two fine maps for the San Francisco Bay Area. In 1957, there was published a map showing the engineering geology of the Oakland West Quadrangle, followed in 1958 by a companion map of the geology of the San Francisco North Quadrangle. The ground surface is shown by contours at

25-ft (7.6-m) intervals above an altitude of 25 ft (7.6 m), and at 5-ft (1.5-m) intervals below that; bedrock contours are at 100 and 25 ft (30.1 and 7.6 m) intervals, respectively. The maps are valuable not only because of the rugged character of the rock surface, which can be observed at the surface, but also because of the concentration of large buildings in areas of erratic geology. Near the Ferry Building, at the foot of Market and Mission streets, bedrock lies nearly 90 m (300 ft) below the present surface, whereas it is exposed at the surface just a few blocks away to the northwest and southeast. The bedrock contours show clearly that this strange variation is due to an old buried river valley, a not uncommon geological feature, but one that can play havoc with foundation designs if unsuspected and undetected (Fig. 4.15). The local section of the American Society of Civil Engineers has also been active in assembling engineering information on subsoils, some of which has been summarized in a useful general paper.

Montreal, Quebec In 1966, the Montreal City Planning Department (Service d'Urbanisme) published as their Bulletin no. 4, a color brochure of convenient size (by the use of folding plates) illustrating very clearly the general bedrock and soil geology of the island of Montreal and the immediately adjacent areas. Relief and drainage characteristics are also shown, as are land slopes and local forest coverage—all the basic material for the initial phases of planning. The city's Public Works Department, jointly with the Geological Survey of Canada, has produced a useful brochure on the local soils and their engineering properties. A fine geologic map accompanies Geological Report no. 152 of the Geological Exploration Service of the Quebec Ministry of Natural Resources. This work has now been even further extended (1986) as a "Cities of the World" paper in the Bulletin of the Association of Engineering Geologists.[7]

Pittsburgh, Pennsylvania On a somewhat different scale is a publication of the Topographic and Geologic Survey of Pennsylvania dealing with the geology of the Pittsburgh area as prepared in cooperation with the Pittsburgh Geological Society. This publication, for example, was prepared "mainly for the interested adult and the secondary school earth science teacher," but it provides, at the same time, an excellent beginning for any site investigation in this city. The four pages of references to publications relating to the geology of the Pittsburgh area alone make this publication worthwhile as a starting point for the more detailed studies required for urban civil engineering work.

Archival Records

All urban areas have some form of records of their past history. Some may be very sketchy and so not of much use in site studies, but others may reveal information of great value. In earlier times, the local public library was the usual repository for old city maps, old prints showing the location before

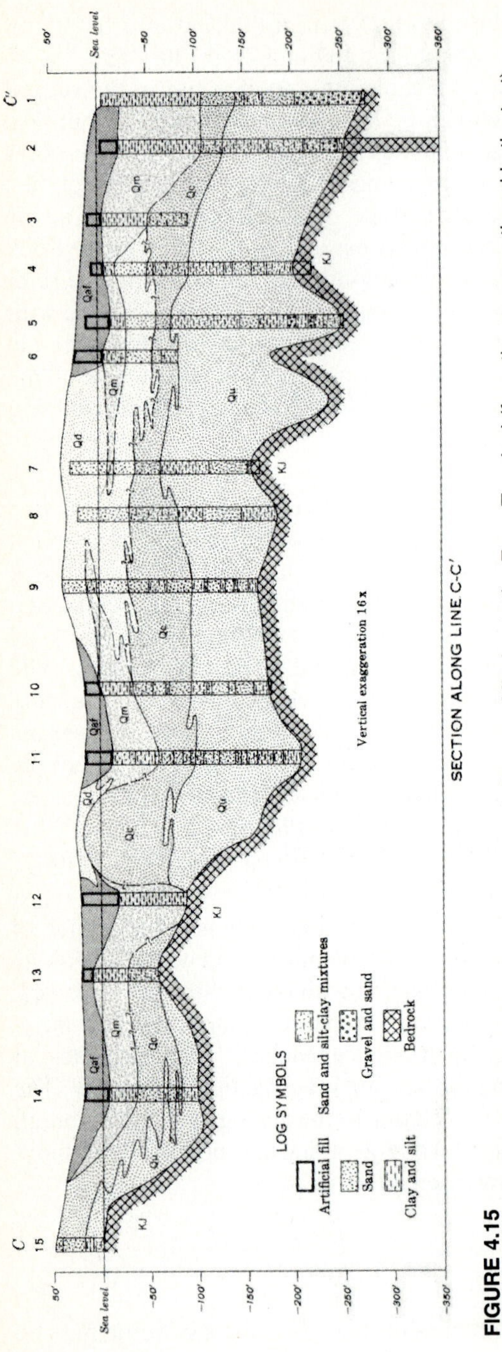

FIGURE 4.15
Section of the geology beneath San Francisco, from 20th Street at Mission to the Ferry Terminal (from the map mentioned in the text). *(Courtesy U.S. Geological Survey.)*

modern growth began, and similar historical records. Today, many cities have their own specialist archival staffs, and in the case of some larger cities, their own archives. Now that archival material is being properly collected, search is also warranted in the archives of the respective state (or province), or even the national archives, since their map rooms often contain unsuspected treasures.

New York, New York The twin towers of the World Trade Center at the foot of Manhattan in New York City are known around the world. Their foundations rest on the well-known Manhattan schist, in an excavation 21.3 m (70 ft) deep, and cover an area of six city blocks (Fig. 4.16). As part of the site investigation archival material was consulted, in the form of maps dating from 1783. It was found that the Hudson shoreline extended as far as the building site and was at that time built up with rock-filled timber cribs. With this, and similar archival information available, detailed studies of the site were made, and the excava-

FIGURE 4.16
Excavation for the foundations of the World Trade Center, New York City, showing tieback anchors at right and subway tunnel crossing the site, the scale of the operation given by the group of visitors. (*Courtesy Port Authority of New York and New Jersey.*)

tion contractors, knowing what would be encountered, were greatly aided in their work. Similar experiences have assisted site investigations, and so foundation design and construction, in other cities in the United States, notably San Francisco and Pittsburgh.

Canada Despite the relative youth of most Canadian cities, the same approach has proved of value there. Records in the 70-year-old city of Edmonton, Alberta, enabled a local consultant to prepare a complete atlas of the old coal workings that still exist under the central part of the city, to the lasting benefit of all concerned with subsurface investigations in this area. A new garage building in the center of Canada's capital city of Ottawa had to be located on a site which, the responsible engineer recalled, might have been an old quarry used by the Royal Engineers and their contractors in the building of the Rideau Canal between 1826 and 1832. The Public Archives of Canada contained not only a plan of the workshop area of the Royal Engineers but also one of the earliest photographs ever taken in Ottawa, both showing the old quarry. With this archival material available, borings were located appropriately, their interpretation greatly assisted by the old records. In a corresponding manner, old maps, one found in the Provincial Archives, and the other in a local city hall, were useful aids to modern investigations for two building projects in Nova Scotia.

Other Examples Neglect of old records, or a failure to search thoroughly for relevant archival material, can naturally have unfortunate results. If a contractor, for example, starts on a foundation-excavation job expecting to encounter nothing but soil and encounters unsuspected rock-filled timber cribs (a very common early form of construction in North America), the work will be hampered and there will be the inevitable, and justified, claim for extra payment. Old unsuspected works even beneath the lower limit of excavation or pile driving may cause trouble to a completed building. In July 1974, serious settlement took place under one of the new buildings of the University of Kent in Canterbury, England. Investigation showed that this was possibly due to the partial collapse of an 1830 tunnel on the Canterbury and Whitstable Railway (the great George Stevenson being the engineer). This tunnel ran immediately beneath the affected building, the pile foundations of which had the effect of transferring the relatively light building loads closer to the tunnel than shallow foundations would have done. The tunnel had to be backfilled as a part of the remedial work.

Even more remarkable was an experience in Zurich, Switzerland. A new multistory commercial building was to be erected on a choice building site to the east of the city's center and about 300 m (1,000 ft) from the edge of the Lake of Zurich. The city is of ancient foundation and it is known that so-called "prehistoric" lake villages existed along the original shores of the lake, the present built-up shoreline being formed by retaining walls founded on reclaimed land. The city archaeological staff indicated that the site might possibly be over part of the original shoreline, requesting that watch be kept for any artifacts in the cores of exploratory borings. The young engineer charged with boring supervision noticed a "piece of stone" in one core; it proved to be a flint arrowhead. Thus alerted, archaeologists of the city were able to arrange for the entire excavation over the building site to be carried out by archaeologists under scientific control while building superstructure was constructed and completed, supported on the surrounding basement walls and specially constructed central supports. Remains from five different civilizations were thus discovered, the oldest dating back to 4,000 B.C., and over 10,000 piles were uncovered (remains of an extensive lake village), to the enrichment of historical and archaeological records.[8]

Utility Records

Most municipalities have good records of the utilities buried beneath streets, even if little is known about the underlying geology. Records of buried utilities are clearly essential, as may readily be appreciated with one look at the maze of pipes and cables uncovered in excavating in a street near any civic center. A common pattern is to have a coordinating committee responsible for the maintenance of all such records, their accurate recording on large-scale maps and, in many cases, having absolute control over any disturbance of streets before a check has been made of the records and formal approval issued. This is the pattern followed, for example, in Philadelphia and Toronto. Local utility records will not show geology beneath streets, but they do have value to the civil engineer in two ways. They must be consulted before any boring is undertaken in urban areas and the necessary permission accordingly obtained. Beyond this, they show how such invaluable records can be prepared by cooperative effort, and at little direct expense, once the will to maintain them is there and the necessary facility for record keeping is available.

The first of these points may be illustrated in the case of Toronto. In 1944, the city started planning the first section of its subway system. It was necessary to supplement the information about the geology underlying the downtown area by arranging test borings to determine accurately soil and bedrock conditions. The route of the first part of the subway lies directly under Yonge Street, the main south-north artery of the city. Utility plans were consulted with the result that only under one of the four corners was there enough space between the pipes and cables shown on the utility plans to permit the sinking of a boring. Even there, one "empty" square area 60 cm (24 in) on the side was all that was available—and it was bounded by a high-voltage electrical cable, a multicircuit telephone cable, and high-pressure gas and water lines. Based on the relevant utility drawing, the hole was located accurately and drilling was satisfactorily completed.

Geophysical Exploration

Despite all urban logistical difficulties, investigation of the actual subsurface conditions at urban building sites is always imperative. The same methods already generally described will be used; however, such work will be more costly than usual and must be conducted so as to cause as little inconvenience as possible in the use of the streets where investigation has to be carried out. It might be thought that geophysical investigations have no place in city work, but experience has shown otherwise.

Investigations in Detroit have shown what can be achieved with relatively simple equipment and a minimum of interference with municipal services. By means of a six-channel seismograph with a suitable control unit giving a continuous record on sensitized paper, it was found that in some cases a hard blow from a sledgehammer was sufficient to give useful records for shallow depths. Inserting part of a stick of 40 percent dynamite in a hole 1.2 m (4 ft) deep was effective for rock depths up to 30 m (100 ft). Proper precautions naturally have to be taken in carrying out such work in city areas, but with accurate plans of underground utilities which have just been mentioned, the placing of small charges or explosive caps can usually be safely arranged. In one test case in Detroit, rock was located at a depth between 37 and 43 m (123 and 142 ft) with an error of only 3 percent. "Signal enhancement" seismographs are now widely used to take advantage of very small charges, caps, or even shotgun shells.

Seismic refraction investigations were carried out in a somewhat similar manner for the start of the Prudential Center in the Back Bay district of Boston. A large area, much of it occupied by railroad tracks, was ultimately cleared for this great development, of which the 55-story Prudential Tower was the first unit. Borings had shown that the geology underlying the site consisted of up to 3 m (10 ft) of fill overlying sediments of the former tidal flats. Sand with some gravel formed the base of the tidal deposits below which was a thick stratum of Boston Blue Clay. A thin layer of lodgement till overlaid the Cambridge Argillite bedrock. The tower was to be founded on caissons carried to bedrock. Seismic refraction enabled borehole deductions to be confirmed and extended over the area to be occupied by the tower. The only unusual feature of the work was the necessity for placing the half-pound charges of dynamite in shot holes drilled into the upper part of the clay stratum, after charges placed in shallow holes in the fill material had proved to produce unsatisfactory wave forms. Good accuracy was obtained, as was shown from caisson inspection records during construction.

UTILIZING EXPLORATION RESULTS

Underground explorations must be correlated without delay with the results of geologic mapping at the site. This requirement is another argument in favor of having one or more resident engineers or engineering geologists present during all exploratory work, since the correlation can best be carried out while the work is in progress. The most useful means of combining the results of the two methods of investigation is to draw tentative geologic profiles based on survey work, along lines connecting borings. The records of these holes can then be plotted to scale on the profile as work proceeds. It may be necessary to use an enlarged vertical scale for this plotting. By means of this simple device, it will be possible to keep constant check on the progress of holes, to stop them when they appear to have gone far enough, and to locate new holes to clear up doubtful points revealed by the profile.

Field management is a three-dimensional problem. Those who are not accustomed to three-dimensional thinking will be greatly aided by an actual model of the subsurface information as it is gradually revealed by a test-boring and drilling program. The use of models for this purpose on all but the very smallest jobs is now well-established practice. In their simplest forms, such models can consist merely of vertical sticks, colored to correspond with the different strata encountered and located on a plan of the works in correct relation one with another.

One of the most extensive models ever developed for study of subsurface conditions was constructed to illustrate the exploration program carried out at the site of the TVA Kentucky Dam, near Gilbertsville, Kentucky. The dam is a concrete and earth-fill structure 2,595 m (8,650 ft) long, founded on siliceous and chert limestone. The complexity of underground conditions called for 817 borings, aggregating 30,189 m (100,631 ft), one of the most extensive investigations ever carried out in North America. Progress of the work was followed by means of a peg model constructed of 5-mm (3⁄16-in) steel pins, suitably colored and arranged in plan. The resulting 1:100 scale model was over 30.5 m (100 ft) long, but it proved invaluable, not only to those engaged in the work but also to the consultants during their regular visits to the dam site (Fig. 4.17).

Plastics are now used to form transparent models, with embedded pegs to illustrate test borings. In this manner a vivid impression of underground conditions is given by means of a solid model that can be handled and viewed from a variety of angles. And in the case of extensive underground excavation, the process can be carried still further and a model of the excavation itself prepared, marked, or colored to correspond with the geological strata to be encountered, as determined from exploratory borings, shafts, and tunnels. With the aid of such models, contractors can readily visualize the character of excavation work and can plan their drilling and blasting operations long in advance of actually encountering the changes in strata with which they have to contend. Computer-imaged, three-dimensional drawings are now gaining in favor; colors are available and may be changed instantly, along with scale and viewer perspective angle.

FIGURE 4.17
Model of the Kentucky Dam site near the mouth of the Tennessee River, Kentucky. The total length of the model is about 30 m (100 ft), every drill hole being represented by a painted rod. (*Courtesy Tennessee Valley Authority.*)

CONSTRUCTION-RELATED ACTIVITIES

When Construction Starts

There is a widespread but mistaken belief that the work of the geotechnical engineer charged with investigation of the subsurface at a projected building site is finished with the report summarizing such studies. That is really only the "halfway mark," since the geotechnical engineer should have a continuing responsibility for studying and recording the subsurface conditions as they actually *are,* when exposed by excavation. It is naturally always hoped, and expected, that conditions will be just as predicted from the site investigation, but about this there can be no certainty. Throughout this volume examples will be found where conditions as revealed when excavation was carried out differed markedly from what had been anticipated. Sometimes these variations can have serious consequences for the progress of the job and even—in extreme cases—necessitate changes in design. Excavation may also reveal unsuspected features of the geology which, while not affecting construction adversely, are of such significant, academic interest that they should be recorded before being hidden again from human sight.

Site investigation, therefore, is not finished until excavation has been completed and work is about to start—for example, on the placing of the foundation structures or on the lining of a tunnel. It may be difficult to persuade owners that such an extension of site investigation work is imperative, but imperative it is, as all too many examples from practice can testify. From the professional point of view, such checking on actual conditions is also of vital importance. It is only in this way that defects in investigational procedure can be detected, and methods and tools improved. When conditions are found to be just as expected, there will naturally be a sense of satisfaction, but when variations are found, variations previously unsuspected, there is an equally important challenge to see where the methods used for preliminary work went wrong. And before any excavation is covered up by permanent structures, there is one further vital operation to be performed.

As-Constructed Drawings

When a structure is built, one of the most important requirements during construction is the careful recording of all changes, even the most minute, from what is shown on the contract drawings. Everyone with construction experience knows well that, of all the tasks on the job, the most tedious and difficult is the recording of such changes on contract drawings which, when so amended, become the *as-constructed drawings*. With construction proceeding all around, there is always so much to be done immediately that there is great temptation to skimp on recording accurately the exact dimensions and details of the structure as built. But as-constructed drawings are *absolutely essential,* not only as a record of what has been built for payment and other current purposes, but also to have a permanent record in case the day should come, as it does so often, when changes to the structure have to be made, or an extension has to be constructed, or there is a geotechnical problem with project performance. At such times, the absence of as-constructed drawings can be costly and even hazardous.

Important as are such drawings for what is built, they are even more important—if such an expression can be allowed—for the ground conditions upon which the structure is to be erected. Preparation of as-constructed drawings for what is revealed by excavation is, therefore, the final responsibility of the geotechnical engineer and the proper conclusion of all site investigation work. Not only is a record of the geology actually encountered in excavation essential for current purposes but, just as with the structure, such records are essential and invaluable if work ever has to be carried out adjacent to the building site, as for an extension. Anyone who has had to carry out such building extension work, where as-constructed drawings of the original excavation were not available, will know how frustrating, costly, and even dangerous is the absence of such simple records. These records, so easy to prepare when excavation is complete, give information that will be impossible to obtain again without new excavation—but they must be made right then and there.

Some qualification, perhaps, should be made to the use of that word "easy." Recording the conditions revealed when excavation is complete is, in itself, an easy operation. But it has to be done at a critical time in the overall construction procedure. With excavation complete and the foundation bed approved by the responsible engineer, there is almost always a keen desire to get the concrete in place, to get started on the "real job" (as so many think of it). Records of the foundation bed often have to be made, therefore, amid an extremely busy construction scene, with the first concrete forms ready to be placed and supervisors anxious to see the first concrete in place. Despite all the pressures that such timing may create, the recording of the foundation beds and other features of excavation must be well and accurately done, if the record is to be fully useful. Careful explanation will usually convince even the most impatient worker on the job that the necessary recording is essential to the well-being of the project.

As impatient as constructors may be, careful attention to as-constructed mapping by the project geologist is one of the most important safeguards available to the owner. It is the owner who pays for this valuable service and evidence of contractor problems may appear in this way for early resolution. And, if not, the as-constructed records and dated photographs are available for use in any regrettable legal actions.

Finally, there is perhaps no one operation that is as vivid a reminder that no two foundation beds are ever identical as is recording what excavation reveals. The fascination of geology lies partly in its infinite variety, and site investigation is yet another way in which the wide scope of geology is brought to light. Whatever the geological conditions revealed, the record of them will add to the steadily accumulating body of information on the geology of the area in which the building site is located.

CONCLUSION

Site area geology must be the starting point for all well-conducted site investigations. Drilling and field testing of soil and rock can only be fully effective when set against a sound background provided by thorough knowledge of the geology underlying the site. Although only rarely appreciated, engineers instinctively rely upon continuity to design their structures on the basis of a relatively few carefully selected boreholes. Consider, for example, the design of a simple bridge pier, measuring 20 by 3 m (65 by 10 ft) in plan. Sinking two 7.5-cm-(3-in)-diameter exploratory borings at the site of the pier would be a reasonable approach to the investigation of this relatively small site. If the boreholes were carried to a depth of 15 m (50 ft) and soil samples obtained, and if it were assumed (for simplicity) that the prism of ground beneath the pier that would be stressed when the pier was loaded had its sides at 45° to the vertical, a simple calculation would show that the volume of ground tested and sampled—ground upon which the design was to be based—was only 0.0007 percent of the volume of ground that would be under stress. This is limited sampling indeed, and yet it would be regarded as "good practice," thanks to the unacknowledged contribution of geology. Fortunately, this uniformity of ground conditions is widespread; study of the local geology will usually, but not necessarily always, suggest when variations are to be expected.

Site investigation is one of the most crucial parts of civil engineering practice. It is, correspondingly, the supreme challenge to the geotechnical engineer. The geology underlying the site provides the basis for all the work that must be done in exploring underground conditions in order to determine how the site can safely be used for the purpose intended. No two sites are ever the same. Two sites may be similar, but even the slight differences that will always exist may prove to be unusually significant. Every investigation of a building site is, therefore, a voyage into the unknown. As buildings get larger and more complex, the investigation of building sites increases in importance and extent. As the more obvious and more convenient sites for larger structures such as bridges and dams are progressively used up, site investigations for these larger projects will become more complex and will often expose conditions that will tax the ingenuity of even the most competent designer.

It has been said, in reply to the discussion of papers describing the design and construction of one of the world's greatest bridges, that "most site investigations prove to be inadequate." This is perhaps an undeserved accusation, although a glance at most of the cases described in this book might appear to support that statement. It is to be remembered, however, that the cases selected for summary description here include, by their very nature, some geological feature that caused difficulties. For most of these cases, site investigation probably was inadequate. There are, however, throughout the world very many times the number of cases herein noted which were designed, constructed, and are now operating with complete satisfaction, without any of the troubles that will be described later. For this vast majority of civil engineering structures, site investigation was entirely adequate. In some cases, it may have been adequate because local geologic conditions were so favorable that they compensated for any deficiency in investigation that may have existed. In general, however, and to an increasing degree, site investigations in civil engineering practice have been satisfactory because they were carefully and thoughtfully planned and conducted. This text assumes the burden of helping to ensure that site investigations of the future will be entirely adequate for the successful design and construction of all civil engineering works needed, as they will be throughout the world, for the steady improvement of the physical standard of living.

REFERENCES

1 "Redesign and Construction of Prettyboy Dam," *Engineering News-Record, 111,* pp. 63–67 (1931).

2 H. B. Gough, "A Wet Shaft Frozen Tight," *Engineering News-Record, 173,* pp. 666–668 (1939).

3 R. F. Legget and F. L. Peckover, "Canadian Soil Penetration Tests of 1872," *Canadian Geotechnical Journal, 10,* pp. 528–531 (1973).

4 I. B. Crosby and S. F. Kelly, "Electrical Subsoil Exploration and the Civil Engineer," *Engineering News-Record, 102,* p. 270 (1929).
5 Crosby and Kelly, loc. cit.
6 D. G. Moye, "Engineering Geology for the Snowy Mountain Scheme," *Journal of the Institution of Engineers of Australia, 30,* pp. 287–298 (1955).
7 L. Boyer, A. Bensoussan, M. Durand, R. H. Grice, and J. Berand, "Geology of Montreal, Province of Quebec, Canada," *Bulletin of the Association of Engineering Geologists, 22,* pp. 329–394 (1985).
8 R. F. Legget and W. R. Schriever, "Archaeology from a Swiss Test Boring," *Canadian Geotechnical Journal, 23,* pp. 250–252 (1986).

FURTHER READING

Subsurface Exploration and Sampling of Soils for Civil Engineering Purposes by J. Hvorslev, although first published in 1948, remains unchallenged as the major reference work on this subject; it was reprinted by the American Society of Civil Engineers in 1976.

If the idea of geology beneath city streets needs further illustration, there is now available, in addition to the series of papers in the *Bulletin of the Association of Engineering Geologists,* a volume entitled *Geology under Cities,* published in 1982 by the Geological Society of America as vol. 5 in its series of "Reviews in Engineering Geology," with R. F. Legget as editor. Papers in it deal with the geology beneath Washington, D.C., Boston, Chicago, Edmonton, Kansas City, New Orleans, New York, Toronto, and Minneapolis–St. Paul, Minnesota.

CHAPTER 5

OPEN EXCAVATION

Open excavation work is frequently regarded as the relatively simple matter of digging a hole in the ground. This same careless regard is often given to earth-work construction. All experienced engineers know otherwise. The problems introduced are sometimes intensified by their very simplicity. The engineering design efforts involved in these works may be more indefinite and consequently more involved than in many other branches of design work. Excavation methods and rates of progress are clearly dependent on the material encountered and its geological structure, the same two factors that determine the finished cross section of the excavation. All too often, these vital factors do not receive the study they deserve; similarly, fill placement is often characterized by unforeseen problems related to geologic factors.

Open excavation consists essentially in removing certain naturally formed material, within specified boundaries and to certain grade levels, in the most expeditious and economical manner possible. This requires accurate knowledge of the nature of the materials that have to be handled, their relative structural arrangement, their behavior when removed, the possibility of meeting water during excavation, and the possible effect of the excavation on adjacent ground and structures. Neglect of preliminary investigation work in open excavation is to some extent understandable. Such work looks easy, and the unit prices generally charged are so low that it is often thought that the consequences will involve only a few cubic meters, more or less.

Preliminary investigations should generally follow the lines already suggested. The resulting geological maps should show estimated rock contour lines ("top of rock") or the same for any stratum so different from the surface material as to affect excavation methods and progress. Geotechnical profiles must assist the engineer in visualizing the structure of the mass of material to be excavated, and the possibility of encountering groundwater. These profiles must also show the relation of the excavation to adjacent strata and structures so that possible damage may be foreseen. As-constructed records of all geological features are necessary as a constant check on the validity of design-basis information and to serve as a warning of future troubles.

An outstanding illustration of these various facets is the first rail subway in Canada of the Toronto Transit Commission. The first section of this railway was 7.5 km (4.6 mi) long. Every boring record that could be found for the central city area was studied as the basis of an exploratory boring program. Boreholes were then put down at about 120-m (400-ft) intervals right through the heart of the city. Soil samples were taken, and groundwater levels were observed until construction commenced. It was determined that the lower 0.8 km (0.5 mi) of the "cut-and-cover" subway box structure would be founded on shale. The overburden constituting the bulk of all excavation was a mixture of lodgement till and the geologically famous Toronto interglacial beds of glacial sands, gravels, silts, and clay. Design and contract documents were prepared on the basis of the information obtained through these preliminary studies. As a result, bids received were remarkably close. Cut-and-cover excavation was geologically recorded in step with excavation; soil samples were taken at intervals of 15 m (50 ft) and later lodged with the Royal Ontario Museum for safekeeping.

The work was unexciting from the engineering point of view, because soil conditions proved to be almost exactly as predicted by the preliminary study. No unusual soil problems were encountered, even in some of the difficult underpinning jobs that had to be done adjacent to the subway. Some minor claims were received by the commission after the work was completed (naturally), but all were settled satisfactorily and without recourse to law.

In general excavation work, the dip of the strata to be encountered is a factor of major importance. If the strata are horizontal, or approximately so, excavation work will be relatively straightforward, and side slopes can be determined with some degree of certainty. If the bedding is appreciably inclined, excavation methods will be affected, and hazards possibly increased. The side slopes must be so selected as to be in accord with the natural slopes of the bedding. Faults that cross the area over which excavation is to be carried out may cause serious trouble through juxtaposing strata of variable hardness or strength. In soft ground it may be difficult if not impossible to detect faults before digging starts, and hazards will thus be increased. Finally, all excavation work is influenced primarily by the nature of the material to be excavated. Broadly speaking, this can be classified as *unconsolidated material* (soil), *weak rock,* and *rock*. It will be convenient if the general problems of excavation are considered under these headings. Contract specifications must adequately define materials to be excavated, since claims most often occur in connection with open excavation.

A MAJOR EXAMPLE OF EXCAVATION

Generally speaking, more claims for extras have been received for alleged incorrect classification of material to be excavated than for any other cause. Perhaps the premier case in point is the St. Lawrence Seaway and Power Project (on both sides of the U.S.-Canadian border). The 8-m (27-ft) waterway, capable of handling oceangoing vessels, extends from the head of the navigable St. Lawrence River at Montreal to Lake Ontario and thence into the Upper Lakes. Construction involved new locks and canals, some deepening of the canal link between lakes Ontario and Erie, with much dredging of channels and the construction of a large number of ancillary works.

By international agreement, the two locks (Eisenhower and Snell) in the international section of the river were located on the United States side. All the earlier, smaller locks had been built by Canada and were located on the Canadian side, where geological conditions are favorable. The interconnecting canal was a major new cutting excavated through dry land, with a bottom width of 133 m (442 ft) (see Fig. 5.1) and over 15 million m^3 (20 million yd^3) of excavation. Site investigations included an extensive boring program, soil sampling, soil testing, and an excavation demonstration pit. On the contract drawings for the canal, *till* is clearly indicated in the borehole logs, although the presence of two tills is not differentiated. In the associated specification it is clearly stated that "the materials to be encountered... will consist predominantly of compact to very compact glacial till." On the contract drawings (and specifications) for the powerhouse works, the terms *sand, gravel,* and *clay* are used in various combinations. Many of the borings are indicated as "wash" borings, taken some years before construction started. The only mention of till is made in reference to a few borings taken along the line of the old Cornwall Canal in Canada, some distance from the location of the powerhouse works.

FIGURE 5.1
Excavation in progress on the St. Lawrence Seaway and Power Project, in the Wiley-Dondero Canal (in the United States). (*Courtesy St. Lawrence Seaway Development Corp.*)

It is doubtful whether the use or nonuse of the term *till* made much difference to the bids received. Bidding was keen from contractors with wide experience in earth moving who clearly wanted the jobs—to some extent, it has been suggested, for prestige purposes. Early unit prices for excavation of soil varied from \$0.31/yd^3 to \$0.58/yd^3; \$1.26 was a later figure. Of the contractors involved with the canal, one went bankrupt, another defaulted, and a third entered a \$5.5 million claim on a \$6.5 million contract.

Work on the United States half also involved large quantities of excavation for the hydroelectric facilities. Here, on contracts with a total value of \$89.1 million, claims for extra payment amounted to \$27.6 million (or 31 percent of the contract price); settlement was for \$4.8 million (or only 17 percent of the amount claimed). The chief complaints were that the "sand and gravel" (actually lodgement till) were "cemented," some of the material requiring ripping or even blasting before being moved by scrapers; and that the marine clay was very difficult to handle with normal equipment because of its stickiness (Fig. 5.2).

As usual, contractors were supposed to have satisfied themselves as to the nature of the work on which they had tendered bids, an understanding clearly reflected in the small amounts given in settlement of claims. The glacial till had been under detailed geological study since 1952, with early maps resulting from this work being on "open file" at the New York State Geological Survey in Albany. MacClintock and Stewart list about 150 papers dating back to 1843 dealing with the surficial geology of the surrounding St. Lawrence Lowland. As early as 1910, Fairchild had suggested two tills. All papers after that date clearly indicate that till is the predominant soil beneath the upper stratum of marine Leda Clay. The upper (Fort Covington) till was well recognized; the lower (Malone) buried till had been suggested, directly or indirectly, by several geologists. Since the lower till had been subjected to submersion and then to the pressure of the ice of the second glaciation, its compact nature was a certainty.

All this geological information was available before the bids for the Seaway and Power excavation were submitted, but it does not appear to have been fully utilized. Immediately adjacent to the Wiley-Dondero Canal lies the Massena

FIGURE 5.2
Exposure of typical lodgement till in excavation for the Eisenhower Lock on the St. Lawrence Seaway and Power Project. (*Photo: R. F. Legget.*)

Power Canal, excavated at the turn of the century to supply St. Lawrence water to a new power station. Men were still alive (in 1957) who had worked on this job and who could therefore have testified that the original contractor for this excavation also went bankrupt, apparently because of exactly the same difficulties with the same soils, and no more than 1.5 km (1.9 mi) away. Four hundred km (250 mi) upstream, the Welland Ship Canal, built in the 1920s, encountered the same problem in excavating tough till. Vast claims ensued, indirectly precipitating the failure of the leading Canadian contractor. A major account of this work, published in 1929, states that

> ...the rock surface was overlaid by glacial till. Over the northern one-third the rock dropped much lower, and was overlaid by very hard boulder clay of the first glaciation, over this again being the ordinary till of the second glaciation.

This experience on the St. Lawrence Seaway and Power Project should provide invaluable insight for those who have to deal with till, especially its excavation.

ECONOMICS OF OPEN EXCAVATION

The basic control over the respective costs of construction of tunnel and open cut is that of geologic structure. For an open cut, the volume to be removed per foot of length is naturally much greater than for an equivalent tunnel, but the unit cost of excavation will be lower, since the work can be done in the open without underground hazards and with more efficient excavating machinery that is not suited to the underground. If the side slopes can be trimmed at a satisfactory angle, no equivalent to the lining of a tunnel will normally be necessary; but if the side slopes have to be restricted, retaining walls at the foot of the slopes may involve a greater cost than tunnel lining. Comparative estimates of cost for tunnel and cut can be calculated. When these are applied to locations with increasing depth to grade, a section can be selected as that at which the change in type of construction can be made most economically. Geologic sections along the center line of the proposed cut and at right angles to it (at several stations) are essential. The proposed cut can then be seen in relation to underground drainage, the dip of the strata, and possible unusual underground structural relationships. In consequence, important modifications of design may have to be made.

Consider such a section as shown in Figure 5.3. In the tunnel, drainage of underground water need not be considered, as no water will reach the tunnel. However, in the cut, drainage will certainly present serious problems. Water will be continually seeping out at contact A, and if the quantity of water is sufficient, it may tend to wash some of the sand out onto the face of the slope. In order to escape, the water must flow over the face of the shale, thus tending to weaken it. There might even be some travel of water along bedding planes of the shale, reducing shear strength and giving rise to possible slope failures.

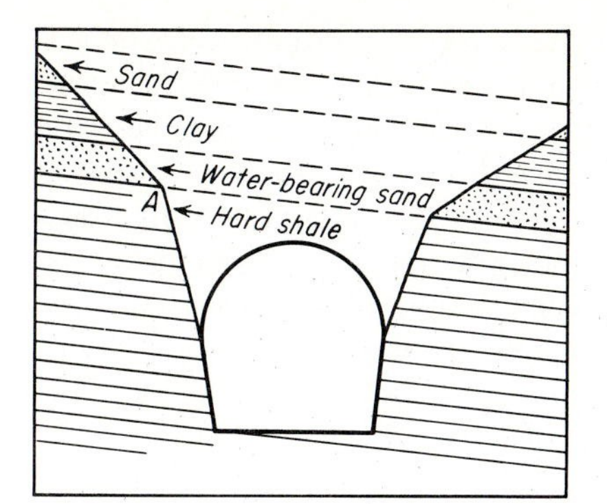

FIGURE 5.3
Simplified diagram showing a possible relation between tunnel and open excavation dependent upon geology.

Alternatively, through this unrestricted drainage of the underground porous bed, unstable conditions might be induced in the overlying clay stratum. For such relatively common cases, standard calculations of relative economy may not present a true cost picture. In the particular case illustrated, it will be advisable, even at the expense of increased first cost, to extend the tunnel construction beyond the theoretical economical limit to avoid the possibility of the difficulties indicated.

More complicated geologic structures may be encountered. Faulted ground is a menace in all material; although it causes difficulties in tunnel work, it may cause greater difficulties in open excavation, owing mainly to the lack of lateral restraint and subsequent failure of large masses on fault planes (Fig. 5.4). Overall dimensions and finished grade depths of open excavations will normally be determined by considerations other than excavation costs. Thorough preliminary investigations are necessary, and these may reveal special conditions which

FIGURE 5.4
Simplified geological section showing how the dip of strata may affect open excavation.

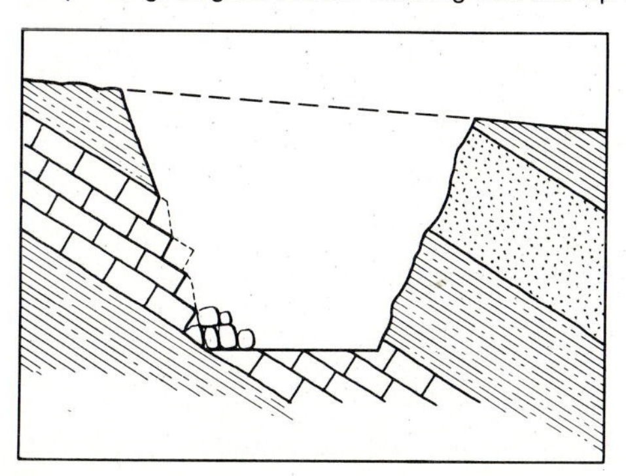

profoundly affect the economic aspect of the work undertaken. Subsurface explorations may show that a slight relocation will avoid troublesome material. Finally, geologic conditions will have an overall control over the economic cross section of the cut, since the structural relationships of the strata to be exposed affect the nature of side slopes quite as much as does the nature of the material. A point of special consideration is internal and surface drainage, which can be relatively simple and inexpensive, yet which leads to appreciable slope steepening or improvement in the factor of safety of the side slopes.

Each of the two great modern water power projects at Niagara Falls provides an illustration of the geologic aspect of excavation economics. For the Sir Adam Beck No. 2 Project of the Hydro Electric Power Commission of Ontario, the two large pressure tunnels lead to a final 3.5 km (2.25 mi) of open canal, designed for a flow of 1,130 cms (40,000 cfs), excavated through Paleozoic sedimentary strata. Careful stratigraphic correlation of the rock strata, which are inclined slightly to the horizontal, showed that if the canal invert was kept in the Grimsby Sandstone, drilling and mucking would be facilitated and a better degree of canal slope stability would be obtained. This vivid example of the interrelation of geologic conditions and civil engineering design was depicted graphically on a very large illustrated sign erected for the information of visitors while the construction job was in progress.

OPEN EXCAVATION IN SOIL

Excavation in soil may vary from work in good, clean, sharp gravel or glacial till to digging in soft, cohesive clay. Between the two extremes lie the many and varied substances classed generally as *soils* or *earth* or by some other such generic name, in nearly all of which cohesion is a property of some significance. In considering the mechanics of these materials, as distinct from *rock,* conditions other than those effects studied in the ordinary strength-of-materials laboratory must be investigated. It is here that soil mechanics has made such a great contribution to the field of civil engineering, such as by the determination of slope stability, based on laboratory investigations of the shearing strength of the soils to be encountered. All of this, however, is of little avail without accurate knowledge of the geology of the site to be excavated. The exact interrelation of the various soil strata, and particularly their relation to local groundwater conditions, must be known with certainty if excavation is to proceed as planned and without trouble. There is no branch of "soil work" in which soil mechanics and geology must be so closely associated as in the determination of slopes to excavation. That this is no new observation is shown by the comments made by Alexandre Collin more than a century ago.

Preliminary Considerations

The bottom grade level of open excavations will be determined either by relative elevation of other parts of the works involved or by general economic considerations. In any event, the total quantity of material to be excavated from side slopes will have a considerable effect on total cost. In all rail and highway cuts, for example, the nonproductive excavation of material to form flat side slopes can be far larger than the volume of material that must productively be moved from that part vertically above the level base of the cut. It is clear, therefore, that side slopes should be as steep as possible, consistent with safety, to minimize the volume of material to be excavated.

Geologic factors should be given the fullest consideration in the preliminary and final determinations of side-slope geometry. This has not always been done; many side-slope designs have been based on the past experiences of those concerned with the work. On the basis of practical experience, charts of angles of repose for materials of various kinds have been made up and are in general use today. Unquestioning reliance on such charts is not advisable, since it is almost impossible to correlate the broad material classifications in these tables with the detailed results of exploratory borings.

The use of assumed, nongeologically evaluated angles of repose in very preliminary designs is understood. A full study, however, should follow of the geological structure of the complete excavation prism to discover the extent of geologic effects, to investigate the effect of removal of the prism on adjoining geologic structure, and to determine what soil mechanics investigations are advisable. Sands and gravels are usually unaffected by exposure to the atmosphere but for clays, special attention must always be paid to the effect of exposure to the atmosphere during excavation. The moisture content of clay is a critical factor in such determinations; it can profitably be studied in the laboratory if undisturbed samples are obtained for testing. The old construction practice of covering up exposed steep faces of clay evolved in order to maintain unchanged the water content of clay. Its use will often result in considerable savings in the course of excavation. The mechanical properties of some clays will profoundly influence the progress of excavation.

Methods Used

The foregoing comments on open excavation in soil will be obvious to all who have had such experience. Those, however, with such experience will also appreciate that the suggestions made are as equally often neglected or not even considered. Geologic knowledge of all the strata to be penetrated in excavation work is absolutely essential before work begins. This is true regardless of the methods to be used for excavation. Increased capabilities of large-scale excavation equipment have been so marked that average unit prices for excavation on large North American jobs have changed little since 1950, despite a threefold increase in the general price index. But even the best of equipment will not operate efficiently in material for which it was not designed. Geological study, therefore, is a prerequisite for getting the best out of earth-moving equipment.

Projects of confined area will often necessitate vertical excavations at the property boundary. The use of vertical retaining walls has grown steadily since

1950. Many such walls consist of sheet-steel piling. The severe vibrations induced by pile driving, although usually unimportant in sand and gravel, may produce strange results in clay, especially if the driving is under water. One such case, a small pier, utilized steel piles driven into overconsolidated marine clay which naturally showed up in the preliminary wash boring as "very hard." When disturbed by the pile driving and consequent churning action, the clay changed character completely. It became almost fluid in the neighborhood of the piles and gave a greatly reduced passive pressure as resistance to lateral movement. While the pile-driving operations were in progress, a large dredge was engaged in excavating a nearby channel in the same clay. The steel cutterhead of the dredge had to be used to break up the clay sufficiently for it to be pumped; but starting at about 120 m (400 ft) from the site of the pile driving, the clay became progressively softer as the neighborhood of the piles was approached until finally the steel cutter was stopped and the clay was removed by direct pumping.

Boulders usually present a serious impediment to the driving of steel or prestressed concrete piling. A geologic interpretation of exploratory borings is always necessary. Knowledge of the glacial nature of strata encountered, for example, suggests the possible presence of boulders even if the boring logs do not show any. Furthermore, if steel piling is to be driven into rock, a study of the local geology will sometimes be the only means whereby a reasonable estimate can be made of the nature of the rock surface and whether or not it will be weathered enough for the piles to penetrate it at least for a little way.

Examples of Open Excavation

Over 230 million m^3 (300 million yd^3) of soil was moved from the bed of Steep Rock Lake in northwestern Ontario before all the valuable iron ore beneath the lake bed was accessible for open-pit mining. This excavation job involved the moving of more soil than was moved in the building of the Panama Canal. In 1943 and 1944 the lake, with an area of about 26 km^2 (10 mi^2), was drained by a major river-diversion project. The natural bed of the lake consisted of varved clays with high sensitivity and a liquidity index always greater than 1, and sometimes as high as 1.5. By a vast dredging operation the soil covering the ore bodies was removed easily, but the surrounding slopes had to be finished off in a stable condition. This involved careful cutting with high-pressure water jets to a final slope configuration that was based on careful geological studies carried out with simple equipment made on the job and closely associated with continuing laboratory studies of soil properties. Basically, the method used was to trim the soil to bench slopes of 1 in 3 and to limit the vertical height of each bench to 6 m (20 ft). A horizontal berm, its width depending on the shear strength of the soil, was then excavated and graded gently away from the face of the slope to obviate drainage onto the erodible face of the excavated soil. The slopes were successfully finished off and vegetation completed the stabilizing process. The most striking example of what could be achieved with this very unstable soil (which, due to its high moisture content, ran like pea soup when disturbed), was the forming of a 48-m- (160-ft)-high natural dam from the undisturbed material to retain an undrained part of the lake bed, while on the other side excavation proceeded down to one of the ore bodies 60 m (200 ft) below the crest level of this unusual barrier.[1]

Almost at the other extreme from the Steep Rock Lake was the excavation for one of the largest automobile parking garages in the world. Beneath a 2-ha (5-acre) peaceful garden in Los Angeles is a three-level garage capable of holding 2,000 automobiles, the lowest level about 9.7 m (32 ft) below street level. A careful preliminary investigation testing had disclosed that the excavation would be in clean, fine brown sand, sand and gravel mixed with a little clay, fine gray sand and silt, and a local siltstone known as the Puente Shale. The area would generally be dry, but some groundwater at a depth of about 6 m (20 ft) was expected in the northern part of the square.

Excavation was carried out over the entire site by 2-cubic-yard shovels that were even able to excavate the shale, with some difficulty. The relatively limited quantity of groundwater was handled easily by intermittent pumping from a deep pump well at the north end of the site (Fig. 5.5). Because of the generally dry site and its great area, an ingenious system was developed for

FIGURE 5.5
Excavation in Pershing Square, Los Angeles, in 1951, for the construction of the parking garage now hidden beneath the central garden. (*Courtesy American Society of Civil Engineers;* from reference 2.)

foundation construction, a system based on the accurate knowledge of subsurface conditions that had been obtained. Excavation was carried out in two 4.8-m (16-ft) lifts. Concurrently, a large drill rig surrounded the entire site with 1.5-m-diameter holes, spaced as 2.7 m (9 ft) centers, going below final grade level, and belled out at their bottoms to serve as footings. In these holes, precast concrete columns were then placed, each weighing 9 tons. Soil was left in place at its angle of repose up to the tops of these columns while the rest of the excavation was completed and the foundations were placed for the whole garage except for the one bay around its entire perimeter.

As the massive concrete structure rose to its finished level, 75 cm (30 in) below finished garden level, it was used to support struts that were then fixed to the individual columns all around the site. With these supports in place, the remaining excavation of the sloping bank could be completed, and the final perimeter bay of the garage constructed, incorporating the precast columns placed when the job began.[2]

The size of the two excavation projects just described is far removed from the normal small excavation encountered on ordinary civil engineering construction. The necessity for due attention to geology, however, remains the same, irrespective of the size of the enterprise. One useful example is, unfortunately, typical. The basement for a large building in Massachusetts involved excavation to a depth of 7 m (23 ft). Only one boring was placed in the area of the boiler house, and only to a depth of 4.3 m (14 ft). Instead of the glacial overburden encountered over the main part of the site, the contractor encountered bedrock at a depth of 4.5 m (15 ft) in this deeper cut. The bedrock was hard metasandstone of Pennsylvanian age (the Rhode Island Formation). Blasting could not be used, but, fortunately, the contractor was able to remove the rock with a large tractor-hauled ripper, although at greatly increased cost. His claim for extra payment led to the usual argument. Although a settlement was eventually achieved, all the trouble would probably have been avoided if a more thorough site investigation had been performed.

SUPPORT FOR EXCAVATION

The most extreme cases of lateral support of excavations are those carried out in narrow trenches, such as for sewer and water main installations. Successful design calls for exactly the same attention to subsurface conditions, and to the geology of the strata encountered, as is necessary for the largest projects. Workers' lives are involved and so the safety of such excavations is of paramount importance. Because human safety is involved, it is in some ways even more necessary than usual to have detailed knowledge of the subsurface before laborers are allowed to work in the confined spaces of narrow trenches. Experienced contractors know better than to allow any workers to proceed into unsupported cuts deeper than 1.5 m (5.0 ft) (U.S. OSHA regulations), especially those excavated in clay, even though the sides may stand vertically when first excavated.

On a larger scale, especially in urban areas, the necessary support of the sides of excavations must be planned and designed well in advance of construction. These designs must be based upon the best possible subsurface information. The ingenious solution used for the Pershing Square excavation in Los Angeles has been mentioned briefly.

The major excavation for Carlton Center in Johannesburg, South Africa, involved four complete blocks in the center of a large city excavated to almost 30 m (100 ft) below street level. A 202.7-m-(665-ft)-high office tower and an associated office building are the main features of the 2.4-ha (6-acre) Carlton Center, located in a 120.9- by 152.4-m (400- by 500-ft) hole carried to a depth of as much as 29 m (95 ft). Subsurface conditions consist of a complex system of igneous and metamorphic rocks—diabase, dolerite, quartzite, and shale—weathered to depths of up to about 30.5 m (100 ft) and overlain by a thin veneer of transported soil. The boundary between transported and residual soils is distinguished by a "pebble layer." Groundwater was present in the weathered rock, requiring dewatering and essential water-level monitoring in all streets around the site. Regular movement monitoring was also carried out to ensure that there was no damage to neighboring buildings. An innovative excavation support involved a ringbeam of prestressed concrete wide enough to be used as a roadway (Fig. 5.6). Accurate knowledge of all subsurface conditions made

FIGURE 5.6
Excavation for the multiblock Carlton Center, Johannesburg, South Africa, the scale given by the size of the men in the right background; the large strut and surrounding beam are of reinforced concrete. (*Photo: R. F. Legget.*)

possible the safe and efficient carrying out of this immense excavation in so critical a location.

The usual method of supporting the sides of both large and small excavations in soil is to drive around the site, in advance of excavation, either sheet-steel piling (the necessary supports being placed as the work progresses) or "king piles" (usually steel H sections) at regular intervals. Suitable lagging is placed between adjacent piles as the hole deepens. Such temporary walls can be supported either by horizontal systems of bracing, such as those used at Johannesburg, or by inclined struts founded either on rock or on the central part of the concrete foundation slab, such as those used on the Pershing Square job in Los Angeles. Since the mid-1960s steel tied-back anchor systems have been used, extending beneath adjacent buildings and placed in tension to compress earth material in the active pressure failure zone. This system operates on the Mohr-Coulomb principle of increasing strength across the most likely failure surface. Naturally, contractors have developed variations of these methods but, until relatively recent years, temporary supports for excavations were usually of this general type.

The geology of some urban areas is so complex that troubles can often be anticipated because of unexpected ground conditions unless a very carefully planned subsurface investigation has been carried out. A published account of one such case begins with these words: "When a contractor digs in New York City, he expects to find almost anything!" The job that prompted this remark was the foundation for an extension to Mount Sinai Hospital in Manhattan, involving excavation over an area of 30 by 60 m (100 by 200 ft). Exploratory borings showed sound rock. When a shovel started excavation in advance of the driving of steel H piles, it encountered rock broken into large chunks. This was then found all over the site, the broken rock apparently being earlier subway muck.

Slurry-Trench Method

The practice of civil engineering construction has long been distinguished by ingenuity. Two excellent post–World War II examples are major new methods of supporting the sides of excavations in soil and weak rock. The success of either requires an unusually accurate knowledge of subsurface conditions. The first is known as the *slurry-trench (Milan) method.* It was developed by an Italian firm in 1950, and its first major use in North America appears to have been in the tightly confined areas of the second section of the Toronto subway. Two 60-cm- (2-ft)-wide trenches were excavated along the alignments of the two sides of the finished subway box structure. The upper 1.5 m (5 ft) were provided with concrete sides to serve as guides for the special excavating buckets (about 60 cm or 2 ft wide), which carried the trenches to a final depth of from 7 to 16 m (24 to 54 ft).

The special feature of this method is the use of a bentonitic clay slurry in the trenches, the density of the slurry being such that it will support the exposed soil safely as the excavation proceeds. "Tremie" concrete is then placed from the bottom upward in the completed trench after steel reinforcing cages have been lowered into place. As the concrete fills the trench, the bentonite slurry is forced out for reuse. Upon curing, a normal reinforced-concrete wall was in place, and excavation proceeded between the two concrete walls.

Accurate subsurface investigation is again an essential part of the slurry-trench method, as is the absolute necessity for consistent soil conditions along the slurry trenches so that there can be no possibility for slurry loss. If, for example, a pocket of gravel or an unsuspected lens of gravel is encountered, the slurry can be lost, with serious damage to the trenches.

Not only have concrete walls placed in slurry trenches come into use for unusually difficult foundation excavations, but they have also been used for other purposes. Cutoff walls had to be constructed, for example, below 1,620 m (5,400 ft) of earth embankment during the construction of Wanapum Dam on the Columbia River, the embankment serving first as a cofferdam and then being incorporated into the main dam structure. Careful advance tests were made before the slurry trenches were constructed, some as deep as 24 m (80 ft) and from 2 to 3 m (7 to 10 ft) wide, the entire project requiring about 500,000 barrels of slurry.

Tieback Method

The second relatively new method consists of providing support for the walls temporarily surrounding an excavation by tying them back into the ground behind that which is actually being retained by the walls. Once again, accurate knowledge of subsurface conditions around the site being excavated is absolutely essential for the safe use of this method. The idea is believed to have originated with Andre Coyne, a distinguished French civil engineer. It was pioneered in the United States by Spencer, White and Prentis, foundation contractors of New York, who patented the idea of prestressing the tieback anchors. The tieback system gives the great advantage of a working site uncluttered by the usual bracing. Figure 5.7 shows a usual arrangement, the inclined ties being well anchored into the adjacent bedrock, the exact location and strength of which must naturally be computed on the basis of site geologic conditions.

Unusually difficult geological conditions were encountered in the close confines of the new Hamilton Plaza building in White Plains, New York. The local gneiss was of such poor quality that the upper 7.5 to 9.0 m (25 to 30 ft) could be excavated without blasting. A sharp fold in the gneiss near the center of the site resulted in smooth and micaceous foliation planes dipping into one wall of the excavation at angles varying from 38 to 70° to the horizontal. The vertical walls were stabilized by the use of steel cable tiebacks, each placed at such an angle to the horizontal as to compensate for the local dip of the rock. With the weak rock secured in this way, 160,000 m^3 (210,000 yd^3) of rock was excavated (as well as 38,000 m^3 or 50,000 yd^3 of soil) to depths of from 19 to 22

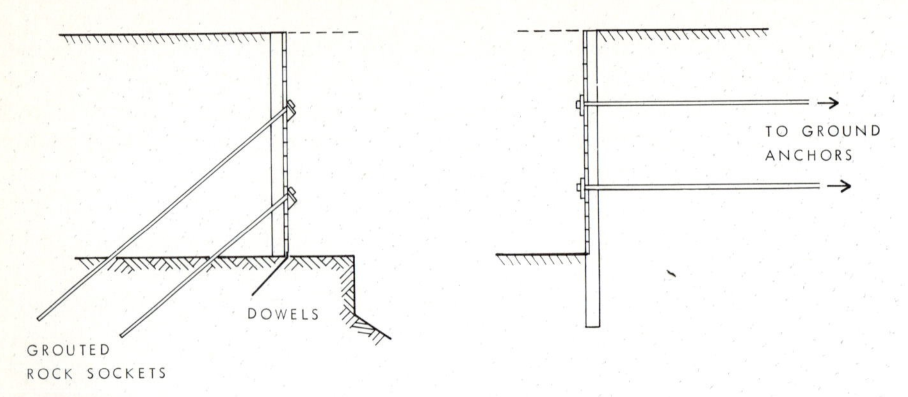

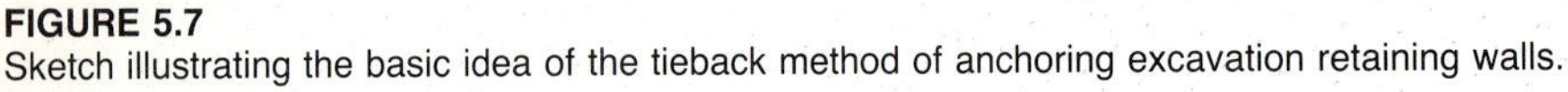

FIGURE 5.7
Sketch illustrating the basic idea of the tieback method of anchoring excavation retaining walls.

m (65 to 75 ft). Even these depths were exceeded in the excavation for the foundations of three modern buildings in Los Angeles which went to depths of 34 m (112 ft) in the local soft, dark gray Pliocene siltstone. All the foundations were successfully completed with the use of tiebacks, 2,700 tiebacks being used for one job alone.

An Example of Combined Methods

Excavation for the World Trade Center in New York revealed a problem of the greatest complexity in view of the size of the area that had to be excavated and the depth to be attained (21.3 m or 70 ft below street level). The solution was found in the use of a combination of both the new methods that have just been described. A slurry trench was constructed and then supported by tiebacks anchored in the Manhattan Schist. A further reason for the use of the slurry-concrete walls was the necessity to exclude groundwater inflow. The excavation spoil was trucked to an adjacent site in the Hudson River and placed inside a sheet pile containment resulting in $90 million of reclaimed land (see Fig. 4.16).

SINKING OF SHAFTS

Shaft sinking is one of the most specialized parts of open excavation; this is a technology shared by civil and mining engineers. The first requirement is of open-cut planning, an accurate knowledge of the strata to be penetrated and the location (and possible variations) of the groundwater level. For shallow shafts, well points may be used, but the depth to which this method will be effective is limited. Pumping from any great depth will usually be impracticable, not only because of the amount of water to be pumped, in view of the size of the cone of depression, but also because of the possible subsidence effects of the pumping operation upon surrounding property, or at least in lowering local groundwater supplies. Accordingly, other means have been adopted for shaft sinking in water-bearing strata; one of the most common in recent years has been freezing of the surrounding ground.

Freezing has been used as an aid in civil engineering work for many years. There are early records of Russian constructors forming cofferdams in rivers by working in winter periods and allowing ice to form in relatively simple enclosures as the water level was gradually lowered. The first patent for the application for artificial refrigeration in engineering work appears to have been taken out in 1883 in Germany, with the first application of that process to shaft sinking carried out in that same year for a mine in Schneidlingen, Germany. The first application in North America may have been around a shaft for the Chapin Mining Company of Iron Mountain, Michigan. Methods are now refined, theories of freezing action in soils have been well developed (notably in the U.S.S.R.), and the number of applications has steadily but slowly increased. The use of the freezing process is costly; its success is wholly dependent upon the certainty and uniformity of ground conditions all down the shaft. Since it is temperature-dependent, there is always an element of risk attached to it.

Some of the most remarkable shafts of recent years have been those sunk to depths up to 900 m (3,000 ft) at the Saskatchewan potash mines. Potash solubility makes leakage into any shaft a matter of unusual seriousness in any case in which the potash bed has been reached. Because of the water-bearing Blairmore Formation, at least seven of the major shafts in this area were sunk with the aid of freezing. Further work with grouting was necessary in some cases even after a permanent lining had been put in place. A good example of civil engineering application is the excavation of the large shafts used during the construction of a vehicular tunnel under the river Scheldt at Antwerp, Belgium. These 21-m- (70-ft)-in-diameter and 26-m- (87-ft)-deep shafts penetrated water-bearing sand with an interbedded turf layer. For the freezing process, 116 holes were bored, one set on a circle of 25 m (86 ft) diameter, and one set on a circle of 23 m (78 ft) diameter. The spacing between the holes was 1.3 m (4.5 ft), and all were sunk to a clay stratum 27 m (90 ft) below ground level. A 15-cm (6-in) pipe was sunk and sealed in each hole and a 5-cm (2-in) pipe, open at the bottom, was inserted in each of the larger pipes to form the necessary circulating system for the brine solution used to lower the temperature of ground and water. It took four months to complete each of the requisite cylinders of ice and frozen ground; they were then maintained in a frozen state until the shaft construction was complete.

Under exceptional circumstances, other more specialized methods can be used to "solidify" soft ground. Several chemical-injection methods have been successfully applied to shaft work but, like all other methods, they depend for their success upon an accurate foreknowledge of the strata to be penetrated.

CONTROL OF GROUNDWATER

The freezing process has also been used to control movements of soil caused by groundwater on a few major cases of open excavation. One of the most remarkable, and most successful, of these isolated examples was the control of earth movement during the construction of the Grand Coulee Dam for the U.S. Bureau of Reclamation. The dam is located on the Columbia River, Washington, and is one of the major dam structures in the world. At the dam site, silt had originally filled the river valley to a depth of about 150 m (500 ft), but the river had subsequently worn this down to a depth of between 12 and 15 m (40 and 50 ft). During this erosive process, the river channel swung from side to side in the valley, with the result that prehistoric silt slides were frequent. Excavation in the dried-out river bed encountered the toes and activated many of these old slides.

This silt is a very fine glacial rock flour, containing from 20 to 25 percent colloidal material. When undisturbed, the material stood up well, but as soon as it was moved, it proved unstable at slopes steeper than 1:4, even when relatively dry. Drainage by wells and tunnels assisted in controlling slides, as did slope correction. Near the center of the east excavation area, however, the bedrock was intercepted by a narrow gorge 36 m (120 ft) deeper than the average bedrock elevation, in which silt created an unusually serious slide. After a 5-cubic-yard shovel could make no headway against the movements, the engineers finally decided to freeze a solid buttress arch behind which excavation could proceed. This was done with the aid of 377 special freezing points and a circulating ammonia-brine-solution refrigerating system which had a capacity of 72,000 kg (160,000 lb) of ice per day. Freezing of the 12-m-(40-ft)-high dam started when the slide was moving at a rate of 60 cm (2 ft) per hour, and saved its own cost alone in the excavation which did not have to be undertaken. In addition, several weeks of very valuable time were saved in connection with work in the cofferdammed riverbed.[4]

Perhaps the most unusual method of groundwater control also has its applicability rooted in the geologic curiosities of silt and clay soils. This is the application of direct electric current over a suitable electrode field, causing the phenomenon of *electroosmosis* in clays and silts. First used in Europe, the method was introduced to North American practice by Professor Leo Casagrande. As early as 1807 Reuss had discovered that an electric potential applied to a porous diaphragm will move water through capillaries toward the cathode, as long as current flows. Helmholtz explained the phenomenon in 1879. Much experimentation had to be carried out before its practical application to soils was possible but this was achieved, with one World War II application by German forces in Norway giving notable results. Well points are usually used as the negative electrodes and steel rods may be driven into the ground to be drained, midway between electrodes.

An early application of this method in North America was in the excavation of the foundation for a power plant at Bay City, Michigan. Excavation was in wet silt, an although a well-point system operated at high vacuum was used successfully for the first stage of excavation (generally in medium sand), excavation for the second stage (generally in sandy silt with seams of sand) proved difficult with the standard well-point installation. Electroosmosis was therefore tried, the well points already installed being used as negative electrodes, the deeper parts of the excavation standing up well.

A more extensive and unusual application was the stabilization of a part of the almost completed earth-fill West Branch Dam in Ohio. The 24-m- (80-ft)-high embankment is about 3,000 m (10,000 ft) long. With 2,700 m (9,000 ft) completed, movement of the central part of the dam was observed just before the fill was finished. Careful studies by the U.S. Army Corps of Engineers revealed that high pore pressures in the clay on which the embankment was founded were responsible. They would have dropped in time, but rapid remedial measures had to be applied to permit the dam's completion. About 1,000 electrodes, roughly one-third of them well-point cathodes and old rails serving as the anodes, reduced pore pressures and the project was completed, although at considerable extra cost.[5]

Both the freezing method and electroosmosis are elaborate to install and relatively costly, so that their use is naturally limited to exceptional cases. When normal methods of water control fail, one of these special methods may prove economical if soil conditions are suitable for its use and if no unusual geological features will be encountered in the subsurface to interfere with its successful operation. Specialists must be engaged for the successful application of these methods, but all civil engineers should be familiar with their potential so as to recognize when to apply one of these methods.

DRAINAGE OF OPEN EXCAVATIONS

Groundwater flow theory is routinely applied to drainage of water at or near the surface of the ground, and especially in open excavations. Pumping will usually be necessary, but should be designed only after local geology has been studied and the most suitable form of pumping arrangement selected.

A useful example is on record from the Salt River valley of Arizona. Excessive irrigation of 82,000 ha (203,000 acres) of farmland brought about waterlogging of sandy and silty clay underlain by clay and caliche. Drainage by open channel proved to be ineffective and the water level began to rise 0.4 m (1.4 ft) per year. Soon 26,000 ha (64,000 acres) had groundwater within 3 m (10 ft) of the surface. Geological investigation led to the decision to pump the water through relief wells, utilizing cheap, off-peak power. Pumping was started in 1923, and the waterlogged area was quickly reduced.

The concept of the *cone of depression* is relevant and important even in determining pump locations and depths. The same idea supports the concept of tile conduits for shallow surface drainage. Long used in European farming, tile underdrainage was introduced into North America in 1835 by a Scottish farmer, John Johnston, with an installation near Geneva, New York, that is still in use. Civil engineers have traditionally used tile drains around building foundations and the rear of retaining walls. Some groundwater which is close to the surface cannot be drained gravitationally and is an important factor affecting foundation stability on clay soils. Geology is critically important in detection of clay soils susceptible to shrinkage when drying. Special study of local rainfall records is in order to determine Thornthwaite's soil-water balance through rainfall and evapotranspiration.

In earlier days, pumped drainage of large excavations was limited by the available capacity and types of pumps. The arrival around 1925 of submersible pumps led to the established practice of using such pumps in gravel-packed wells. Sometimes two or three sizes of gravel are needed to provide the equivalent of a filter around the intake to the pump to avoid undue erosion of the surrounding soil. In some cases, submersible pumps are left in place after the completion of construction to provide for permanent drainage. A notable example is the Albert Canal in Belgium.

Graving (or "dry") dock construction is almost always an unusually difficult operation. Areas are always large and immediately adjacent to deep water. Even though these notable structures are not numerous, the record of each case is well worth careful study. The King George V Graving Dock, at Southampton, England, required ten installations of 35-cm (14-in) pipes, gravel-packed in 60-cm (24-in) holes. Permanent groundwater relief pipes were installed to deal with artesian water at the dock.

The great World War II west coast dock of the U.S. Navy at San Diego, with an entrance sill 13.5 m (45 ft) below mean low water, was designed to accommodate the largest naval vessels of the time. Careful soil studies and a full-scale field test confirmed that it might be possible to build the dock in the dry. The site was therefore dredged to grade and then protected by a rock-fill seawall, behind which sand fill and an impervious silty, clayey sand blanket were placed.

Thirty-six 9,000 lpm (2,000 gpm) deep-well pumps were then installed, together with three rows of well points which circled the construction site and stabilized the soil slopes. In all, 1,600 well points yielded up to 9,000 lpm (2,000 gpm). In 23 days, the entire site had been dewatered to 18.6 m (62 ft) below sea level; the average pumping rate was 165,000 lpm (36,000 gpm). A continuous pumping rate of from 90,000 to 113,000 lpm (20,000 to 25,000 gpm) was necessary for eight months until completion. The deep-well pumps and motors became the permanent pumping installation. Settlement of adjacent ground was also anticipated and accommodated.[6]

A very different use of deep-well pumps accompanied the late 1960s construction of the 13-km (8-mi) Welland Bypass of the Welland Canal in southwestern Ontario, close to Niagara Falls. Limestone bedrock lay about 30 m (100 ft) below ground level and was overlain by varied deposits of lodgement till, sand, and gravel, and an upper stratum of heavy clay. Water under subartesian pressure provided domestic water for many wells. Any interference with the groundwater level would cause unacceptable local inconvenience. A major test-pumping program using two deep wells was therefore carried out prior to the start of construction.

Groundwater observations were taken over an area of 650 km^2 (250 mi^2) around the route of the bypass canal. These showed a drop of level of 30 cm (12 in) even 19 km (12 mi) away from the pump location, a vivid reminder of the flow of groundwater through permeable ground. Test results also permitted development of a deep-pumping program for dry construction of the two tunnels under the bypass canal. The consequently great savings are a reflection of the same principle as that used by Robert Stephenson at the Kilsby Tunnel, now carried out much more easily with modern deep-well pumps. Temporary arrangements were made for supplying water to all homes with interfered well supplies and limited, permanent pumping continues in the vicinity of the two tunnels.

WELL POINTS

Most construction drainage is handled by the use of *well points,* which are simple to install (by experts), effective, and economical. Basically, the system consists of a special pump drawing on a number of manifold-connected well points for lowering groundwater to an elevation below the lowest excavation level. Material to be excavated is thereby predrained and converted from its wet, treacherous state to a dry condition, thus facilitating excavation.

Well points are jetted to a depth 1.5 m (5 ft) or more below bottom grade level and are located closely outside the area to be excavated. They are spaced at such distances apart that the cones of depression around each well point intersect. Thus, the water level midway between the points is lowered below any level at which it might cause trouble. All the points are connected with a header pipe leading to the pump. The system can be extended and modified in many ways, notably in the excavation of deep trenches; successive surfaces of depression are indicated in Figure 5.8. Bold and simple in conception, the system can be fully effective only with a sure and certain knowledge of the geological conditions at the site, since the success of its operation depends on the travel of the groundwater in the strata encountered.

Construction of the Victoria Park pumping station, a part of the Toronto, Ontario, water-supply system, covers an area 26 by 89 m (87 by 297 ft). Construction excavation had to be taken to a depth about 10 m (35 ft) below the level of Lake Ontario, the shore of which was only about 30 m (100 ft) distant. Excavation proceeded through stiff clay for about 9 m (30 ft) of the total depth of 14 m (46 ft); small water boils then made their appearance. Auger borings found a few previously undetected, thin water-bearing sand seams. Well points were jetted down to full depth all around the area to be excavated, and the excavation was then completed without encountering water.

In other cases the facility of working in the dry—made possible by use of the well-point system—has permitted the simplification of foundation design. The large Holland Plaza Building in New York City, for example, was designed to be supported by a concrete-pile foundation placed in a clayey sand. A well-point system dried up the excavation, and a heavy reinforced-concrete mat foundation was used instead.

The well-point system can be designed to give effective results even in some

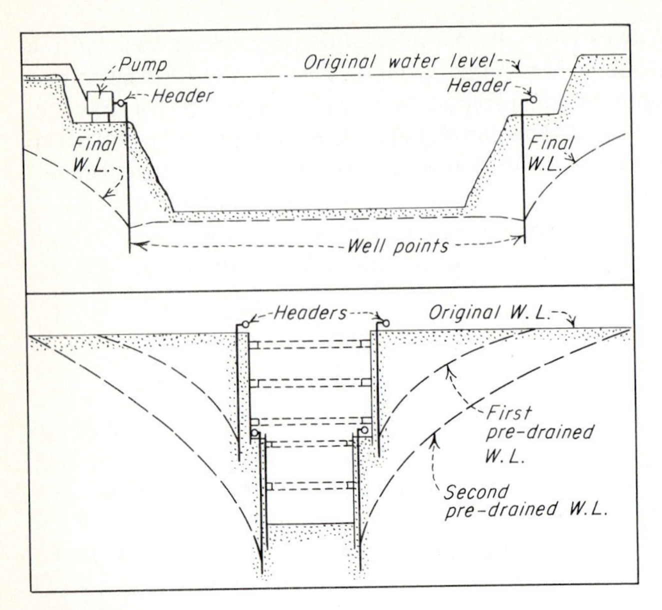

FIGURE 5.8
Diagrams illustrating the use of well points in open excavation and for trench excavation in water-bearing ground.

types of clay. The quantity of water pumped out of clay is naturally very much smaller than that pumped out of more permeable material. Careful study has suggested that in such cases the suction caused by the pumping action results in a slight vacuum at the base of each point. This leads to an unbalanced pressure at the nearest exposed excavation face, and atmospheric pressure outside prevents water from leaving the clay and, indeed, tends to hold the material at a slope far in excess of that at which it would naturally stand. At the other extreme, well points have been successfully used for dewatering weak rock. If the points can be driven into place, or placed in suitably drilled holes, they can frequently change a wet and difficult job into one that is as "dry as a bone," provided always that careful study of the strata to be drained has been made in advance.

Use of well points in sand strata is an obvious application. One of the most extensive of early installations was for the construction of an outfall sewer into Jamaica Bay, New York City, where excavation was carried out in an artificial peninsula of sand fill that was made of dredged sand deposited over some of the original muddy, marginal bay bottom. This extensive installation required three levels of well points in operation at once, delivering as much as 60,000 lpm (800,000 gph). During the construction of the Denison Dam by the U.S. Army Corps of Engineers on the Red River between Texas and Oklahoma, an extensive well-point installation was used to dry up not only an area in which the closure section of the dam was to be constructed but also an area in which slumping had occurred so that excavation had to be carried on under rigid control. In still another instance, water-bearing coarse gravel was successfully dried out by well points over an area of 76.2 by 60.9 m (250 by 200 ft) during the construction of a power plant of the Pennsylvania Electric Company at Warren, Pennsylvania, immediately adjacent to the Allegheny River; on this job four *permanent* bronze well points were installed before construction was complete.

Deep excavations in the Key West district of Florida were traditionally avoided by contractors because of the groundwater difficulties encountered in the local oölite, a soft white rock composed of organically formed calcium carbonate. In 1954, the initial attempt was made to drain this unusual material by well points at the Key West sewage lift-pump station 12 m (40 ft) square and 7 m (22 ft) below the surface, with groundwater standing at only a depth of 60 cm (2 ft). Open pumping reduced the water level by 2.1 m (7 ft) and well points were successful in pulling the water down the remaining distance. Special attention was paid to the spacing of the points to ensure proper approach velocities so that screen clogging would be avoided.

Figure 5.9 illustrates an unusual geological condition that was encountered during the building of the west anchorage for a cable-stiffened suspension bridge over the Lempa River in El Salvador. Well points here had three functions: (1) to dry up the area in which excavation was to proceed, (2) to reduce hydrostatic pressure in the underlying soil strata in order to eliminate any possibility of blowing in the bottom of the excavation, and (3) to stabilize the soils adjacent to the large area excavated. Artesian water had been noted to flow up unplugged exploratory boreholes that had penetrated the underlying strata, causing sand boils. This is a reminder of the necessity of plugging all such holes, when preliminary investigation is complete, lest the same holes cause unsuspected trouble when encountered again after construction begins.

Problems encountered in well-point dewatering generally do not arise from defects in the well-point system but because the geology of the ground being

FIGURE 5.9
Simplified cross section of the west anchorage of the Lempa River suspension bridge, El Salvador, showing the soil conditions which necessitated the three-stage well-point installation. (*Courtesy American Society of Civil Engineers.*)

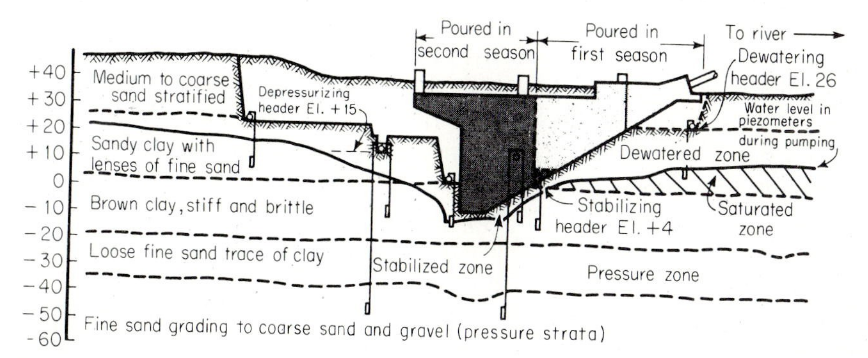

dewatered proved to be different from what had been expected. Geology was, therefore, the culprit. As has been so rightly stated, "The greatest mistake in the Miami area is to take one boring (only) or to evaluate the subsurface on the basis of a previous block several blocks away." This is wise advice for any site to be dewatered by well points.

EFFECTS OF DRAINAGE

The fact that construction drainage is effective over great distances is all too often forgotten, sometimes with unfortunate results. One notable example occurred 14.5 km (9 mi) east of Harrisburg, Pennsylvania, in the Hershey Valley, a pleasant area underlain generally by an Ordovician limestone of the Beekmantown Formation. For many years, the Annville Stone Company had been mining the high-calcium Annville Limestone at a location 2.4 km (1.5 mi) northeast of the factory of the Hershey Chocolate Company. The stone company had pumped 16,000 lpm (3,500 gpm) out of its workings without serious effect. In May 1949, the company suddenly increased the pumping rate in its lower workings to 30,000 lpm (6,500 gpm). Within a short time, groundwater levels were affected to varying degrees throughout an area of 2,600 ha (10 mi^2). Over 100 sinkholes developed, some with serious effect. Since the Hershey Company was affected, it started recharging groundwater levels at its plant. It began with 16,000 lpm (3,500 gpm), but soon increased this to 45,400 lpm (10,000 gpm). Groundwater levels immediately started to rise again, but the pumping rate of the stone company also increased from 29,500 to 36,500 lpm (6,500 to 8,000 gpm). After attempting to have the recharging operation stopped by the courts without success, the stone company adopted a program of grouting around their operations. With the recharging continuing, groundwater conditions in the area around the mine were thus effectively restored by May 1950. The field investigations carried out in connection with this case constitute a splendid example of a good groundwater survey.[7]

More serious is the example of Johannesburg, center of the world-famous gold-mining operations of the South Africa Rand. The location of active mining has moved from within city limits to new areas to the west of the city. Operations have gone deeper, to follow the gold-bearing reefs. In 1960 a major dewatering program was initiated in the Far West Rand mining district some 65 km (40 mi) west of the city. A thick bed of Transvaal Dolomite and dolomitic limestone (up to 1,000 m or 3,280 ft thick) overlies the Witwatersrand gold-bearing beds. Thick vertical dikes cut across the dolomite and divide the great bed into compartments, one of which, the Oberholzer, is at the center of the new mining area. Large springs flowed in the vicinity of these dikes prior to 1960, but the big pumping program soon dried these up and then started to lower the water level in the compartment.

Between 1962 and 1966 eight sinkholes larger than 50 m (165 ft) in diameter and deeper than 30 m (100 ft) had appeared, together with many smaller ones

(Fig. 5.10). In December 1962 a large sinkhole suddenly developed under a crushing plant, taking the entire plant and 29 lives. In August 1964 a similar occurrence took the lives of five people as their home dropped 30 m (100 ft) into another suddenly developed sinkhole. The mechanism by which these catastrophic subsidences develop is not yet fully understood. It appears, however, to be related to the prior existence of caverns in the dolomite, the roofs of which are in such unstable equilibrium that the lowering of the groundwater changes ground conditions sufficiently to reduce the necessary support for the roofs, which then collapse with dire results.

Serious effects may even result from simple gravity drainage if the local geology has not been carefully studied in advance. The White Pass and Yukon Railway, linking the sea at Skagway, Alaska, with Whitehorse, capital of the Yukon Territory 176 km (110 mi) away but 900 m (3,000 ft) higher, was built in the closing years of the last century in response to the Klondike gold rush. The rail line drops 870 m (2,900 ft) from the top of the White Pass to sea level and includes only one short tunnel. Although the engineers generally followed the dictates of local geology, they disregarded them at one point. At what used to be Lewes Lake, 130 km (81 miles) from Skagway, the route originally selected ran along the lakeshore but this was so broken by coves and bays that it was decided to lower the water level of the lake by 4 m (14 ft) to use an earth bench

FIGURE 5.10
Sinkholes which developed suddenly in the West Rand area near Johannesburg, South Africa, after deep pumping had started for mine drainage. (*Photo: R. F. Legget.*)

for the right-of-way. A channel was therefore excavated through the earth ridge that clearly controlled the water level of the lake. It would have been too much to have expected a preliminary geological study in those times in this wild and inhospitable area. It turned out that the excavation penetrated a glacial moraine which was so easily eroded by the increased velocity of the lake discharge that an immense gully soon formed, draining the lake by 21 m (70 ft) instead of the intended 4 m (13 ft). Only about a dozen small pools remain of the original lovely lake. This is an unusual case, yet it remains a cautionary tale and a reminder of what *should* be done in the way of preliminary geologic investigation.

These examples come from large projects, but exactly the same influences of pumping or even of uncontrolled or undetected drainage can happen on the smallest job. In urban areas this can be very serious, especially if timber piles are present in support of historic structures. Prior to about 1930 timber piling was commonly used in the clay soils of coastal cities, on the assumption that the water would remain constant so that rotting of the wood, due to alternate wetting and drying, need not be considered. Groundwater control during urban construction operations is, therefore, of vital importance. It can be approached only on the basis of detailed knowledge of the local subsurface.

The great tower of Strasbourg Cathedral, designed in 1439, was built with stone footings supported on timber piles. A 1750 drainage system lowered the water level appreciably and caused the tops of the piles to decay. Serious settlement of the tower resulted. Some years ago, the tower was jacked up after its columns had been encased in concrete, and new reinforced-concrete foundations were installed. Many other European buildings of historic interest have been similarly affected.

In the United States a similar problem occurred with the Boston, Massachusetts Public Library. About the year 1929, cracks were noticed in the building. Consulting engineers found that the tops of many of the timber piles had decayed, some having rotted completely. About 40 percent of the building had to be underpinned; affected pile heads were cut off and replaced with concrete (Fig. 5.11). The foundation strata in this part of Boston consist of lodgement till, blue marine clay, and silty sand outwash in an upward sequence above bedrock; pile foundations are therefore usual in this city area. An investigation revealed that groundwater was leaking away into a low-level sewer constructed in about 1912. A test dam was used partially to block the sewer, and groundwater levels rose enough to warrant leaving the dam as a permanent feature. Further remedial measures were taken in 1955; at the northeast corner of Copley Square, perforated metal pipes were installed to serve to recharge groundwater in the affected area with sewer water. It is thought that the nearby Boylston Street (Green Line) tunnel of the Boston subway also has groundwater influence and is under continued careful observation.[8]

The Boston incident represents a case of unexpected and quite uncontrolled drainage, but it is so striking an example of what *can* happen when engineering

FIGURE 5.11
Upper sections of timber piles under the Public Library Building, Boston, Massachusetts, showing deterioration attributed to varying groundwater levels. (*Courtesy Boston Public Library.*)

works are in direct contact with groundwater that it carries a most useful message. Not only timber-pile foundations are affected by lowering of the local groundwater surface. The bearing power of soils, particularly of clays, may be appreciably changed by an alteration in their moisture content. Allied troubles include the shrinkage of clays as they dry out and the possibility that the finer particles will be washed out of waterlogged sand as groundwater recedes or is drawn from it. All these conditions have been responsible for notable foundation troubles. Any drainage project should therefore be based on careful study of the local geology and the known groundwater conditions, to ensure that the effects of the cone of depression around all pumping units will not have an objectionable impact on neighboring property or structures. Legal suits for damage claims due to neglect of this simple guide are common in urban areas.

FILL FOR EMBANKMENTS

Some of the most notable advances in soil mechanics have been in the way in which soil is "borrowed" from its natural location and used as an "engineered material" in construction. Laboratory studies determine the compaction characteristics of soils, maximum densities at optimum moisture content now being a regular requirement for the placing of all fill material other than the smallest amounts. Regretfully, another of the semantic differences between the practices of geology and civil engineering is in the use of the words *compaction* and *consolidation*. They are here used in the engineering sense, i.e., *compaction* is the act of compacting soil with mechanical devices at a specified moisture content to achieve maximum density; correspondingly, *consolidation* is the process by which a fine-grained soil will compress under any applied load. Laboratory tests indicate the strength properties of the remolded soil as the basis of accurate designs for the dimensions of fill structures. If the fundamentals of soil compaction are properly applied and based upon adequate geotechnical tests, there is no excuse for any appreciable settlement. The State of California, for example, regularly places finished pavement on top of highway fills up to 27 m (90 ft) in height, shortly after completion of the fill.

PROBLEMS WITH FAULTS

Many construction projects are faced with open-excavation work in material that varies from rock to soft unconsolidated deposits. As long ago as 1925 such unusual excavation work was encountered on the Cofton Tunnel on the Birmingham to Gloucester section of the (then) London, Midland and Scottish Railway (now British Railways) in England. The 1838 and 1841 tunnel had to be opened up and reconstructed. In the north, the tunnel was in soft, false-bedded sandstones with thin beds of marl, while in the south the strata were sandstone and tough marl. For the civil engineer, therefore, they can be classified generally as both rock and soil. Original construction records were not available, but the geological survey map of the district showed the presence of a major fault. Troubles due to earth movements were therefore anticipated, and some of a serious nature were encountered very early in the work (Fig. 5.12). Before the final 73 m (80 yd) of tunnel was demolished, a careful geological survey was made to locate the failure-prone Longbridge fault. The evidence tended to show that a tunnel instead of an open cut had been chosen because of a realization of the troubles related to the fault plane to be encountered. The open cut was completed satisfactorily, but after great difficulties. A member of the Geological Survey of Great Britain was in constant attendance; the paper describing the work is a masterly account by a civil engineer of what was essentially a major geological problem.[9]

EXCAVATION OF ROCK

Successful excavation of solid rock necessitates preliminary determination of rock units and their boundaries or contacts, the nature of discontinuities to be encountered, the possible presence of water, and the presence of any weathering or alteration. The structural arrangement of rock units will often control the cut slopes to be adopted in design. When the general dip is in the direction of the slope and at a lesser angle than the slope, an *adverse* condition of

FIGURE 5.12
Opening-up of the Cofton Tunnel, England, showing a hanging-wall slip in the west slope, caused by a fault plane. (*Courtesy British Railways, Midland Region.*)

daylighting exists, in which there is no lateral restraint to slope movement. Structural arrangement will also affect underground water problems. The arrangement of rock strata will determine the general lines of water movement, and in connection with the porosity or impermeability of beds, the levels at which water may be encountered.

Of great importance is the *nature of the rock* to be excavated. By "nature" is meant not merely the general lithologic classification of the rock but a vivid conception of the actual properties of the particular type of rock to be encountered all over the area to be excavated. The value of general classification is by no means unimportant—indeed, this is most essential—but each class of rock can vary so considerably that, for civil engineering purposes, it must be further described by some indication of its physical properties. Thus, *sandstone* can vary from a hard and dense rock to material that is little better than well-compacted sand. *Granite* can prove to be one of the hardest of all rocks, and yet it may be found in such a state of decomposition as to be excavatable by handshovel. All preliminary geologic information must therefore be checked thoroughly with rock logs, compared to the physical state of surface rock samples and the core itself. Similarly, it is essential to know with some degree of accuracy before work begins how the rock to be encountered will stand up after long exposure; this is necessary to forestall possible future accidents and to keep maintenance charges to a minimum.

If due allowance is made for exceptional variations in physical properties, the following notes will serve as a general guide. The stability and permanence of igneous and metamorphic rocks when they are exposed to the atmosphere can generally be relied upon unless the rocks are badly weathered. If they are highly jointed, sheared, or shattered by local faulting, the greatest caution must be observed in finishing off slopes. Caution is especially required for work done during severe winter weather, as in Canada, where frost action may temporarily slow the rate of surface disintegration. Sedimentary rocks must be considered with great caution and judgment in such work. Clay in any form, even as a shale which may be exceedingly hard when first exposed, must be regarded with suspicion, particularly if the bedding planes are dipping at any appreciable degree. The action of the atmosphere may soon reduce this material to an unstable state, with consequent extra trouble and expense. When clay is known to be present, ample allowance should be given in the design of cut slopes, and the economics of protecting the exposed faces with light retaining walls should be investigated. *Flysch* deposits, alternating sequences of shale and sandstone, nearly always cause troublesome stability problems, especially when the present cut slope axis is about parallel to strike. Sandstone and limestone, if in a firm and solid state, will stand with vertical or almost vertical faces; but since both types of rock vary from a sound, solid state to material that can be crumbled in the hand, every case must be considered on its own merits.

Adequate *drainage facilities* are naturally an important feature of all rock-excavation work. They will be provided by one of the standard methods available in civil engineering practice, but geological considerations may affect them to some degree. For example, the 1920 construction of a new railway in El Salvador and Guatemala encountered a weak conglomerate called *talpetate*. All cut slopes had to be left as steep as possible to carry off the torrential tropical rains before they led to erosion of the conglomerate. Similarly, drainage in all types of shale must be arranged to remove surface water as quickly and surely as possible to minimize the swelling-based disintegration of the rock. Limestone may also be affected by excessive contact with water.

The usual method of drilling and blasting may not always be suitable when the exact *condition of the rock* to be moved is known. Such a case occurred during construction of the main spillway for the Fort Peck Dam in Montana, which is founded on local Bearpaw Shale. This weak rock can be excavated without the aid of blasting. To excavate the 150,000 m^3 (200,000 yd^3), three different types of cutting machine were used: an auger which drilled holes 1.5 m (5 ft) in diameter, and two special adaptations of electrically operated coal-cutting machines, one used for cutting vertical slots between holes and the other for horizontal undercutting. The Bearpaw Shale disintegrates when exposed to the atmosphere, and for this reason a 1.2 m (4 ft) cover layer was left in place over the excavated area at the start of excavation. The finally finished surface was sprayed with a bituminous paint against which the concrete was immediately placed. Not only does this rock disintegrate when exposed, but it exhibits a slow rebound after excavation, owing to the release of internal stresses previously resisted by the superincumbent rock now removed.

As had been predicted, excavation for the powerhouse at Wheeler Dam on the Tennessee River in the United States encountered an "alternating series of nearly horizontal layers of pure limestone and a cherty (siliceous) rock, the latter occurring in much thicker strata than the former." More than 380,000 m^3 (500,000 yd^3) of rock had to be removed to depths varying to 17 m (57 ft) below the existing stream bed, and owing to the naturally fractured nature of the rock, it was planned to avoid blasting close to the faces of the deep cuts. Close-line drilling was therefore carried out with wagon drills all around the site (Fig. 5.13). The undisturbed rock face was important in the structural design of the powerhouse, and the elimination of overbreak made the method economical. Such examples are now extremely common; all confirm the importance of geological features in rock excavation.

Methods of Rock Excavation

Rock excavation is an elementary construction operation, and it is not surprising that meticulous attention to detail—coupled with steady advances in the equipment used for drilling and mucking, in the procedures used in drilling,

FIGURE 5.13
Line-drilled excavation for turbine pits in the powerhouse area of the foundations for Wheeler Dam, Tennessee River, Tennessee. (*Courtesy Tennessee Valley Authority.*)

and in the nature of explosives used in blasting—have greatly improved the efficiency of rock removal. In Sweden, rock-excavation practice (as well as tunneling) has been outstanding, especially in the adoption of small, mobile, one-person drilling rigs. Correspondingly, the wide adoption of removable, long-wearing bits has led to great advances in the operation of drilling in general. Innovations in blasting techniques should be considered whenever conditions warrant; one unusual example is the blasting of overburden and rock at the same time.

A new sea-level canal was constructed to cut off a sharp bend in the Motala River, some 145 km (90 mi) southwest of Stockholm. The Lindo Canal has a length of 6.5 km (4 mi); the first section is 3.2 km (2 mi) long. Its depth is 9 m (30 ft) and its bottom width 54 m (180 ft); thus, it is wider than some of the major ship canals of the world. The project involved removal of about 3.8 million m^3 (5 million yd^3) of material, most of it by dredging; 20,000 m^3 (26,000 yd^3) was rock. At the entrance to the canal, lodgement till with overlying clay occurred on top of the bedrock to depths of 9 m (30 ft); the clay was saturated and thus awkward to remove. Special drilling rigs with 16-m (52-ft) feeds were adapted for driving a casing pipe simultaneously with sinking drill steel through the overburden, to be ready for drilling as soon as rock was encountered. Drill steel was withdrawn when the hole had been drilled to depth, and the casing pipe was left in place. Plastic pipe was then set in place, the casing pipe and the holes loaded with 40 to 50 cartridges. One major blast used 12,000 detonators and 55,000 kg (123,000 lb) of dynamite in 4,750 holes to move 27,500 m^3 (36,000 yd^3) of rock and 54,000 m^3 (96,000 yd^3) of overlying clay and till. The high ratio of dynamite to rock moved was employed to fragment the rock to a finer degree than usual to facilitate its removal by dredging.

Most geologists or civil engineers will not usually be concerned with the details of blasting technique. But even the best of blasting techniques will be only partially successful if the geology of the work site has not previously been studied so that it is known with certainty. Major rock excavation projects can often make beneficial use of comparative tests of the effects of blasts set off at varying distances. As such experience is accumulated, vibrations due to blasting will gradually cease to be a hazard. Such precautions should include noting all preexisting damage to buildings. Cracking especially should be noted; if carefully recorded, it will often be found to coincide exactly with the damage reported after blasting has taken place. In one instance, a large rock-excavation job was carried out in the vicinity of an important hotel. At the end of the job, the "usual" claim for damages caused by blasting was received. The claim, however, was not pressed when it was divulged that the contractor had rented a room in the hotel where he had maintained a vibration record throughout blasting operations. This record showed that the vibrations caused by the blasting were less serious than those caused by heavy traffic.

Modern civil construction contains many fine examples of rock excavation illustrating how accurate geologic knowledge facilitated the work. One of the most unusual rock-excavated structures is the foundation basin for radar and radio telescope at Arecibo, Puerto Rico. A reflector of 261-m (870-ft) radius had to be constructed to a high degree of accuracy. A geological siting study led to the selection of a location in coral limestone country, characterized by unusually large sinkholes. The amount of excavation was minimized by utilization of a weathered natural depression between pinnacles of sound rock. Despite the advantages of this natural site, the contractor still had to remove 210,000 m^3 (270,000 yd^3) of rock, generally by blasting except for final trimming, and to place 150,000 m^3 (200,000 yd^3) of compacted fill to shape up the final surface so it would conform as closely as possible with the spherical reflector.

In great contrast, modern rock excavation methods applied to hard, dense granite gneiss involved excavation of 1.6 million m^3 (2.1 million yd^3) for Interstate Highway 240 through Beaucatcher Mountain, North Carolina (Fig. 5.14). Nearly 500 structures lay within 600 m (2,000 ft) of the cut, including a 65-year-old water reservoir, while Beaucatcher Tunnel (on U.S. Route 70) lies only 48 to 69 m (160 to 230 ft) from the big cut. Controlled blasting was conducted on the basis of test shots. Slopes were successfully completed with

FIGURE 5.14
The Beaucatcher Tunnel, North Carolina, showing its proximity to the deep cut here required for Interstate Highway 240, which is being completed in stages. (*Courtesy North Carolina Department of Transportation, Raleigh, N.C.*)

presplit holes (7.5 cm; 3 in) on 75-cm (30-in) centers; 11.2-cm (4.5-in) holes were used for production blasting, sunk 7.5 to 9.0 m (25 to 30 ft) on a 2.4- by 2.4-m (8- by 8-ft) pattern.[10]

QUARRYING IN CIVIL ENGINEERING

Building stone is usually obtained from established quarries. The two most common types of work requiring special supplies of quarried rock are the construction of rock-fill dams and the use of rock for the construction of rock embankments or special structures such as breakwaters. If the quantity of rock required for projects such as these or any other kind is appreciable, the civil engineer will be well advised to obtain the services of a quarry expert.

After the site of the quarry has been stripped, the exposed rock surfaces should be studied carefully. Special attention should be paid to the dip and strike of the strata, to the presence of any unusual features such as folds and, particularly, to jointing. The engineer will then be able to open up a working face in the most advisable way, taking advantage of the orientation of geologic structure to facilitate both the blasting and the removal of rock. The location and depth of shotholes and the necessary kind and amount of explosive to be used are matters that will have to be determined accurately for every new quarry face opened up. The problems involved are similar to those involved in the normal excavation of rock in grading work. The Federal Highway Commission studied 71 different highway grading and excavation projects and concluded that over half were poorly blasted and that only 15 were classed as "good." The consequent average increase in operating time for shovels was 42 percent; and as the volume of rock handled was less than the dipper capacity, the average overall efficiency was only 50 percent. As rock drilling is an expensive operation (1 ft of hole is roughly equivalent to about 1 lb of dynamite in North America), efficient blasting practice will tend to minimize shothole drilling.

Only infrequently do civil engineering projects call for quarrying on a major scale, apart from the construction of rock-fill dams. One job that did need a lot of rock was the construction of the Hiwassee Dam of the TVA. All the aggregate necessary for this great project, including even the sand, had to be obtained from a specially opened quarry, as the rock from excavation was unsuitable. Good graywacke rock was found at a suitable location; in all, 1.3 million m^3 (1.7 million yd^3) of rock was quarried. Interesting experiments were carried out to check the use of 22.5-cm (9-in) drill holes as compared with 15-cm (6-in) holes. The rate of drilling the two sizes was found to be about the same, although the 22.5-cm (9-in) holes filled with explosive produced twice as much fragmented rock per foot of hole as did the 15-cm (6-in) holes.

Far more extensive was the rock-quarrying operation called for by the construction of the rock-fill embankment across Great Salt Lake to replace the old trestle of the Southern Pacific Railroad. What was described at the time as "the biggest controlled explosion in history" involved the use of 800,000 kg

(1.79 million lb) of explosive in one blast; this resulted in over 1.5 million m^3 (2 million yd^3) of suitably fragmented rock. The location was Promontory Point on a peninsula extending into the lake from the north. The total quantity of rock required for the complete job was 10 million m^3 (13 million yd^3), all of which was a hard quartzite.

CONCLUSION

Some open-excavation work is called for on almost every civil engineering project. It is small wonder therefore that practice in this field is so diverse and so interesting despite the overall simplicity of the basic operation. Improvement of well-accepted techniques may be expected to continue; new features will be added in matters of detail. It is significant that, despite the steady worldwide rise of prices in practically every field of activity, the cost of excavation on civil engineering projects is still the "best bargain in construction."

The handling of water incidental to excavations, be it surface water or groundwater, whenever it occurs on civil engineering projects is of the utmost importance and is always dependent on geology. Without an appreciation of subsurface conditions—both the character and arrangement of bedrock strata and soil and the presence, character, and potential movements of groundwater—the planning of the best and most economical drainage or dewatering facilities may be a difficult achievement. Accordingly, the carrying out of site investigations for civil engineering works should always be conducted so that the investigation leads to the clearest possible picture of subsurface conditions, especially of any possible unusual interrelation between groundwater and the strata in which it occurs, always mindful that annual variations in the level of the groundwater surface are probable. It is strange that water, the most common of materials while at the same time one that is essential for life, should so often be neglected, and not only in construction operations. In the laboratory study of soils, it was not until the vital importance of the exact moisture content of soils in situ was fully appreciated that real progress in understanding the mechanics of soils was possible. Dr. Karl Terzaghi has said: "On a planet without any water, there would be no need for soil mechanics." There *is* water on this planet, and as it encroaches on engineered works, it is one of the greatest challenges to the civil engineer; geology can render invaluable aid in meeting that challenge.

REFERENCES

1 R. F. Legget, "Soil Engineering at Steep Rock Iron Mines, Ontario, Canada," *Proceedings of the Institution of Civil Engineers, 11,* pp. 169–188 (1958).
2 C. A. McMahon, "Los Angeles Constructs 2,000 Car Underground Garage," *Civil Engineering, 21,* p. 689 (1951).

3 M. S. Kapp, "Slurry-Trench Construction for Basement Wall of World Trade Center," *Civil Engineering, 39,* pp. 36–40 (April 1969).

4 G. Gordon, "Arch Dam of Ice Stops Slide," *Engineering News-Record, 118,* p. 211 (1937).

5 "Electro-Osmosis Stabilizes Earth Dam's Tricky Foundation Clay," *Engineering News-Record, 176,* pp. 36–45 (June 23, 1966).

6 D. R. Warren, "Novel Construction Plan for Graving Dock Suggested by Soil Studies," *Civil Engineering, 14,* pp. 323–328 (1944).

7 R. M. Foose, "Ground-Water Behavior in the Hershey Valley," *Bulletin of The Geological Society of America, 64,* pp. 623–646 (1953).

8 B. F. Snow, "Tracing Loss of Groundwater," *Engineering News-Records, 117,* pp. 1–7 (1936).

9 R. T. McCallum, "The Opening-out of Cofton Tunnel, London, Midland and Scottish Railway," *Minutes of Proceedings of the Institution of Civil Engineers, 231,* p. 161 (1931).

10 "Artful Blasting in New Cut Protects Old Tunnel Nearby," *Engineering News-Record, 201,* p. 24 (September 28, 1978).

CHAPTER 6

TUNNELS AND UNDERGROUND SPACE

Tunneling must have been one of the earliest human construction activities. It may be that the natural caverns frequently found in limestone first suggested to early man the idea of habitable passages in rock. Excavated underground dwellings and temples, the first tunnels, are to be found in many of the ancient civilizations. In early times, excavated passages beneath the earth were utilized for purposes other than residence. Tunnels soon came to be constructed for drainage of mines and quarries. Water supply and road construction also necessitated the digging of tunnels at early dates in human history.

Today, tunnels are used for the same purposes. They facilitate transportation, are used in the generation of water power, for transport and water supplies. Other exceptional uses are to be found, but tunnels generally can still be classed as they were 2,000 years ago—either as aqueducts or as viaducts. It is indeed strange to reflect that this ancient branch of construction retains some of its original character. A tunnel is and can be only a tunnel, having a certain length and a certain cross section.

Underground space is a natural result of the extension of tunneling technology, the term being used here to indicate excavated (or natural) openings beneath the ground surface for human activities. Methods of excavation are similar to those used in tunnels, but the dimensions are considerably larger than even the largest tunnel cross sections. Underground power stations have been in use throughout the twentieth century and all bulk oil storage in Sweden is now located underground for safety and economy.

The use of underground space for manufacturing and even for residential purposes is a reversion to a practice rooted in antiquity. These modern uses preserve the visual environment and have the further advantage of conserving energy. With the world's energy situation steadily growing ever more critical, all aspects of energy conservation are becoming increasingly important. Underground space is yet another example of the ultimate dependence of socially desirable developments upon geology.

So little appreciated is the potential that the use of the underground presents for energy conservation, that some background to design is herein presented. Normal ground temperature increases with depth below ground surface. Diurnal effects cease close to the surface and annual variations terminate at about 10 m (33 ft) in temperate climates. This constant ground temperature is close to the local annual average air temperature. In northern parts of the United States and in southern Canada, this temperature is about 10°C (50°F). If this temperature is compared with the comfort temperature required in buildings (say, 22°C or 72°F), it will be clear at once that the heating load required to bring underground space to a comfortable human state will be constant throughout the year and will be appreciably less than that for aboveground structures. Considerable economies result in the capital cost of heating equipment, since air conditioning will not be required in summer. The annual energy consumption will be drastically reduced. Between one-quarter and one-third of all the energy used in North America is for the heating and cooling of buildings. The significance of the use of underground space thus becomes crystal clear.

Fortunately, accurate figures from modern installations support these suggestions. Bligh and Hamburger have reported that Kansas City with 10 percent of all cold storage in the United States experiences facility preparation costs at $30 per square foot above ground, and $8 to $18 below ground; operating costs are $0.12 above ground but only $0.01 for underground installations, the reduction in energy consumption being comparable. Many underground installations have been made for reasons other than the conservation of energy (such as underground power stations located below ground for convenience, safety, and economy in construction cost). Each use illustrates what can be done with good design and if the local geology is satisfactory.

HISTORICAL NOTE

The Jordanian desert, about 480 km (300 mi) due east of Cairo, is the site of the largely underground city of Petra. Temples and halls were hewn out of the local red sandstone of the great Rift Valley that runs from the Jordan depression to the Gulf of Aqaba. The land had been the traditional home of the Horites, cave-dwelling predecessors of the Edomites of the Old Testament. Specific biblical reference to Petra is uncertain but the city was certainly in use in the sixth century B.C. This major trading settlement peaked in activity around A.D. 100, and lasted until the Mohammedan conquest of 629 to 632. Its temples and carvings are a salutary reminder that modern humans were not the first to think of gaining security by "going underground."

Petra was not alone. The river Nile has long been regarded as one of the cradles of civilization. There, too, are to be found temples and other underground spaces carved laboriously for human use. Probably the best known today are the temples of Abu Simbel, if only because of the international effort that went into their preservation. Located 280 km (175 mi) upstream of the site of the Aswan High Dam, the structures in the pink sandstone contain four immense, pre–1200 B.C. figures of Rameses II (20 m or 66 ft high) and the great cliffside temple, 16.2 by 17.4 m (53 by 57 ft), with a corridor connecting it to an inner sanctuary that extends 60 m (200 ft) into solid rock. This masterpiece was physically raised above the rising waters of the Nile on completion of the High Dam.

Malta contains underground temples and gathering places at least 5,000 years old, hewn out of solid sandstone with flints which must have been brought to the island from the mainland. These early tunnels were generally built through the softer types of solid rock, but some were excavated through unconsolidated materials and so required immediate lining for stability. Another notable example of ancient underground construction was the tunnel under the river Euphrates, probably the first submarine tunnel of which any record exists. It was 3.6 m (12 ft) wide and 4.5 m (15 ft) high and was built in the dry, since the river was temporarily diverted.

The Romans were preeminent in early tunnel construction. They appear to have introduced the use of fire into tunnel construction, utilizing the principle (known to others earlier) that a heated rock, if suddenly cooled, will crack to some extent and so make excavation easier. They also probably employed vinegar instead of water as a cooling agent and as an acidic disintegration agent when working in limestone. It is known also that the Romans utilized intermediate vertical shafts and even inclined adits in the construction of their longer tunnels, notably of the tunnel built for the drainage of Lake Fucino. Volcanic tufa was another of the softer rocks pierced by these intrepid builders; one notable tunnel through this material, that which gave the road between Naples and Pozzuoli passage through the Ponlipio Hills, was 900 m (3,000 ft) long and 7.5 m (25 ft) wide.

Even more remarkable are the vast and numerous caves of Cappadocia, Turkey. The early underground Christian city here (at Derinkyu) with a population of 50,000 was connected by an 8-km (5-mi) tunnel with another large underground city. The chalk caves bordering the river Seine of northern France near Vetheuil are honeycombed with caves, many of which have been used for the shelter of humans and animals for hundreds of years. One of the greatest of Norman chiefs made his home in a group of caves known as Le Grand Colombier (the great dovecote), which includes one room over 100 m (328 ft) long. The great caves at Pommeroy Park, used by the Romans as a building-stone quarry, are famous today for their storage of maturing champagne (Fig. 6.1). In the south of France, close to the

FIGURE 6.1
Wine storage in the Cave de Vouvray, typical of the way in which underground storage is used in France for this vital purpose. (*Courtesy La Reveu Vinicole, Paris.*)

valley of the Rhone, is the greatest of all these underground wonders, the Bramabian, with its underground river and 9.6 km (6 mi) of galleries. The underground river of Labouiche is said to be the longest in the world, more than 3.2 km (2 mi). Visitors can sail more than 1.6 km (1 mi) by boat. The underground torrent of the Cigalière, with its waterfalls 18 m (60 ft) high, is in the same area. Then there are chasms such as the *gouffre* Martel, which extends downward 420 m (1,400 ft), only 270 m (900 ft) of which have been explored.

The Middle Ages saw no advance in the technique of tunnel building. Even the introduction of gunpowder, first used in tunnel work during the period from 1679 to 1681 at Malpas, France, had little immediate effect. It was the extension of canals and the introduction of railways that finally initiated the great advances of the last 200 years, during which most of the tunnels now in use were constructed. Although rudely built and simple in conception, ancient tunnel works were inevitably dependent upon geological considerations—not only in design but also in construction. Simultaneously with advances in tunneling technique has gone the development of geology, and so today the two are intimately associated.

GEOLOGY AS THE UNDERPINNING OF DESIGN

Of all the activities of the civil engineer, without question it is to tunneling and underground space that geology can most usefully be applied. Geological problems alone affect design and construction methods once the general location and basic dimensions of a tunnel are determined. Accurate location and construction methods therefore depend on the rock through which the tunnel is to be driven. The necessity of lining must usually be determined by the short-term deformation of the host rock or the behavior of the mineral stability of the rock when exposed to the air and possibly to inflowing water. A thorough geological investigation before construction begins is, therefore, of paramount importance in all tunnel work.

Basically, accurate geologic profiles along all possible routes available for the tunnel are the first requirement. Knowledge of the engineering characteristics of the host rock is the second and is of no less importance. With this information at hand, the engineer can decide whether or not the construction of the proposed tunnel is a practical and economical possibility. If designing a pressure tunnel in which water is to be conducted under a pressure head, the engineer can determine with some accuracy what cover of rock must be allowed between the tunnel line and the ground surface. The necessity for, and the design of, an artificial lining for the tunnel generally can be determined, and the likelihood of having to grout the rock adjacent to the tunnel can usually be foretold from information gained in a geological investigation. Finally, such information will give the engineer some indication of the percentage of overbreak likely to be encountered when construction is under way and for which allowance must be made in estimates. To the contractor for tunnel work, geological information is equally vital. The entire construction program and construction methods depend on the material to be encountered, and the main hazards, such as underground water, are determined solely by geological conditions.

There are some general geological observations about tunneling which first demand attention. Clearly, the ideal condition in tunneling is to encounter one easily excavated material only, a material which contains no water-bearing discontinuities and which is unaffected when exposed to air. Rarely is anything approaching such an ideal material found; the London Clay which most of the London tube railway tunnels penetrate in part is perhaps as close to the ideal as is found anywhere. Normally, tunnel locations can be changed only for economic consideration, and so rock conditions at a particular location have to be accepted by the engineer and explored thoroughly in order to identify an accurate design and suggest a range of acceptable construction techniques to the bidding contractors.

In general, Archean or Precambrian rocks, the oldest types geologically, are difficult to excavate; construction in such formations is consequently relatively expensive. Paleozoic rocks, on the other hand (geologically younger), are usually the most simple to excavate and, consequently, they afford more economical construction. Formations of Tertiary to Recent origin increase construction difficulties and are often considered as weak-rock. Tunneling becomes generally more difficult as the formation becomes younger, and Recent sand and gravel deposits are particularly awkward to drive through. Novel methods have been adopted in tunnel work for penetrating such materials; the practice of freezing the water in water-bearing strata is perhaps one of the most ingenious.

As in every other type of civil engineering work, economic considerations generally predominate in tunnel design and construction. Geologic predictions should be made of not only the materials to be penetrated but the actual cost of construction as it is affected by the rate at which excavation progress may be anticipated. From estimates, detailed comparisons of the costs of various alternative routes for the work proposed can be prepared. In the Kinlochleven hydroelectric scheme in Scotland (25,000 kW with a head of 300 m or 1,000 ft), such a study led to the adoption of a plan calling for an open reinforced-concrete conduit following the contour of the hillside as the main aqueduct. In the neighboring Lochaber scheme (90,000 kW with a head of 240 m or 800 ft), the main aqueduct is a tunnel route based on a similar and extensive economic study utilizing preliminary geologic information.

PRELIMINARY WORK

Geologic sections along possible tunnel routes form the basis for tunnel engineering planning and design. The sections are compiled on the basis of general survey methods. Tunnels are usually located in such a way that accurate preliminary correlation of the nature of the strata to be expected in the tunnel with surface conditions will be difficult. Subaqueous tunneling ground, for example, can be readily checked only from indications on adjacent dry land; tunnels under cities will pierce strata covered by built-up city areas; and tunnels through high mountain ranges will often lie at significant depths below the surface—over 2,100 m (7,000 ft) for the Simplon Tunnel in Switzerland. In many such cases, it will be impracticable, if not impossible, to put down the usual exploratory drill holes, which generally serve to confirm the deductions of geological survey work. It is for this reason that tunnel-construction records (to be mentioned later) are of vital importance.

A serious result of incomplete preliminary information occurred during the construction of the Lötschberg Tunnel in Switzerland, from Kandersteg to Goppenstein, between October 1906 and September 1911. The tunnel is 14.5 km (9.04 mi) long, and it was believed that it would penetrate solid rock (granite) throughout. Unfortunately, about 3.2 km (2 mi) in from one portal, the drilling broke into an ancient glacial gorge, now filled with detritus and followed by the Kander River, and in an instant 6,100 m^3 (8,000 yd^3) of material rushed in and 25 men lost their lives. The heading had to be bulkheaded off and the course changed; only in this way was the tunnel eventually finished, finally being 0.8 km (0.5 mi) longer than had been anticipated. Had modern geological investigation techniques or geophysical methods of exploration then been available, this trouble might have been anticipated.

Alpine tunneling has given the engineering profession varied and invaluable experience sorely tested in complex geology. Even with the most meticulous preconstruction investigations, nothing can be certain about driving Alpine tunnels other than that the work will be difficult. Together the tunnels in this region constitute probably the world's greatest concentration of major tunnels. In particular, the 11.5-km- (7.1-mi)-long Mont Blanc highway tunnel from Hameau des Pélérins in France to Entrèves in Italy provided a fascinating comparison of different methods of tunnel driving between the Italian and French ends and of differences in rock types—schist and granite at the French end and calcareous schist, quartzite, and granite at the Italian end.

A variety of North American and European drilling and mucking equipment was used. Much loose and rotten rock was encountered, necessitating the extensive use of rock bolting and steel supports. The tunnel was started in 1958 and opened for use in 1964. Even the Mont Blanc Tunnel has now been surpassed in length by the 15.4-km (10.2-mi) St. Gotthard vehicular tunnel, roughly paralleling the 60-year-old railway tunnel, also with varying rock conditions and much trouble with inflows of groundwater. Started in 1970, the St. Gotthard Tunnel was opened for use in 1980.

Problems can also be experienced, however, even in quite short tunnels, if preliminary subsurface investigation is inadequate. One example will illustrate what has already been said about the absolute necessity for relating geological studies to the engineering problems to be faced—i.e., the necessity of developing and applying engineering geologic data and not just academic (or "pure") geology.

The Broadway Tunnel through the Grizzly Peak Hills on California Highway 24 in the outskirts of Oakland was designed in the midthirties to provide improved highway connections to the east but for reasons that will shortly be obvious the problems associated with its excavation were not publicly described until 1950. The two parallel bores, about 900 m (3,000 ft) long and 11 m (36 ft) wide were complete in the midthirties. The special highway district formed to build the facility engaged a prominent geologist. His favorable report was mentioned in the contract documents (and made available to contractors) but not officially endorsed or quoted. A consortium of six experienced contractors was awarded the job and accepted this preliminary report without recourse to a geological study of their own. Very little in the way of supports or lining was anticipated, but a pilot drift showed that none of the ground penetrated could be left unsupported prior to lining. Two bad cave-ins took place, one taking three lives. The original contractors withdrew from their contract and sued, unsuccessfully, for over $3 million. The project was completed by others. There are few more telling examples of how inadequate geological advice, unrelated to proper geotechnical investigations, can be so misleading.[1]

Another example of serious trouble in tunnel construction was that on the 4,470-m (14,885-ft) Las Raices Tunnel, Chile, for the Transandian Railway. Surface indications in the few rock exposures above the tunnel suggested a uniform granite porphyry without troublesome water inflow. Driving encountered generally sound rock with some fracture zones and clay-filled shear zones, and some infiltration of water. With little warning, a break-in occurred at about 525 m (1,750 ft) from one face, pouring 2,000 m^3 (2,616 yd^3) of mud into the tunnel and trapping 42 men. They were eventually rescued in a remarkable way through a special rescue tunnel. Through the failure chimney to the ground surface above the tunnel, it was discovered that the tunnel had tapped and drained an ancient glacial channel which had been filled in with river deposits, yet another example of the trouble that can result from underground conditions caused by glacial action of the past.

Core drilling along a proposed tunnel alignment is made to supplement preliminary geological investigation whenever possible. This exploration will give a partial validation of the surficial geologic mapping. These two lines of evidence have prefaced the construction of many important tunnels, among them several which serve the great metropolitan area of New York.

TUNNELS FOR NEW YORK CITY

Some of the early New York tunneling experiences were not very fortunate. The first Hudson River tunnel was started in 1874, apparently without explorations. The work was interrupted several times and was completed only in 1908. Ten exploratory pipe soundings were utilized for the East River Gas Tunnel but apparently not with geological correlation. Construction between 1891 and 1894 to take three gas mains across the East River to the city of New York anticipated solid rock river bottom throughout. Trouble with water was soon encountered, however, and eventually the tunnel ran into a shear zone of soft decomposed rock. Construction methods had to be radically changed, calling for compressed air and a shield.

Just after the turn of the century, the Board of Water Supply of the City of New York initiated one of the world's greatest water-supply undertakings, the Catskill project, which even in this day of great projects is still a monumental piece of engineering (Fig. 6.2). Two large reservoirs in the Catskill Mountains to the north of New York were linked to the city by a 176-km (110-mi) aqueduct. Two large pressure tunnels were also constructed beneath the city for the distribution of the Catskill water; City Tunnel No. 1 was completed in 1917, and City Tunnel No. 2 in 1936. The successive chief engineers to the Board of Water Supply have been aided by three distinguished geological consultants, Professors W. O. Crosby, J. F. Kemp, and C. P. Berkey, appointed in 1905 and 1906 to advise on geological matters affecting the works. From the very start of the work in 1905, this great enterprise was characterized by the closest possible cooperation between geologists and engineers. In the words of Dr. Berkey:

> In this project virtually nothing was taken for granted. Every new step was the subject of special investigation with the avowed purpose of determining the conditions to be met; and, when these were determined, the plan of construction and design of the structure were brought into conformity with them. In this manner specifications could be drawn with sufficient accuracy to avoid most of the dangers, mistakes, and special claims commonly attending such work. Very few features or conditions were discovered in construction that were not indicated by the exploratory investigations, and such as were found proved to be of minor significance and were cared for at moderate expense.

When it is considered that the country traversed by the aqueduct (as is shown so clearly by the geological section along the whole route shown in Figure 6.2) "exhibits so great a variety of natural features and physical conditions that virtually every individual section of this aqueduct presented special geological problems" and that the dam sites were generally marked by great deposits of glacial debris with ancient river valleys thus covered from view, this remarkable result will be more fully appreciated.

Fortunately for the engineering profession, the results of this cooperative work were disseminated through engineering literature. The paper by Dr. Berkey and J. F. Sanborn, presented to the American Society of Civil Engineers in 1923, is already an engineering classic. It is safe to say that the practice it describes is responsible for some of the change in attitude of the civil engineering profession toward geology which has taken place since the turn of the century, certainly in North America. For those interested in a brief account of the geological features affecting engineering work in and around the city of New York, reference may be made to Guidebook 9, prepared by Dr. Berkey for the Sixteenth International Geological Congress held in 1933. A more recent summary of New York City geology has been compiled by the late T. W. Fluhr.[2]

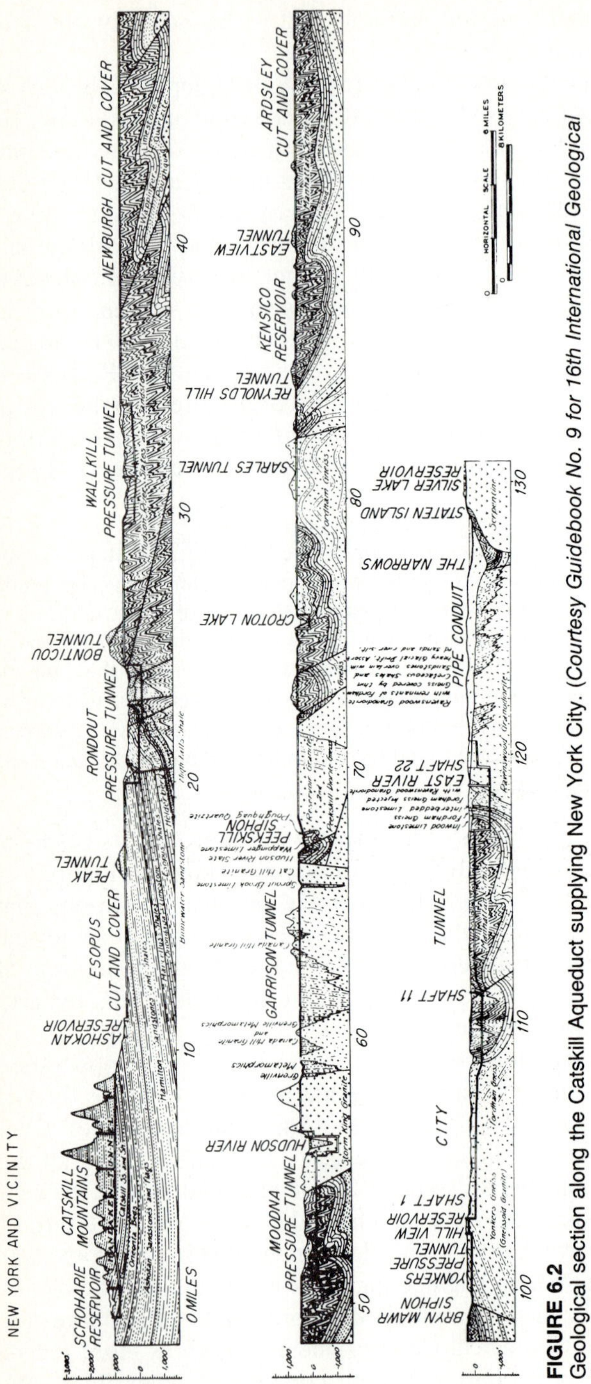

FIGURE 6.2
Geological section along the Catskill Aqueduct supplying New York City. (Courtesy Guidebook No. 9 for 16th International Geological Congress.)

Extensive exploratory work is typical for the Board of Water Supply. On the 29-km (18-mi) City Tunnel No. 1, for example, 13,800 m (46,000 ft) of exploratory borings were put down, and more than 15,000 m (50,000 ft) were drilled for City Tunnel No. 2. The same exploration pattern was continued and experience gained on the earlier tunnels was applied when New York undertook the much greater Delaware project, which involved the most extensive tunneling work ever carried out on any project. From the Rondout Reservoir west of Poughkeepsie, the main aqueduct runs as a deep pressure tunnel 136 km (85 mi) to the Hill View Reservoir of the Catskill scheme; together with the connecting City Tunnel No. 2, the total length of this one tunnel is almost 170 km (105 mi). The East Delaware Tunnel is 40 km (25 mi) long and connects the Pepacton and Rondout reservoirs, and the West Delaware Tunnel (70 km or 44 mi long) connects the Cannonsville and Rondout reservoirs. The latter was holed through in January 1960, and marked the beginning of the last stage of this vast water-supply project, designed to satisfy the water demands of New York City until the end of the century.

The vehicular tunnels so essential to the flow of traffic to and from Manhattan are also marked by unusual geological conditions. Possibly the most difficult was the Queens-Midtown Tunnel, holed through in 1939. Figure 6.3 shows the geological complexity involving tunneling by shield in soft ground, by cut-and-cover in artificial ground (formed by the clay blanket), and in solid rock; in some mixed-face sections the lower parts of the shields were in rock and the upper in loose soil. Compressed air was used, at a maximum pressure of 104 kg/cm^2 (37 psi). At one time, one of the working faces encountered 2.4-m (8-ft) natural lumps of coal debris; then 2.7 m (9 ft) of sand, clay, gravel, and boulders; and finally 2.4 m (8 ft) of sound rock in the invert. Although air was lost at surprising rates, even breaking up and eroding the clay blanket in the riverbed, the face was never lost. Successful completion is a tribute alike to the contractor's skill and to the very detailed geologic foreknowledge.

Equally complex geology was also found in the Wards Island sewer tunnel, constructed during the period 1935 to 1937. Preliminary borings did not disclose the extremely soft altered rock that finally necessitated a complete change of grade. The original plans were made for a depth of 89 m (297 ft) below Manhattan Island. The tunnel was finally put through at a depth of 153 m (510 ft) below water level (Fig. 6.4). One of the original river borings had passed through this material but it had not been possible to secure a reliable sample. The cross section illustrates how the secondary drill holes disclosed satisfactory conditions at a lower grade. Work was resumed in March 1937, and the headings were holed through two months later.[3]

UNDERWATER TUNNELS

Visitors who have occasion to use Wapping Station on the East London line of the London Underground may notice a simple plaque noting the involvement of Sir Marc Isambard Brunel (1769–1845) and Isambard K. Brunel (1806–1859).

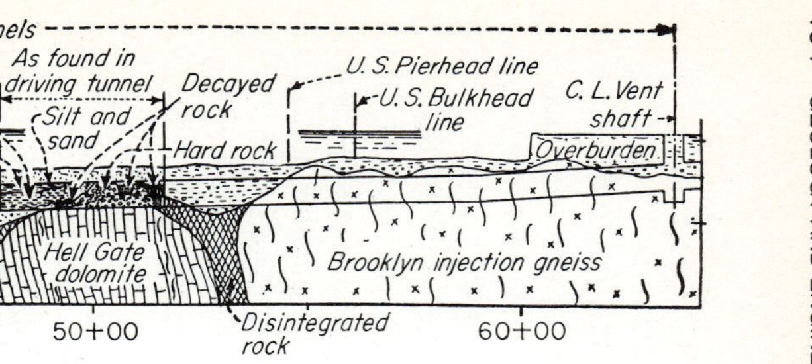

FIGURE 6.3
Diagrammatic section across the East River, New York, showing geological conditions encountered during the driving of the East River vehicular tunnel. (*Courtesy Engineering News-Record.*)

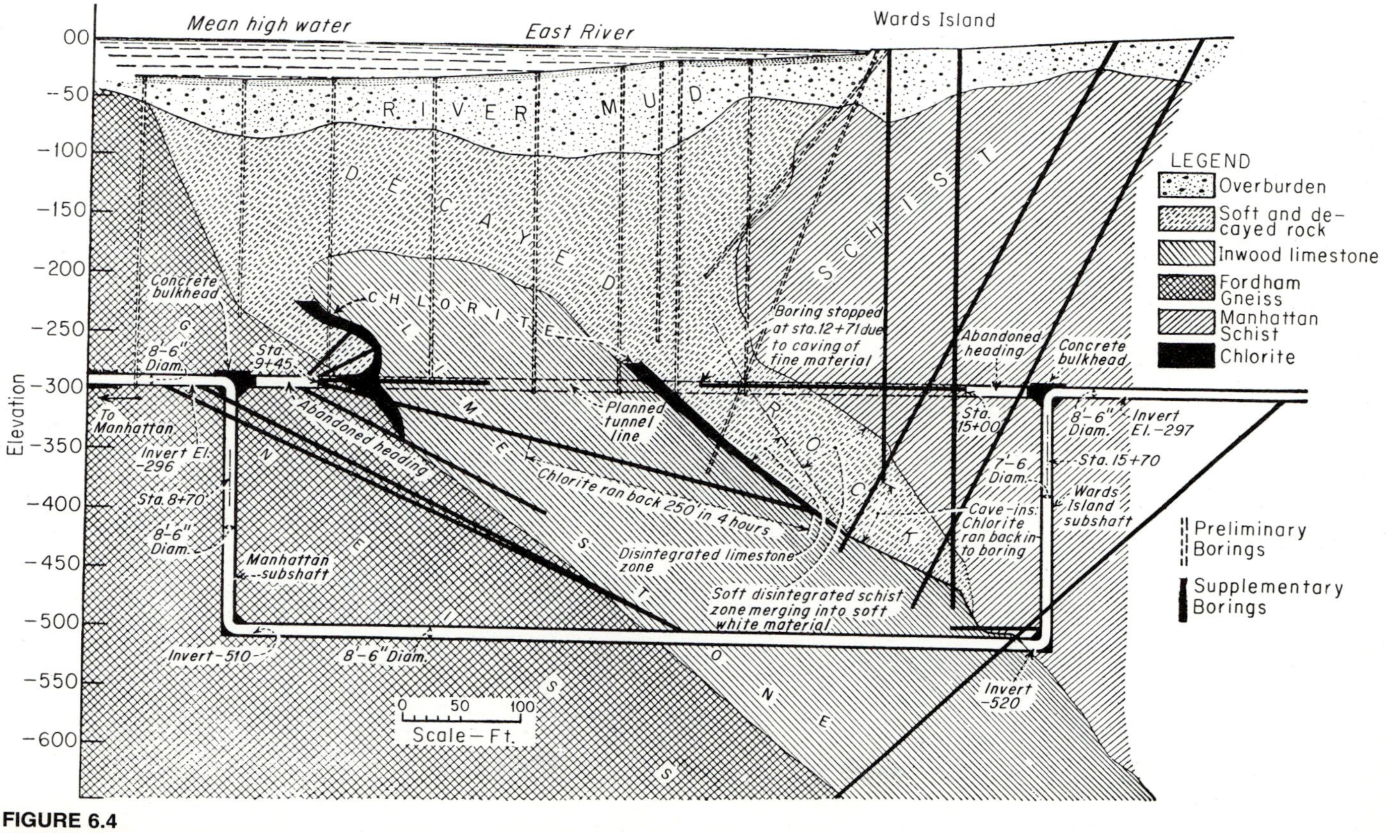

FIGURE 6.4
Geological section on the center line of Ward's Island sewer tunnel, New York, as revealed by test drilling, showing the necessary relocation of the tunnel because of chlorite seams. (*Courtesy Engineering News-Record;* from reference 3.)

This first underwater tunnel was completed in 1843. The younger Brunel constructed the tunnel by shield. This idea came to him while he was examining how the boring mollusc *Teredo navalis* worked in a piece of oak, noting how it bored its way into the wood under the protection of its own shell. A company was formed in 1824 with Brunel as engineer. Borings were placed across the river. "Eminent geologists" proclaimed that the tunnel should not go lower beneath the riverbed than was absolutely necessary. This is one case in which geological advice was less than helpful. Trouble was encountered soon after construction commenced, owing to inadequate cover, and on May 18, 1827, a bad blow occurred and the tunnel was flooded. The works were shut down for six years because of lack of funds. Brunel's personal enthusiasm eventually led to the resumption of operations in 1836, and the tunnel was successfully completed in 1843.

The first underwater railroad tunnel in North America appears to have been that for the Grand Trunk Railway (now Canadian National Railways) beneath the St. Clair River from Sarnia, Ontario, to Port Huron, Michigan. Chief engineer Joseph Hobson was appointed after the general alignment had been selected, and he chose the final detailed location in 1884. He then proceeded to have exploratory borings put down at 15-m (50-ft) intervals throughout the length (1,800 m or 6,000 ft) of the tunnel. Even today, this would be regarded as exemplary site investigation. The tunnel lay in lodgement till with a minimum clearance below the riverbed (4.8 m, 16 ft). The 6.0-m (19-ft 10-in) tunnel now carries a single track, and is lined with the original cast iron segments (another innovation).

Underwater tunnels always involve construction hazards. Such an example is the Almendares vehicular tunnel, connecting two populous residential districts of Havana, Cuba. Intensive subsurface investigations revealed the Miramar approach to lie in several layers of peat, overlying clay, and silt with sand and gravel lenses. At about the midway point, coral limestone was encountered and continued as an available foundation bed for most of the remainder of the structure. In situ testing revealed that water could be controlled in the soil deposits by the use of well points, and the concrete approach structure was thus constructed in the dry. The porous limestone was intrusion-grouted in advance of construction. Thus, 51,000 m^3 (67,000 yd^3) of rock was "solidified," permitting the driving of steel-sheet piling to form a cofferdam in which excavation proceeded in the dry to a depth of 11.4 m (38 ft) below river level.

Notable underwater tunnels link the islands of Japan. Two tunnels link the islands of Kyushu with the main island of Honshu: a railway tunnel completed in 1944 and the Kammon vehicular tunnel started in 1939 but completed in 1957 as a postwar venture. The Kammon Tunnel is 3.45 km (2.16 mi) long and penetrates diorite and porphyrite. Although the host rock is generally hard, the alignment crosses many faults and requires associated cement grouting. Extensive decomposed rock was used in the vicinity of the faults and the work was completed without trouble.

These important tunnels pale to near insignificance when compared with the double-track Seikan railway tunnel between Honshu and the northern island of Hokkaido. The 21-km (13.5-mi) underwater section lies 99 m (330 ft) below Tsugaru Strait. Long approach ways (52 km, 32.3 mi) make this the longest railway tunnel in the world. Access shafts were sunk to 219 m (730 ft) below the seabed, from which tunnels were bored, probing the geology. No summary can do justice to the Seikan Tunnel. Suffice it to say that major construction concerns were related to a long section of weak-rock volcanic ash, also harboring bad ground, which created major inflows of water. The 11.2-m- (37.6-ft)-wide and 8.5-m- (28.4-ft)-high cross section is being excavated by drilling and blasting to create one of the great tunnels of the world.

PRESSURE TUNNELS

Pressure tunnel design always presents three separate problems. The engineer must make sure that the host rock will be impervious, the internal unbalanced hydrostatic head must be resisted, and frictional resistance of the lining must be reduced. The design must also take into consideration the possibility that the tunnel will be empty at times, when groundwater will tend to exert considerable pressure on the outside of the lining. The engineer could naturally resort to the questionable economy of a pressure conduit (of steel or reinforced concrete) that would meet all the requirements called for by the foregoing conditions and neglect entirely the existence of the surrounding rock. Good engineering, however, involves economic solutions to problems faced; and so the civil engineer seeks the cooperation of the geologist so that rock strength can be utilized to keep the lining used to an economic minimum.

Almost all important pressure tunnels are lined today, mainly with concrete. Mass concrete has almost negligible tensile strength and when under tension, even when reinforced, it will open up in minute cracks. To be impervious, the lining must be made from the finest quality concrete placed carefully and well compacted in the forms. To reduce potential deformation, a geological classification of rock-mass suitability must be made and then confirmed during construction. The classification deals with the overall modulus of the host rock and its ability to resist deformation that could lead to fracturing of the liner. The long-term portland cement-concrete-rock bond must be considered, as well as a prediction of blasting overbreak as an aid in estimating lining costs.

Lining design is governed by the anticipated maximum interval water pressure, from which can be calculated a lining bursting pressure. This pressure must be resisted by the lining–host rock system. At tunnel portals and where only a minimum cover of rock is present, the lining must be designed to resist all stresses. Various methods are used to compute the strength of the host rock. A conservative assumption is that the weight of the column of rock vertically above the tunnel is all that can be assumed to resist liner deformation, where unusual tectonic stresses are not present. Lining design is an exercise in engineering judgment. The laws of mechanics can suggest certain dimensions but all calculations depend ultimately upon the accuracy of the assumptions made as to the strength of the host rock. Information required includes determination of geologic units along the tunnel axis, determination of intact rock properties, the rock mass classification, and an estimate of the probable location and quantity of any groundwater inflow.

This essential approach was employed in design of the Sir Adam Beck No. 2 development at Niagara Falls, completed in 1956. This new 1.2-million-hp powerhouse was placed adjacent and upstream from the old one but required different intake and main conduit arrangements. Instead of using an open canal skirting the city of Niagara Falls, Ontario, the Hydro Electric Power Commission of Ontario decided, after exhaustive engineering and geological studies, to use twin 13.5-m- (45-ft)-diameter lined concrete tunnels to convey the water from the intake along 8.8 km (5.5 mi) of its journey to the powerhouse. Figure 6.5 illustrates the geologic stratigraphy at Niagara Falls and its relation to the twin tunnels. The shale, for example, is basically incompetent and disintegrates upon exposure to the atmosphere. Tunnels were driven, however, mainly in competent rock, with the excellent Irondequoit Limestone used to form the crown of the tunnel. Core samples were taken as part of the exploration program. When tested, these gave a range of rock properties so that a design could be prepared with confidence.

During construction, a careful observational program was carried out to detect possible rock movements. Observations confirmed design expectations; by extrapolation, a maximum inward horizontal displacement of the walls of the first tunnel was predicted at 2.7 cm (1.08 in). Almost 2.5 cm (1 in) of this deformation was recorded within 180 days of excavation. The corresponding inward movement for the second tunnel was only about one-quarter to one-third of that in the first; this is an indication of the early stress relief in the rock due to excavation of the first tunnel.

Pressure tunnels occasionally fail. Critical review of the circumstances indicates the hazards associated with such tunnels. Leakage from a partially lined pressure tunnel was blamed as the cause of a massive landslide which severely damaged the Whatsam power plant building and transformer station in British Columbia in 1953. Water was conveyed through two 3.6-m- (12-ft)-diameter tunnels, 3.2 km (2 mi) long, to the small power station in which 25,000 kW was generated. Only the ends of the tunnels had been provided with concrete as an economical feature of design.

Rather more serious was the Sydney (New South Wales) pressure tunnel constructed from 1923 to 1928. This water supply tunnel is 16 km (10 mi) long and was designed initially with an internal diameter of 3 m (10 ft). It is constructed throughout in Hawkesbury Sandstone of Triassic age, of varying geologic character, fissured and jointed, and marked by current bedding. Before the completed tunnel was put into service, sections of it were tested. During the test, the lining ruptured to varying degrees in all but one of the sections tested. In general, the failures occurred (1) where rock was poor and

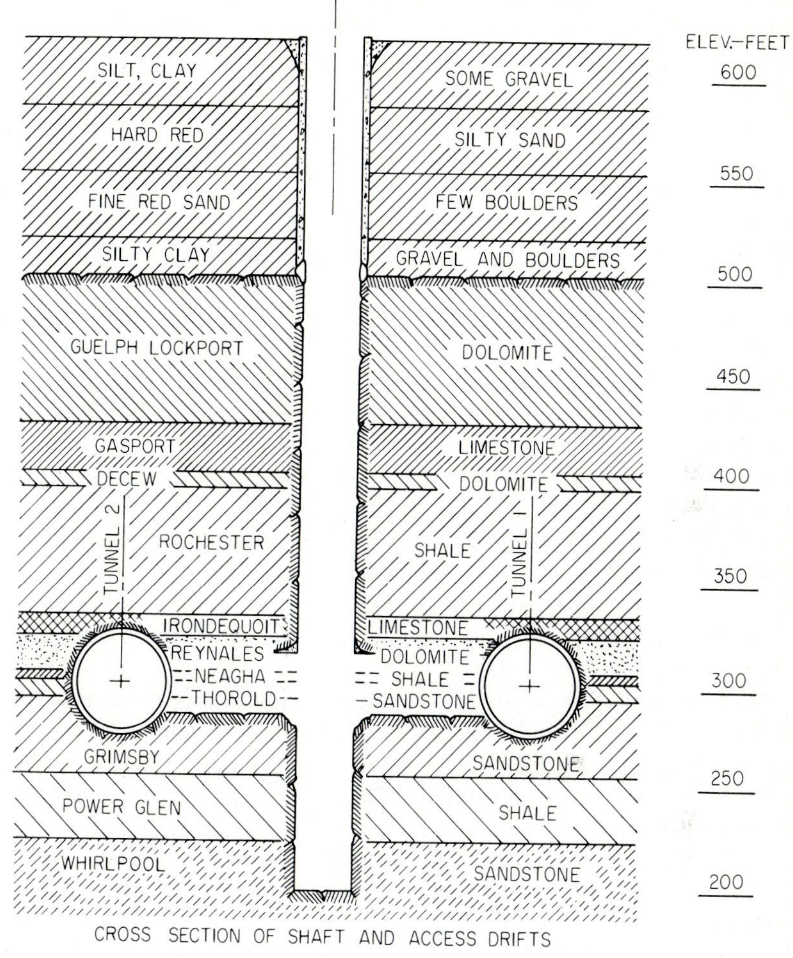

FIGURE 6.5
Cross section of rock strata at Niagara Falls, Ontario, showing the location of the twin tunnels for the Sir Adam Beck plant (near construction shaft no. 1). (*Courtesy Ontario Hydro.*)

joints had been encountered, and (2) where the rock cover above the tunnel was of theoretically inadequate thickness. The test water broke forth at the ground surface and instrumentation showed that part of the lining and the surrounding rock had moved laterally 1.3 to 1.9 cm (0.50 to 0.75 in) in the direction of the axis of the tunnel and at an inclination conforming to that of the sandstone bedding. As a remedial measure, the whole tunnel had to be lined with a continuous steel pipe. Although the failure may have been due to faulty structural design, it was possibly associated with a lack of appreciation of the true nature of the sandstone host rock.

TUNNEL SHAPES AND LININGS

Tunnel lining is of two classifications, corresponding to the two major tunnel uses. Road and rail tunnels often require lining to resist the pressure exerted by the host material. A lining may likewise be necessary to protect the wall rock from atmospheric influences and exhaust gases and vapors. Water tunnels, on the other hand, must often be lined to present a smooth, slow friction surface to the flow of the water as well as to provide an impervious barrier to leakage of water into the surrounding rock.

Determination of the final tunnel cross-sectional shape will also be affected by the host material. Circular water tunnel sections are the most economical from the hydraulic point of view, but considerations of construction methods may result in a horseshoe section with an inverted arch bottom. In weak rock, structural considerations of the design of lining will generally lead to the use of a full circular section. Tunnel roofs are usually designed as semicircular arches even in high-quality rock; the remainder of the section (in traffic tunnels) is proportioned on the basis of economic considerations. An interesting variation is the Roman road tunnel through the Ponlipio Hills. This tunnel has a pointed arch 7.5 m (25 ft) wide, 6.6 m (22 ft) high at the center of the tunnel, but 22.5 m (75 ft) high at the ends; the increase was intended to bring light into the tunnel. As the total cost of a tunnel varies almost directly with its cross-sectional area, the importance of the correct determination of this sectional area will be clear.

Isambard K. Brunel designed and directed construction of the Box Tunnel on the Great Western Railway about 1835. Built on a gradient of 1 in 100, the Box Tunnel is today a leading feature on the main line of the Western Region of British Railways, and is still largely unlined. At the time of construction, Brunel had to put up with adverse comments from eminent geologists. The most severe critic was the famous Dr. William Buckland of Oxford, who maintained that the unlined portions of this limestone tunnel were dangerous; the "concussion of the atmosphere and vibration caused by the train," he said, "would make the rock fall away." Brunel was unimpressed by this so-called "academic advice" and went ahead with the tunnel as he had planned it. The equally famous Hoosac Railway Tunnel in northwestern Massachusetts, completed in 1873, then and still the longest railroad tunnel east of the Mississippi and the first tunnel in which nitroglycerine was used for blasting, was not lined and is still in excellent condition after continuous use since its opening.

Lining design is not within the scope of this book. The technology of soft-ground tunneling, however, has led to the development of segmental cast-iron or pressed-steel plates or precast reinforced-concrete panels that are now in almost universal use as lining. In solid rock tunnels, the timber framework of construction is still left in place as a semipermanent lining. In bad ground, it may be replaced by masonry or surrounded by a solid concrete lining. Shotcrete or reinforced-concrete linings are now used almost universally to cover exposed surfaces liable to disintegrate. Concrete-placing methods follow standard practices, but there is at least one case in which geology may be said to have contributed significantly even to the placing of concrete lining.

Lining may amount to as much as one-quarter of the tunnel cost, so the importance of this feature of design will be obvious. Geological input will be of great value in determining the design of the tunnel cross section. High-quality igneous and metamorphic rock will probably not require lining for traffic tunnels. Zones of low-quality or sheared or jointed rock may have to be countered not only by permanent linings but they may also require rapid temporary installation before deformation can occur. In the Simplon Tunnel, difficulty was experienced at a distance of about 4.4 km (2.75 mi) from the Italian portal because of pressures exerted by soft, altered calcareous mica schist; even 40-cm (16-in) rolled-steel beams used as temporary strutting buckled. Quick-setting cement provided the final solution to the problem. It was applied quickly after excavation, although at a cost of $1,100 per meter of tunnel length. In other places, rock in the tunnel floor buckled and was forced upward. On the other hand, in sections driven through high-quality rock, even at the maximum depth below the surface, no deformation occurred; the lightest type of masonry lining proved adequate.

For a variety of reasons, sedimentary rocks produce most tunnel-lining problems. As a result of their mode of origin, sedimentary rocks are likely to change in character over relatively short distances, especially in tilted strata. Original depositional characteristics make these rocks susceptible to changes in stress. Furthermore, sedimentary bedding planes naturally affect the physical stability of any tunnel section bored through them, depending on the angle made by the tunnel invert and dip and strike of the beds. The underground water carried in shear zones and along bedding planes will also affect the design of a lining. All these factors, however, can be estimated from preliminary geological investigations.

OVERBREAK

Tunnel contract documents must be carefully drawn up with full regard for all the geological conditions anticipated. Payment for overbreak is almost certain to be one of the matters that will cause the most trouble at the conclusion of the contractor's operations. *Overbreak* is used to denote the quantity of rock that is actually excavated beyond the design perimeter (''A'' line) previously fixed by the engineer as the finished excavated tunnel outline. In the case of a lined tunnel, not only does the contractor not receive payment for excavation, but also additional concrete backfill will be necessary to cover more than a stated minimum beyond the ''A'' line (usually only 15 cm or 6 in).

A study of records and descriptions of early tunnels suggests that overbreak was not seriously considered in the early days of modern construction, possibly because many tunnels were then paid for on the basis of a unit length of completed tunnel. In modern work, payment is more usually made per unit volume of excavation and per unit volume of concrete or other lining. Under such circumstances, it will be readily seen that overbreak must be kept to a minimum, and as the quantity will depend primarily on the nature of the rock penetrated, given equally good driving methods, preliminary geological investigations are needed to estimate the acceptable degree of overbreak.

Soft-ground tunnels can generally be excavated exactly to the neat line of the drawings, and construction equipment (shields, lining plates, etc.) can be designed to fit to this line. Fine-grained igneous rock will usually break closer to the design section than sedimentary or metamorphic rock, which are characterized by a planar grain of weakness (bedding or foliation). Some types of chalk and dense sandstone also give good excavated sections. Tunnel engineers and geologists will usually be able to make a close cooperative estimate of the way in which the rocks will break. It will then be possible to prepare specifications so that they will conform as closely as possible to the situations that develop as the work proceeds. The following list represents a typical specification dealing with payment for excavation.

1 No points of solid rock must project beyond the "B" line, called the *neat line of excavation,* fixed at a distance from the inside perimeter ("A" line) of the finished tunnel section equal to the minimum thickness of lining required.

2 No flat surfaces of exposed rock of more than 3 m^2 must occur within a specified minimum distance from the finished perimeter, usually 10 cm (4 in) more than the minimum lining thickness.

3 All excavation and concrete lining will be paid for up to, but not beyond the pay line ("B" line), although all cavities beyond this line must be carefully filled, generally with concrete of an inferior mix.

The fixing of the pay line is usually the most difficult part of the design of the tunnel cross section. The location of the pay line will depend upon the nature of the rocks to be met; the way in which they will tend to break; the necessity of temporary timbering, and whether this will have to be incorporated in the finished lining or not; and the anticipated strike and dip of the rocks. The overbreak is the volume of excavation that actually has been taken out beyond the "B" line. Clearly, the engineer will want to fix the line so that a contractor will not be faced with the possibility of excessive overbreak, an uncertainty that will be reflected in the unit prices tendered for excavation by those contractors who appreciate the significance of geology and in claims made by those contractors who do not.

Two photographs are reproduced showing finished cross sections of the Lochaber water power tunnel in Scotland, as excavated. The first (Fig. 6.6) shows the shape of section obtained when the shot hole drilling was almost perpendicular to the strike of the mica schist. The marks of the drill steel are clearly visible; the overbreak here is very small. The second view (Fig. 6.7) shows the shape of section obtained when drilling was approximately along the strike. The section is almost rectangular owing to the way in which loosened

FIGURE 6.6
Main tunnel of the Lochaber water power project, Scotland; a view near heading 3E showing the effect of drilling at right angles to the strike of the mica schist. (*Courtesy Sir William Halcrow & Partners, London.*)

blocks of the schist have fallen away. These views illustrate more vividly than any words the dependence of overbreak upon structural geological conditions as well as upon the nature of the rock penetrated.

This concern for overbreak may seem to be overemphasized. Not only does the extra rock represented by overbreak represent an unwarranted cost but, in most cases, this space must be filled with concrete. Excessive overbreak can lead to great increases in cost over initial estimates. Overbreak can be experienced with almost all types of rock and is even a factor when bores are made using tunneling machines, if sections of loose rock are encountered. Typical was the case of a small drainage tunnel excavated in Toronto through Ordovician (Dundas) shale for which blasting was necessary. The concrete lining, instead of being 0.3 m (12 in) thick as designed, ranged actually in thickness from 0.6 to 1.2 m (24 to 47 in) in thickness—and this was in easily excavated shale.

CONSTRUCTION METHODS

Tunnel construction methods are as complex and varied as the imagination of the worldwide fraternity of specialists who work in underground construction. Yet it is important to encapsulate herein the elements of drilling, blasting,

FIGURE 6.7
Main tunnel of the Lochaber water power project, Scotland, a view near heading 5W showing the effect of drilling along the strike of the mica schist. *(Courtesy Sir William Halcrow & Partners, London.)*

excavation, and tunnel boring, since construction methods depend primarily on the geologic conditions encountered. Broad classification of the material will determine generally the type of construction to be adopted. The exact nature of the material, together with local practice, will determine the detailed variations of the main method in use—for example, the use of top, bottom, or central headings in rock tunnels and the use of the English, Belgian, German, Austrian, or other schedule of excavation in soft-ground tunnels. Geologic conditions govern the choice of methods although operational details will be determined by experienced construction personnel, either engineers or geologists. Special construction techniques will be required to handle unusual conditions that may be encountered.

In very deep tunnels, elevated ground temperatures may adversely affect the progress. Temperature variations are difficult to predict accurately. Ground temperature inside the Lochaber Tunnel in Scotland (during construction) rarely varied from 12°C (54°F). Mine workings provide some useful data. At the

Lake Shore gold mine in northern Ontario, tests show that the rock temperature there rises 1°C (1°F) with every 50 m (163.4 ft) of depth below surface. In Great Britain, a corresponding figure has been found to be 1°C (1°F) for every 32 m (60 ft) of depth. The most serious difficulties due to abnormal temperatures have been encountered during the construction of the long tunnels that pierce the European Alps.

Other construction problems arise when tunnels are in close proximity to other underground works. For example, construction of the Breakneck Tunnel on the New York Central Railroad in New York State required enlargements of an existing double-track tunnel, extensively used and only partly lined. A new double-track tunnel had to be constructed parallel to the existing bore, separated by 9 m (30 ft) of hard granite gneiss. The northern portals are close to one of the siphon shafts carrying the Catskill water supply from the deep-water tunnel under the Hudson River. Clearly, the excavation had to be accomplished in such a way that the siphon shaft would not be disturbed. The dense gneiss would transmit explosion vibrations without much damping. Rock excavation was therefore carried out by perimeter drilling of 60-cm (2⅜-in) holes, 1.8 m (6 ft) deep, around the tunnel section, spaced at 3-in centers. The 16-cm (⅝-in) gaps between each hole were then broached after which the isolated core of rock was broken up and removed by pneumatic methods. The operation was successful (Fig. 6.8).

FIGURE 6.8
Breakneck Tunnel of the (former) New York Central Railroad, New York, showing the portal of the new second tunnel on the right; note the proximity of the shaft house for the Catskill water-supply tunnel to the right of the new tunnel. (*Courtesy New York Central Railroad.*)

Slaking ground occurs when weak rock is affected by exposure to air. Certain types of clay and shale are particularly susceptible to the influence of the atmosphere, and a common practice (as in Toronto tunnel work) is to spray the shale with a thin coat of shotcrete as soon as it is exposed. More elaborate methods were utilized in the case of the construction of the P. L. M. Railroad tunnels on the line between Nice, France, and San Dalmazzo, Italy. Almost pure anhydrite was encountered in the Col de Braus Tunnel for a length of about 1 km (1,000 yd) and in the Caranca Tunnel for a few hundred meters. Water reaching the anhydrite in the former tunnel from adjacent Jurassic limestone caused the exposed rock to increase about 30 percent in volume. The trouble became more serious during the time when work was halted from 1914 to 1919. After construction, undrained groundwater led to the rupturing of a portion of the masonry lining. Aluminous cement concrete and an elaborate drainage system were installed; the results have been satisfactory.

This experience was utilized in the later construction of the Caranca Tunnel. At completion of mucking, the rock was covered with coal tar. The lining was placed with the least possible delay, using aluminous cement; and after its completion, coal tar was injected under pressure behind the lining. The tunnel was put into use in 1921 and has given satisfactory service since.

Methods of excavation have naturally advanced in recent years in keeping with general progress in mechanization and in equipment controls. Remarkable developments have been made in tunnel-boring machines (TBM). Even though Colonel Francis Beaumont in 1883 used a simple type of boring machine for the pioneer exploration tunnel for the long-discussed Channel Tunnel, it was not until after World War II that the modern TBM experienced its phenomenal development. Excavation of Missouri River diversion tunnels for the Fort Peck Dam (1934) saw unsuccessful trials of two coal saws adapted for tunnel work. Trials were resumed (1949) at Fort Randall Dam, using a jumbo-mounted, circular-operating ring saw, with some success. At the Oahe Dam (1954), four different types of specially designed TBMs were used with real success; rates of advance in faulted shale were twice as fast as by conventional methods. A geologist of the U.S. Army Corps of Engineers was in constant attendance throughout this work at Oahe Dam, mapping every advance; predictions based on his detailed geological mapping proved of value to the tunnel contractor. These machines were the forerunners of the elaborate TBMs of today (Fig. 6.9).

Nonrotary excavating machines based on coal-working machines were successfully tried as early as 1953 on a 1,200-m (4,000-ft) sewer tunnel in Euclid, Ohio. Thereafter, advance was rapid. One of the latest developments has been the invention in England (later refined in Japan) of a "slurry mole," a TBM using bentonite slurry at the working face in poor ground, by an ingenious application of the same principle as that used for slurry-trench construction. High-pressure water jets have been incorporated in some shields, and elaborate controls have reduced the labor necessary for smooth operation. No less than 382 TBMs were at work in Japan in 1979, and were said to be carrying out more tunneling work than in the rest of the world put together. A

FIGURE 6.9
A large modern tunnel-boring machine breaking through one of the major Chicago drainage tunnels. The scale can be gauged from the timbers in the foreground. (*Courtesy Metropolitan Sanitary District of Greater Chicago.*)

variety of types of TBMs are now available. They are costly machines and so are economical only for tunnels beyond a critical length (about 1.5 km). TBMs weighing 850 tons, with cutter heads operated by 9,850-kW motors, have been successfully excavating 10.5-m- (35-ft)-diameter tunnels in Chicago.

The vaunted efficiency of tunnel-boring machines tends to obscure the fact that their success depends on accurate knowledge of the geology (including groundwater) of the tunnel axis. There are already cases in which machines have been trapped underground by unsuspected squeezing ground conditions or due to unnecessarily slow operating progress. All that has so far been said, therefore, about tunnel geology and the importance of the best possible preliminary investigations must be reemphasized for all cases in which the use of TBMs is contemplated. One failure can more than wipe out all the savings possible with the ordinarily high efficiency of machine operation. One example, reportedly caused by the unsuspected presence of a "buried valley" resulted in the collapse of ground around a *mole* in the 8,475-m (27,800-ft), 3.9-m- (12-ft, 11-in)-diameter sewer tunnel on the island of Montreal. It took more than a year (using compressed air) for the mole to be rehabilitated.

There are other important aids to tunnel construction. Freezing, as described for open excavation, may occasionally be the only solution in very poor ground. In driving a double-track mass transit tunnel beneath a riverbed in Tokyo, Japan, a Japanese contractor froze 37,000 m^3 (1.3 million ft^3) of soil and then used a coal-machine excavator to dig out successfully the frozen earth. Rock bolts are now a valuable tool for stabilizing weak rock faces. *Spiling* is inserted by reinforcing bolts in holes drilled inward from the working face.

GROUTING

Grouting has become a powerful aid in tunnel work, especially when used to reduce the flow of groundwater. Its use depends on the geological nature of the strata to be encountered. Tunnel grouting methods are traditional; a cement mixture is usual, although sometimes special chemical additives are injected first to ensure cementation of fine cracks.

Fractures must be present if the rock is to be grouted satisfactorily. Grouting is a preferred method to counter the effects of excessive inflow of water, structurally weak rock incapable of supporting the tunnel arch, and material liable to disintegrate on exposure to the atmosphere. A careful geological investigation can usually anticipate the need for grouting in tunnels in the above-cited conditions and in water-bearing strata such as coarse gravel.

Leakage of river water into the famous Severn Tunnel of the (old) Great Western Railway in England was greatly diminished by specialized cementation many years after the tunnel's construction. The construction of the new Mersey vehicular tunnel was greatly facilitated by the same process, and precementation in water-bearing gravel facilitated the construction of some of the inclined escalator shafts for the London tube railways.

Chemical grouting, as opposed to portland cement grouting, has also played a part in assisting with the control of groundwater in tunnels. The Pennsylvania Turnpike adapted the seven partially driven 50-year-old tunnels of the never-completed South Penn Railway for use on the highway. The completed tunnels vary in length from 1,053 to 2,035 m (3,511 to 6,782 ft), representing about 60 percent conversion in enlargement of the old railroad tunnels. Detailed geological studies were made under the direction of A. B. Cleaves. The local geology was unusually complex, involving 19 shale and sandstone formations and groundwater pH values of 3.4 (highly acidic) to 8.3 (highly alkaline). Vitrified-clay tile drains were widely used for drawing off the highly acidic waters to prevent damage to the concrete lining. After a decade, seepage in some of the tunnels was proving to be a nuisance, and chemical grouting with expanded shale aggregate backfill was used to seal the tunnels.

UNUSUAL GEOLOGICAL PROBLEMS

Almost all the tunneling problems so far mentioned are of a geologic nature; this would be true no matter how many examples were cited. There are many problems no less serious than bad ground. The relative space devoted to each problem is proportional (roughly) to its incidence in practice.

Groundwater

Groundwater is often the main source of trouble in tunnel construction. Drainage facilities from the headings may be required and when necessary invert grades are not available, the additional trouble and expense of pumping is necessitated. Water in any appreciable quantity will impede construction work. If the host ground is soft and liable to a loss of strength by pore pressure buildup, the use of compressed air may become essential. It is essential to have accurate preconstruction assessments of groundwater conditions. No major civil engineering operation should be initiated in the underground before something is known about the level and flow of groundwater.

It is equally important to check carefully on all water encountered during construction. This will lead to determination of the source of the water and therefore a method to seal the water off in some way. It will also make sure that no serious undermining or cavitation occurs. In deep tunnel work, plutonic water (water that has not come from the atmosphere or the surface of the earth) may be encountered. During the construction of the Simplon Tunnel, alleged plutonic water was encountered, such that both a hot and a cold spring of water issued into the tunnel within 1 m (3.3 ft) of each other. During the construction of the Moffat Tunnel in Colorado, calcium chloride was dumped in a lake 420 m (1,400 ft) above the tunnel line, with traces of the chemical entering the tunnel no more than two hours later.

Two tunnels lie beneath the river Mersey at Liverpool, England: the Mersey Railway Tunnel, completed in 1885 and which still provides an important electric railway service, and the 1934 Queensway Vehicular Tunnel, at that time the largest subaqueous road tunnel in the world. The host rock is the Bunter Formation, a red sandstone of the Triassic system. The tunnels lie almost wholly in the middle of three Bunter members; the porous, massively bedded, and jointed host rock bedding dips only 2 to 5°. Throughout the district the bedrock is covered with glacial deposits, lodgement till, sands, and gravels, and these deposits extend across the riverbed. In 1873, T. Mellard Reade, a distinguished Liverpool civil engineer and geologist, predicted that a buried preglacial valley would be found under the present bed of the Mersey.

Construction of the Mersey Railway Tunnel started in 1879. Boreholes were sunk and Sir Francis Fox, the engineer for the work, took heed of Mr. Reade's forecast, which later proved to be correct. Sir Francis, though, had seen that the tunnel had been lowered to a safe elevation. The contractors had encountered the buried channel suddenly, but only in the crown. Crews were

FIGURE 6.10
Queensway Vehicular Tunnel under the River Mersey, Liverpool, England, a view showing the final phase of rock excavation with some permanent lining in place. Scale may be judged by the man in the center background. (*Courtesy Mott, Hay and Anderson, London.*)

withdrawn from the workings, but the channel-bottom clay cover held, and work was resumed and successfully completed.

When the second tunnel was started in 1925, Professor P. G. H. Boswell undertook extensive geological investigations for designing the increased tunnel diameter (13.2 m or 44 ft) in its new seaward location and to help achieve construction economy. The anticipated bottom level of the buried valley (Fig. 6.10) hinged on clarification of the fact that the old drainage system flowed in the direction opposite to that of the present river. It was desirable to keep the invert level of the new tunnel reasonably high. The actual minimum separation of tunnel crown and the bottom of the valley was only 90 cm (3 ft) at one spot. The old valley, although filled with water-bearing sand, was covered with impervious lodgement till, and so safety could be assured. It was not practical to put down boreholes in the river at that time, owing to the range of tide and to the amount of shipping. Exploratory borings were placed ahead of the working face to lengths of 45 m (150 ft). The total cost of this extensive preliminary and exploratory work amounted to less than 0.2 percent of the total cost, a remarkably good investment.

The railway tunnel was not made watertight (even today about 22,600 lpm or 5,000 gpm have to be pumped out of its drainage sumps), and the water has the composition of the Mersey and comes through the bedding planes and joints in the sandstone.

Groundwater problems in tunneling are best emphasized by reference to the "hottest and wettest tunnel ever driven," the Tecolote Tunnel of the Cachuma project of the U.S. Bureau of Reclamation. The tunnel, 10 km (6.4 mi) long and 2.1 m (7 ft) in diameter, was built to convey water from the Cachuma Reservoir on the Santa Ynez River to the California coast near Santa Barbara. It penetrates the Cretaceous sedimentary rocks of the Santa Ynez Mountains in which it was known in advance that a major fault would be encountered. One of the special precautions taken was a survey of all springs and wells within 16 km (10 mi) of the new tunnel in the event that they dried up. Water flows, methane and hydrogen sulfide gas, and heavy ground had been encountered in two earlier bores in the area. Gases were found in small quantities and caused some trouble; deformation of temporary shoring also occurred. But it was a steady increase in the quantity and temperature of water entering the tunnel that led to a temporary shutdown of the work in 1953. Work was resumed the next year, and the tunnel was successfully holed through early in 1955, but only after further and almost incredible difficulty. The flow of water reached a maximum of 34,000 lpm (9,000 gpm) and the temperature reached 47°C (117°F), making working conditions at the face almost unbearable (Fig. 6.11).

Localized inflow occurred along faults and attempts were made to grout

FIGURE 6.11
Tunneling crew at the Tecolote Tunnel returning to their working heading in water-filled mucking cars since the temperature at this location was 97°F (36°C) when the photo was taken. (*Courtesy U.S. Bureau of Reclamation.*)

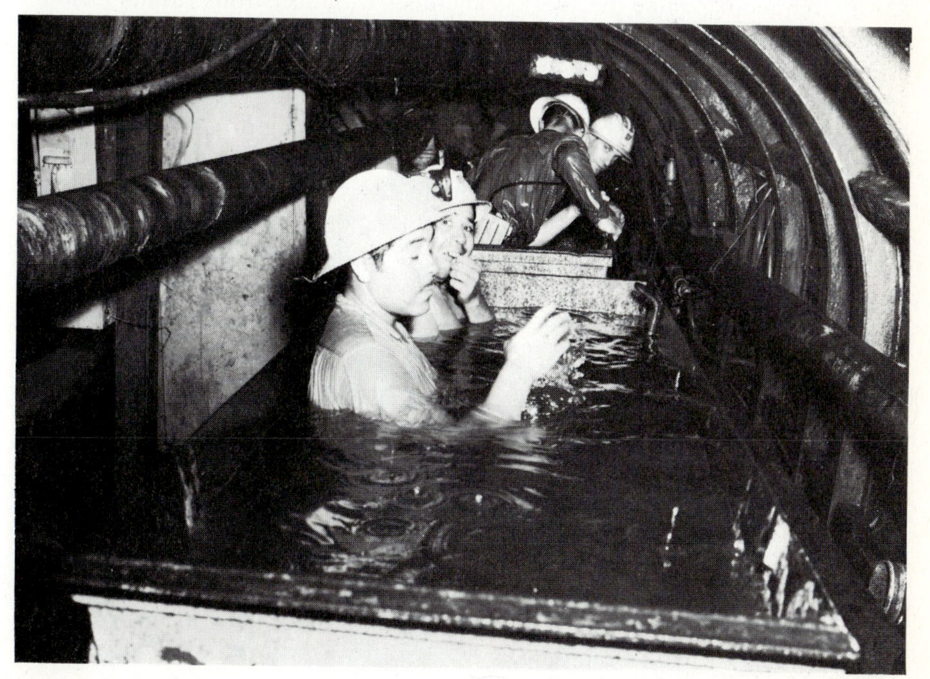

around the tunnel to slow up the flow of water, but special pumping equipment was needed to deal with the 34,000-lpm (9,000-gpm) flow. Men going to work at the face had to travel through the "hot" section of the tunnel, immersed in water up to their necks; they rode in dump cars specially provided for this job. It can be seen from this example that even the most expert geological advice is often unable to detect the potential construction impacts of local variations in geologic structure. It does point to the necessity of continual attention to every detail of the geology encountered in tunneling until the job is complete. This adversity, however, was turned to some good effect. The excessive flow from the tunnel was piped to the Santa Barbara area to relieve a serious water shortage; thus, the tunnel trouble proved to be the salvation of the local water authority.

Presence of Gases

Methane gas, not uncommon in sedimentary rock of petroleum-producing regions, may, if not detected, result in fire or explosion. Methane is also lethal if inhaled. If, therefore, study of the local geology gives any suggestion that methane may be encountered, the most stringent precautions must be taken, such as the use of spark-proof equipment and the presence of gas detectors near the working face. Shale in the Toronto area sometimes contains methane. Contractors on the first section of Toronto's subway were warned of this possibility and took all necessary precautions. Some recent tunneling jobs have encountered methane unexpectedly, sometimes with disastrous results including the loss of life. Shales, clays, and any soils of organic matter should be suspected to harbor methane.

Methane, carbon monoxide, and carbon dioxide should be anticipated in the vicinity of former landfills under which tunnels are being driven; these gases have been responsible for deaths in tunnel workings. Even more remarkable was the presence of nitrogen in shaft sinking for a new London tube railway station, thought to have originated from the "dead air" in the voids of intervening gravel (the oxygen having long been dissipated) seeping into the open working, with a potentially dangerous result.

Weathered Rock

In situ weathering of bedrock is another problem which can lead to serious consequences if encountered in tunnel work. In areas of residual soils, bedrock may be weathered to saprolite and superficially appear to be rock, displaying hard-rock structural features but with negligible shear strength. This soil-like material was the cause of delay and trouble on the Wilson Tunnel in Honolulu, an 832-m (2,775-ft) vehicular tunnel. The first 570 m (1,900 ft) of the tunnel was driven without difficulty, but caving and loss of life occurred in the volcanic rock which had turned to saprolite.

In another case, borings had suggested that solid chalk would be encountered for sinking an access shaft to a new cable tunnel to be constructed under the river Thames between Tilbury and Gravesend. When bedrock was reached, it proved to be highly disintegrated to a depth of 6 m (19.7 ft), probably as a result of exposure to earlier permafrost conditions.

CONSTRUCTION RECORDS

Tunnel construction work requires maintenance of accurate, complete, and up-to-date geological records from the start of construction. This can be done most conveniently by preparing a peripheral geologic map of the tunnel walls and frequent geologic sketches of the geologic contacts, bedding, joints, shear zones, faults, and weathered or altered zones. These features must be observed after every round has been fired, and the strike and dip of the rock must be recorded regularly, along with overbreak zones. All this information should be carefully and continually compared with the design-basis geologic section along the tunnel as predicted from preliminary investigations. In addition, it must be compared with surface topography and outcrop details in the vicinity, in order to continue to compile as complete a picture as possible. A close check can now be kept on the relation of observed to anticipated geologic conditions. Any unusual departures can be checked on discovery, and implications can be incorporated into the tunneling methods. Alternatively, underground geologic difficulty may serve as a guide to similar troublesome spots that may be encountered along the tunnel axis.

This "as-constructed" geological record is of great interest and importance to the geologist from another standpoint entirely; it gives information obtainable in no other way and, if the tunnel is to be lined or used as an aqueduct, at no other time. Such records can therefore be of supreme value as a basis for later negotiations, arbitration, or court proceedings in the regrettable event of contractor claims for additional cost reimbursement. Should any trouble develop in the future operation of the tunnel, the cause may often be traced to some unsatisfactory condition of the surrounding ground, if this information can be obtained from the records available. When tunnels, especially pressure tunnels, are being examined after continued periods of use, a geologic map or section will always help by indicating the stations that need special attention. If a railway tunnel is to be increased in cross section, an accurate geological record of the original work will be of inestimable value.

All civil engineers in charge of tunnel work should also invite the director of the local geological survey to inspect and map the tunnel and should furnish to the survey a copy of their final peripheral geologic maps. This courtesy, involving no expense and but little trouble, should likewise be extended when possible to the head of any university geology department in the immediate neighborhood. The discretion of scientists is such that no engineer need ever hesitate to take this step.

There are many examples of tunnels that have yielded invaluable geological information, thanks to such cooperation. The most important known to the writers is in the moving autobiography of the great German geologist, Hans Cloos. Entitled *Conversation with the Earth* and available in English translation, it is recommended to all engineers interested in geology; it is a most human and stimulating document, penned shortly before the death of Dr. Cloos as a "valve and a means of salvation from the confinement, brute force, and untruth of the times."

Dr. Cloos was greatly interested in the Rhine Graben of western Germany, about which controversy over its origin had raged for years. He tells of the interest to him, as possibly providing a clue to this geological riddle, of the proposal to build a new approach to the main railway station of the lovely city of Freiburg in the Black Forest near the Swiss border. Although the original contract for the work was signed in 1910, World War I and postwar conditions delayed activity until August 1928. The tunnel was to be 509 m (1,696 ft) long and to pierce the Lorettoberg, below the line on the Gunterstal. Its western portal would be cut into sandstone and its eastern into gneiss, so that it might reveal the exact interrelation of the young sediments of the Rhine Valley (limestone, sandstone, marl, and clay) and the ancient gneiss of the Black Forest and the Vosges. Dr. Cloos visited the work regularly, and in April 1929 he was able to see the excavation completed and to view, first-hand, the hitherto hidden Rhine Graben fault.

> There it was right before my eyes...a furrow ran up the wall into the ceiling and down the other side, as neatly as if it had been done with a knife...dipping 55°away from the mountain and downward below the plain....I took pencil and paper and drew as much of the three-dimensional picture as I could reduce to lines and planes. For the wet spongy stone would not be there very long. It was to be sheathed in, and only a little window would be left open through which posterity may catch a glimpse of the extraordinary phenomenon that I have been privileged to see.

The main tunnel of the Lochaber Water Power scheme is an example of tunnel construction in which close attention to geology during construction was an important feature. This Scottish tunnel extends from the main reservoir for the project, Loch Treig, to the side slope of Ben Nevis (Great Britain's highest peak) overlooking Fort William, on Loch Linnhe; it is slightly more than 24 km (15 mi) long and has an effective average diameter of about 4.5 m (15 ft). The tunnel passes through one of the most complicated parts of Great Britain's rock strata. Many types of rock were encountered; granite, shattered mica schist, and schist of excessive hardness. The route was selected for topographic reasons, but prior to construction, a geologic survey was made by E. B. Bailey (later director of the Geological Survey of Great Britain). The preliminary information was confirmed to a remarkable degree during Dr. Bailey's as-constructed detailed mapping.

Several interesting facts not shown by surface outcrops were discovered and appreciated by most of the resident engineering staff (one of the writers was a member for all too short a period). At the Loch Treig end of the tunnel, excavation proceeded in soft mica schist, much disturbed by faults and shear zones. There were seams of clay in this section, and the rock adjacent to them was soft; timbering and cast-iron lining had to be used. After two of these disturbed zones had been passed, it was noticed that they corresponded to the cliff lines on the hill above the tunnel. Accordingly, succeeding cliff lines were related to the presence of faulted and sheared ground, which was predicted to within a few feet.

Two unfortunate cases may now be cited in conclusion. The absence of construction records for the first long railway tunnel ever to be constructed, the Woodhead Tunnel on the line connecting Manchester and Sheffield in the Pennine district of northern England, later proved serious. Twin single-line tunnels, one completed in 1845 and the other in 1852, are 4,772 m (15,906 ft) long, making this structure the fourth longest tunnel in the United Kingdom. Disintegration of the brick lining in the 1940s made maintenance difficult, and daily traffic of over 80 trains on each line left little time available for the necessary repairs. In 1948, when electrification was pending, it was decided that a duplicate double-track tunnel, parallel to the existing twin tunnels, would be constructed. Owing to a necessary slight realignment, the length of the new tunnel is 4,811 m (16,037 ft), making it now the third longest. It was excavated at 9.2 m (30 ft, 6 in) wide and 7.9 m (26 ft, 4 in) high; a mass concrete lining reduced the effective width to 8.1 m (27 ft).

Since no technical account of the planning and execution of the old tunnels could be traced, no firsthand record of the behavior of the shale and sandstone rock that had to be penetrated was available. On the other hand, there was discovered a large and detailed longitudinal section along the original tunnel and through the five construction shafts. This was the section made in 1845 by the original resident engineer and deposited in the museum of the Geological Survey. It showed the strata encountered, the fossils found, and the limits of all geological features. In the report of the government's inspecting engineer (of 1852), reference was made to "bulging of the sidewalls" for a distance of about 15 m (50 ft), a condition that was corrected before the tunnel was approved for public use.

The new tunnel was completed in 1953. There had been a great deal of trouble with rockfalls and even with the original excavation method, owing to the character of the dominant black argillaceous shale host rock. At about 270 m (900 ft) from the Woodhead end of the tunnel, just at the slabby sandstone shale contact, a very large rockfall took place, involving 30 m (100 ft) of tunnels. A cavity formed to 21 m (70 ft) above the invert. The fall crushed steel-rib supports, as shown in Figure 6.12, and delayed work by six months. The tunnel was eventually completed and lined. Construction records from the earlier tunnels would almost certainly have disclosed similar trouble. This case is a useful warning of the troubles that may be encountered with shale. Any available record should be consulted by all who have major tunnel work to carry out in shale.[5]

The second case is the railway tunnel at Clifton Hall, England, a short

FIGURE 6.12
Roof collapse between 270 and 291 (900 and 972 ft) from the portal of the new Woodhead Tunnel, England, built for British Railways. (*Courtesy Sir William Halcrow & Partners, London.*)

branch line between Patricroft and Molyneux Junction, not far from the Woodhead tunnels. The tunnel was only about 1,170 m (3,900 ft) long, straight in plan, lined with brick and of horseshoe shape, 7.4 m (24 ft, 9 in) wide and 6.7 m (22 ft, 3 in) high. It was constructed in 1850 and regularly and carefully inspected. On the morning of April 13, 1953, the "ganger" responsible for this stretch of line noticed a small amount of brick rubble on the track; a swift inspection by the railway engineering staff followed. The brick lining at the location noted was seen to be under stress; steel ribs were made up but the tunnel crown failed on the morning of April 28. A mass of wet sand and rubble from an old construction shaft fell into the tunnel and a crater formed suddenly at the surface. Three houses had been built over the top of the unknown and completely hidden shaft; all collapsed into the shaft and five residents of two of the houses were killed.

The official inquiry brought out the fact that the tunnel records had been destroyed in a 1940 air raid and a subsequent 1952 fire. Records found later disclosed the fact that eight construction shafts had encountered water-bearing sand and timbering had decayed with tragic results.

TUNNELS UNDER BOSTON

Metropolitan Boston lies on the complex bedrock geology of the Boston basin (Fig. 6.13). This bedrock is divided into the Roxbury Conglomerate and the Cambridge Argillite. To the north, the latter is in contact with the Lynn volcanic rocks. From Figure 6.13 it will be seen that all the tunnels marked lie within the Boston basin, except for the Malden Tunnel that crosses its northern boundary. All the tunnels other than the Dorchester Bay Tunnel (built in the 1880s) were projects of the Metropolitan District Commission (MDC) of the Commonwealth of Massachusetts. Staff geologists and special consultants carried out preliminary investigations, so that all recent tunnels were completed without undue difficulty, all by experienced contractors.

The Main Drainage Tunnel (11.4 km, 7.1 mi long) lies largely under the harbor. The 3-m-(9.8-ft)-in-diameter tunnel was constructed between 1954 and 1959 and was finished with a 60-cm (24-in) concrete lining. The rocks penetrated were Cambridge Argillite, which proved to be the most competent, requiring steel supports for only 11 percent of its exposure; and Dorchester Shale of the Roxbury Formation, the least competent, requiring supports for 87 percent of its exposed length. A short stretch of Roxbury Conglomerate was sound, but the shales and argillites required almost continuous support. Much jointing and many minor faults were encountered, but these did not seriously influence the need for support.

The City Tunnel Extension, constructed between 1951 and 1956, is almost the same length as the Main Drainage Tunnel (11.4 km or 7.1 mi). It also lies in solid rock for its entire length, its invert being between 30 and 117 m (98 and 384 ft) below mean low sea level, with ground level above the tunnel rising to 60 m (196 ft) above this datum. The tunnel has a finished diameter of 3 m (9.8 ft) in an excavated diameter of about 4.05 m (13.3 ft). It conveys water from the Chestnut Hill Reservoir in western Boston to the southwestern part of Malden. Host rocks, the same as along the Main Drainage Tunnel, proved to be much better for tunneling; only 5.6 percent of the total length required steel supports and one-third of this in weak argillite. Another third was related to dikes, the remainder to shear zones, faults, joints, and bedding-planes. The tunnel cut diagonally across one of the major folds within the Boston basin and so revealed geological information of much interest. One hundred and six faults were mapped, most of them with apparent displacements of only a few meters, but in 17 cases the displacement amounted to more than the height of the tunnel.

Construction of the Malden Tunnel took place in 1957–1958. Its slightly less than 1.6-km (1-mi) length (Fig. 6.13) was known in advance to pass through the major north bounding fault of the Boston basin. The location of the contact was determined with reasonable accuracy by exploratory borings. Excavation

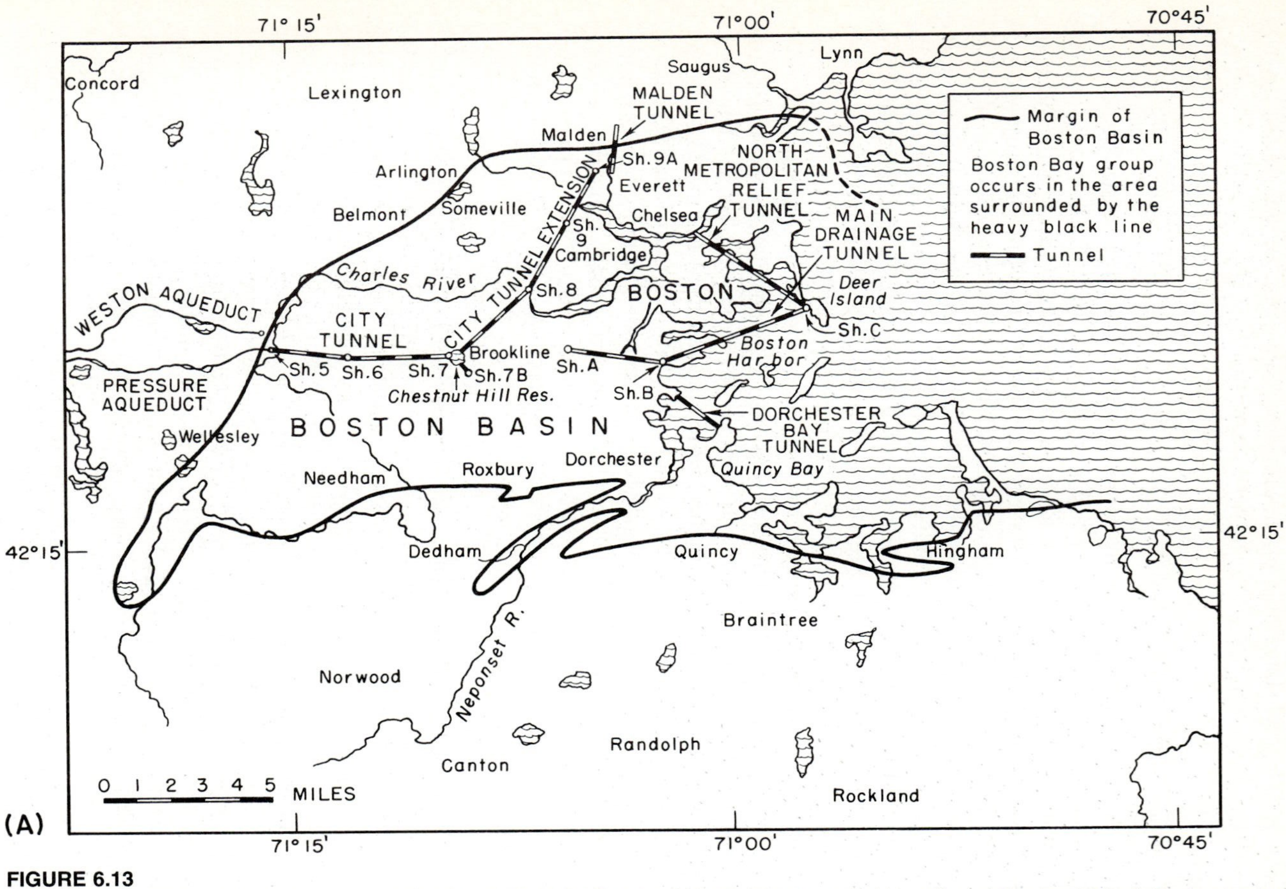

FIGURE 6.13
Index map showing the location of the principal tunnels under Boston, Massachusetts, and the outline of the (geological) Boston basin. (*Courtesy Boston Society of Civil Engineers.*)

proved the accuracy of the predictions, for the existence of the fault was clearly demonstrated by difficult excavation. Fifty-two percent of the tunnel length required steel supports, all in the vicinity of the fault. The actual contact of the Lynn Felsite, to the north, and the Cambridge Argillite, to the south, was tight on the west side of the tunnel but had a gouge-filled separation of about 2.5 cm (1 in) on the east side. This faulting had sheared the rocks to either side and a great deal of water was encountered. Pumping rates varied but reached an average of 5.5 million liters (1.45 million gallons) per day for February 1958, in contrast to the relatively dry state of the other tunnels. The tunnel has a finished internal diameter of 3.75 m (12.3 ft) and lies at an invert depth of as little as 84.3 m (276 ft) below the local datum.

Boston basin structure has been intensively reported by Dr. L. LaForge and by Professor Marland P. Billings and students of Harvard University. The Billings work appears in a notable series of journal papers of the Boston Society of Civil Engineers, the oldest engineering society of the United States. This intensive mapping recorded the complete tunnel sections to a scale of 1:240. The resulting folios are archived at the MDC and at Harvard University, and can be studied by all. Small-scale sections are part of the BSCE papers. These Boston tunnel papers are recommended as splendid examples of the sort of record that should be prepared and published for all major tunnels and especially for those exhibiting unusual geological features.

USE OF OLD MINES AND QUARRIES

Unused mines or underground quarries of urban areas should not be overlooked for purposes other than for mining. Wartime has always demanded such "safe" space for storage. World War II saw great activity in this direction. The salt mines at Hungen, north of Frankfurt, were used to store the choicest German art treasures during the war years, including the world-famous bust of the Egyptian queen Nefertiti. One hundred tons of treasures from the British Museum were also stored in a disused tunnel near Aberystwyth in Wales, while the famous Elgin Marbles were stored in one of the London underground railway tunnels. The Polish salt mines at Wieliczka near Cracow are used permanently as a viewing gallery. The famous Bath stone mines of England were put to use as wartime workshops. The even temperature, good ventilation, steady relative humidity, and absence of sunlight combine to make these mines ideal for mushroom culture. At one time, 5.2 ha (13 acres) underground were used for this purpose, producing between 113,000 and 136,000 kg (250,000 and 300,000 lb) of mushrooms per year. This same area of Wiltshire, where excavations started in Roman times, housed many secret British wartime operations, including a complete aircraft factory. Even today one large area is still in use for a large classified governmental operation.

Quite the most remarkable use of "mined-out" space, however, is to be found in and around Kansas City, Missouri. This important urban area is in the middle of a 225-km- (140-mi)-wide belt of exposed Pennsylvanian sedimentary strata. A subunit, the Kansas City Group, consists of alternating limestones and shales. Many rock escarpments in the area, up to 15 m (50 ft) high, expose different members of the group, the most important of which is the Bethany Falls Limestone, an excellent building stone. Quarrying and mining of this and associated beds of limestone have steadily increased. The value of such nonmetallic mining is now estimated to be at least $25 million per year. The limestone has been used not only as building stone, but also for lime and portland cement, concrete aggregate, and road construction.

The general mining method leaves large bedding pillars of rock to serve as roof supports. Horizontal bedding results in level floors in the mined-out areas (see Fig. 6.14). Groundwater is encountered only occasionally, as these large underground areas are generally dry, with the usual constant temperature and steady relative humidity. The total thickness of the Bethany Falls Limestone, from 6 to 7.5 m (19.5 to 24.5 ft), results in mined-out space that is almost ideal for underground operations. About 15 active mining operations excavate about 4 to 5 ha (10 to 12.5 acres) of space each year. There are an estimated 1,160 ha (2,870 acres) of mined space beneath the area around the city, with 4.7 million m^2 (5.6 million yd^2) of finished space available for warehousing and other purposes in 21 different mined areas.

Today, almost all limestone mining in Missouri is being carried out with eventual use of the mined-out space in mind. The pillars are now geometrically

FIGURE 6.14
Inside one of the disused limestone mining areas in Kansas City, Missouri, showing the naturally level floor and competent rock roof. (*Courtesy Missouri Division of Geology and Land Survey, Rolla.*)

spaced to give efficient cleared spaces (Fig. 6.15). Floors are graded, and entrances are left with sufficient roof cover to give safe portals. Almost 60 ha (150 acres) have now been put to use for office, manufacturing, and storage purposes in the Kansas City area alone. Ventilation must naturally be provided along with minor adjustment of air temperature and humidity. Rock bolting to secure safe roof rock is occasionally necessary, but in the areas that have been well mined, little extra work is usually found to be necessary prior to the finishing of the space for use.

In all urban and regional planning, therefore, close attention should always be paid to any unused space beneath the ground surface of the area being planned. It may provide economical and satisfactory solutions to problems of space, its use will be energy-conserving, and the visual environment will benefit. There are problems to be faced in utilizing such space, as outlined later in this chapter, but they are capable of engineering solutions, always assuming the local geology is suitable.

FIGURE 6.15
One of the entrances to the old limestone mine workings in Kansas City, Missouri, showing the rock formation involved and one of the railroad tracks that leads to areas now used for storage. (*Courtesy Missouri Division of Geology and Land Survey, Rolla.*)

UNDERGROUND POWERHOUSES

The construction of underground power stations (almost always water power or pumped storage stations) is now a standard industrial design concept. Their use was pioneered at the end of the nineteenth century by one of the first Canadian power stations at Niagara Falls, in which turbines were located at the bottoms of rock excavations. The first completely underground station was the Snoqualmie Falls station of the Puget Sound Light and Power Company, located east of Seattle. Built in 1898 and still in operation with much of its original equipment, this powerhouse was excavated in lava flow basalt, only 60 m (200 ft) long, 12 m (40 ft) wide, and 9 m (30 ft) high, located 81 m (270 ft) below surface level and with a tailrace tunnel.

In Europe, the first such plant was that built in 1911 for the Porjus project in Sweden. The first Italian underground station was that at Coghinas in Sardinia, built in 1926. Today Italy is one of the leaders in this phase of hydroelectric design, having over 100 underground power stations already in use. Despite the fact that the local geology must possess solid rock of excellent quality for such large excavations, there are even underground power stations in the Highlands of Scotland. This area is famed in geological circles for some of the world's most complex hard-rock geology, sometimes exposed to view as in the Ceannacroc station. Together with the associated Glenmoriston station, it was constructed in meta-sedimentary rocks, mainly schists and granulites. Rock bolting was used to reinforce the exposed roof of the station. Much grouting was used to combat water inflows during construction, and reinforced concrete barrel-vault roofs were installed, but much wall rock was left exposed to view.

The main problems associated with this specialized use of the underground are again the necessity for an accurate foreknowledge of bedrock conditions, close inspection as excavation proceeds, preplanned means for dealing with water inflows, and readiness to use rock bolts, either directly or with prestressing and other means of rock support. Reinforced concrete may be used, but, more frequently, bolt-supported wire mesh with a sprayed shotcrete cover is employed. Only rarely has the adequacy of preliminary information been in question, although frequently actual rock conditions have not been exactly as had been hoped for from the best of the preliminary studies.

A few examples from around the world will show what varied host rocks have been used for underground powerhouse sites. In Sweden more than one-half of all water power stations are underground. Two early examples were the Kilforsen power plant (285,000 kW) and the Stornorrfors plant (520,000 kW) which was the largest Swedish power station when completed. Sweden is fortunate in being underlain by competent Precambrian granite and gneiss. Over 2 million yd^3 had to be excavated for the Stornorrfors tailrace tunnel alone, but this is 4 km (2.5 mi) long with a cross section 26.5 by 16.0 m (87 by 52 ft) wide. Many innovations in rock drilling and excavation and in shaft sinking were featured on this project, which was a milestone in the steady development of underground construction (Fig. 6.16).

FIGURE 6.16
Excavation for the tailrace tunnel of the Stornorrfors water power station on the Ume River, Sweden. (*Courtesy Statens Vattenfallsverk, Valigby, Sweden.*)

Similar use of the underground was being made at about the same time (the mid-1950s) in France, notably in the Montpezat powerhouse. This station utilizes water obtained in a complex scheme involving an ancient volcanic crater used as the storage reservoir for the flow from three tributaries of the upper Loire. A head of 900 m (3,000 ft) generates 105,000 kW.

In the United States, the important Haas station on the Kings River project in California was built by the Pacific Gas and Electric Company, this being of excellent massive granite. Here it was possible to substitute 600 m (2,000 ft) of unlined tailrace tunnel for steel penstocks above ground.

Improved practice of the last three decades has led to completion of the Swiss Hongrin underground pumped storage plant on the shores of Lake Geneva, at Veytaux. Dogger Limestone and calcareous schist host rock called for careful excavation procedures and extensive use of special prestressed rock anchors.

Over 1,000 prestressed rock anchors and almost 4,000 rock bolts were used on the Waldeck II underground pumped storage plant, near Hemforth, West Germany. This plant was built in mixed strata of slate and coarse graywacke, and sandstone. So vital was rock reinforcing that a borehole TV camera (7.5 cm, or 3 in, in diameter and 1.5 m, or 5 ft long) was used by project geologists to survey about 10 percent of the anchor holes.

On the other side of the world, the Tumut project of the Australian Snowy Mountains scheme contains two underground stations. Tumut 2 was built entirely in granite and granitic gneiss, which was most thoroughly explored before designs were completed. No major changes were necessitated during the progress of the work.

In great contrast, due entirely to geologic competence quality, and not to the preliminary engineering and geological studies, was the Belgian Coo-Trois-Ponts pumped storage plant in the Ardennes, near Liège. Topographically, the site was ideal, but the local geology consisted of thin, almost horizontal quartzite and phyllite strata of unusual brittleness. A major exploratory boring program was carried out before construction started, and an inspection tunnel 295 m (984 ft) long was driven, but when excavation started, host rock faulting was far worse than expected. Unusual difficulties in construction were successfully overcome with the aid of all the devices already mentioned. Some of the rock bolts were as much as 21 m (70 ft) long, some supporting as little as 10.8 m^2 (116 ft^2) of exposed rock.

Canada has consistently had some of the largest underground power stations of the world. The Chutes des Passes station for Aluminium, Ltd., on the Péribonka River in the Lake St. John district of Quebec generates 750,000 kW. This station was followed by Outardes III station, 21 m (70 ft) wide, 42 m (140 ft) high, and 124 m (415 ft) long, built in the late 1960s; the Portage Mountain station in British Columbia, built at about the same time, with a main hall 25 m (85 ft) wide, 267 m (890 ft) long, and 30 m (100 ft) high; and the Kemano power station in British Columbia, 342 m (1,120 ft) long, 24.6 m (80 ft) wide, and 41.7 m (137 ft) high. This power station has now been eclipsed by the Churchill Falls station in Labrador, and this in turn by the LG-2 station, the first of four vast projects which form the great James Bay scheme of Hydro Quebec. The LG-2 station is 26.5 m (87 ft) wide, 47.3 m (155 ft) high, and 483.4 m (1,580 ft) long, and excavated in Precambrian granitic gneiss cut by numerous pegmatite and diabase dikes (Fig. 6.17).

UNDERGROUND FUEL STORAGE

World War II led to developments in the use of specially excavated vaults for a variety of storage purposes—initially for defense, but now for civilian use also. The engineering and geological features of such works are no different from those discussed in connection with tunnels; the shapes and sizes of the excavations alone are different. The unusual character of some of these storage arrangements makes their success highly dependent upon geological factors.

In view of all that Pearl Harbor meant to the defense of the United States, it is not surprising to find that one of the first large underground storage installations was constructed there during the late war years. Naval diesel and fuel-oil supplies were stored in 20 cylindrical vaults, each 100 ft in diameter and 250 ft high, with domed inverts and crowns. These were constructed as steel containers and set in a concrete lining within excavations in thick basaltic lava flows. Grouting was carried out to fill interstices in the rock adjacent to the lining and also to place the concrete lining in compression before the tanks were filled. When the war ended, it was found that vast underground storage

FIGURE 6.17
Excavation in progress for La Grande underground water power station in northern Quebec, east of James Bay (Hudson Bay). (*Courtesy Societé d'Energie de la Baie James, Montreal.*)

facilities had been constructed also in Germany, Japan, Italy, Czechoslovakia, and Sweden. In some cases, these large excavations were used not only for storage but also for munitions manufacturing. About 5 million m^3 (6 million yd^3) of rock was excavated from one part alone of one of the largest German installations.

With the imperative of war removed, other ideas for underground storage were developed. In the lead were studies by Harald Edholm of the Swedish State Power Board, who wondered if the immiscibility of oil and water could be utilized as the basis for underground oil storage facilities. Using an abandoned feldspar mine on the east coast of Sweden, Edholm developed a system which worked exactly as he had predicted. The mine was cleaned out and boreholes were installed to ensure a constant supply of water at the bottom of the vast storage cavern. The oil floated on top of this layer, and suitable arrangements were made for getting the oil into the "mine" and out again when wanted. Oil is now stored to a depth of 78 m (260 ft); the extent of the storage area is reflected in the fact that a rowboat is used for firsthand inspection of the subterranean oil lake. It was estimated that a comparable "standard" oil-storage depot would have cost about five times that of the full operation of the abandoned mine (Fig. 6.18).

FIGURE 6.18
Typical rock cavern for the storage of oil in Sweden, the scale indicated by the human figures. (*Courtesy Skanska Cementjguriet, Sweden.*)

The use of abandoned oil and gas wells for the storage of refined petroleum products is a regular practice and has led to specially mined underground cavities for the same purpose. Notable among modern installations is that of the Imperial Oil Company of Canada, in the backyard of its large refinery at Sarnia, Ontario. A vast salt bed lies 660 m (2,200 ft) below the plant and extends under much of this part of Canada and the adjacent United States. Rotary boreholes were sunk into the salt bed; two sets of pipes were installed in the hole, and dissolution of the salt began. In three months, 45 million L (384,000 gal) of water was used, and 10.5 million kg (23 million lb) of salt was dissolved. This created a natural storage cavern with a capacity of 30,000 bbl, which was later expanded to 50,000 bbl; an adjacent cavity was similarly formed with a 40,000-bbl capacity. Butane and propane were then pumped into the caverns and stored; brine is pumped back as the lighter petroleum products are withdrawn from storage. The economy of this method of storage will be obvious. Even the saving in the cost of the ground necessary for "normal" storage tanks is itself enough to pay for much of the subterranean work.

More than a million bbl of fuel was soon being stored in North America in this way with probably more than 25 million bbl in abandoned oil and gas wells. The vast underground storage project of the Standard Oil Company of New Jersey at Linden, New Jersey, employs the Triassic red shale that characterizes so much of the geology of New Jersey. Through a 90-m (300-ft) shaft only 105 cm (42 in) in diameter, 107,000 m^3 (140,000 yd^3) of rock was removed after being mined out to form a vast honeycomb pattern of underground tunnels. These tunnels gave a total storage capacity of 675,000 bbl, all to be used for the storing of liquefied petroleum gas.

The Standard Oil Company of New Jersey, in a search for economic oil storage, found two adjacent abandoned slate quarries at Wind Gap, Pennsylvania. The company connected the two quarries and rehabilitated them to store 42 million bbl. At the time of its completion, this was the largest single oil-storage facility in the world. Because of the clean lines to which the slate had been excavated, it was possible to fit the necessary pontoon floating roof; the tightness of the slate formation provided a nonleaking containment. As oil is pumped out of one quarry, water is supplied from the other in order to keep the top level approximately the same at all times. The project again employs the principle of the immiscibility of oil and water.[6]

Since 1960, the pioneer storage installations have been completely eclipsed in scope and size by developments such as that of Sun Oil Company, a million-bbl liquefied propane gas oil-storage installation, excavated 127 m (425 ft) below ground level at a major tank farm 32 km (20 mi) southwest of Philadelphia. Excavation was in solid granite and consisted of 2,100 m (7,000 ft) of storage tunnels 7.5 m (25 ft) wide and 10.5 m (35 ft) high on a rectangular layout, with 19.8-m- (66-ft)-square pillars being left in place for support. All access to the workings was through a single 1.5-m (5-ft) -diameter shaft. Rock temperature is 16°C (61°F) and the gas will be stored at this temperature, its corresponding pressure being 6.3 kg/cm^2 (9-psi).[7]

The safety, economy, and visual effectiveness of underground storage for bulk fuels naturally led to the concept of freezing to form suitable underground unlined cavities in soil, where no bedrock was available. Some of these storage units were in successful operation for a time, but trouble with them has been experienced, notably in New Jersey and in England. Two tanks constructed in this way at Canvey Island on the river Thames in the late 1960s were abandoned in 1975 after high operational costs rendered surface storage more economical. Despite the fact that the tanks were used to store liquefied natural gas at–162°C (−248°F), which might have been expected to maintain the frozen cylinder of soil forming the tank, fissures developed which resulted in leakage and subsequent frost heaving at ground level.

This relative failure in underground storage is a rarity since when good bedrock is available, underground storage of both gas and oil is now well-accepted practice. So extensive now is underground storage in Sweden and elsewhere in Scandinavia that Swedish engineeers convened the first international conference on underground storage in 1977. Over 1,000 people attended, 57 countries and every continent being represented. Reports were presented indicating major underground storage facilities for oil, gas, liquefied gas, and even of molasses, in 24 countries.

Typical of these underground oil storage facilities is that at Nynäshamn, 50 km (31 mi) south of Stockholm on the Baltic Sea, which holds 900,000 m^3 (31.7 million ft^3) of oil. Groundwater in the Precambrian host rock has the effect of confining the lower-density oil in its caverns; minor seepage is pumped from the bottom of the caverns. Figure 6.18 effectively shows the appearance of a typical Swedish oil-storage cavern before being filled with oil, the pipes being those required for the pumping of water and oil. Underground storage is now a proven technology, when local geological conditions are favorable.

UNDERGROUND SPACES FOR HUMAN USE

The remarkable use of worked-out underground mining space in Kansas City suggests that such abandoned mines can be used for purposes other than bulk storage. Our present concept of the use of space below ground arose out of wartime and postwar requirements for civil defense, notably (again) in Sweden (Fig. 6.19).Under the revised Swedish Civil Defense Law of July 1, 1960, all communities of over 5,000 people must be equipped with properly designed "normal shelters." These are usually provided in basements and similar cellars; the legal requirement shows clearly the attention given to civil defense in this strategically located Scandinavian country. Ten percent of the urban population is required to control emergency operations and are to be accommodated in deep rock shelters. More than 100 of these unusual facilities have been constructed below a rock cover of at least 15 m (50 ft), but some have as much as 30 m (100 ft) of rock above them. Quite a number of these excavations have been designed for peacetime uses such as the pumping installations for the water-supply system of Goteborg, on Sweden's southwestern coast.[8]

The most notable of comparable underground installations in North America is the command combat operations center of the North American Air Defense Command (NORAD) near Colorado Springs, Colorado. Completed in 1962, this fully equipped underground installation is capable of accommodating a total staff of 700, with all services necessary for their living, such as water supply, power supply, and sewage disposal. The center lies almost completely in the Pikes Peak Granite of Cheyenne Mountain. Preliminary borings had indicated that the granite was sound enough to support the large chambers, but as-constructed geologic observations would be necessary as excavation progressed. The two approach tunnels are 789 and 366 m (2,580 and 1,200 ft) long, respectively, 8.7 m (28.5 ft) wide, and

FIGURE 6.19
Swedish destroyer in a completely underground marine dock, the scale given by the man near the bow of the vessel. (*Courtesy Fortifikationsfirvaatningen, Sweden.*)

6.75 m (22.1 ft) high. Where the tunnels meet, the great complex of underground chambers starts. Rock pillars between chambers are as much as 30 m (100 ft) wide. Within the excavations, three-story spring-mounted steel "buildings" are designed to resist blast shocks and to protect the crews and communications systems housed inside.

A geologist of the U.S. Army Corps of Engineers mapped the granite exposed during excavation. This mapping revealed two major joint systems and the axis of the chambers was reoriented to a more favorable direction. Close geological observation at one of the most vital intersections of two main chambers revealed in advance highly weathered rock and close jointing. This problem was dealt with by construction of a reinforced-concrete dome at the base of the intersection chamber and its jacking up into place below the imperfect rock roof (Fig. 6.20) These details of structural design were developed during the course of the work and were based on geology exposed by excavation.[9]

FIGURE 6.20
Construction of the NORAD underground command center at Colorado Springs, Colorado, in progress, showing framework and reinforcing steel for the reinforced-concrete dome used at the main tunnel intersection. (*Courtesy Commanding Officer, NORAD, Colorado Springs.*)

Archival material must be kept at constant temperature and humidity; underground storage provides, therefore, ideal conditions. State archives in Norway and Sweden are so located in massive Precambrian rock. The Church of Jesus Christ of the Latter-Day Saints (Mormon Church) has chosen the Wasatch Front granite. Six storage caverns, located 210 m (700 ft) below ground surface, provide 6,000 m^2 (65,000 ft^2) of storage at a constant temperature of about 15°C (59°F) and relative humidity of 40 to 50 percent.

Libraries can also be located below ground with advantage. This has been demonstrated at the University of Illinois in Urbana, where a new undergraduate library was constructed entirely below ground level, excavation being in lodgement till. A 7,000-m^2 (75,000-ft^2) extension to the famous Radcliffe Science Library of Oxford University was cut-and-cover constructed entirely below ground level, thereby preserving the visual delight of the University Museum building. Tiebacks were used to support the vertical sides of the finished excavation of coarse-to-medium gravelly sand, 5.1 m (17 ft) overlying stiff fissured clay with silty zones.

Public garages are now an urban necessity, but represent an unproductive use of above-ground land. The major downtown public garage in Los Angeles was built below ground level, and there are a growing number of smaller garages now following the same pattern. Congestion in the historic center of Quebec City was diminished by construction of a three-story underground garage excavated in hard, fine-grained middle Ordovician limestone, shale, and intervening beds of limestone conglomerate.

POTENTIAL PROBLEMS

As in all branches of civil engineering work, there are special geological problems in the use of the underground. Groundwater conditions must be established with certainty and for at least a full year's cycle before designs can be completed, although the presence of groundwater around unlined oil-storage caverns is an advantage in preventing oil loss into surrounding rock discontinuities. "People space," however, must be dry, and there must be no possibility of even minor flooding. Once recognized, this potential trouble can be mitigated by design provisions.

A second potential source of trouble is the possibility of slaking or disintegration of underground walls on exposure to air. The only real problem that has had to be faced so far in the extensive use of underground space in Kansas City, for example, is due to the exposure in some areas of the Hushpuckney Shale. When exposed to air and the moisture that air normally contains, this material undergoes slow, chemically induced expansion, and in consequence, slight problems with floor movement. This does not detract from the remarkable use of the underground at Kansas City but it is, again, a very valuable reminder of the absolute dependence upon geology of this use of space beneath the surface of the ground.[10]

CONCLUSION

Many more examples could be easily be quoted from worldwide tunneling and underground construction. The large number of examples is probably due to the fact that every tunnel job is unique at least in some one respect and often in several. Descriptions of tunnel construction are common to civil engineering literature, a fortunate reinforcement of the application of geology to all tunnel work. Notable among available records is the compilation by the California Department of Water Resources of complete records for 99 tunnels of various size, shape, character, and geology. Charts prepared from this study provide one of the most comprehensive estimating guides of this kind available in civil engineering literature.

Great tunnels continue to be built, and the story of human tunneling activities is an ever-widening one. Even as this volume goes to press, approval has been given yet again to begin pilot bores for the under-the-English-Channel tunnel. As early as 1800, a proposal to construct a tunnel was made; in 1867, definite plans were advanced. Geological investigations of 1875 produced 7,000 soundings. Almost 4,000 samples were obtained from the seabed and exploratory shafts were sunk. The consensus of general geological opinion is that the proposed tunnel (63 km or 39 mi long) should be constructed throughout in the Lower Chalk Measures, although the continuous existence of this formation across the channel can only be surmised.

In addition to its value in promoting the use of underground space, geology can assist also in conservation of energy through development of underground space. There is an unusual amount of activity in this area. The American Underground Space Association has its own professional journal, *Underground Space,** and it will be found that The Geological Society of America entitled one of the sessions at its 1978 annual meeting (in Toronto) "Geology beneath Cities," a title later applied to a GSA volume containing the papers given at that meeting as well as other papers. The geology beneath modern cities is, indeed, of vital importance.

REFERENCES

1 B. N. Page, "Geology of the Broadway Tunnel, Berkeley Hills, California," *Economic Geology, 45,* p. 142 (1950).

2 T. W. Fluhr and V. G. Terenzio, "Engineering Geology of the New York City Water Supply System," New York State Geological Survey, Open File Rept. 05.06.001, Albany, N.Y., 183 pp. (1984).

3 "Tunnel Looped Under a Fault," *Engineering News-Record, 119,* p. 220 (1937).

4 E. R. Crocker, "Hottest, Wettest Tunnel Holed Through," *Civil Engineering, 25,* p. 142 (1955).

5 P. A. Scott and J. L. Campbell, "Woodhead New Tunnel: Construction of a Three-Mile Main Double-Line Railway Tunnel," *Proceedings of the Institution of Civil Engineers,* pt. I, *3,* p. 506 (1954).

6 "Roofed-in Slate Quarry Now Stores Oil," *Engineering News-Record, 153,* p. 24 (September 16, 1954).

7 "Largest Mined Gas Storage Cavern Carved from Granite," *Engineering News-Record, 196,* p. 24 (January 1, 1976).

8 O. Albert, "Shelters in Sweden," *Civil Engineering, 31,* p. 63 (November 1961).

9 T. O. Blaschke, "Underground Command Center," *Civil Engineering, 34,* p. 36 (May 1964).

10 R. M. Coveney and E. J. Parizek, "Deformation of Mine Floors by Sulfide Alteration (Kansas City)," *Bulletin of the Association of Engineering Geologists, 14,* p. 313 (1977).

* Known as *Tunneling and Underground Space Technology* beginning in 1986.

CHAPTER 7

BUILDING FOUNDATIONS

All engineering structures must be supported in some way on geologic materials; this is the inevitable connection between geological conditions and foundation design and construction. Dams, bridges, and transportation structures are usually of considerable size and so achieve an undue importance in the popular mind, since the major part of engineered construction is made up of a large number of smaller projects. This great group of miscellaneous structures can cause just as many trials and difficulties as large projects, especially difficulties related to foundation problems. The supporting elements of this great miscellaneous group of structures can be broadly described as "building foundations."

Building foundation work was, for many years, largely a matter of rule of thumb, restricted only by such building regulations as were relevant. All too often, construction methods, which are of vital importance in many foundation operations, were left entirely to building contractors, sometimes even to the final detail. In the hands of a competent construction company, this practice may not have been too objectionable, but in other cases it led almost certainly to unsatisfactory foundation work. Today, the foundation of structures is generally recognized as an important part of building design. On many projects, large and sometimes small, architects are now asking for and funding more thorough preliminary investigations, and building plans are frequently prepared in association with a definite scheme of construction. Contractors are thus protected against many of the uncertainties that they might otherwise have to face. Simultaneously, the science of foundation design has brought about correlation of theoretical design methods with the results of their application in practice. Along with improvements in field and laboratory soil-testing methods, better and safer advantage is being taken of both the flaws and possibilities that many building sites present. This care is especially important in view of the generally decreasing quality of available urban building sites.

Foundation design consists of three essential operations: (1) determining the exact nature of the foundation beds that are to act as a support, (2) calculating the loads to be transmitted by the foundation structure to the strata supporting it, and (3) designing a foundation structure to fit the conditions ascertained as the result of operations (1) and (2). Building loads may be affected in a general way by the local geology, but for any particular region they will as a rule be related only to structural design. Design of the foundation, however, is completely dependent on the nature of the ground underlying the building site. Determination of ground conditions is essentially a geological problem, leading to selection of the type of foundation as a matter of engineering judgment. Soil mechanics theory is employed in this design but accuracy may be invalidated if a full study of the geology of the underlying strata has not also been made.

This broad outline provides the necessary background for consideration of the application of geology to building-foundation work. The fact must be emphasized that geological conditions alone are to be discussed—a warning necessary in view of the incomplete picture of foundation engineering that the following pages display. Adequate determination of geological conditions at a building site is but one of the three essential components of complete foundation design, and its isolation from the other two parts, and particularly from the final detailed correlation of loading and ground conditions, necessarily distorts its significance in foundation work. Foundation failures are rarely due to faulty structural design; they are usually related to failure of the load-bearing foundation beds.

INFLUENCE OF GEOLOGICAL CONDITIONS ON DESIGN

Ground conditions at a building site may be classified as one of three general types according to foundation design possibilities:

1 Bedrock may lie at ground surface or so close to the surface that the building may be founded directly upon it (Fig. 7.1).

2 Bedrock may occur at a distance below ground surface which makes possible direct transfer of building loads by the foundation.

3 The nearest rock stratum may be so far below the surface that the structure will have to be founded upon the unconsolidated material overlying the rock.

Geological processes of the past directly influence this broad classification. The borough of Manhattan, in New York City, typifies the first classification;

FIGURE 7.1
Crown Center development on Signpost Hill, Kansas City, Missouri, under construction, foundation beds of alternating strata of shale and limestone being exposed, the latter used for all major foundations. (*Courtesy Missouri Division of Geology and Land Survey, Rolla.*)

some of the buildings are founded directly on the Manhattan Schist which appears as outcrops over part of that famous island. Much of Montreal is of the second type; Paleozoic bedrock is overlain by unconsolidated materials of the Pleistocene and Recent periods which vary in thickness, but which in some places reach more than 30 m (100 ft). London is an example of the third type; deposits of London clay, sands, and gravel overlie the chalk bedrock, which occurs too deeply to serve as a bearing horizon. It would be of interest to study the influence that ground conditions of these broad types have had on architecture, but engineers will readily appreciate the significance of the groups without the aid of this comparison.

FOUNDATIONS ON BEDROCK

Determination of site conditions in the case of the first of the three main types of foundations, i.e., when bedrock is at or near the surface, will be a relatively simple matter. Conditions of this type are rare in urban areas, possibly because cities are usually located on riverbanks where unconsolidated deposits are the rule rather than the exception. Here, geology first serves to estimate the soundness of the rock, along with determination of the significance of structural features, and possibly with a determination of seismic risk. Rarely will there be a question of whether the bearing capacity of rock is sufficient to withstand building loads, but if compression tests of the rock are made, great care should be taken to see that they are made with specimens loaded in a direction corresponding to that at the site. Sedimentary rocks are notorious for their strength anisotropies. When tested parallel and at right angles to stratification, compressive strength may vary as much as 50 percent. As sedimentary rocks include many weak rock types which have low bearing capacities, this point is often of importance. Typical figures for slate are 500 kg/cm^2 (7,220 psi) at right angles to bedding, and 360 kg/cm^2 (5,180 psi) in the direction of bedding.

These are average figures obtained in connection with the foundation of the Boston Parcel Post Building, on Cambridge Argillite. Values for the usual bearing stress utilized for various rock types will be found in engineering handbooks and treatises on foundation design. Independent tests for all but the smallest projects are usually advisable. In situ tests of rock will be necessary in the case of unusual loading or weak rock (see p. 235).

Bedrock covered with a thin mantle of soil will be seen only when the surface has been cleaned off. When this was done at the site of the Clinton Engineer Works in Clinton, Tennessee, following an exploratory boring program, two differing rock conditions were found. On one side of a small valley was shale of the Conasauga Formation, which had weathered in the upper few feet to a soft brown clay. Weathered shale, with some clay seams, started at about 1.5 m (5 ft) below the surface. Only at depths of about 3.6 m (12 ft) was unweathered shale encountered, but it was sound and adequate for the contemplated foundation loads. On the opposite side of the small valley (the site involving 175 structures), quite different conditions were encountered. Here was dolomitic limestone of the Knox Formation, inclined away from the center of the valley with a dip of 55°. Erosion had taken place in many of the exposures of the upended beds, the crevices so formed being generally filled with soft residual clays, and with many boulders.

Locations of individual buildings were dictated by operational needs, and these site conditions had to be accepted. After cleaning of the rock, grouting was carried out, and concrete footings were built over this grouted area. Further complications were found at one of the process buildings, since the "rock" encountered by the test borings proved to be large, flat-topped, irregular boulders, some 6 m (20 ft) below the surface, so closely packed that normal excavation methods for removal of the soil around could not be used. The whole area was washed down and then entirely covered with a heavy, solid concrete mat which safely spread foundation loads across the embedded boulders and the intervening limestone.

Another bedrock-related problem was encountered at the St. Vincent Hospital in Portland, Oregon. Although a comprehensive site investigation program had been completed, excavation revealed two narrow, buried, rubble-filled channels crossing the site. After detailed study, it was decided to cover over the two rubble-filled channels with concrete plugs, to cover the entire site with a 1.5-m (5-ft) -thick blanket of compacted rockfill, and then to construct a concrete mat foundation to support the main central tower structure of the hospital.

FOUNDATIONS CARRIED TO BEDROCK

Where rock does not outcrop at the surface in the vicinity of a building site, underground exploratory work is doubly essential. If bedrock lies within a reasonable depth below the surface, exploratory drilling and, if necessary, geophysical surveying should be utilized to determine accurate contours of top-of-rock surface, as well as the groundwater surface. The word *reasonable,* although frequently objectionable, is here used to express the many local variables that may affect this economic depth. The maximum foundation depth so far utilized appears to be 75 m (250 ft) at the Cleveland Union Terminal Tower, Cleveland, Ohio.

A variety of engineering methods is available for transferring the building load down through the overburden to bedrock. Choice of the method will depend upon economics and feasibility of alternatives. If the depth is not too great and if the soil is free of boulders, end-bearing piles are frequently used. If steel piles are to be used, the possibility of corrosion must be studied; if concrete piles are to be used, the possibility of high sulfate content in the groundwater must be investigated. Cast-in-place concrete piles will require a good bearing stratum with depth. Wooden piles require consideration of possible variations in groundwater level, since most wood deteriorates under alternating wet and dry conditions.

For greater depths of overburden, for ground that contains boulders, and for carrying very heavy building loads, the use of some type of caisson or cylinder of concrete, either cased or uncased, will often provide an appropriate solution. For soft blue clay, as is found in the Chicago area, a simple type of open caisson (the "Chicago well") has proved satisfactory for depths well over 30 m (100 ft). Some of the most difficult foundation jobs yet undertaken have had recourse to deep foundations placed on sound rock.

Experience in the glaciated Detroit area shows how important it is to know also the full character of the underlying bedrock. The fissured limestone bedrock is overlain by clay strata of varying types, with a layer of very hard glacial lodgement till ("boulder clay" or "hardpan" being local names) immediately above the rock. Depths to this impervious layer are as much as 36 m (120 ft). If the till is not pierced, foundation design and construction are not unusual. Shallow foundations may bear on to clay strata, and piles or concrete piers on to the impervious layer. If the glacial till is pierced, thin, or fractured, subartesian water may rise as high as 30 m (100 ft) above the rock. Such conditions were encountered during the construction of the Greater Penobscot Building, 169.5 m (565 ft) high, having a total foundation load of 91 million kg (200 million lb). Compressed air was applied to construct the bell-mouth, open concrete caissons, which were carried right to rock. Site groundwater has an unusually high sulfur content, which is injurious to concrete, making it necessary to pregrout the surrounding host rock.

The Great Lakes region harbors problems associated with bearing strata encountered at intermediate depth. A typical Buffalo profile shows up to 4.5 m (15 ft) of miscellaneous fill, then 6.0 m (20 ft) of clay and silt, with compact sand, silt, and gravel extending to bedrock at depths usually over 18 m (60 ft). Excavation in this combination of material is not as easy as in Chicago, and steel end-bearing piles are driven to rock, though boulders frequently interfere with pile driving.

Geologically related flaws can led to failure of concrete caissons. Blocks of cohesive soil may fall into the caisson boring during placement of concrete, later leading to failure under load. A single deviation from the anticipated soil profile, or one lapse in inspection procedures, can create a defect in the design member of reinforced concrete. A new three-storey, reinforced-concrete building in Montreal had to be demolished because of the failure of cast-in-place concrete piles. An unsuspected stratum of week organic material, not strong enough to withstand the pressures exerted upon it while piles were being driven, was the cause of the failure.

Bedrock surface contours must also be known with certainty. Even a relatively flat bedrock surface must be verified by one or two borings. If an irregular bedrock surface is suspected, then an adequate characterization must be carried out over the whole of the building site. One glance at Figure 7.2 should be enough to establish this point in the mind of the most skeptical of readers. This is a model of the foundation of a modern Oslo office building close to the famous town hall. One corner of the building is carried on concrete piers resting directly on rock; the other is supported on 50-m (164-ft) steel piles,

FIGURE 7.2
Cutaway view of model of modern office building on Fridtjof Nansen's Plass, Oslo, Norway, showing variation in bedrock levels. (*Courtesy Norwegian Geotechnical Institute.*)

driven to rock through the soft, sensitive clays that distinguish this part of Scandinavia (as they do the St. Lawrence Valley in Canada). Despite the almost phenomenal foundation-bed conditions, maximum settlement has amounted to no more than a tolerable 3.8 cm (1.5 in).

Mined areas such as Pittsburgh present many examples of foundation difficulties due to the presence of abandoned workings. A notable solution is that of the Veterans Administration hospital, a 10-story building with two basements, located 30 m (100 ft) over an abandoned coal mine. Details of the mine were not available, apart from some surface cavings. Subsurface exploration employed 7.5-cm (3-in) NX core borings, supplemented by three 75-cm (30-in) -diameter Calyx holes. Cost estimates for subsurface treatment led to the decision to grout up the partially backfilled mine workings. All the old workings were economically consolidated as an artificially "reconstructed" rock foundation. In Zanesville, Ohio, a 3.5-million-L (800,000-gal) elevated steel water tank was constructed over old mine workings for which no adequate plans were available, and supported on one main central and 12 circumferential concrete columns, all carried about 10.5 m (35 ft) into rock below the mine workings. These columns were reinforced only through the upper, weak sandstone; the lower parts in hard shale were plain concrete.

Finally, a contribution from the famous island of Manhattan. Figure 7.3 shows a section through the foundation excavation for the Chase Manhattan Bank. Water-bearing silt and "hardpan" boulders of lodgement till made the final stages of excavation most difficult. The Joosten chemical solidification system was applied to reduce permeability, and the excavation was completed without loss of soil or distress to adjacent structures.

Sloping bedrock surfaces require special design and construction precautions, since gravity-induced slippage of loose overburden soils can be aggravated by pore water pressure. This is the plausible explanation for serious movement of the Cahuenga Pass multiple-arch retaining wall constructed in Los Angeles in 1925. This wall is 135 m (450 ft) long and its height varies up to 18 m (60 ft); the centers of buttresses are 9 m (30 ft) apart. The wall was founded on bedrock at its ends but not in the central portion. Rock was here 12 m (40 ft) below ground levels. Spread footings were therefore used and carried to sloping rock, at a depth of 6 m (20 ft). Movement of this central section was observed before all the fill had been placed. It continued even after the loading on the wall had been reduced and in July 1927, it amounted at one place to 43 cm (17 in) outward and 38 cm (15 in) downward. The load on the spread footings was limited to 39,000 kg/m^2 (8,000 psf), but the underlying rock sloped steeply away from the wall, so the conclusion was that soil strata and wall were moving down on the rock surface as a unit.

An allied problem is to have stratified bedrock available as a foundation stratum but dipping at such an angle to the horizontal that the stability of the upper layers is doubtful. In parts of New York City, the Manhattan Schist is so inclined; it is a feature that is occasionally encountered elsewhere. Local details will determine the best solution of the problems thus presented, but a usual method is to drill the surface rock and anchor it by dowels to layers that are far enough below the surface as to be beyond the range of possible movement.

FOUNDATIONS ON SOIL

Soil foundations are the most common of all load-bearing situations. Adequate subsurface investigations are imperative. The general rule calls for exploratory work to a depth of at least twice the width of the structure and greater than this if possible, especially if the presence of relatively soft strata is suspected. The extra cost of penetrating a meter or more is negligible compared to the value of the results. Soil structure interaction transmits foundation loads to unconsolidated strata in two general ways: (1) by constructing a continuous raft, or spread footing, which rests directly on a near-surface stratum; and (2) by driving piles or casting piers into the soil or by constructing piers therein. The

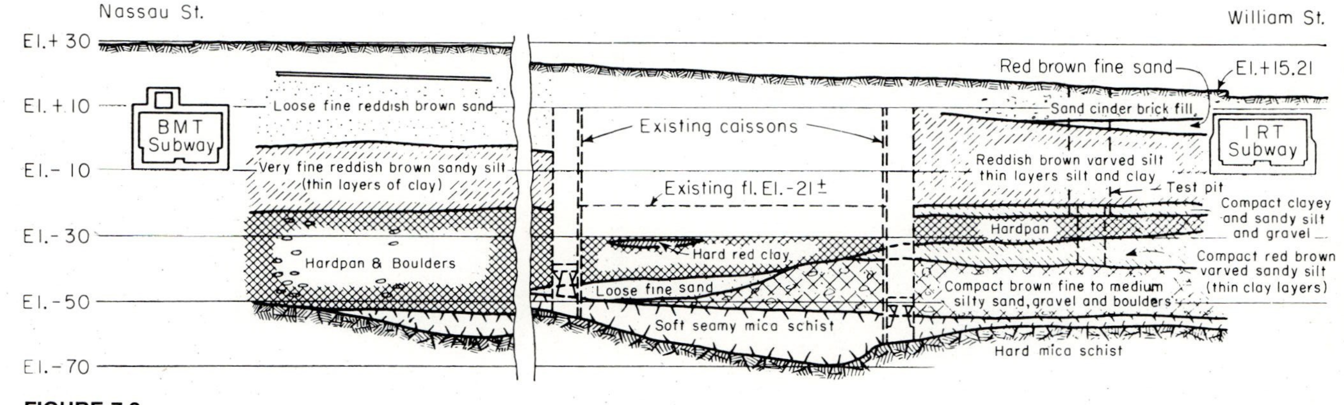

FIGURE 7.3
Simplified cross section through the foundations of the Chase Manhattan Bank, lower Manhattan, New York. (*Courtesy American Society of Civil Engineers.*)

first method is commonly varied so that the load is distributed through a number of isolated footings. The strength of the second method may depend on end bearing at a reasonably hard stratum, or it may be the result of skin friction. Other minor variations have been developed, but these methods include generally all leading foundation classes for unconsolidated strata.

Figure 7.4 is a graphical illustration of the way in which building loads are dissipated by transfer of stress to geometrically increasing volumes of soil as the distance from the foundation structure increases. The lines of equal pressure indicated in Figure 7.4 are colloquially known as "bulbs of pressure," a convenient though inexact term. The proportional reductions of stress shown are typical values; exact figures for any set of assumptions made can be calculated for any given set of conditions—one of the great contributions of soil mechanics to foundation design. This simple diagram shows the vital necessity of having an accurate knowledge of subsurface conditions to a depth of at least twice the width of the structure, for the bulb will actually vary in shape as a result of soil strength factors.

When this precaution is not taken and a building is erected without accurate subsurface information, trouble may develop if, for example, there is a buried stratum of weak soil beneath the site. There are on record all too many cases

FIGURE 7.4
Simple "bulb of pressure" diagram, showing the effect of size of a loaded area on the stress distribution beneath it.

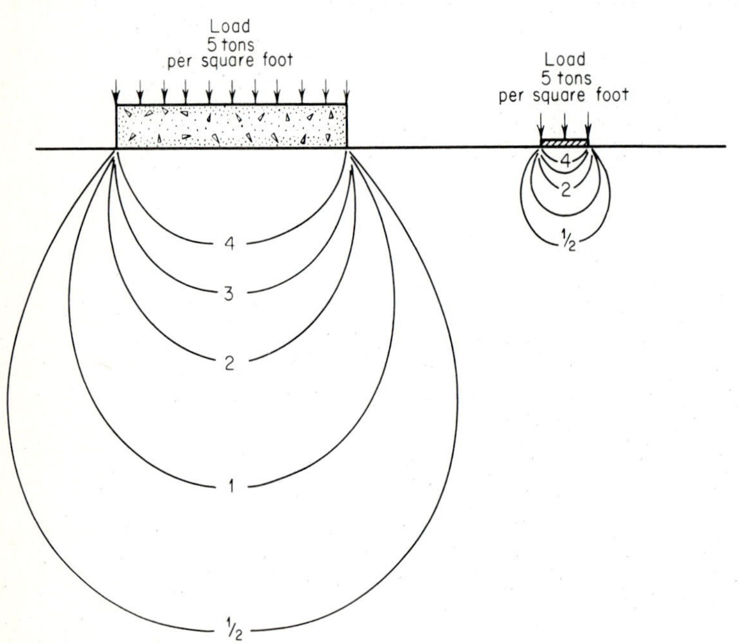

in which serious settlement has occurred from this cause. The following account is typical. A large building was planned and designed on the basis of borings. Prior to completion, the building slid laterally 10 cm (4 in) and settled 10 cm (4 in) at one place in its front wall. New borings were put down to a depth of 26 m (87 ft); 13 and 9 m (42 and 31 ft) deeper than initially explored. A stratum of "very soft black clay and fine sand" was found at 16.5 m (55 ft) below the surface; the upper strata were clay, with some sand of varying consistency. The front of the building was about 27 m (90 ft) wide. Underpinning and pile driving were required to correct the unsatisfactory condition, and at "probably four times what the cost would have been had the piles been driven before the wall was built."

A further and rather ordinary example was that of the Westinghouse Electric and Manufacturing Company Building in Philadelphia, constructed on the basis of boring records supplied by the previous site owners. These showed 2.4 to 3.0 m (8 to 10 ft) of loose fill and then uniform clay and sand to rock level, 13.5 m (45 ft) below the surface. Bearing-pile foundations were chosen for the 10-storey building and test rods were driven to determine the length of piles, fixed at between 7.5 and 9.0 m (25 to 30 ft). When eight of the ten storeys had been constructed, the building was found to have settled 10 cm (4 in) on one side. New core borings and an open test caisson were put down to rock, a mica schist at 17 m (57 ft). The borings passed through 8 m (27 ft) of fill, 1.2 m (4 ft) of peat, 30 cm (1 ft) of gravel, 4.5 m (15 ft) of silted peat, 2.4 m (8 ft) of silt, and 60 cm (2 ft) of coarse sand and gravel instead of the fill, clay, and sand expected. An elaborate and ingenious underpinning operation had to be undertaken immediately, and the footings had to be supported on 37.5- and 40.0-cm- (15- and 16-in)-pipe piles, concrete-filled and carried to rock.

Accurate knowledge of subsurface conditions at a building site includes determination of soil properties as obtained by careful laboratory testing of undisturbed soil samples. With these basic soil properties, it is possible to calculate the rate and probable total settlement of the building due to the increase of stress in the supporting soils. Correspondingly, safe bearing capacities for soil can be calculated, safe against excessive settlement and against overstressing of the soil immediately under the foundations. The responsibilities of the engineer emphasize the absolute necessity of accurate subsurface information at building sites.

Foundation failures will steadily decrease in number and magnitude. Study of past failures can be of real assistance to modern design. An extreme example of the difficulties caused by ignorance of soil behavior is provided by the National Museum Building in Ottawa. This massive red sandstone structure of 1910 now has a somewhat unusual appearance, having lost the elaborate upper part of the entrance-porch tower shortly after building was begun. Serious settlement was noted before the building was completed. The main porch had separated from the rest of the building by 35 mm (14 in) at roof level and was just at the point of actual shear failure when the tower was removed. Until 1975 it was possible to see stark evidence of at least 30 cm (12 in) of settlement. The building was founded on over 30 m (100 ft) of the sensitive marine clay already mentioned, a clay which was not strong enough to carry the heavy stress concentrations induced by this splendid example of Victorian "wedding-cake" architecture.

GROUNDWATER

Groundwater conditions govern the success of many building sites. The problem of falling groundwater levels is not peculiar to Boston; it is one of the most widespread problems of civil engineering works. In San Francisco, a 14-storey building near Sansome and Bush streets had to have its entire load transferred from 1,100 wooden piles to 210 concrete piers as early as 1930 because of water level variations. In Milwaukee, Wisconsin, and in all too many cities, the same problem has, unfortunately, had to be faced.

How can such foundation troubles be prevented and remedied? One method, followed in some building regulations, is to make sure that all piles are either cut off or embedded in concrete at a level well below the minimum elevation possible for the local groundwater. An alternative method is to use "composite" piles of protected sections within the range of water-level variation and timber below this level. The Northwestern Mutual Life Insurance Building in Milwaukee, Wisconsin, lies on glacial strata, with bedrock 75 m (250 ft) below ground surface; the water level in 1930 was 5 m (17 ft) below the surface. Timber piles were used, and lengths of 10-cm (4-in) pipe, fitted with screw caps and strainer bottoms, were built into the foundation footings. Water level readings are taken periodically and it has proved necessary to add pile-moisture water to about one-fifth of the pipes in order to maintain satisfactory foundation conditions.

Britain's noble Cathedral Church at the ancient capital of Winchester is famous also in the annals of foundation engineering. Groundwater comes close to the surface in the vicinity of the church. An early 1900s scheme of reconstruction involved shoring the outside of the building; supporting the great arch vaulting inside to prevent collapse; using steel tie rods when necessary, in association with an extensive use of high-pressure grouting in all the damaged masonry; and then underpinning all of the exterior walls down to a bed of gravel. This last operation was carried out in complete darkness created by the black groundwater since the available pumping equipment would probably have removed some of the silt and possibly disturbed the foundation beds in other ways. The work involved excavating from beneath the walls and piers in successive pits, down to the gravel, and then placing concrete by hand up to the underside of the masonry of the walls. The entire job was done in a period of five years by expert diver William Walker, and regularly inspected by the consultant, Sir Francis Fox, also in a diver's suit. The building was reopened at a great service held on July 14, 1912. A statue of William

Walker was later placed in the cathedral in grateful recognition of his notable and unique engineering work.

Fifty years later, a new hotel was built within 51 m (167 ft) of the cathedral. The record of the earlier work was available, and the same groundwater condition had to be dealt with, so as not to cause damage to the cathedral. Well points, all connected into a larger pipe "header," were used in closely spaced rows to surround the area that had to be pumped dry (Fig. 7.5). This dried out the excavation for the hotel. Naturally, careful groundwater records were maintained; the most important observation well was inside the cathedral. With excavation complete, and the building of the superstructure proceeding, the groundwater was allowed to regain its normal position.

Groundwater may also cause a foundation to "float" upward. At the new general care building of New York's Harlem Hospital, the volume of water

FIGURE 7.5
Excavation for the foundations of a new hotel in Winchester, England, showing the header for the well-point dewatering system (a horizontal pipe, left of center) and the great cathedral in the background. (*Courtesy James Drewitt & Son Ltd., Bournemouth, England.*)

handled by the well-point system was so great as to require operation until 12 storeys of structural steel and the first three concrete floors were in place. This was done in order to give sufficient deadweight to counteract the calculated hydrostatic uplift pressure.

When similar circumstances are encountered with structures that do not have sufficient intrinsic weight to offset anticipated hydrostatic pressure, other measures have to be taken. When the Ley Creek sewage-treatment tanks at Syracuse, New York, have to be emptied for maintenance or repair (about once every three months), the superintendent must first set in operation permanent well-points that lower the groundwater level around the tanks to a sufficient depth to avoid floating the empty tanks. Soil conditions in this area include a thick stratum of what is locally known as "black sand," a medium-to-medium-coarse sand, underlying a sandy silt and clay blanket at the surface.

More unusual was the problem faced during the construction of five New York City Housing Authority 14-storey apartment houses adjacent to the boardwalk at Coney Island. Figure 7.6 shows the local soil conditions and the proximity of two of the adjacent buildings, which could not be disturbed during construction. Laboratory study showed that the organic silt of both beds had been overconsolidated in the past by sand dunes. Hence, the foundations of the large apartment blocks would have to be so located as to give some degree of buoyancy, in order to reduce the additional pressure they would exert on the silt beds. This meant excavation below the existing groundwater level which, in turn, meant unusually careful groundwater control operations to ensure the stability of the adjacent buildings. Detailed pumping tests were undertaken, and a large number of groundwater observation wells were installed. The job was completed successfully by the combination of a well-point system pumping water out of the excavation for the apartment blocks and a recharge system pumping water back into the ground in the vicinity of the existing buildings. System operation showed no evidence of any settlement of the existing structures or of any damage because of the manipulation of the groundwater conditions in the area.

PILED FOUNDATIONS

The four main methods of transmitting light loads to soil strata can be outlined as: (1) individual footings under columns; (2) continuous raft, or spread footing, under either a complete building or one section of it; (3) end-bearing piles driven to depths of safe load transmission; and (4) excavated cylindrical shafts in which concrete piers (suitably reinforced) can be formed to transmit the loads from footings to a safe bearing stratum.

With the wide variety of pile types, available choices are usually driven by economics. Bearing piles, when circumstances require their use, can serve as excellent foundation units; buildings all over the world testify to this. In some exceptional cases, the use of piles has, however, actually weakened the foundation bed intended for use. If, for example, site investigations show that

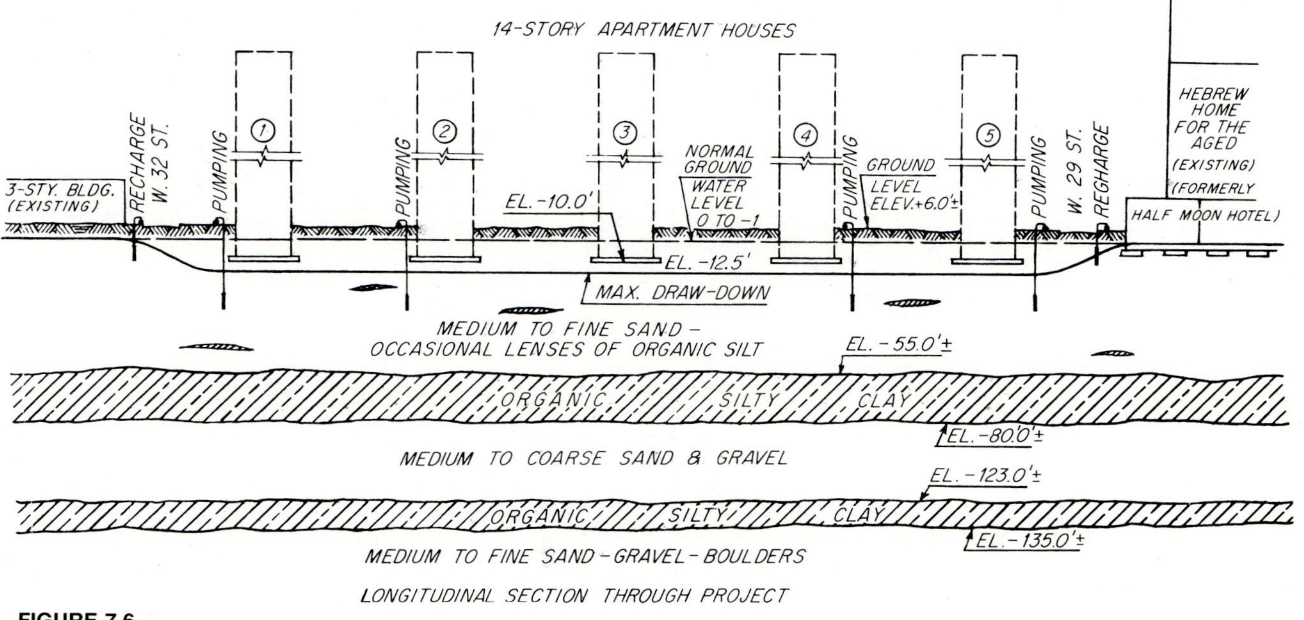

FIGURE 7.6
Cross section through foundation strata for buildings in Coney Island, New York, in relation to groundwater conditions. (*Courtesy American Society of Civil Engineers.*)

the building area is underlain with sensitive clay, then pile driving must be viewed with extreme caution, since in many of these clays the vibrations set up by pile driving are enough to reduce their effective bearing capacity by partial "liquefaction." In the province of Quebec a large church had to be completely underpinned after having been founded on piles in complete disregard of the extremely sensitive clay on which it was founded; piles had proved satisfactory for supporting a church in a neighboring village which happened to have quite different subsurface conditions.

The fundamentals of pile action in soil must be fully appreciated before piles are even considered as possible foundation units. The "bulb of pressure" (using the common term) naturally applies to groups of piles. If the length of piles is great in comparison with the width of the foundation (as is often the case), then the soil will be stressed to a considerable depth beneath the bottom tips of the piles. The neglect of this fact has led to serious settlements of buildings founded on piles, below the bottom tips of which existed soil strata of insufficient strength to resist, without undue compression, the loads thus imposed so far below ground surface.

The rule of thumb for exploration depth (knowing with certainty geological conditions to a depth of at least twice the width of every structure) should be extended if piles are to be contemplated. The best of load tests carried out on single piles will fail to indicate the danger of using a fully piled foundation where there is a weak stratum below. From the many examples of neglect of this sort that can, most unfortunately, be cited, there may be mentioned the case of a 20-storey hospital founded on a Recent alluvial deposit 30 m (100 ft) thick. The deposit was a soft clay with interbeds of firm-to-dense sand. More than 10,000 end-bearing piles were driven to about 14 m (46 ft) below ground surface to a 2-m (6-ft) -thick sand stratum. Within a year of completion, differential settlement had begun; within twenty years the settlement at the center of a dish-shaped area was nearly 1 m (3 ft), all due to consolidation of clay beneath the sand stratum to which the bearing piles had been driven.

Bearing piles, both end-bearing and friction-bearing, are excellent foundation units when properly used in appropriate ground conditions. It is unfortunate that so many "horror stories" of the misuse of piles are to be found in the annals of civil engineering. If the failures are described with constructive intent, some advantage is gained, since the cases can then be cautionary tales. Some failures of piled foundations are directly due to lack of appreciation of pile action in relation to the geological conditions at the site; others are caused by mistakes in pile design or driving.

A particular case of pile failure due to geology involves the early 1930s construction of a large grain elevator at Durban, South Africa, on a site underlain by a surface stratum of sand overlying a thick stratum of clay. The properties of the clay were not fully appreciated, because of an inadequate sampling from the ten boreholes at the site. As a result, the length of the bearing piles for supporting the structure was reduced from that originally intended. The clay was a normally consolidated deposit with low shear strength and high sensitivity, and it was therefore susceptible to considerable loss of strength by such disturbance as the driving of piles.

Three hundred and eight piles for an annex were driven first, and the concrete foundation structure with tunnels was completed about two months after pile driving. Settlement, even without further loading, started as soon as the foundation structure was complete and accelerated after the building of the superstructure, a settlement of almost 65 cm (25 in) taking place in less than one year. Similar settlement took place under the working house. Work had to be stopped and an attempt made to underpin the structures to rock, but even this proved difficult because of the character of the clay. A new foundation was designed, carried to rock, and the elevator was eventually completed but at considerable loss of time and great extra expense. The published records of this case are vivid reminders of the vital importance of a full understanding of subsurface conditions before piles are included in any foundation design.

"FLOATING" FOUNDATIONS

Structures to be constructed on weak or sensitive soils may also employ what are colloquially called "floating" foundations, an accurate description even though the first impression it gives is slightly misleading. Even weak soils manage to sustain their own weight without settlement. If, therefore, a depth of soil can be removed, the weight of which is equal to the weight of the structure to be built on the site, then the weight of the structure can replace the weight of the soil, and no settlement of the resulting sunken structure should result. This is sound in theory, but the actual execution of such designs involves considerable ingenuity in construction methods. Figure 7.7 shows such ingenious means. The New England Mutual Life Insurance Company building in Boston's Back Bay district is 10 storeys high, with an 87-m (290-ft) tower. Ground conditions from the surface down were 6 m (20 ft) of man-made fill; 6 m (20 ft) of alluvial silt; 3 m (10 ft) of oxidized, hard yellow clay (on which the building was founded); 21 m (70 ft) of glacial clay; and finally 9 m (30 ft) of compact sand and gravel above bedrock. Groundwater was only 3 m (10 ft) below the surface. The final design consisted essentially of a 12-m (40-ft)-deep concrete box, the excavation for which just about balanced in weight the total building load of 130,000 tons, thus inducing no basic change of stress beneath the final foundation.

A corresponding example from Great Britain is the Grangemouth oil refinery of Scottish Oils, Ltd., on the Firth of Forth. Here the subsurface conditions consist of a very soft (and weak) clay for a depth up to 9 m (30 ft), underlain by a 2.7-m (9-ft) layer of stony clay, below which gray silty (estuarine) clay extends more than 30 m (100 ft). A previous structure, with bearing pressures as low as 5,468 kg/m^2 (1,000 lb/ft^2) had settled at the site, and a 33-m (110-ft) chimney had settled 60 cm (2 ft) out of plumb. Even with cast-in-place, bulb-ended concrete piles, the new structures also settled.

For the latest structures, therefore, and after careful soil testing, a total load

FIGURE 7.7
"Floating box" foundation structure being sunk in position near Georgetown, Guyana. (*Courtesy Soil Mechanics Ltd., London, and the Demarara Sugar Terminals Ltd., Guyana.*)

of 37,000 tonnes was carried on 15 separate cellular caissons, which were sunk to depths up to 7.5 m (24 ft), forming, in effect, a set of "floating" foundations for the refinery. The largest of the caissons was 51 by 51 m (170 by 170 ft). They were made with thin reinforced-concrete walls constructed at ground level, resting on reinforced-concrete pads which were removed when sinking started. Soil within the cells was excavated by grab; a concrete plug sealed the bottom of each cell when the correct depth to give the required buoyancy was reached. Looking at the new refinery today, with its complex equipment and highly instrumented operations, one finds it hard to believe that all the structures are supported on such weak soil, so successful has this foundation design proved to be. There is little need to emphasize that "floating" foundations can only be designed on the basis of the most accurate possible knowledge of the subsurface geology.

CAISSON FOUNDATIONS

Japanese engineers have developed a pressure-balanced raft foundation that depends to an unusual degree upon complete and accurate knowledge of subsurface conditions. Over 20 such foundations have been installed in the Tokyo district. Under suitable subsurface conditions, a cutting edge is built on the surface of the ground and upon this the basement of the building is constructed as a ''superstructure'' above ground. When the resulting structure has the requisite weight, sinking is started by excavating beneath it; control is exercised by the progress of excavation under the cutting edge, and sinking is induced by the weight of the structure (see Fig. 7.8). This idea has long been used for the construction of bridge piers. Boulder-free soils are obviously desirable; the most appropriate conditions are those given by alluvial or lake deposits of uniform soft or medium clay.

The largest building yet to be constructed in this way is the Nikkatsu International Building in Tokyo, which has a four-storey basement, nine storeys above ground, and a three-storey penthouse; it measures 97.5 by 66.0 m (325 by 220 ft) and occupies a full, triangular city block. Six borings and five

FIGURE 7.8
Sections through the foundation caisson for the Nikkatsu International Building, Tokyo, Japan, showing the way in which the caisson was transformed into the permanent foundation structure. (*Courtesy American Society of Civil Engineers.*)

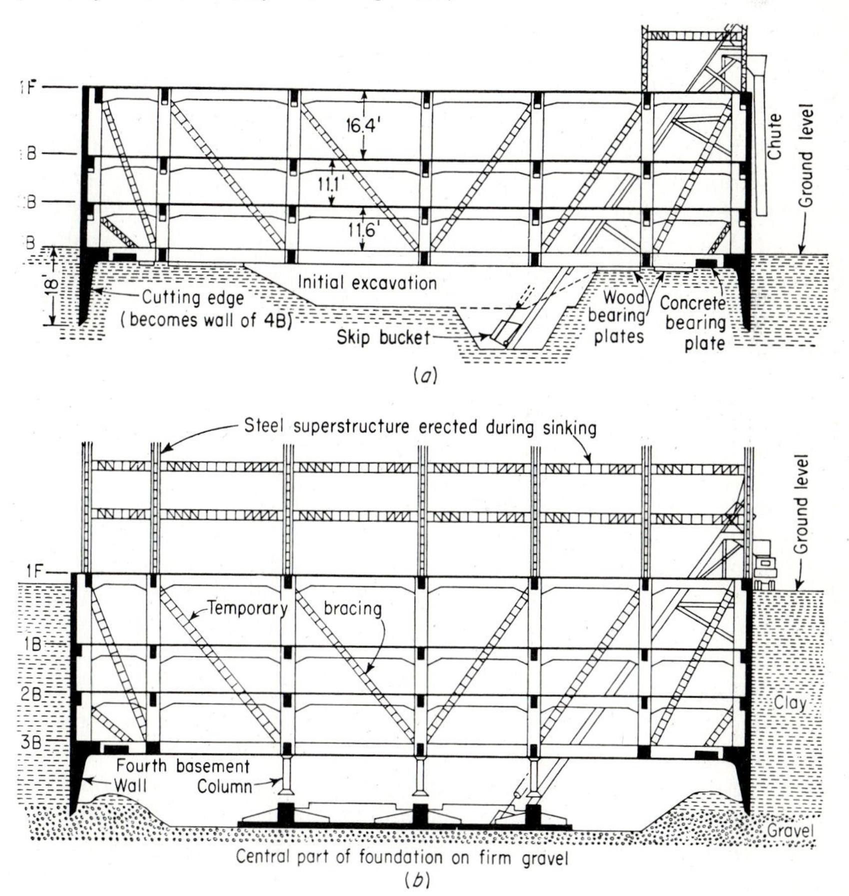

wells were sunk to depths of 45 m (150 ft) in determining subsurface conditions. Beneath the subsoil, a blue-gray clay extends for 15 m (50 ft), underlain by gravel and sand beds. Groundwater was encountered at a depth of 12 m (40 ft), but not in any quantity until a gravel bed was reached. These are almost ideal conditions for the sinking method. Three basement floors were built above ground before sinking started; the massive caisson then weighed 25,000 tonnes.

A European version employs the versatile bentonitic clay as a low-shear strength lubricant, injected in an annular ring around the caisson formed by a slightly wider diameter cutting shoe. A 530-car parking garage, with seven levels below grade, has been constructed in Geneva, Switzerland, on 7.5 m (25 ft) of water-bearing sand and gravel overlying 21.3 m (70 ft) of soft clay that grades into a good, firm clay on which the caisson was founded. Suitable soil conditions and an absence of boulders were an absolute prerequisite.

A different type of caisson foundation was the only economical solution for an eight-storey-high storage warehouse in the port of Stockholm. The only site available was a spoil fill of miscellaneous excavation waste, including large blocks of rock up to 2 m^3 (2.6 yd^3) in size, underlain by lodgement till, providing a geological combination of unparalleled difficulty. Heavy loads from the building were transmitted to rock through a drydock-constructed basement of 47 by 54 m (158 by 180 ft) in plan. The basement was floated to the site and sunk on specially leveled sand pads. The second "storey" was then constructed, holes being left in the structure through which heavy steel pipes, 95 cm (37 in) in diameter, could be lowered, being rotated by special equipment, serrated cutting edges penetrating the miscellaneous material encountered before rock was reached. Special arrangements were made at the foot of each shaft for transmitting the heavy loads that each would have to carry to the underlying bedrock. Construction was completed successfully, all on the basis of a thorough exploration of the unusual subsurface conditions.

PRELOADING OF FOUNDATION BEDS

Many buildings to be founded on weak soil do not require deep basements. An adequate knowledge of site geological history, along with laboratory tests on good soil samples can be used to calculate induced consolidation of the soil by in-place preloading. Temporary loads can be removed when necessary settlement has taken place. The permanent structure can then be built without excessive settlements; actual settlement will be controllable to a reasonable degree. This method is traditional practice for approach fills to bridges and similar earthworks. The concept of preloading was used in 1819 by Thomas Telford during the building of the Caledonian Canal in Scotland, and by Sandford Fleming in Canada in 1870.

A prime example is the Cathedral of Mary Our Queen in Baltimore, a monumental structure 112.5 m (375 ft) long and up to 28.2 (94 ft) wide, with two stone towers 39.9 m (133 ft) high. Here, preloading in the form of a pile of earth 9 m (30 ft) high, spread over an area of 36 by 21 m (120 by 70 ft), was carefully designed to obviate the differential settlement which the extra load imposed by the two towers would have caused in the underlying saprolitic soil formed of weathered mica schist.

An unusual example was provided by the rehabilitation of a large World War II warehouse at Port Newark, New Jersey. Built in 1941 by the U.S. Navy, the building (measuring 165 by 76 m or 510 by 256 ft) was essentially a structural steel frame supported on 25.5-m (84-ft) wooden piles. The Port Authority of New York and New Jersey acquired the building in 1963 and found that the immediate foundation soils were inadequate to support live loads. Newark Bay is here underlain by red shale bedrock over which there is a fairly uniform stratum of red clay 21.3 m (70 ft) thick. Over this is about 7.8 m (26 ft) of black organic silt, and on this the usual heterogeneous collection of miscellaneous fill material so often found around the shores of long-established harbors. At the site of the building, the fill consisted of cinders.

In keeping with the Port Authority's experiences in normal preloading of outside areas, it was decided to apply the same procedure to the floor inside this building. The entire building was therefore filled to a depth of 3.6 m (11.8 ft) with soil brought in by trucks, giving an additional floor load of 6,350 kg/m^2 (1,300 lb/ft^2). Left in place for 14 months, the fill was all removed by scrapers in January 1966, when the floor was found to have settled 0.43 m (17 in). A new floor was then built that, careful calculations suggested, might settle an additional 38 cm (15 in) in the ensuing five years, an amount that could readily be arranged for in the final floor design.[1]

BUILDINGS ON FILL

It is sometimes necessary to build on sites that are not naturally level, requiring either fill or excavation. Correspondingly, it is sometimes necessary to make additions to existing buildings, some preparation of the adjacent areas being first necessary for this purpose. In times past, both these situations frequently led to differential settlements. Today such undesirable effects can be avoided. Soil studies, and especially the understanding of soil compaction at optimum moisture content, can now result in the placement of accurately compacted soil as fill in any area and to any normal depth, with the certainty of minimum settlement.

An early, major use of site grading fill was the immediate post–World War II construction of the great Clinton Engineer Works in Tennessee. A large U-shaped building area of 750 m by 120 m (2,500 by 400 ft) was to be prepared. Optimum moisture contents of all fill were determined and engineered fill was placed in 15-cm (6-in) layers by sheepsfoot-roller compaction. At the time this was a revolutionary solution which gave completely satisfactory results.[2] Today, such use of compacted soil under strict control is standard civil engineering practice. Its successful use always depends upon known and acceptable subsurface conditions.

SETTLEMENT OF BUILDINGS

All buildings settle, as does every other structure erected upon the surface of the ground. It is strange how this simple fact puzzles the layman, but every civil engineer and geologist knows well that foundation beds, whether rock or soil, are just as susceptible to the imposition of loads as are all other solid materials. The stress induced by building loads will be reflected in the strain in the supporting materials. With properly designed foundations, the amount of compression, or consolidation in clay soils, will be so small as to be unseen by eye. When "settlement of building" is discussed, therefore, it is almost always *excessive* settlement that is under review or, in some cases, unacceptable *differential* settlement between parts of a building which causes structural damage. Settlement ultimately is dependent upon some inadequacy of the foundation beds. Only very rarely has settlement gone beyond the point of excessive vertical movement, bringing soil or rock to shear failure. There is, however, one famous case where this did happen; the lessons to be learned from it, and the ingenuity displayed in the correction of the failure, make it an essential example.

Construction of a large grain elevator at Transcona, Manitoba, was started in 1911. The site chosen was adjacent to one of the world's largest railway yards just outside Winnipeg, to facilitate the rapid handling of grain from the Canadian prairies. The drying house was 18 m (60 ft) high and measured 5.6 by 9.0 m (18 by 30 ft). The adjoining workhouse was 54 m (180 ft) high and 21.0 by 28.8 m (70 by 96 ft); and the bin structure was 30.6 m (102 ft) high and 23.1 by 58.5 m (77 by 195 ft). The 65 circular bins arranged in five rows were capable of storing 36,000 m^3 (1 million bu) of grain. A reinforced-concrete raft foundation 60 cm (2 ft) thick supported the bin structure, which weighed 20,000 tonnes when empty. Storage of grain started in September 1913. On October 18, when 31,500 m^3 (875,000 bu) were in the bins, a vertical settlement of 30 cm (1 ft) occurred within an hour after movement was first detected. The structure began to tilt to the west and within 24 hours was resting at an angle of 26°53′ to the vertical; its west side was 7.2 m (24 ft) below its original position, and the east side had risen 1.5 m (5 ft) (Fig. 7.9).

Fortunately, because of the monolithic nature of the structure and its sound construction, no serious damage was done to the bins, apart from their displacement. Accordingly, and through an outstanding underpinning operation, the bins were forced back into a vertical position, supported on concrete piers carried to rock, and the entire elevator was rehabilitated. It has been in steady use ever since. The site is underlain by Ordovician limestone upon which rest lodgement till, sand, and gravel. Then come 12 m (40 ft) of glacial lake clays in two distinct layers of equal thickness, clays deposited in glacial Lake Agassiz. Three meters (10 ft) of Recent alluvial deposits and outwash complete the subsurface profile.

How much was known of the subsurface conditions when the elevator was constructed is not clear, but recent investigations of the failure leave no doubt

FIGURE 7.9
An unusual photograph of the failure of the Transcona grain elevator, Winnipeg, Manitoba, showing the upheaval of the clay caused by the soil shear failure. (*Courtesy Foundation Co. of Canada Ltd.*)

that the failure was caused by overloading of the glacial lake clay. Ultimate-bearing capacity, determined on the basis of laboratory tests on samples, was found to be 31,300 kg/m^2 (6,420 psf), as compared with a calculated bearing pressure at failure of 30,300 kg/m^2 (6,200 psf), reasonably close agreement that confirms the value of studies in soil mechanics for foundation design when associated with accurate knowledge of subsurface geological conditions.[3]

Admittedly, the failure of the Transcona elevator was an extreme case. Closer to the normal occurrence was the slight tilting of a reinforced-concrete water tower at Skegness in England. The 33-m (100-ft) tower was founded on a reinforced-concrete mat; it tilted 60 cm (2 ft) out of plumb when first filled with water. Borings revealed about 9 m (30 ft) of water-bearing sand with irregular interbedded peat, all underlain by lodgement till which, if it had been investigated, could have been used as the foundation. Restoration of the tower employed a ring of steel-sheet piling, driven to varying depths, deep on the "low" side and shallow on the "high" side. Open, large-diameter boreholes were then drilled on the high side, to an elevation well below the steel piling. This side of the tower base was then loaded still further with sandbags, until the excess pressure started to force some of the subsoil from beneath the footing up into the boreholes. Settlement was controlled by removing varying quantities of the extruded soil from the open boreholes. In six weeks, the tower had righted itself; and the steel piling was all driven down to the glacial till, thus enclosing the remaining subsoil beneath the tower to give all the bearing capacity required.

All records of building subsidence are reminders that the earth's crust is not the solid, immovable mass so popularly imagined. When solid rock outcrops at the surface, it provides a foundation stratum of material that is as susceptible to stress and strain as any other solid matter. If unconsolidated material constitutes the surface layer, it provides a foundation bed even more liable to movement under load, and long-time settlement may be anticipated if some clays are loaded beyond a certain limit. Modern foundation design aims not at eliminating settlements (which is impossible) but at so controlling them that the structure supported will exhibit no undesirable effects.

A few other famous examples are worthy of mention. All of these historic buildings were erected long before foundation engineering had even been recognized as an important aspect of construction. The Leaning Tower of Pisa naturally comes to mind first (Fig. 7.10). Located in northwestern Italy, the tower was started in the year 1174 but not completed until 1350. It has continued to tilt since then; the present displacement is over 4.8 m (16 ft) in its

FIGURE 7.10
The leaning tower of Pisa, Pisa, Italy. (*Courtesy Ente Nazionale Italiano per il Turismo, Rome.*)

FIGURE 7.11
The Taj Mahal, Agra, India. (*Courtesy High Commissioner for India, Ottawa, Canada, and Lt. General Sir Harold Williams.*)

total height of 53.7 m (179 ft). Its foundation consists of a circular slab 19.2 m (64 ft) in diameter, with a central hole 4.5 m (15 ft) in diameter. Whether or not there are piles under the slab is still uncertain. Foundation strata consist of a bed of clayey sand 3.9 m (13 ft) thick, underlain by 6.3 m (21 ft) of sand resting on a bed of brackish clay which is of unknown thickness. Pre–World War II investigations and laboratory tests suggested that the tilting is attributable to the clay layer 8.4 m (28 ft) below the foundation slab.

Remedial work was begun by strengthening the foundation slab and ring wall masonry with cement grout under low pressure. As a final measure, the sand stratum surrounding the foundation was consolidated by the use of a chemical method employing a single gel medium injected at a low viscosity. This treatment may not have been wholly effective, since movement of the tower continues at the rate of 1 mm (0.04 in) per year. It is estimated that it will take 200 years at this rate of movement before the tower is in real danger. With a bearing pressure of some 87,500 kg/m^2 (8 tons/ft^2), it is a wonder that the tower has not settled more.[4]

The Taj Mahal in Agra, India, regarded as one of the world's most beautiful buildings, betrays cracks in the cellar immediately beneath the plinth terrace on the north side and in the eastern superstructure of the mausoleum. Examination of the foundations of the north wall some years ago revealed the condition shown in Figure 7.11, which may add further to appreciation of this masterpiece of Indian building and architecture.

Even so famous a North American symbol as the Washington Monument, in the heart of the capital city of the United States, has had settlement troubles. This great masonry shaft, 167 m (555 ft, 5 in) high above its foundation and weighing 81,120 tons, settled about 14.5 cm (5.75 in) in the fifty years after its completion to full height in 1880. Further settlement is improbable, provided the ground around the monument is not disturbed. Started in 1848 by public subscription, the monument had risen to a height of 45 m (150 ft) by 1854 when funds ran out. In 1876, before Congress appropriated funds for completion of the shaft by the Corps of Engineers, the shaft was 4.5 cm (1.75 in) out of plumb but was brought back into plumb by an ingenious system of loading in connection with the underpinning. Borings were sunk to a depth of only 5.4 m (18 ft)—still in the water-bearing sand and gravel stratum upon which the monument rests, an oversight and indication of the shortcomings in foundation engineering at that time. Deeper borings of 1931 revealed a complex pattern of subsurface conditions; a layer of soft blue clay overlies bedrock, which was reached about 27 m (90 ft) below the monument. Since the thickness of the clay stratum varies considerably beneath the footing, the initial differential settlement is not surprising.[5]

One of the most elaborate New World buildings to serve the arts has achieved fame also because of its settlement. The Palace of Fine Arts in Mexico City has already settled 3 m (10 ft), and its elevation is still falling in relation to streets outside the influence of the building. Begun in 1904 and completed 30 years later (see Fig. 7.12), its structural-steel frame is covered with Italian Carrara marble. The building measures roughly 80 by 117 m (267 by 390 ft); its total weight of 58,500 tons is carried on a concrete-mat foundation which weighs 46,000 tonnes. Subsoil conditions at the site of this building consist of about 50 m (165 ft) of volcanic ash-rich "clay," with some interbedded layers of sand and sandy clay. The high moisture content of the main stratum is generally responsible for the remarkable settlement. The building was designed by an architect who spurned engineering advice (a recorded saying of his was that "if the structure is pleasant to my eye it is structurally sound"); it has been stated that he did engage an engineering consultant to design the foundations, although no records appear to exist of the design assumptions.[6]

PREVENTION OF EXCESSIVE SETTLEMENT

Today, the fundamentals of good design have produced innumerable buildings all over the world which are performing quite satisfactorily with no excessive or differential settlements. Immediately across the Plaza de la Alameda in Mexico City from the Palace of Fine Arts is the modern Torre (Tower) Latino Americana, completed in 1951 and performing with complete success, exactly as was

FIGURE 7.12
The Palace of Fine Arts and the adjacent Torre (Tower) Latino Americana in Mexico City, the differing foundations of which are described in the text. (*Courtesy Departamento de Turismo, Mexico D.F.*)

anticipated in design. An extensive investigation included borings to a depth of 70 m (240 ft), undisturbed samples of soil being taken at every change in soil type. In a typical boring, over 30 distinctly different strata of soil types were sampled and tested. They included volcanic ash, pumice, sand, and even some gravel. The results of laboratory tests, along with study of the local groundwater situation, led to one of the most ingenious foundation designs yet applied to a large building.

This foundation achievement was based on detailed knowledge of every aspect of the underlying geology. The resulting foundation consists of a concrete slab supported on end-bearing piles driven to good bearing on a stratum of sand 33.5 m (110 ft) below ground surface. These piles extend through the uppermost and thickest stratum of volcanic clay, the cause of most of the famous settlements. An excavation was made for two basements and for the foundation slab structure, to a depth of 13 m (42 ft, 6 in), considerably below the groundwater surface. The site was therefore surrounded with a solid wall of wooden sheet piling, swelling in contact with water, which gave tight enclosure. Anticipated settlements, caused by the total weight of the building (only 24,000 tonnes as compared with the 58,500 tonnes of the Palace of Fine Arts) to the piles, from the piles to the sand-bearing stratum, and through it to the strata below—were carefully calculated. The building has been under close observation and settlements have been exactly as calculated over 30 years ago.[7]

Victoria, capital of British Columbia, is graced by the Empress Hotel which opened in 1908. When chosen, the site was a muddy bay (James Bay) completely flooded at high tide. The City of Victoria ceded the land and built a seawall across the bay to enclose the desired site, pumping from the harbor an average of 3.9 m (12.8 ft) of fill over the entire area to bring it up to the level required for building. Borings in 1904 and 1913 along with exploratory shafts revealed a blanket of silty clay (the Victoria Clay) over the entire area, to depths up to 30 m (100 ft).

Origin of the "hard yellow clay" is found in the geologic history of the site. During the 2,000 years following deposition, sea level dropped to 9 or 12 m (30 or 40 ft) below its present level, thus exposing the newly deposited clay. The clay became dessicated near its surface, leaving a characteristic weathered crust about 4.6 m (15 ft, 6 in) thick. After that, the sea level rose to its present elevation, and a new layer of sediment was deposited over the weathered surface of the Victoria Clay, illustrating how geology can explain a soil profile that would otherwise be puzzling. Typical boring logs show "hard yellow clay" over "blue clay on gravel and sand." The "blue clay" is the Victoria Clay and "hard yellow clay" is its weathered crust. Timber piles were driven to the underlying sand and gravel on the western side of the hotel, but piles over most of the site were driven only into the blue clay. Settlements were soon observed. Newly appointed foundation engineers excavated large quantities of soil from under the south end of the building. These voids were enclosed by concrete and settlement was arrested. Settlement observations were then started and, in 1970, assurance was given that future settlement would be very small indeed.[8]

BUILDINGS OVER COAL WORKINGS

Old coal workings are common hazards beneath some urban areas. Few cities are so fortunate as Edmonton, Alberta, for which an *Atlas of Coal Workings* was produced by a local geotechnical consultant, a publication based on an exhaustive study of all available early records of this relatively new city of the Canadian West. In older cities (especially in older countries), the fact that coal workings do exist in building areas will usually be known, so that geologic investigations can be made to take care of unexpected settlements. An ingenious example is that of the special foundations for four circular, re-inforced-concrete water tanks at Heanor, Derbyshire, England. Each of these 10.8-m (36-ft) -diameter tanks was built to replace a reservoir that had failed because of subsidence over colliery workings. The same site had to be used. The foundations were carried to a stratum of hard shale, and each tank is supported at three points (instead of the usual four or more) to eliminate indeterminate stresses at any stage of irregular settlement. The feet rest freely on the supporting foundations, and the tanks may be releveled if subsidence takes place.

To the east-northeast of Prague in the lovely land of central Bohemia are important lignite deposits worked by surface mining. The thirteenth-century town of Most sits right on top of 27.3 m (90 ft) of lignite that contains 90 million tonnes, only 1.5 to 15.2 m (5 to 50 ft) beneath the narrow streets. The town has been razed and replaced with modern apartment blocks, about 3.2 km (2 mi) away. A central point of the old town had been the sixteenth-century gothic Church of the Ascension. Its singularly beautiful interior has three naves; it is 59.3 m (195 ft) long, 26.4 m (86 ft) wide, and 31.3 m (102 ft) high. The Czechoslovakian government decided to save this beautiful church, minus its badly fractured bell tower. Carefully "caged" with a steel framework, the 12,000-tonne church was moved 870 m (2,850 ft) and downhill by 10.5 m (34 ft) to a newly prepared site (Fig. 7.13).

UNDERPINNING

Some of the most difficult of all foundation problems are encountered when column loads of especially valuable old buildings are transferred to new and reliable foundations. Wooden piling decay and progressive settlement can also often be corrected by underpinning. Successful underpinning is always dependent upon accurate knowledge of every detail of the subsurface.

FIGURE 7.13
The church at Most, Czechoslovakia, while being moved from its original location as part of the relocation of the town to permit further coal mining. (*Photo: J. Skopek, Prague, C.S.S.R.*)

All visitors to Stockholm will recognize the great Royal Palace located in what is known as the Old Town, immediately opposite the Grand Hotel on the inner harbor. The palace, built over the last two centuries, has its wings founded on the site of an older palace that burned in 1697, and is said to have 700 rooms. The two wings have suffered from serious settlement, as much as 0.6 m (2 ft) in the last half-century. Some remedial work was done in the 1920s but trouble continued, due possibly to the increased load of the concrete then placed, the heavier and increased traffic on roads adjacent to the palace, and slowly dropping groundwater levels. (This drop in groundwater levels is possibly associated in part with the known rise in the level of all of Scandinavia due to rebound following the last glacial period.)

The two palace wings rest partially on the remains of the former building that burned. A timber raft near the main building and timber piles were used as foundation units, the mat resting on 3 to 6 m (10 to 20 ft) of miscellaneous fill. Beneath the fill is a thin stratum of clay with some sand, then a bed of dense gravel from 9 to 23 m (30 to 75 ft) thick overlying bedrock. Using the concrete mat of the 1920s as a working platform, the specialist contractor for the underpinning operation sank 29 cylinders though all these strata to bedrock. These were cleaned out, filled with concrete, and used as the new foundations for the two wings. This difficult work was successfully carried out in the confined quarters of the old stone-arch basement, where a number of archaeological items were discovered and given to the Stockholm City Museum.[9]

PRECAUTIONS ON SLOPING GROUND

There is still another geological problem with building foundations: the possible instability of sloping ground. Sloping sites would not normally be chosen for building, but the mounting value of real estate in urban areas has increased the use of such sites, especially as they offer beautiful vistas. The west coast of North America immediately comes to mind, but other locations have had similar troubles. In the area of greater St. Louis, Missouri, where the local loess is a soil most susceptible to water, many houses built too near the tops of slopes have shown evidence of damage after heavy rains or when drainage was interfered with. In the suburban municipality of Bellefontaine Neighbors, for example, the 1957 slide of a steep slope into a borrow pit damaged 10 houses on St. Cyr Drive; three were complete losses.

In Los Angeles, however, this type of trouble has been most widespread and serious, In 1952, when heavy rains followed seven very dry years, extensive damage was caused to many hillside properties. Many lawsuits resulted, since damage on one site often affected adjacent properties. Before the following winter, the City passed a grading ordinance and placed some 60 percent of the city's area under special regulations. Rains in more recent years have caused further damage. Local studies have continued and public interest has been aroused. In 1957, for example, a Geological Hazards Committee of the City of Los Angeles was formed voluntarily by local geologists and engineers to assist civic authorities. Most dramatic and personally tragic are the losses of homes due to movements caused usually by the action of uncontrolled water. But such slope movement also destroys useful land and disrupts normal municipal services.

Along coastal roads in the Los Angeles region there is the added hazard of undercutting of weak rock cliffs by the sea, but even here it is the effect of uncontrolled surface water and groundwater that causes most of the trouble. The application of geological principles to the problems encountered in constructing buildings on sloping sites is probably the most perfect example in engineering work of the old tag that "a stitch in time saves nine." For the expenditure of a very small sum, an examination of any site for a proposed small building can be made by a competent professional adviser. If good advice so received is followed, damage that might run into tens of thousands of dollars can often be avoided. A few simple hand borings, examination of the local geology, study of groundwater conditions in the vicinity and of surface drainage arrangements, and above all, a close scrutiny of the records of local weather, can lead to expert advice on how a sloping site can be used or, alternatively, whether it should be avoided.

SMALL BUILDINGS

The examples cited in this chapter are almost all large structures, inevitably so since they create the heaviest loads to be imposed upon the ground. Any failure of these structures to perform as designed involves considerable financial loss. Small buildings, however, can be subject to all the problems that affect larger structures and so must at least be mentioned because of their relatively greater numbers. Reference again to Figure 7.4 will be a useful reminder that the loads from all buildings, large and small, stress the ground beneath them, hence, the supporting ground must be strong enough to sustain the increase of stress caused by the building loads. The loads from small buildings such as residences will usually be so small that almost all normal soils will sustain them.

Sometimes, however, small buildings have to be erected on soil that is clearly not strong enough for the normal type of foundation structure. Typical was a new two-storey firehouse in San Francisco. Its location, selected on the basis of fire-fighting logistics, was a site underlain by 6 m (19 ft) of fill material overlying bay mud extending about 30 m (100 ft) to bedrock. Any normal design would have imposed loads on this material which would have resulted in major settlement. It was therefore decided to float this small structure on a 1.2-m (4-ft) -deep, ribbed, reinforced-concrete foundation mat. Lightweight concrete was used for the superstructure and other design features were included to keep the dead load to the minimum possible. The dead load amounted to 1.22 million kg (2.7 million lb), but this was almost exactly the same as the weight of the fill removed. Only the smaller live load (122,500 kg or 270,000 lb) will therefore increase the stress in the underlying soil. Even with this care settlement is estimated at no more than 30 cm (12 in) in the next 50 years.[10]

This example shows how foundation schemes generally used for large structures can, when necessary, be adapted for much smaller buildings. Groundwater problems will be the same but can usually be guarded against by having open-tile drains surrounding the lowest part of the foundation structure (house basements, for example). Building on floodplains is to be avoided at all costs. Any chance of slides from adjoining sloping ground must be thoroughly investigated. The possibility of an innocent-looking site being actually located over filled ground, and possibly over an earth-covered site used earlier as a refuse dump, must never be forgotten. The investigation of the building site for smaller structures is just as necessary as that required for major structures, although on a much more modest scale.

Fortunately, there are many available guides to site selection for the use of the small builder and prospective homeowner. An excellent one was prepared as a service to the community by the members of the Pittsburgh Geological Society. Entitled *"Lots" of Danger,* it is a simple quarto-sized document written in nontechnical terms and amusingly illustrated to be easy reading for everyone. Its subtitle is "Property Buyers' Guide to Land Hazards in Southwestern Pennsylvania"; its contents are just that. The title and the reference to hazards were deliberately selected to bring home to the average person that even small buildings may be subject to unsuspected dangers. In a similar way, a careful study has been made by geologists in Texas of real estate brokers' and builders' knowledge of geology and other physical site determinants. It was found that, although the "brokers and builders are aware of and interested in the geologist's contribution...they indicated that the contribution must be both informative and intelligible." This is useful advice for all who have to deal with the problems of foundations for small buildings.

Powerhouse Foundations

GENERAL CONSIDERATIONS

The current energy situation is a crisis that will not pass. The peak in world crude-oil production may occur around the year 2000, while the peak of world coal production is estimated to occur around the year 2150, not a long period ahead in terms of human history. Against this general background, the requirements of the immediate future must be met by the best possible use of fossil fuels, the maximum use of water power, and the development of tidal and solar power.

So vital is this assured supply of centrally generated power that safety in the design of power stations and efficiency in their construction are paramount. Even though the problems associated with their design and construction are similar to those encountered in other civil engineering work, there have been some unusual problems that have required proper attention to geology in all aspects of site investigation.

Severe environmental limitations are now imposed upon site selection for power stations, sometimes to the actual elimination of sites that may be suitable from purely engineering and economic points of view. All thermal stations, no matter what fuel is used, must be close to a large body of water for cooling purposes. Geothermal stations must be located where ground temperatures have been found to be suitable. Tidal power stations must be built in the sea at locations where the tidal range makes such development possible, and water power stations must naturally be sited in accordance with the topography that provides sufficient fall. Add to these restrictions the problems of design presented by the unusually heavy and dynamic loads to be found in power stations and one can see an engineering challenge of unusual complexity.

The necessary location of thermal stations adjacent to bodies of water such as rivers and lakes usually involves the use of riverside or lakeside sites, locations which almost automatically present geological conditions with some undesirable features. Only rarely will any source other than lakes or rivers be available, apart from occasional use of seawater for cooling purposes for stations located on seacoasts. A reverse problem of thermal stations is the disposal of fly ash, when the fuels being used result in large accumulations of this waste material. Efforts are being made to find uses for the large quantities of this material produced by some power stations, but in an increasing number of cases, some means of disposal in bulk must be found. British engineers have suggested the use of "littoral drift in reverse," the utilization of the constant movement of seabed material along most coasts propelled by the action of alongshore currents. The idea was conceived that, if fly ash were deposited in the sea, it would be moved naturally away from the point of deposition, all finer materials eventually being deposited in ocean deeps.

GEOTHERMAL POWER

Both the United States and Canada have completed national surveys of all locations at which geothermal energy may be tapped. Hot springs have been used at least since the time of the Romans and probably long before that, the hot water naturally showing the high ground temperature of bedrock at the location of all such springs. The use of this thermal energy for commercial purposes appears to have started at Larderello in Italy in 1777, when borax was recovered from natural steam which comes out of the ground at this location, its condensation leading to the descriptive name *fumaroles suffaroli*. It was at this same location in Italy that power was first commercially developed from natural steam in 1905. In the years following World War II more than half of all geothermal power (389,000 kW) came from the 14 Larderello installations situated not far from Pisa.

New Zealand was then second only to Italy in the use of this natural source of power, with 192,000 kW produced in its Wairakei plant. Development of this plant followed an intensive study of geothermal power in general and in relation to the active volcanic region of New Zealand, in which heat loss from the ground surface is from 20 to 100 times greater than the normal rate. Smaller plants are now operating in Iceland, Japan, Mexico, and the Soviet Union, with studies anticipating such plants being developed in a number of other countries such as France, Turkey, and Chile.

In the United States, The Geysers, 45 km (90 mi) north of San Francisco, emit rare dry steam. In 1922, a 20-kW generator was driven by the use of natural steam. In 1955 the Magma Power Company leased extensive areas, and by 1957 six wells had been successfully sunk to depths of several hundred feet. Pacific Gas and Electric Company now has ten installations at The Geysers, generating 421,000 kW, with further expansion in view, so that it is now the world's leader in the use of geothermal power. The natural steam is run through centrifugal separators at the well head, to eliminate mineral particles, and then piped directly to steam turbines which drive electrical generators in exactly the same way as in a normal thermal station. Geological problems are subsidiary to legal difficulties involving environmental disputes.

Hot springs are much more common throughout the world than flows of steam and are usually developed as health spas which utilize such warm or hot water for medicinal purposes. Flows are usually limited in quantity, but there are a number of locations where naturally hot water is utilized for commercial purposes. Quite the most extensive use is in Iceland, where nearly one-half of the total population (just under 200,000) live in houses warmed by natural heat. At the capital of Reykjavík such heat comes from groundwater at 187 to 266°F.

TIDAL POWER PLANTS

Of all the sources of power in nature, the tides probably appeal to the imagination more than any other. In locations such as the Bay of Fundy in eastern Canada, where tides range up to 13.5 m (45 ft), it is tantalizing to see huge volumes of water being moved twice a day in each direction. Several natural forces control complex times of variations in tidal range, rendering essential the coupling of any tidal power project either with major pumped storage or, more desirably, with a power network which will enable other stations (such as thermal plants) to supply the extra power to meet power demand.

Design involves no more than a development of concepts already successfully used in water power stations. Problems of construction are more serious in view of the unavoidable location of tidal power plants, i.e., in the sea, with water levels changing continually every day. Unusually careful site investigations must be carried out in the open sea, with constantly varying water levels. France and the Soviet Union both have constructed modern tidal power developments. The French plant is located at the mouth of the Rance River on the northwest coast of France, close to the mediaeval city of St. Malo. Electricité de France, the governmental power agency, operates 24 10,000 kW units. The powerhouse is flanked at one end by a rock-fill embankment and by a concrete spillway structure with a ship lock at its other end, the total length of the combined structures being about 720 m (2,400 ft). Site investigation showed that competent gneiss was continuous across the site at depths up to 12 m (40 ft) below low tide level. Fortunately, the bedrock surface was generally exposed, with only occasional deposits of sand and gravel which had to be cleaned using a compressed-air caisson. Concrete slabs were then cast within the caissons as bases for the two necessary cofferdams which were floated into position and sunk to the bedrock floor. It was estimated that the cost of the cofferdams represented about one-third of the total cost of the plant.

Profiting by the French experience, Soviet engineers designed their first, although much smaller, tidal power plant as a prefabricated concrete structure floated into place, thus eliminating the need for cofferdams. The plant is located on the Arctic coast at Kislogubsk, near the mouth of the Ura River not far from Murmansk. High tides are experienced all along this coast, so that further major installations are planned. This first plant, equipped with one 400-kW unit, was placed in commission in 1970 (Fig. 7.14). Site conditions were favorable. Bedrock was exposed, requiring some underwater blasting to give a level bed, on which loose rock was removed by clamshell bucket. Sand and gravel was placed as the necessary bearing surface. The two installations, so different in size and in execution, illustrate well the significant impact of site

FIGURE 7.14
The Kislogubsk tidal power plant of the U.S.S.R., on the northern coast of the Soviet Union, west of Murmansk, facing the Barents Sea; the complete power station was constructed in a temporary dock near Murmansk and then towed to its final location. (*Courtesy Press Counsellor U.S.S.R. Embassy, Ottawa; Novotsi Photo.*)

conditions for tidal power plants. For such plants of the future, site investigations are bound to be challenging indeed.

WATER POWER PLANTS

No two water power plants are ever the same. Each plant exhibits some design feature of special interest. The powerhouse itself, unlike the case with thermal and nuclear plants, may be only a small part of the entire project. At least one dam will almost always be necessary along with tunnels and pipelines to supply high-head plants, the location and installation of which sometimes encounter geological problems which may be critical until construction is complete. And when water power plants are in operation, the change in hydrogeological conditions which they cause may sometimes create postconstruction problems. For this short chapter, the authors have selected half-a-dozen cases known to them, each of which illustrates at least one major problem that may be encountered in water power development.

The Arapuni Development

The Arapuni water power scheme is located on the North Island of New Zealand, about 190 km (120 mi) south of the city of Auckland which it supplies with power (Fig. 7.15). It was constructed between 1925 and 1932 by the national Public Works Department. It is essentially a curved gravity-type main dam 58 m (192 ft) high, with a crest length of 91.5 m (305 ft) on a 75-m (250-ft) radius, which diverts the water of the river Waikato into an open headrace canal 1.2 km (¾ mi) long, finishing with a spillway dam and penstock intake. Steel penstocks built into rock tunnels lead to the power house, the capacity of which is 150,000 kW operating under a head of 52.5 m (175 ft).

Waikato Valley geology is complicated; the valley has been the scene of repeated volcanic activity, and as a result the course of the river and its gradient have varied widely and often. Old river courses have been filled with volcanic debris and several old river channels occur at varying heights above the existing riverbed. One of these was utilized as the main part of the headrace canal.

Host rocks are chiefly a welded tuff and flow breccia. The tuff is believed to have been ejected as incandescent dust from vents in the ground and then to have flowed and fused together as it cooled. The tuff collected at the bottom of river valleys, where it was water-cooled quickly and is highly fractured and unconsolidated. The solid tuff may contain up to 30 percent water in ultramicroscopic pore spaces, as the rock is durable and moderately hard. Underlying the tuff are alternating beds of tuff and breccia, including a pumaceous breccia on which the main dam is founded. Below this last stratum, softer tuffaceous material extends to great depths. These volcanic deposits are uneven; they also proved to be so elastic that "the absence or presence of 10 feet of water in the

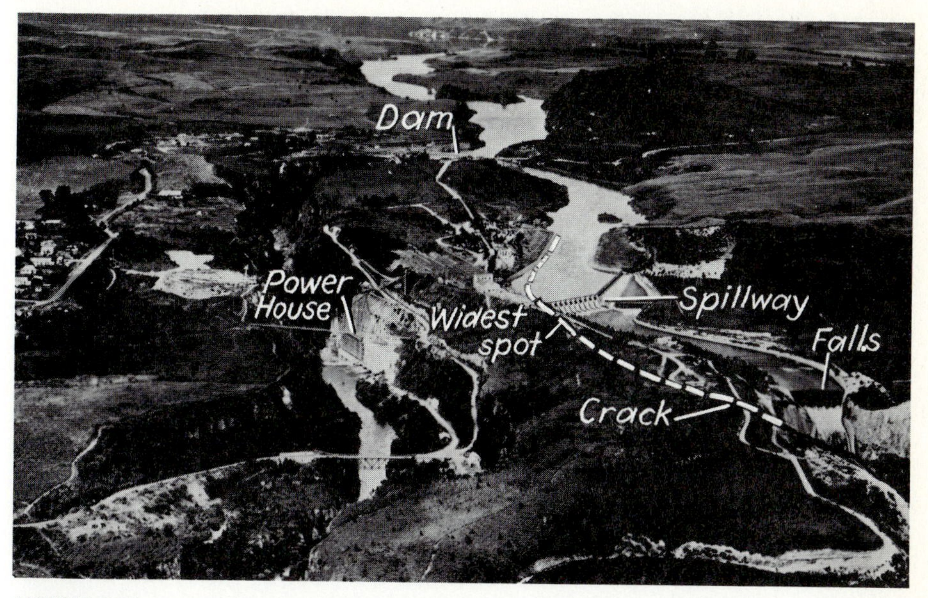

FIGURE 7.15
General aerial view of the Arapuni water power project, North Island, New Zealand, showing location of the crack which is described in the text. (*Courtesy F. W. Furkert, D. P. W. New Zealand.*)

gorge as the diversion tunnel was opened or shut, caused decided and opposite tilts to be registered on the seismograph in the powerhouse."

On June 7, 1930, while the dam was retaining water and the powerhouse was under load, a crack occurred in the local country rock, roughly parallel to the flow of the river (Fig. 7.15). It was widest (5 cm, or 2 in, across) where the spillway joined the penstock intakes, and it extended for about 600 m (2,000 ft). Observations then made showed that "the whole mass of country, about 600 m (2,000 ft) long, 45 m (150 ft) thick, and 120 to 240 m (400 to 800 ft) wide, was bent over towards the gorge." The power station was shut down, and the lake drained; "as the lake fell, the ground recovered its position, the cracks closing except where jammed by drawn-in debris, and the powerhouse, suspension bridge, etc, regaining their original positions."

This most unusual occurrence was naturally investigated and studied closely by geological and engineering experts. The consensus was that the movement was due to leakage of water from the headrace canal, which affected the adjacent volcanic rock to such an extent that movement took place. Remedial measures centered on waterproofing the headrace lining, resulting in satisfactory functioning today.

While the remedial works were in progress, many unusual features developed. Thus, as the water level behind the dam was being lowered, gas consisting of 96 percent nitrogen escaped from the rock in the headrace. The gas was unusual in that it contained no oxygen, and its volume may have reached the surprising figure of 1,400 m^3 (50,000 ft^3) instead of the 700 m^3 (25,000 ft^3) mentioned in the published description. Remedial grouting experienced considerable lost water circulation, including the entire volume of a fire hose turned into one hole.

The Bonneville Development

Quite different were the problems faced in the 1934 development and expansion of the Bonneville power project on the Columbia River (Fig. 7.16). This 1,076-MW (megawatts) dam and powerhouse is interesting, as it lies athwart the Cascade Slide, a great landslide which occurred about 700 years ago, after the river had cut a gorge estimated to have been at least 60 m (200 ft) deeper than the modern channel. The slide deflected the river and altered the topography all around the site. Preliminary investigations, therefore, had to be carried out with unusual care. The unusual part of this work was to obtain a stable location for a displaced railroad on the Oregon shore. The slide area is about 2.4 km (1 mi) long and 900 m (3,000 ft) wide; its slow movement is

FIGURE 7.16
Bonneville Dam and powerhouses, as now completed looking east; the new powerhouse is on the left. (*Courtesy U.S. Army Corps of Engineers, Portland District Engineer.*)

attributed to the "lubrication" by groundwater of an underlying stratum of shale. Solution of the problem included extensive drainage works and a major retaining wall.

The accompanying aerial view (Fig. 7.16) shows full utilization of the river. The only possible location for the still-larger second powerhouse was on the northern shore, where the ancient landslide had occurred. Extensive subsurface investigation revealed boulders of 30-m (100-ft) diameter, in a matrix of sand. Below this heterogeneous slide material lies the original alluvium, which has been the cause of seepage around the end of the spillway dam. Beneath this is slaking bedrock which loses strength when exposed to air. It was decided to surround the entire 16-ha (40-acre) area for the new 558-MW powerhouse with a slurry-trench cutoff wall. Four and a half million m^3 (6 million yd^3) of excavation was then removed down to bedrock, in some places 30 m (100 ft) below mean water level in the river above the existing powerhouse.

The railway had to be relocated through a single-track, concrete-lined tunnel constructed through the old slide material, involving some of the most difficult tunneling anywhere in recent years. A pilot bore was driven through sections that were known from preliminary investigations to be in unusually difficult material. The 420-m (1,400 ft) tunnel main cross section (7.5 by 10.5 m, or 25 by 35 ft, and horseshoe-shaped) was then driven, and with very little blasting, was sprayed with a 7.5-cm (3-in) layer of shotcrete. Steel ribs, spaced on 75-cm (2.5-ft) centers, had to be used in much of the tunnel. Every phase of the second part of the Bonneville development faced geological problems created by the old slide material. The full record of this remarkable construction operation will be an engineering classic.

The Cheakamus Development

A very different ancient landslide complicated construction of the 140-MW Cheakamus water power development in British Columbia. Water reaches the turbines through steel pipes under a head of 337.5 m (1,125 ft), flowing to the pipes through a 10.4-km- (6.5-mi)-long tunnel. The unusual aspect was a requirement that the lake level be raised by 19.5 m (65 ft), through construction of a 26.4-m (88-ft) dam at the only possible location for such a dam, which consisted clearly of the rock debris of an ancient slide. Geological studies defined a century-old rock slide, probably involving 15.2 million m^3 (20 million yd^3) of rock debris. Dr. Karl Terzaghi studied the site in his own inimitable way, and decided that the dam could be built. His account of the problems became the last publication of his lifetime. Dr. Terzaghi visited the site of this 450-m- (1,500-ft)-long and 26.4-m (88-ft)-high dam no less than 18 times during the course of construction, adjusting the design as excavation revealed newfeatures of the complex geology of the heterogeneous rockslide material. The dam stands today, performing quite satisfactorily, as one of the supreme examples of adapting most unfavorable geological conditions to the service of humanity.

The Kelsey Development

On the banks of the Nelson River off Hudson Bay stands the 155-MW Kelsey power development. Built to supply power to the nickel plant at the new town of Thompson, Manitoba, the powerhouse operates under a head of from 15.0 to 15.5 m (50 to 55 ft). Local bedrock is a medium-to-dark-gray paragneiss, thought to have been formed from granitized sedimentary rocks. Local soils are ablation till so there was ample, good aggregate available for the main concrete powerhouse and spillway structure, the associated rock-fill dam, and the dozen dikes needed to complete the project. There was, however, one complication. Two of the dikes (Nos. 2, east and west) were confirmed to lie on permafrost which would thaw under the latent heat available from the reservoir water. This would then result in subsidence of the ground surface, and so of the dike structures. These dikes were therefore designed to settle as the ground warmed, over a matter of years. These compacted sand-fill embankments, with sand drains installed in the ground beneath each dike in advance of dike building, have performed exactly as anticipated. Maximum settlement after seven years of operation has been 2 m (7 ft), all of which was made up by the regular placement of new fill in accordance with the original design.

The Foyers Development

In Scotland, on almost the same latitude as that of the Kelsey station but with greatly more temperate climatic conditions, stands the 300-MW Foyers pumped storage project of the North of Scotland Hydro-Electric Board. The complexity of the bedrock geology in the Scottish Highlands is well known. Unusual care was therefore taken with the site investigations for this project. Fourteen boreholes were put down along the 3-km (1.8-mi) line of the power tunnel and seismic surveys assisted in interpretations. Assumptions made for purposes of design were confirmed in the field, and a representative of the general contractor had this to say:

> From the Contractor's viewpoint the appraisal of ground conditions was unusual in that greater reliance was placed on the trial shaft by Loch Ness and on the geological survey than on conventional site investigation tests and boreholes. More borehole information in the vicinity of both upper and lower control works would have been valuable, particularly in relation to the nature and depth of the overburden, but it is unlikely that additional boreholes elsewhere would have justified the extra cost. Rock conditions within the zone of the Great Glen fault adjacent to Lock Ness were too variable to permit detailed prediction, but elsewhere the geological survey provided an accurate forecast for construction purposes.[12]

This helpful and constructive comment underscores the fact that civil engineering works have not only to be designed but must be also constructed. The general contractor did encounter real difficulty in constructing his temporary works, as steel piling was selected for forming a circular cofferdam 29 m (95 ft) in diameter. It was impossible to drive the sheet piles to depths greater than 7 m (23 ft) because of the toughness of the lodgement till and the presence in it of cobbles and boulders. The steel piles were again driven in an artificially formed gravel ring. It is to be noted that it was, again, the properties of the lodgement till which caused the difficulty. Again, it cannot be emphasized too strongly that excavation for pile driving in glacial lodgement till must rely on preliminary subsurface investigations to determine the actual properties of the till.

The Kootenay Development

Exactly the reverse type of problem had to be faced in the construction of the powerhouse of the Kootenay Canal project in British Columbia. Following the downstream construction of Libby Dam in the United States, an increased flow was available for power generation. The best site for the powerhouse involved 1.9 million m^3 (2.5 million yd^3) of hillside excavation in water-bearing glacial silt. The danger of liquefaction due to blasting and the passage of heavy equipment had to be faced.

It was eventually decided to utilize electroosmosis to drain-stabilize the soil while excavation proceeded (Fig. 7.17). The work was carried out in six stages as excavation worked up into the hillside. Thirty-cm- (12-in)-diameter, open-ended steel pipes were driven at the head of each slope, on 6-m (20-ft) centers, down to bedrock, which in places was over 30 m (100 ft) deep. Pipes were cleaned out by water under high pressure and then backfilled with sand. In each such sand column a 32-mm (1.25-in) well point was inserted, encased in a 50-mm (2-in) pipe. Another parallel set of 50-mm (2-in) pipes was driven to the same depth, 3 m (10 ft) away, to act as the anodes. Direct current at 150 volts was then applied, the well points were activated, and groundwater flow increased to 150 lpm (40 gpm). The work was completed without difficulty.

THERMAL PLANTS

Thermal plants, also known as oil- or coal-fired utility stations, have grown in size during the worldwide switch from fuel-oil dependency for power generation. From the common pre-1970s size of 50 to 250 MW, today's giants are energy centers made up of one or more units of 500- to 750-MW capacity. Many of the stations are "mine-mouth" operations, located within an easy truck haul from a surface coal mine. Foundation loads at the stations are generally heavy and are often placed at variable elevations over plant areas of several hectares. Among the various allied foundation considerations are placement of coal handling facilities capable of receiving and storing unit train-loads of 100 rail cars per week, and the transfer and disposal of fly ash and fluegas desulfurization wastes in the range of 5 to 25 percent of that volume as the coal is consumed.

FIGURE 7.17
Electrodes for the application of electroosmosis for slope stabilization on the Kootenay power canal of the British Columbia Hydro and Power Authority. (*Courtesy B.C. Hydro and Power Authority, Vancouver.*)

Considering their huge demands on cooling water flow, choice plant sites are generally picked of necessity not on the basis of superior foundation geology, but on the basis of water supply, along with coal supply and power transmission needs. Many of these sites are underlain by complex and relatively weak and unconsolidated floodplain or coastal soils. These factors, combined with frequently high groundwater conditions, present a variety of geological challenges in the selection and design of appropriate foundations.

Problems with Uplift Pressure

Problems in construction of steam power plants are often related to nearby cooling-water sources such as rivers or lakes. The TVA Shawnee steam plant on the Ohio River downstream from Paducah, Kentucky, is founded on sand and gravel overlain by an impervious blanket of loess. Construction dewatering was similar to the permanent drainage control necessary to keep uplift pressures under rigid control. TVA designers chose a system of 84 relief wells of 20-cm (8-in) diameter, connected to a common collection header and 9,500-lpm (2,500-gpm) pumps. An alternative method of controlling uplift pressures is featured at the Johnsonville TVA station. The station is founded in water-bearing chert, with the possibility, therefore, of a 15-m (50-ft) head of water acting on the lowest foundations. This could have been balanced by the addition to the foundation structure of 50,000 tonnes of extra concrete as deadweight. As an economical alternative, the designers arranged to tie down the foundation by means of reinforced-concrete "cells" drilled into the chert bedrock. Two hundred and twenty holes, generally 75 cm (30 in) in diameter and penetrating 5.7 m (19 ft) into the rock, were drilled and anchored.

Problems with Weak Foundation Strata

Placement of the 700-MW South Bay power plant of the San Diego Gas and Electric Company of Chula Vista, California, was on a low, flat, marshy area on San Diego Bay, previously occupied by saltwater evaporation ponds. Site investigation showed soft surface soils extending to 6 m (20 ft) below the surface. Older and firmer soils then followed, but bedrock was not reached by a boring terminating at a depth of 63 m (210 ft). Friction piles were considered after preliminary studies had shown that a concrete mat foundation would be subject to too great a settlement and that end-bearing piles or caissons were not possible. An innovative solution adopted—removal of the soft surface soil down to an elevation 3 m (10 ft) below mean sea level and its replacement with carefully compacted soil suitable for supporting the powerhouse loads. Select, compacted fill was a beach-deposited, wind-blown dune sand obtained from the Coronado side of San Diego Harbor.

Continuous and accurate field control of all backfill operations was naturally essential for the success of this unusual foundation, a tribute to the contribution that soil mechanics can make to a geologically difficult situation. There are at least three other power plants in California alone successfully founded on compacted fill over weak foundation strata. Preloading is common practice with smaller structures such as bridge piers, as will be seen in Chapter 8, but the same principles can also be applied to the large areas required for power stations.

This was well shown in the design and construction of the 250-MW steam power plant of the Cajun Electric Power Cooperative, Inc., located 35 km (22 mi) upstream of Baton Rouge, Louisiana, in the west alluvial floodplain of the Mississippi River. This site is underlain by 1.8 to 2.4 m (6 to 8 ft) of medium clay, overlying 22.5 to 30.0 m (75 to 100 ft) of soft clay and loose silt, with occasional pockets of organic clay, geologically typical for floodplains of major rivers. Dense sand suitable for end-bearing piles exists only at or below 30 m (100 ft) beneath the surface. Predicted settlements, if no treatment were given to the site, were so high as to be unacceptable. Friction piles therefore presented what might be called the conventional solution, but preloading of the upper soil strata in order to reduce significantly the expected settlement was a possibility. The contractor was allowed to use as surcharge the topsoil removed from the entire site, giving a surcharge load of twice the uniform design load.

This preloading would give twice the settlement to be expected from the actual loads, thus reducing the period during which the surcharge had to remain in place. Surcharge was left in place for a period of nine months, during which a maximum settlement of about 30 cm (12 in) took place.

Problems with Subsidence

Power plants nearly always present unique foundation design considerations. Plants located in coastal areas of soft, compressible, unconsolidated deposits may further encounter potential subsidence. The Sam Bertron steam power plant of the Houston Lighting and Power Company is one of five plants located on the Houston Ship Channel. The Houston area has been subsiding for some years. When built, the plant was at an elevation of 6 m (20 ft) above mean sea level. By 1972, the plant had subsided to 3.8 m (12.8 ft) above mean sea level. The plant is located about 6.4 km (4 mi) inland from the Baytown Tunnel where, in 1961, a hurricane tidal surge of 5.0 m (16.8 ft) was recorded. The combined threat of local ground settlement and future tidal surges demonstrated that remedial measures were essential. Soil conditions at the site consisted of a bottom channel deposit of Recent soft, organic, gray-clay muck, with shear strength as low as 488 to 976 kg/m^2 (100 to 200 psf), overlying reasonably strong Pleistocene clay (shear strengths of about 4,882 kg/m^2 or 1,000 psf), the contact being generally at about 6 m (20 ft) below sea level. A protective levee would offer the most reasonable protection, but further borings revealed that the channel had been overdredged, replacing the Pleistocene clay with compressible muck to depths of up to 7.5 m (25 ft). An alternative design was prepared, consisting of a permanent cofferdam of sheet pile cells driven into the channel bed (Fig. 7.18).

NUCLEAR PLANTS

Nuclear power plants are essentially thermal generating stations, but with special features because of the use of enriched uranium as fuel and the unusual precautions that have to be taken to ensure the safety of all operations. Although in the late 1980s there has been a downturn in the number of such plants being built, and although the future possibilities for wider use of nuclear plants are still uncertain, some plants will still be constructed. Stringent regulations covering both the design and construction of nuclear plants already in force will certainly remain and may be strengthened. Geological conditions at the sites of all nuclear plants are more fundamentally important than for any other type of civil engineering works.

One of the earliest major nuclear plants was the 580-MW Sizewell plant of the Central Electricity Generating Board of Great Britain, commissioned in 1965, and then the largest such plant in the world. It is located just over 160 km (100 mi) northeast of London on the Norfolk coast. One reason for the selection

FIGURE 7.18
The Sam Bertron steam power plant of the Houston Lighting and Power Company, showing the special cofferdam, the purpose of which is described in the text. (*Courtesy Bernard Johnson Inc., Houston, Texas.*)

of its location was its simple site geology. Exploratory borings encountered only well-graded, compact sand to depths of at least 48 m (160 ft). So satisfactory was the sand found to be that the reactors are founded on a 2.4-m- (8-ft)-thick concrete-matte foundation. Geology also directly affected construction of the two 3.3-m- (11-ft)-diameter tunnels for the cooling water intakes. These subsea tunnels were driven under compressed air, in sand, to intake caissons 540 m (1,800 ft) offshore. Similar in concept are the two 5.8-m- (19.0 ft, I.D.), 5.03- and 5.23-km (16,500 and 17,155 ft)-long cooling water tunnels of the Seabrook Nuclear Station, New Hampshire, driven in quartz diorite and high-grade schist.

The first nuclear generating plant of the TVA was that at Brown's Ferry in karst terrane on the north shore of Wheeler Lake, one of the TVA-controlled reservoirs. The geology of the site was intensively studied, first in a regional sense and then in detail, with 34 borings in the power block area and 46 elsewhere on the site. In only 11 of the 80 holes was any evidence of dissolution cavities found, and then only a total of 5.1 m (17 ft); 2.7 m (9 ft) were found in the 34 holes beneath the main areas to be loaded. It was concluded that the local bedrock would serve well as the foundation bed for the plant, even with the high loads that would be involved.

Another example, with differing conditions, is the Salem nuclear generating station of the Public Service Electric and Gas Company of Newark, New Jersey. This Delaware River location is on Artificial Island, about 63 km (39 mi) south-southwest of Philadelphia. The island was formed of dredged material deposited here around a natural sandbar in the river at the end of the lastcentury by the U.S. Army Corps of Engineers. The site for the two-unit, 1095-MW station was explored by 35 borings 10 cm (4 in) in diameter, all carried to a depth of 60 m (200 ft) below grade, which was about 2.7 m (9 ft) above mean sea level. Subsurface conditions were found to be reasonably uniform under the site.

Hydraulic fill and alluvium were generally found immediately below the surface to a depth of 7.5 to 9.0 m (25 to 30 ft), followed by 1.5 to 3 m (5 to 10 ft) of coarse sands and gravel, below which the local Kirkwood Formation extended to about 21 m (70 ft) below grade. Kirkwood strata consisting of gray silts and clays with some basal sands were deemed to be unsuitable for supporting the loads from the plant. From the 21-m (70-ft) level to about 42 to 45 m (140 to 150 ft) below grade is the Vincentown Formation, a silty sand stratum that is in places well cemented. It is dense and was considered suitable for the powerhouse loads. Beneath it are dense sand formations down to crystalline bedrock, at 540 m (1,800 ft) below the surface.

Seismic design considerations indicated that possible shear forces would overcome the bearing capacity of piles socketed in the Vincentown Formation. It was therefore decided to excavate down to the top of this formation for the area required for Class I (most sensitive) loads. As suitable fill was not to found in the vicinity, lean concrete was chosen as structural fill to the base level of foundation structures. Excavation of this 1.2-ha (3-acre) area, 21 m (70 ft) below grade, was accomplished by circular steel sheet-pile cell units with interconnecting arcs, and dewatering by 300 wells, each 27 m (90 ft) deep.

Nuclear power plants in steadily increasing numbers have been built in many oil-dependent countries since the early pioneer plants of the late 1950s. There has been a corresponding growth in the number and rigor of the regulations which must now be satisfied before a license to build and operate a new plant can be issued. Little is to be gained, therefore, by any further descriptions of geologically interesting installations. Attention will rather be directed to regulations current in the United States, one of the world leaders in nuclear plant technology. Even this treatment is difficult since, under mounting public pressure, regulations are repeatedly revised.

Two generalities are common to regulations for the siting of nuclear power plants. First, careful steps that are enumerated for site selection for nuclear plants are nothing new to civil engineering practice. The need for extra certainty at nuclear plants, especially with regard to seismic risks and the presence of active faults at building sites can naturally be understood, although even these matters are regularly considered for many "critical" civil engineering projects, such as major dams. Second, the necessity for preparing regulations in codified form necessarily creates a subdivision for the disciplines involved in site selection, whereas in practice, as all civil engineers and engineering geologists know well, the approach must be an interdisciplinary one with the closest cooperation between all those involved. In current U.S. regulations, for example, there are separate sections on geology, hydrology (much of it actually hydrogeology), seismology, meteorology, and even geography. While regulations must naturally be followed meticulously, watch must be kept against any rigid subdivision in the application of the several disciplines involved in site studies to ensure the absolutely essential coordinated approach to final site selection that must be made.

A construction permit must be issued by the Nuclear Regulatory Commission (NRC) before site work on a nuclear power plant can commence. Issuance is based upon consideration by the NRC of the applicant's Preliminary Safety Analysis Report (PSAR). Full description of the site proposed must be given in this report, including meteorological, hydrological, geological, and seismological characteristics. An Environmental Report (ER) must be submitted with the PSAR. Both submissions are independently reviewed for the NRC and the independently chartered (by the Congress) Advisory Committee on Reactor Safeguards (ACRS).

The major geological considerations that have to be fully reviewed in an application include the possible presence of active faults and an assessment of the probability of earthquake-induced displacement taking place on them, and the possibility of the occurrence at the site of such hazards as landslides, land subsidence, uplift, harmful groundwater movements, and volcanic effects. Construction methods and materials planned for use in construction must be reviewed. The static and dynamic stability of the structures proposed, with special emphasis on seismic probability, must be considered. Detailed seismic investigation, geological and geophysical, are mandatory. Possible flooding of the site is a matter that must be investigated under the heading of hydrology, whereas "terrain analysis" comes under the heading of geography and includes such geomorphic features as terrain ruggedness.

CONCLUSION

Readers may notice, and be surprised, by the fact that this long chapter deals with such an apparently simple subject as the foundations for buildings. Even at that, the treatment of this vast subject has been restricted to the geological factors affecting the founding of that large group of structures generally designated as "buildings." Unfortunately, the very familiar nature of most building foundation projects often leads to a lack of attention to subsurface conditions. As a matter of record, in 1977 the U.S. General Services Administration (GSA) admitted that in the previous ten years they had spent more than $10 million to repair faulty foundations beneath public buildings for the construction of which they had been responsible. One of two studies of this situation carried out for the General Accounting Office (GAO) showed that of 28 buildings studied, difficulties with foundations and "unanticipated soil conditions" had been experienced in 15, or more than 50 percent. It was said that inadequate geotechnical information was an industrywide problem and one not peculiar to GSA.

Nevertheless, excellent work in the building foundation field is now being done. There is still progress to be made, however, before failures due toinadequate subsurface information for building foundations are eliminated from the construction scene. There is no excuse whatsoever for the neglect of full investigation of the subsurface conditions beneath the building site and for some distance around, with special emphasis on groundwater conditions. One more famous example of what trouble can result from inadequate site investigation is that of the foundation of one of the great buildings that now dominate the famous Raffles Quay in the city of Singapore.

The Asia Insurance Building is a steel-framed 18-storey structure, founded on a site reclaimed from the beach during the last century. The beach formation beneath the site varies from 3 to 4 m (10 to 14 ft) in thickness; it was thought to be underlain by fill and weathered shale and sandstone. Four borings were sunk on the site in 1949 and couldn't go further than from 9 to 13 m (30 to 42 ft) below the surface. A strong chisel was used to obtain chips from the sandstone encountered at the bottom of the holes, since there was no diamond drilling equipment then available in Singapore for coring rock. A caisson foundation was then designed, with bearing intended for the hard layer encountered, and construction was started by a local contractor.

It was soon found that the presumed bedrock was really boulders of weathered Triassic shale as well as sandstone common to Singapore, but the boulders had originated in ballast deposited on the beach from ships using the port many years before. Lack of control over the cylinder sinking added to the difficulties by permitting soil boiling and consequent loss of ground in the adjacent area. Construction had to be stopped, and attention turned to core drilling equipment by then available. Hard clay was found to depths exceeding 30 m (100 ft). Caissons already placed were therefore grouted up and then underpinned. The bottoms of the new caissons were belled out at appropriate depths to give the calculated bearing values determined safe for the clay on the basis of careful laboratory tests of the samples obtained during the diamond drilling operations. The bearing pressure had to be reduced from the original value of 109,360 to 38,300 kg/m^2 (10 to 3.5 tons/ft^2).

Today, at a minimal cost, the equivalent of a small insurance policy, a study of the urban geology at a building site can be made, archival records that may have any bearing on the site can be studied, foundation records of neighboring buildings can be examined, and the necessary verification borings (and possibly core drilling) can be carried out. As part of this survey, the groundwater situation at the site also can be determined and, if time permits, arrangements made for its observation throughout the annual cycle. Geotechnical investigations can follow, all coordinated to give the most accurate picture possible of the subsurface beneath the site. Uncertainties will remain, as always, but study of the local geology will give a reasonable indication of what these may be, so that they can be provided for in contract documents. And when the record of subsurface conditions as actually revealed by excavation is carefully recorded and filed with civic authorities, the general picture of the urban subsurface will

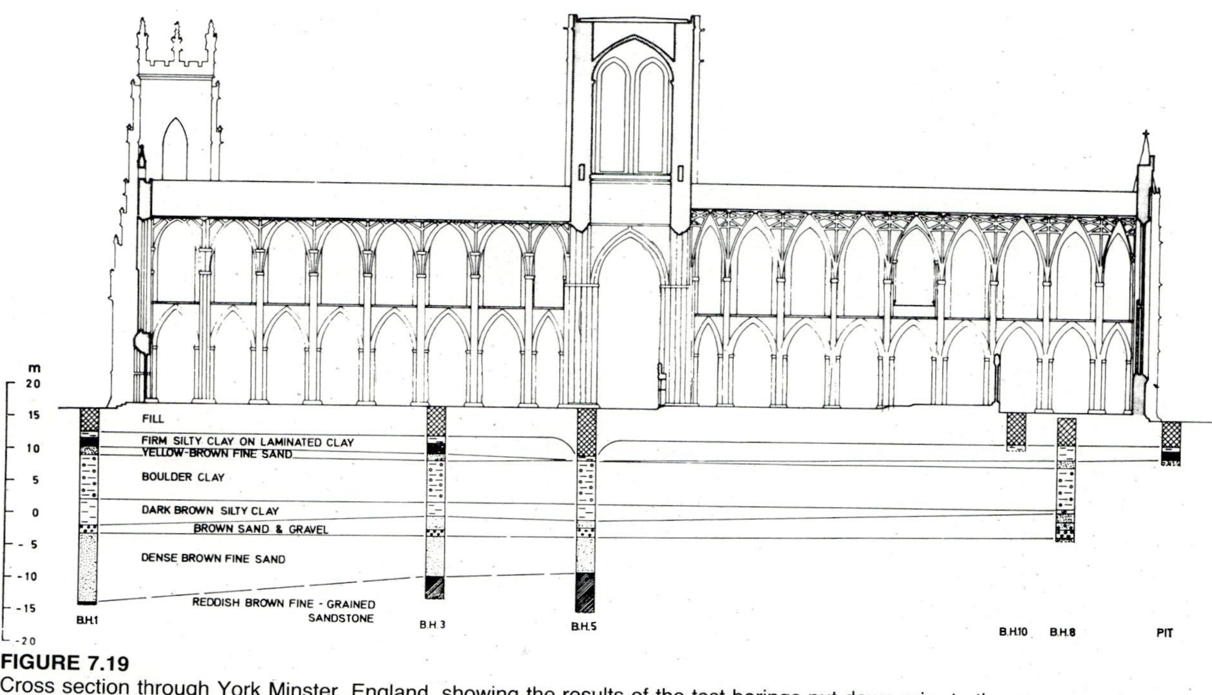

FIGURE 7.19
Cross section through York Minster, England, showing the results of the test borings put down prior to the reconstruction in 1970. (*Courtesy The Institution of Civil Engineers and Ove Arup and Partners, London.*)

be enhanced and made more valuable for building projects of the future. Admittedly, this is the ideal situation, but it is an ideal that has already been realized in some cities. It is an ideal that should be the goal of all concerned with building foundations in every city.

Once the importance, occasional complexity, and invariable interest of building foundations are appreciated, the study of older buildings takes on new significance and value. Interested guests of the Cumberland Hotel in London can discover, for example, that beneath this vast, city-block structure, a tributary of the famous Tyburn Ditch still flows. All who admire the restored majestic beauty of York Minster in England will know that by far the most important part of the reconstruction of this great church was the entire rebuilding of the foundations of the central tower, now open for inspection by an imaginative arrangement of the lower crypt (Fig. 7.19). All observers of the White House in Washington will know that, following its reconstruction, it is as stable and solid as its fine appearance suggests. Practically all major old buildings have a story attached to their foundations if it can be found. Even the building often described as the most beautiful in the world, Chartres Cathedral in France, has its story. The justly famous West Front, with its exquisite carved figures, can be seen at a glance to be unusually placed, almost flush with the front of the two great towers instead of set back somewhat as would be expected. And the explanation is that there was a foundation-bed failure in its original, inset position, back in the fourteenth century. As the reader can see, building foundations have indeed a long and interesting history (Fig. 7.20).

FIGURE 7.20
"...and we can save 700 lire by not taking soil tests." (*Courtesy © Engineers Testing Laboratories, Inc., Phoenix.*)

REFERENCES

1 "Surcharging a Big Warehouse Floor Saves $1 Million," *Engineering News-Record, 176,* p. 22 (March 3, 1966).

2 J. D. Watson and O. R. Bradley, "Compacted Fill Equals Natural Ground," *Engineering News-Record, 135,* p. 810 (1945).

3 A. Baracos, "The Foundation Failure of the Transcona Elevator," *Engineering Journal, 40,* p. 973 (1957).

4 J. T. Mitchell, V. Vivatrat, and T. W. Lambe, "Foundation Performance of the Tower of Pisa," *Proceedings of the American Society of Civil Engineers, 103,* GT 3, Paper No. 12814, p. 227 (1977).

5 D. H. Gillette, "Washington Monument Facts Brought Up to Date," *Engineering News-Record, 109,* p. 501 (1933).

6 J. H. Thornley, C. B. Spencer, and P. Albin, "Mexico's Palace of Fine Arts Settles 10 Ft.: Can It Be Stopped?," *Civil Engineering, 25,* p. 257 (1955).

7 L. Zeevaert, "Foundation Design and Behaviour of Tower Latino Americana in Mexico City," *Geotechnique, 7,* p. 115 (1957).

8 C. B. Crawford and J. G. Sutherland, "The Empress Hotel, Victoria, British Columbia: Sixty-Five Years of Foundation Settlements," *Canadian Geotechnical Journal, 8,* p. 77 (1971).

9 "Sweden's Royal Palace Gets Underpinning to Stop Settling," *Engineering News-Record, 175,* p. 48 (September 23, 1965).

10 "Firehouse, Built on Mud, Is Designed to Float," *Engineering News-Record, 190,* p. 7. (May 31, 1973).

11 F. W. Furkert, "Remedial Measures on the Arapuni Hydro-Electric Scheme of Power Development on the Waikato River, New Zealand," *Minutes of Proceedings of the Institution of Civil Engineers, 240,* p. 411 (1935).

12 D. D. Land and D. C. Hitchings, "Construction (of the Foyers Pumped Storage Project)," *Proceedings of the Institution of Civil Engineers, 64,* pp. 119-136 (1978).

CHAPTER 8 BRIDGE FOUNDATIONS

Design of bridge piers and abutments is an important civil engineering task that will fall at some time to the lot of most practicing civil engineers. History is replete with many examples of substantial bridge works such as the piled foundation of the Roman bridge across the Rhine, and London Bridge, over the Thames. According to a third-century Roman writer, there was a bridge across the Thames just above its mouth as early as A.D. 43. On its arrival in A.D. 1014 to aid King Ethelred of England against the occupying Danes, the fleet of King Olaf (St. Olaf) "rowed quite up under the bridge and then rowed off with all the ships as hard as they could down stream (having secured ropes to the piles supporting the bridge). The piles were then shaken at the bottom and were loosened under the bridge," which gave way, throwing all the defenders ranged upon it into the river.

The London Bridge so well known through illustrations in history books appears to have been completed in the early part of the thirteenth century. The waterway was so reduced by this multiarched structure that swift rapids developed and many persons lost their lives in passing through. The old saying was that "London Bridge was made for wise men to go over and fools to go under." An act of Parliament in 1756 ordered all the buildings on the bridge to be removed and the two central arches rebuilt into one arch. This work inevitably diverted the main flow through the opening and set up serious scouring, which eventually led to demolition of the bridge and replacement with a modern structure.

This is but one of the ancient bridges in which piers have caused trouble. Records are scarce, unfortunately, but it can safely be said that scouring out of the foundation beds adjacent to bridge piers has been a major cause of trouble in the past. The piers of ancient bridges rarely failed because of excessive loading on the foundation beds, if only because of the limitation of span length imposed by the structural materials available. The two defects mentioned can be regarded as the two main possibilities of failure to be investigated in the design of bridge piers. Both are essentially geological in character.

IMPORTANCE OF BRIDGE FOUNDATIONS

However scientifically a bridge pier may be designed, the whole weight of the bridge itself and of the loads that it supports must ultimately be carried by the underlying foundation bed. Although piers and abutments may be relatively uninteresting to structural engineers, the careful consideration of foundation materials is as challenging as the determinate mathematical calculations relating to the arrangement of steel, reinforced concrete, or timber to be used for the superstructure. Sometimes it is assumed that the cost of foundations, compared with the total cost of a bridge, is relatively small. Actual cost records, however, show that the cost of foundations (piers and abutments) often almost equals the cost of superstructure, even on large bridges.

It has not always been fully recognized that concern should always be given to the pier- and abutment-bearing surfaces and whether they can support the structure without fear of any serious movement in the future. Dr. Terzaghi once said that:

> On account of the fact that there is no glory attached to the foundations, and that the sources of success or failure are hidden deep in the ground, building foundations

> have always been treated as stepchildren and their acts of revenge for lack of attention can be very embarrassing.

Of no group of foundations is this more true than of those for bridges.

SPECIAL PRELIMINARY WORK

The first considerations in bridge location are generally those of convenience and economy. Foundation conditions usually take a subsidiarv place, for the prime requirement of a transportation route is that it connect its terminal points by the shortest convenient route consistent with topography. For crossings of deep canyons, considerations of cost usually limit the choice to the site requiring the shortest possible structure. The bridge engineer must therefore often fit design to available foundation conditions. The limitation of site selectivity necessitates acquisition of the most complete geologic information possible.

A still more compelling reason for obtaining full geologic information is that once the construction of bridge piers is started, their respective locations cannot be changed except in most unusual circumstances. More than the usual degree of certainty must therefore be attached to the design and anticipated performance of bridge piers and abutments. There is yet a further reason for this special care in preliminary investigations. Bridges, as a rule, are constructed to cross river or other valleys—topographic depressions that generally exist because of departure from normal geologic structure. Terrain covered by glacial debris may now conceal an older riverbed or other depression well below the existing riverbed. Such conditions are common, and even if known in advance can have serious effects on design.

Riverbeds contain many types of deposits, including boulders, and if preliminary geologic work is not done carefully, an extensive boulder deposit can easily be mistaken for solid rock. A telling example is that of the Georges River Bridge, Sydney, in New South Wales, Australia. Construction of a toll highway bridge to replace existing vehicular ferries was begun in 1923. Three possible sites were explored by an experienced drilling foreman. The borings at the site finally selected showed solid rock at depths below bed level varying between 10.5 and 14.1 m (35 and 47 ft) at regular intervals across a river section about 450 m (1,500 ft) wide, the rock at the sides of which was known to dip steeply. On the basis of this information, a through-truss bridge of six main spans supported on cylinder piers was designed, and a lump-sum contract was awarded. During construction, rock was found at only two of the seven main piers. Additional borings taken to depths up to 39 m (130 ft) failed to disclose any solid rock at all at the other pier sites, and what is even more strange, they disclosed no stratum harder than "indurated sand." Construction had to be stopped and designs changed; in consequence, the bridge took five years to build instead of two and cost 27.6 percent more than the contract price.

Discussion of the paper in which this work was reported to the Institution of

Civil Engineers naturally emphasized the rigid necessity of having borings most carefully watched by a trained observer. The absence of geological references in both paper and discussion suggests that neglect of geologic features may have been a contributory cause of the trouble experienced.

Although this is an unusual and possibly exceptional example, the construction of the Georges River Bridge is a telling reminder of the supreme importance of preliminary geological information in bridge design and of the vital necessity for professional supervision of test boring work.

Another reason for devoting unusual care to geologic investigations at bridge sites in all cases of river crossings is the fact that so much of the ground surface involved is hidden below water. The results of the underwater borings must be correlated with geologic observations secured at the adjacent shores. Where sound rock is encountered, this calls for no unusual attention, provided the exposed surfaces of the rock show no signs of weathering or frequent fracturing, but if any part of the foundation bed consists either wholly or partially of clay, then it is desirable—in most cases imperative—to obtain samples of the clay in as undisturbed a condition as possible. Suitable boring sampling devices can obtain undisturbed samples of clay and other unconsolidated materials, even through great depths of water.

The cities of San Francisco and Oakland are separated by the entrance to San Francisco Harbor. Yerba Buena Island stands in the center of the harbor and divides it into the East Bay and the West Bay. For many years, transportation across the harbor was restricted to ferries, but a bridge reached the construction stage in 1933, being officially opened on November 12, 1936. Early in their planning, the engineers decided upon a program of borings and soil testing to enable them (1) to determine the nature of the subsurface materials, (2) to ascertain the most desirable location for the center line of the bridge, (3) to determine the best location for individual piers, and (4) to select a logical basis for the design of the piers. Preliminary jet borings provided the basis for contouring the top of the rock surface of the harbor. With the aid of additional wash-pipe borings and diamond core borings into the rock, they prepared a final design for the West Bay crossing. Piers were located and designed; all were founded on solid rock and constructed by means of caissons, the behavior of which could be accurately foretold.[1]

The East Bay crossing presented quite distinct problems. Since rock was not found by borings at practicable depths, it became necessary to rely on the overlying unconsolidated material. Cores were obtained and hermetically sealed in the sampling tube, right on the deck of the drill barge; they were soon tested at the University of California. When the containers were opened, perfect cores were generally found, although in some cases a slight swelling was noticed, possibly due to the change in internal pressure in the sample as it came up to the surface. Material was obtained in this way from depths of 82 m (273 ft) below water level, as shown in Figure 8.1.

This unique example still has many ordinary features of preliminary investigations. Adequate borings, not only along the line of the selected bridge site,

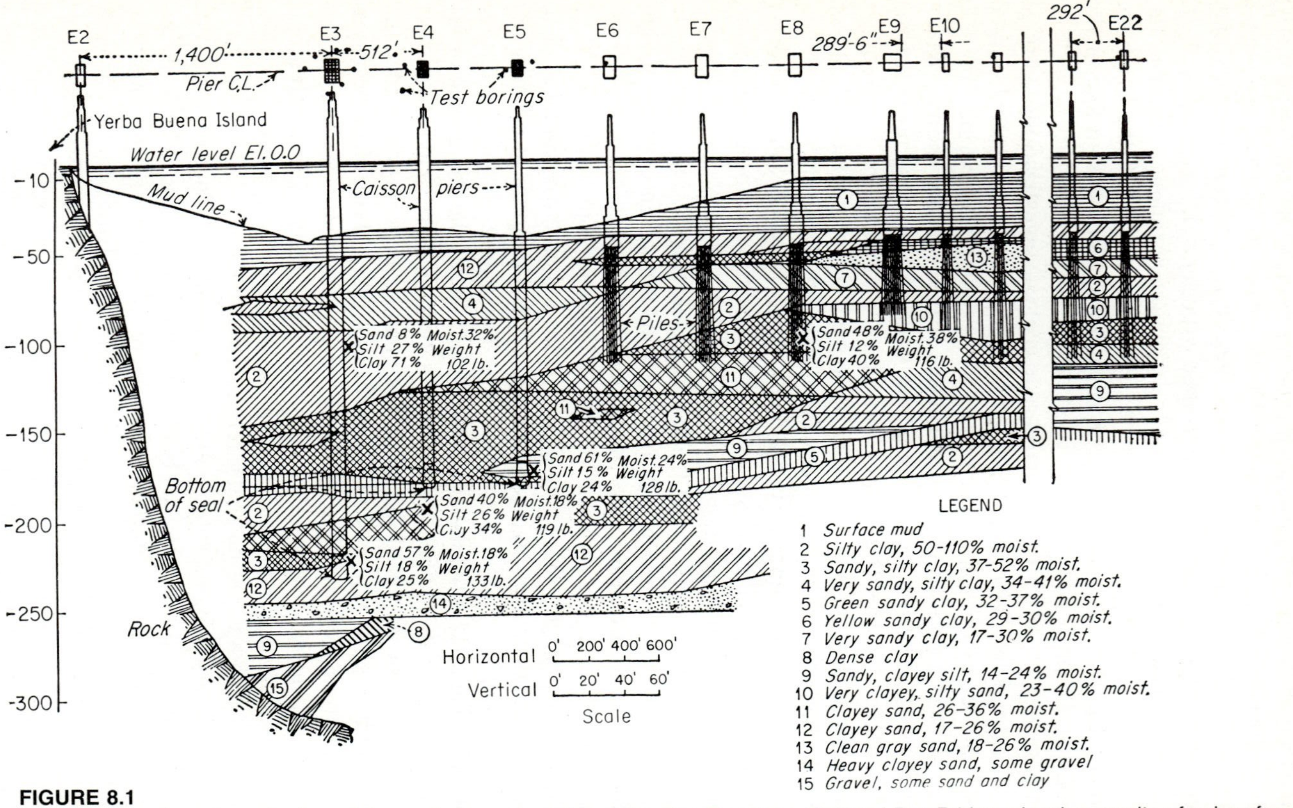

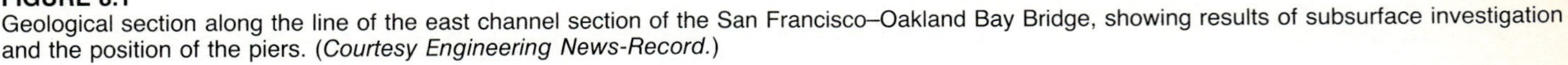

FIGURE 8.1
Geological section along the line of the east channel section of the San Francisco–Oakland Bay Bridge, showing results of subsurface investigation and the position of the piers. (*Courtesy Engineering News-Record.*)

but also on either side of it; careful study of core samples; and correlation of this information with the geologic structure of the adjoining dry ground should present a reasonably accurate structural picture of the foundation beds. This information will enable the designing engineer to locate and design accurately the bridge abutments and piers (Fig. 8.2).

Finally, the necessity of taking all borings deep enough below the surface of solid material (and especially of unconsolidated material) must be stressed. Loadings from bridge supports are always relatively concentrated and often inclined to the vertical. It is therefore doubly necessary to be sure that no underlying stratum may fail to support the loads transmitted to it, even indirectly, by the strata above.

An interesting example of trouble due to this cause is the failure of a highway bridge over the La Salle River at St. Norbert, Manitoba. The bridge was a single reinforced-concrete arch, with a clear span of 30 m (100 ft); the spandrels were earth-filled. The roadway was about 9 m (30 ft) above the bottom of the

FIGURE 8.2
A pictorial representation of the central dual anchorage pier of the San Francisco–Oakland Bay Bridge, showing subsurface conditions. (*Courtesy California Department of Public Works.*)

river, and the height of the fill placed in each approach was about 6 m (20 ft). The bridge abutments were founded on piles driven into the stiff blue clay exposed at the site and thought to overlie limestone bedrock, as shown by preliminary auger borings and the record of an adjacent well. Failure occurred by excessive settlement. The north abutment dropped 1.2 m (4 ft), and bearing piles were bent and broken. Subsequent investigations disclosed the existence of a stratum of "slippery white mud" (actually bentonitic clay) about 7.5 m (25 ft) below the original surface; this material failed to carry the superimposed load.

Local soils were formed in an ancient glacial lake and usually overlie lodgement till, under which is limestone carrying subartesian water. The existence of this water complicated the underpinning of the bridge foundation, but the work was successfully completed, and the bridge was restored to use. The bentonitic clay was previously unknown in the vicinity, and illustrates the uncertainty of glacial deposition. The occurrence has a special interest for engineers; although the bearing piles were driven into the lodgement till ("hardpan"), settlement of the abutment occurred as the result of failure of soft material underlying this.

DESIGN OF BRIDGE PIERS

Generally speaking, there are four types of bridge-pier loading, one or more of which may have to be provided for in design: (1) vertical loads, possibly of varying intensity, from truss or girder spans or suspension-bridge towers; (2) inclined loads, again of varying intensity and possibly varying direction, for arched spans; (3) inclined tensions, from the cables of suspension bridges; and (4) horizontal thrusts due to the pressure of ice or possible debris, the flow of water impinging on the piers, and the wind acting on the bridge superstructure and piers. In earthquake regions, allowance must also be made for seismic forces that may act upon the piers. Combinations of these several loads will give rise to certain maximum and minimum unit pressures to be taken on foundation beds. From considerations of these results and of the nature of the strata to be encountered, the type of foundation can be determined.

Estimation of foundation load at the site of a bridge pier is generally similar to the same operation for other foundation work. Aside from concern for weak strata below the surface, there are two unusual features that may require reductions of the calculated net load on the base area. The first is the allowance for the natural material excavated and for the displacement by the pier of water; and the second is the reduction for skin friction on the sides of the pier because of the usually large surface area exposed as compared with the base area. These two factors are obviously dependent on the nature of the foundation strata. Estimation of the first is straightforward, but that of the second is generally a matter of experience or of experiment during pier sinking, tempered by the results of careful laboratory soil tests.

Weaker strata may even dictate the use of hollow piers to reduce unit loads or of such unusual structures as the open reinforced-concrete framework abutment supports adopted for the Mortimer E. Cooley Bridge across the Manistee River in Michigan. This singularly beautiful bridge, consisting of two 37.5-m (125-ft) deck truss steel cantilever arms supporting a 15-m (50-ft) suspended span and balanced by two 37.5-m (125-ft) anchor arms, has its deck level about 18 m (60 ft) above the level of the ground on either side of the river. The foundation of varying strata of unconsolidated materials was accurately explored and foundation loads were kept to a minimum through use of the open framework design (Fig. 8.3).

When preliminary investigations indicate foundation material of poor bearing capacity, consideration may be given to the use of artificial methods of consolidating such material to improve its bearing capacity. This is no new expedient. The account given by Leland (antiquary to King Henry VIII) in 1538 concerning the Wade Bridge in England revealed that "the foundations of certain of th'arches was first sette on so quick sandy ground that Lovebone (Vicar of Wadebridge) almost despaired to performe the bridge ontyl such tyme as he layed pakkes of wolle for foundation." Although this use of wool has been disputed, the record demonstrates that some artificial means was used to improve bearing capacity. Modern methods (described elsewhere in this book) include grouting chemical consolidation, or leaving the steel piling of the pier cofferdam in place to confine the foundation-bed material and thus prevent lateral displacement. In this way bearing capacity will be increased to some extent.

The foundations for the Tappan Zee Bridge that carries the New York Thruway across the Hudson River for a distance of 4.5 km (2.8 mi) between Nyack and Tarrytown, New York, provide an even more unusual approach to the problem of minimizing loads on weak strata. The bridge site selected through careful studies had to be accepted even though the bedrock drops off under the bridge to depths as great as 420 m (1,400 ft) below water level—a depth too great to be reached by end-bearing piles. The approach spans are therefore carried on friction-bearing piles, driven into the silt, sand, and gravel that form the riverbed.

The main piers carrying the 363.3-m (1,212-ft) cantilever main-channel span (Fig. 8.4), however, are founded on buoyant, reinforced-concrete boxes, carrying about two-thirds of the dead load of the superstructure. The remaining part of the dead load, and the live load, are taken by 75-cm (30-in) concrete-filled pipe piles for the four main piers and by 35-cm (14-in) steel H piles for the four other buoyant boxes; in each case piles and boxes are ingeniously connected together. The steel piles had to be driven to depths up to 52.5 m (175 ft), but the concrete pipe piles went as deep as 102.0 m (340 ft) below water level, being driven through clay and then gravelly clay after the sand and gravel had been penetrated. The hollow piles were mucked out to full depth by waterjet and airlift techniques and then grouted into preplaced aggregate. The grouting consolidated the sheared gneiss and decomposed sandstone bedrock.

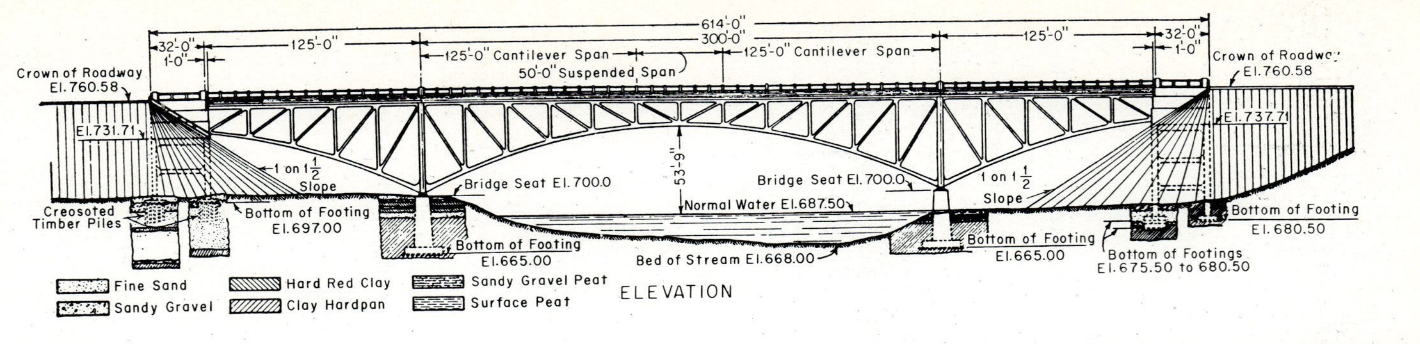

FIGURE 8.3
Elevation of the Mortimer E. Cooley Bridge over the Manistee River, Michigan, showing the foundation strata and special abutment structures. (*Courtesy American Society of Civil Engineers.*)

FIGURE 8.4
The Tappan Zee Bridge over the Hudson River, New York, looking east, showing the spans which are supported by the special piers described in the text. (*Courtesy New York State Thruway Authority.*)

Details of the piers are admittedly an engineering matter, but it was the geology of the site that dictated such a bold design.[2]

Geologic information can be applied to predict settlement of loaded piers. What happens when uneven settlement does take place is well illustrated by the failure of piers 4 and 8 of Waterloo Bridge, London; the whole bridge had to be taken down and a new structure erected. Described by Canova as "the noblest bridge in the world worth a visit from the remotest corner of the earth," Waterloo Bridge was constructed from 1811 to 1817. Timber rafts on timber piles bearing on gravel were designed to protect the pier foundations against scour. Progressive settlement became serious in 1923; the total settlement of pier 4 exceeded 75 cm (2 ft) and naturally caused an arching action between piers 3 and 5.

Settlement may occur from one or more of the following causes: (1) displacement by scour, (2) lateral displacement due to lack of restraint, (3) consolidation of the underlying material, or (4) failure of an underlying stratum. Only condition 3 can be controlled; the other three types are of a nature that may cause serious trouble to the structure. All types can be predicted on the basis of adequate preliminary geologic information.

Provision against unequal settlement of piers has assumed considerable importance in recent years owing to the development of the rigid-frame type of structure, requiring "unyielding" abutments and uniform settlement of piers. Rigid-frame structures founded on clay require isolation of the bridge foundation from the bearing piles and load transmittal through a tamped layer of crushed rock (employed at a Canadian National Railways bridge at Vaudreuil in Quebec). Uneven foundation-bed loading, especially that caused by irregular construction scheduling, must also be carefully considered in design. During the 1932 construction of the Broadway Bridge, Saskatoon, Saskatchewan, concreting of the six arches proceeded in varying stages. As a result, the piers tilted when carrying the dead load of only one adjacent arch rib. A maximum deflection of 15 mm (0.6 in) was recorded as anticipated.

Inclined tensions of the third type of loading are generally transmitted to anchorages in solid rock. This design approach provides for shearing resistance in the rock, which, together with allowance for the dead weight of the anchorage, will be sufficient to balance the tensile forces in the bridge cables. Some inclined tension in bridge cables is taken up wholly by concrete piers, as at the Ile d'Orléans (suspension) Bridge, in Quebec. The suspension span is carried into the long approach structures and secured in anchor piers, one of which is founded on rock, but the other on sand. The stability calculations for these piers had to keep the unit toe and heel pressures within the limit for the foundation-bed material. Frictional values of concrete-to-rock and concrete-to-sand provided the basis for this anchorage. Inclined H-beam piles were driven into the sand underlying one of the anchor piers and were left to project outwardly, in counterforte, into the concrete of the finished pier to give the necessary increase in stability.

As a final example, the Burford Bridge across the river Mole in Surrey, England, was designed to accommodate an unusual geological condition. The bridge is a single reinforced-concrete arch span of 24 m (80 ft), 30 m (100 ft) wide between parapets, with specially selected brick facing. The Mole Valley, located some 40 km (25 mi) to the south of London, flows through chalk formation, which is highly susceptible to groundwater dissolution. Underground cavities here are so large as to receive the whole normal flow of the stream. Borings were put down to see if any such "swallow holes" in the chalk were revealed. Two soft spots were located which proved to be dissolution channels having almost vertical sides and filled with alluvial matter. Concrete domes were constructed over each of the holes, domes founded on circular ledges cut in the chalk around the tops of the excavated channels; the largest dome was 17.4 m (58 ft) in diameter with a rise of 2.4 m (8 ft). The holes were filled up to the undersides of the domes, and the filling was then covered with waterproof paper and used as the lower form for concreting the domes (Fig. 8.5). Each dome was furnished with an access shaft connecting to a manhole at road level by means of which engineers may inspect the swallow holes from time to time to see that no dangerous undercutting or further erosion of the chalk is taking place.

FIGURE 8.5
Foundations for the Burford Bridge on the Mickleham bypass road, Surrey, England; a reinforced-concrete dome covering a "swallow hole." (*Courtesy County Engineer, Surrey, England.*)

DESIGN OF BRIDGE ABUTMENTS

In addition to having to support at least partial structural loads, the abutments of a bridge may have to resist earth pressure against the face and wing walls of the abutment structure. Design may be a complicated matter, considering the difficulties of both bridge-pier and retaining-wall construction. Abutments serving as approach embankments require careful placing of fill, working away from the abutments; adequate provision for internal drainage by cross drains along the lower part of the inner faces of abutment structures; connection of drains to suitable weep holes which cannot become plugged up; and monitoring of all fill settlement.

Abutments located on sloping ground may present different problems, for in addition to having to retain the pressure of earth backing, often with considerable surcharge, they may be subjected to forces set up by the instability of the whole hillside slope; or, alternatively, their foundations may be made insecure by earth movement on slopes below. Again, the forces acting on them are far from symmetrical, so that balancing (especially during construction) is a matter which often calls for great ingenuity in design and dependence on foundation conditions.

Solid rock abutment support is the ideal condition and one which is essential for certain types of arch designs. A famous English bridge illustrates foundation accommodation of a slight variation from the ideal situation—one of the abutments of the Grosvenor Bridge across the river Dee at Chester is founded partially on rock and partially on sand. The bridge, built in 1833, has a clear span of 60 m (200 ft) and a rise of 12 m (40 ft). The construction that had to be adopted at the north abutment, where the outcrop of solid rock was discovered

FIGURE 8.6
Combined pier, anchorage, and abutment structure for viaduct (overhead), suspension span, and steel arch—all component parts of the Thousand Islands Bridge over the St. Lawrence River at Ivy Lea, Ontario. (*Photo: R. F. Legget.*)

to terminate, being succeeded by a deep stratum of loose sand, was to transmit the arch thrust to the sand by means of timber piles.

With arch bridges, abutments are critical parts of bridge design. Figure 8.6 shows an unusual example, demonstrating the possibilities in design when rock conditions are excellent, such as for one of the supporting structures for the Ivy Lea International Bridge that crosses the St. Lawrence River in the Thousand Islands region between Ontario and New York State. Here gneiss bedrock was at the surface; the Thousand Islands are exposures of the Frontenac Axis, the projection of the Precambrian Shield extending into New York and forming the Adirondack Mountains. With such sound rock available the engineers were able to combine in this one structure a main abutment for a long steel arch, an anchorage for cables from one of the main suspension spans, and the necessary support for vertical columns under that part of the bridge joining the arch and the suspended span.

A graceful steel-arch bridge spans the Volta River in Ghana; it is a two-hinged structure with a clear span of 259.5 m (865 ft), making it the seventh longest such bridge in the world when built. Eight months were spent on preliminary geological studies; rocks at the site are metamorphosed Paleozoic sediments (indurated shales and quartzites). Highly folded quartzites form most of the bedrock in the area. Even with this advance knowledge, however, more difficulty was experienced than had been anticipated when the rock was exposed for the abutment on the west bank. Interbedded shale and quartzites were found to an unexpected degree; the sawtooth design prepared for the rock excavation had to be modified to take into account the actual dip and strike of the rock, the depth of which was greater than expected, since some of the few auger borings used had suggested bedrock where only boulders were found. With modification of the design, however, the west abutment was successfully completed, and this fine bridge is now serving as a major transnational link.

When no rock is available, abutment designs must follow methods suggested in the case of pier design. Subsoil conditions must be thoroughly investigated, and the foundation-structure design determined after a full consideration of test results. In many cases of abutment designs on unconsolidated material of low bearing capacity, bearing piles will have to be employed to carry the anticipated load. Such piles, unless driven at a batter, will offer small resistance to lateral movement. For this reason, quite a few abutments suffer inward displacements from their original positions by the excess of earth pressure behind them over the stabilizing forces.

One would not imagine that bridge abutments would ever move inwards against approach embankment fill, but this did happen with seven overpass bridges built at the extreme eastern end of the Macdonald-Cartier Freeway in Ontario. All were excellently designed and constructed bridges supported on end-bearing piles driven to rock, but within three years of completion slight movements of abutments away from the road had occurred. The explanation appears to be that the approach embankments were founded on sensitive Leda Clay. Settlements of this marine deposit were anticipated in design. Settlement would be at its maximum under the full depth of fill but small at the edges of the fill slopes. The "dishing" configuration of the ground beneath the fill would initiate slight movements of the fill toward the point of greatest settlement. The mechanics of the resulting "system" account for the corresponding movements of the abutments.[3]

Not only the design of abutments but even that of bridge superstructures may be affected by the geological structure of the abutment site. Nowhere else has this been better demonstrated, perhaps, than in the case of the Kohala Bridge carrying the Rawalpindi–Kashmir Road across the Jhelum River on the rugged mountainous boundary of the Punjab and Kashmir, India. Originally constructed as a three-span girder bridge having a 39-m (130-ft) center span and two 27-m (90-ft) approach spans, the unstable condition of the hillside at the Punjab abutment caused lateral movement of the girders toward the Kashmir side. The Punjab approach span was seriously damaged in 1929 and finally

FIGURE 8.7
The Kohala Bridge across the Jhelum River on the Rawalpindi–Kashmir road, Pakistan; the "floating span" may be seen on the right, leading from the abutment on the sloping river bank which caused so much trouble. (*Courtesy Northwest Railway of India.*)

wrecked in 1931 through serious landslides triggered by heavy rains. A large fissure was discovered in the hillside about 60 m (200 ft) above the level of the bridge. It extended down into a slope of loose soil and rock detritus and was indicative of future movement.

Reconstruction employed cantilevering a 21-m (70-ft) span back from the first main river pier and connecting it by means of a light suspended span to the hillside road level. Suitable drainage work was also installed as a stabilizing measure. Although the potential for future earth movement has been reduced, the intent is that, should a landslide occur, it will displace the suspended span, which can safely ride up and over the end of the cantilevered span. Thus, little damage will be done to the main bridge structure, and the span can be easily replaced after movement has stopped (Fig. 8.7). Engineering ingenuity has here countered geological instability in a particularly imaginative manner.

PRECAUTIONS AGAINST SETTLEMENT

If settlements due to future mining operations are known to be a possibility at a bridge site, superstructure designs can be prepared accordingly. A notable example of this type of "controlled" design is that of the Clifton Bridge at Nottingham, England, across the river Trent. Coal mining is an important local activity, and exploration had outlined valuable coal seams directly under the bridge location. The structure was therefore designed to take into account a possible settlement should mining take place. The resulting structure is a prestressed-concrete main span of 52.5 m (175 ft), with two end spans of 37.5 m (125 ft) and three 27-m (90-ft) end viaducts, all built on a skew of 24°. The main span had to be designed as two cantilevers with a central suspended span, all statically determinate because of the anticipated settlement. Arrangements were included in the design for the spans to be jacked up off the tops of the main piers whenever this became necessary. The piers are founded on the Keuper Marl, which extends to depths of over 150 m (500 ft) at the bridge site. The marl was badly weathered to a depth of 8 m (27 ft). Consolidation tests showed that it was really a highly overconsolidated clay, indicating that it would swell if exposed and would deteriorate if unprotected from water.

Just as the complex design for the bridge was complete, it was found that a geologic fault between the colliery and the bridge site made profitable working of the coal seams improbable. The local city council therefore bought out the rights for the coal seams for a nominal sum and thus eliminated all possibility of settlement from this cause. It was too late to change the design, so the bridge was built as planned.

Gold-mining subsidence is a constant consideration in Johannesburg, famous for its deep mining. In earlier days, the ore was found almost at the surface and within the limits of what is now the great modern city. Today, the old workings remain unused. They necessitate severe restriction of modern building in a wide strip of land (the old reef) running through the city from east to west immediately to the south of the main city center. Ground movements still take place, and so this area has limited capability of development. It has, however, become the site of a great system of freeways planned in the early 1960s, serving central Johannesburg, and acquired at relatively low cost.

Design and construction of this modern highway presented unusual problems. Ground movements over the old working had to be considered as definite possibilities, as some gold still remains in some of the old, shallow workings. The engineers responsible were able to meet considerable restrictions: a 5-cm (2-in) change in grade in 30 m (100 ft), longitudinal movements of up to 95 mm (about 3 ft) in 30 m (100 ft) in any individual span, and a maximum cumulative movement of 25 cm (10 in) in a group of spans. The engineering solution employs movable support piers, which weigh as much as 600 tonnes, mounted on bearings that can be jacked either vertically or horizontally to compensate for any ground movement (Fig. 8.8).

EARTHQUAKES AND BRIDGE DESIGN

A significant portion of the earth is subject to periodic shocks of earthquakes large enough to bring about structural damage to bridges. In such regions seismic forces have to be carefully considered in bridge design, especially when bridge piers are tall. Among the highest bridge piers yet built were those for the Pitt River Bridge that carries the Southern Pacific Railroad and U.S. Highway 99 over a part of the reservoir formed by the Shasta Dam. The piers have a maximum height of 108 m (360 ft), of which over 90 m (300 ft) is now submerged beneath the waters of Shasta Reservoir.

Because of high-level and recurring seismicity in California, an engineering earthquake study was therefore carried out by the U.S. Bureau of Reclamation.

FIGURE 8.8
Fixed viaduct section of a major bypass road through the city of Johannesburg, South Africa, showing construction of the adjustable piers to allow for future settlements. (*Photo: R. F. Legget.*)

This showed that, under earthquake loading and when submerged, these piers would no longer act as elastic structures fixed at their bases but would act as rigid structures, rotating at their bases. This conclusion led to a seismic design criterion for the piers which were then designed to carry static loads without considering earthquake effects. The anticipated earthquake forces were then added to this design so that the resultant force would be resolved to concentrate at the edge of the pier base.

The Pajaro River Bridge, 148 km (92 mi) south of San Francisco on the Southern Pacific Coast line to Los Angeles, was heavily damaged by an earthquake on April 18, 1906, which destroyed much of San Francisco. The bridge with five simple deck girder spans had a length of 138 m (460 ft). All five spans were moved vertically by the earthquake and all but one horizontally. The motion was enough to throw all the spans off their supports, one span being left hanging precariously over the edge of one of its piers. When the bridge was reconstructed in 1944, advantage was taken of the knowledge that the bridge spans the San Andreas fault. The redesign featured three-span continuous deck girders with one 25-m (86-ft) side span, resulting in a total length of 135 m (450

FIGURE 8.9
The Southern Pacific Coast Daylight Express crossing the Pajaro River Bridge, showing in the lower center of the photo the special bearing support at the top of the concrete pier (as mentioned in the text). (*Courtesy Southern Pacific Transportation Co.*)

ft). Continuity of the central supports in the river bed close to the fault is designed to accommodate any serious vertical movement of the piers or of horizontal movement of the continuous girders. Special rockers and concrete projections will uphold the girders should they be moved off their supports (Fig. 8.9). A heavy pendulum safety trigger has been mounted on one of the piers and designed so that horizontal movement of the center of the bridge of more than 19 mm (¾ in) will cause automatic block signals to prevent access to the bridge. Abutments and the ends of the girders are similarly connected; they will set off danger signals for movement over 25 mm (1 in).[4]

Seismic forces were naturally considered in the design of the majestic Golden Gate suspension bridge, with a main span of 1,260 m (4,200 ft). The northern pier and anchorage are founded on diabase and basalt, and on Marin Sandstone, respectively; the southern pier and anchorage on serpentinite. A horizontal static force of 0.10 g was allowed for the design of the main piers. For the anchorages and superstructure there was a corresponding allowance of 0.075 g.

SCOURING AROUND BRIDGE PIERS

The consequences of bridge construction across a waterway involve three predictable effects: (1) piers and abutments will generally decrease the effective cross-sectional area of the stream and thus inevitably increase the velocity and raise the water level (usually very slightly) upstream from the bridge; (2) piers are obstructions to streamflow and will set up eddies around the piers and may possibly institute crosscurrents in the stream, tending to change it from its normal course below the bridge site; and (3) the combined effect of increased average velocity and eddies may disturb the equilibrium of the bed material between piers and so lead to scouring. All these results represent interference with natural conditions and therefore call for the application of geological information in the engineering solution of the many problems they bring up.

A classic example of the serious effect of a change in permanent water level is the pier-foundation scouring of the old Westminster and Vauxhall bridges over the Thames, in London. This early nineteenth-century scour followed the removal of old London Bridge and the consequent lowering of the water level in the stretch of the river immediately upstream. Normal flow conditions were restored at the site of the remaining bridges since their foundations had not been designed for these conditions; both structures in the course of time had to be completely rebuilt.

Geological factors may appreciably affect runoff calculations prepared to show what hydraulic conditions a bridge will have to withstand. In broad tropical river valleys, bridges may be designed to be completely submerged in flood periods. Naturally, the geological stability of the pier, abutment, and approach foundations must be assured. An interesting example is the Nerbudda River bridge near Jubbulpore in India. It is 366.6 m (1,222 ft) long, consisting of six 29.4-m (98-ft) and eight 13.8-m (46-ft) spans, all reinforced concrete arches; it is founded throughout on basalt. During floods, the bridge is completely submerged.

River-training works are frequently incorporated into bridges on rivers with relatively unstable beds. Without such "training," the volumes of water passing between the several piers may not be the same, and may actually be reversed between pairs of piers. Riverbed scouring may happen even on small streams if bridge piers are placed where they interfere with the normal flow regime in midstream. Scouring of bed materials from around foundation structures supporting bridge spans has been responsible for more bridge failures in the past than any other cause.

SOME CONSTRUCTION REQUIREMENTS

Bridge-foundation design must consider the specific problems of pier construction. Geologic factors affect construction restriction of the watercourse navigation requirements, water depths and tidal range, and the depth below water to foundation-bed level. Pier construction is generally carried out by one of three methods: use of open cofferdams (working either in the dry or in water), use of open dredging caissons, or use of compressed-air caissons. Bedrock foundation conditions will be a potent factor in choosing the specific construction method. In the case of cofferdam work, for example, subsurface conditions will suggest the length of piles required and will indicate what resistance to penetration will have to be overcome in driving. Should top-of-rock occur at a variable level, possibly a combination of two of these methods, especially of the second and third, will be advisable, but such should be chosen only if subsurface details are accurately known.

Recent developments in construction techniques have opened up new possibilities for economical bridge-pier design in cases of poor foundation conditions. Bridge piers can readily be constructed using long steel H piles driven to rock, with the piles possibly encased between bed and water level with sheet-pile cofferdams in which aggregate can easily be placed and converted into solid concrete by specialized grouting techniques. Such piers give all necessary support above water level, provide requisite protection down to riverbed level, and derive their bearing capacity from the column action of deeply buried steel piles transferring loads to bedrock. A growing number of Canadian bridges successfully utilize long steel piles; one of the most notable is the Canadian National Railways bridge over the Kinojevis River in northwestern Quebec. Steel piles up to 52.5 m (175 ft) long were used to form bridge piers in material that could best be described as "soup."

In contrast, there was the dilemma facing Isambard Kingdom Brunel when he designed the Royal Albert Bridge at Saltash over the river Tamar on the Great Western line to Penzance. Divers revealed poor foundation conditions in 1847; thus, Brunel knew he must learn more about the riverbed. A wrought-iron cylinder, 1.8 m (6 ft) in diameter and 25.5 m (85 ft) long, was therefore fabricated, towed out to the site of the central pier, and sunk through the water and mud to bedrock. This cylinder was repositioned for each of 175 borings made into the rock. A top-of-rock profile was plotted; a column of masonry was even built on the rock and carried up to the level of the riverbed. Caissons had been used as early as 1720 by Gabriel, but this caisson designed by Brunel must surely be one of the pioneer structures of construction practice. It measured 11.1 m (37 ft) in diameter, 27.0 m (90 ft) in height, and weighed 300 tonnes. It was floated out into place and fitted with an inner cylinder to give access to the "diving-bell" compartment that eventually extended to rock level.

Caissons must overcome the skin friction associated with sinking through foundation strata. Geologically-based estimates of skin friction should be guided by whatever experience is available for strata which will be penetrated. Frictional resistance can sometimes be overcome by pipe jetting around the cutting edge of the caisson. Caisson trouble is often experienced because of uneven settlement, which may be due to purely mechanical causes, but often it is due to variations in the underlying strata. Several cases of bridge-pier caisson tilting show that such occurrences are generally due to a combination of causes. Ground conditions usually are the most important, either because of uneven settlement due to varying strata or because soil does not yield to the caisson cutting edges in the way anticipated. For example, tilting of the 19,000-tonne, reinforced-concrete open caisson for the east pier of the Mid-Hudson Bridge at Poughkeepsie, New York reached 48°. A stiff stratum of clay and sand was overdredged by more than 3 m (10 ft) below the cutting edge. Sudden collapse of the unsupported bed material seems to have caused the tilting; righting of the caisson constituted an unusual and ingenious construction operation.

Knowledge of subsurface conditions may even suggest innovations such as the "sand-island method," originally evolved for construction of the Grey Street Bridge, Brisbane, Australia, in November 1927. Three reinforced-concrete arch spans, each 71.4 m (238 ft) from center to center of piers, were to be founded in depths of water up to 15 m (50 ft), with bedrock varying from 24.9 to 32.1 m (83 to 107 ft) below high-tide level. Reinforced-concrete cylinders were to be used for the two central piers, and two caissons for the main pier on the south bank. Preliminary information revealed soft clay at the surface stratum, which led to the construction of piled "islands" filled up to above water level with sand. Cylinder and caisson sinking was then carried out under constant control from these islands. Steel piling later salvaged was shown to be well founded into riverbed bedrock. Similar methods have been applied at the Suisun Bay bridge of the Southern Pacific Railroad in California, the Mississippi River bridge at New Orleans, Louisiana, and at the Walt Whitman suspension bridge at Philadelphia.

Some means must also be found to inspect and, if necessary, clean the surface of the foundation beds upon which bridge piers or abutments are to be founded. For bridges which will use low unit-bearing pressures and which will be founded on unconsolidated materials, such inspection is not so important as in rock foundation cases. Visual inspection is always a necessity, for even with the best preliminary core borings available, the engineer must verify rock structure and possible surface disintegration. The rock surface must also be properly cleaned of all unconsolidated material so that good bond may be obtained between concrete and rock. A case in point is the Union Pacific Railroad bridge over the Colorado River in the United States, in which the piers had to be rebuilt only 15 years after their original construction (1925) because the rock-bearing surface had not been properly cleaned prior to concreting.

COFFERDAM CONSTRUCTION

Cofferdams are now such a common feature of bridge piers and abutments and general foundation work that they are generally accepted without much thought being given to their geologic environment. Although frequent reference is made to their "early" use on the river Thames for the construction of Waterloo Bridge (from 1809 to 1817), the following extract from Vitruvius, originating about 20 B.C., is therefore of special interest.

> Then, in the place previously determined, a cofferdam, with its sides formed of oaken stakes with ties between them, is to be driven down into the water and firmly propped there; then, the lower surface, inside, under the water, must be levelled off and dredged, working from beams laid across; and finally, concrete from the mortar trough—the stuff having been mixed as prescribed above—must be heaped up until the empty space which was within the cofferdam is filled up by the wall. . . . A

> cofferdam with double sides, composed of charred stakes fastened together with ties, should be constructed in the appointed place, and clay in wicker baskets made of swamp rushes should be packed in among the props. After this has been well packed down and filled in as closely as possible, set up your water screws, wheels, and drums, and let the space now bounded by the enclosure be emptied and dried. Then dig out the bottom within the enclosure.[5]

Most cofferdams are designed to be pumped dry and to perform the single function of providing an exposed dry area of river, lake, or seabed on which construction can be carried out. The retention of earth materials after excavation is not a difficult matter; it is a special case of the retaining structures already considered. Sealing against outside water pressure above bed level is also a relatively simple matter when the cofferdam piling has watertight joints. One main problem remains, however, that of preventing the inflow of water around the lower edge of the piling—in other words, that of preventing "blows," or piping.

The usual cofferdam structure consists of a single continuous wall of interlocking sheet piling, driven into the foundation strata. The distance is sometimes determined by the resistance offered to pile driving, or is based on previous work, but almost always it is determined in an empirical manner. Experience in the site region must always be a guide in such matters, and the actual evidence presented by the driving of the sheet piling into the host strata will be a telling indication of the underground conditions encountered. Cofferdams generally have to be designed prior to construction; materials have to be ordered before they can be used. Estimates of piling penetration are therefore an essential preliminary, and in this estimating work accurate information on the strata to be encountered will be of value.

Permeability is a key soil property in such considerations, but it cannot be considered alone. Sands and gravels, when revealed by test borings, will naturally suggest the necessity of deep penetration. If sufficiently permeable, they may even dictate the use of a clay blanket all around the outer face of a cofferdam as a means of sealing the direct path that would be taken by water when pumping inside begins. Clay strata, on the other hand, will generally provide an impermeable barrier to seepage, provided the piling is driven below any influence from surface disturbance. In cofferdams constructed in tidal water, or under other fluctuating water levels, the greatest care must be exercised in clay soils. Under certain maximum water-level conditions, the clay will be compressed by one face of the piling and may possibly deform permanently, leaving a gap between piling and clay when the water level changes to the other extreme. This type of water passage is a far easier course for the flow of water to follow than even the pores in sand and gravel, and if this condition is allowed to develop unobserved, serious damage may result and "bad blow" seepage may occur.

These notes indicate the importance of subsurface conditions even in this detail of construction, and above all else demonstrate the necessity of knowing as accurately as possible the various soil types to be penetrated. The writers know of at least one case in which a material that to the untrained eye appeared to be clay actually proved to be compacted limestone flour (with practically no clay content at all). The material was assumed to be clay in cofferdam design, and yet it proved in practice to be as porous as a sieve, leading to most serious trouble with "blows."

Geological advice can be of critical assistance in cofferdam work in foretelling with some degree of accuracy the potential for boulders in the ground to be pile driven. Even one boulder (as in coarse alluvium or lodgement till) can bend or deflect a pile and compromise the groundwater sealing ability of an entire coffer cell. Borings may possibly show the existence of such boulders, but unless the boulder-bearing formation is closely packed, it is probable that only a few borings at the actual site of a bridge pier may not detect boulders. On the other hand, knowledge of the geologic origin and history of the foundation strata will at least suggest whether or not boulders are to be expected. In such cases, extra care must be exercised in preliminary boring work. Foundation records for neighboring works have often been of great value in detecting these geological considerations; the value of such records cannot be overemphasized.

Cofferdam construction hampered by boulders was an important part of the building of the three new bridges across the Cape Cod Canal in Massachusetts in 1934 and 1935. Wash borings had disclosed the existence of glacial deposits of sand and granitic gravel containing numerous granite boulders. Specifications were made to call for foundations to be constructed inside steel sheet-pile cofferdams driven to predetermined depths. What was revealed only by the foundation operations were the variable amount of boulders, their relatively large size, and their distribution. The contractor for the work estimated that 10 percent of the piling might strike boulders that would have to be removed; actually, as much as 40 percent of the piling in one cofferdam was so obstructed. Serious difficulties were encountered at all the main cofferdams (Fig. 8.10). Soil structural disturbance caused by pile driving and boulder removal so disturbed the sand that it went "quick" and thus resulted in large and frequent boils. Ordinary pumping methods to facilitate the removal of the boulders failed to keep pace with the flow of water into the cofferdams, and eventually an extensive well-point installation had to be used both inside and outside the sheeting. The presence of boulders had been revealed during the original construction of the canal 20 years before, but whether or not this earlier construction indicated the presence of boulders in the amount encountered was a matter of controversy.

Cavernous limestone comes up again for mention in connection with an unusual bridge across the Barren River at Bowling Green, Kentucky. The structure was a four-span, continuous-plate girder, with each span 33 m (110 ft) long. Bedrock at the riverbed is limestone, extensively honeycombed below its surface for several meters and filled with mud pockets. Up to 3 m (10 ft) of the cavernous limestone had to be excavated at the two pier sites, which were constructed within sheet-pile cofferdams. Steel sheet piling could not be driven into the limestone, so pile driving and excavation had to be carried out

FIGURE 8.10
Cofferdam for the Bourne south channel pier, Cape Cod Canal highway bridges, showing irregular driving of sheet steel piling due to the presence of boulders, some of which may be seen between tracks and cofferdam. (*Courtesy Fay, Spofford and Thorndyke, Boston.*)

concurrently. The rock blasting so increased the fracture permeability of the rock that strong flows of water penetrated the cofferdams. Various water-sealing expedients were attempted, but eventually 9,000 bags of cement grout were consumed for this purpose around the two relatively small cofferdams.

An entirely different groundwater problem was encountered during the construction of a 1,424-m (4,745-ft) bridge over the Illinois River at Peoria. The Peoria Water Works Co., a private utility, obtains its supplies from wells extending into a sand and gravel stratum, separated from the riverbed by layers of clay and shale of varying thickness. Some of the H bearing piles for the main piers supporting the three-span, continuous-through bridge over the main channel were designed to pass through this water-bearing stratum, and the water company was naturally concerned over the possibility of pollution of its supply. Extensive studies were made and an unusual schedule was worked out to control the water level in the cofferdams from which the steel piles were to be driven. The water level was held down to riverbed level while pile driving was in progress and allowed to rise only when pile driving was complete and thick concrete seals had been placed.

Geology assisted with one of the most important solutions to the "cofferdam problem" known to the writers. Two 8.4- by 18.0-m (28- by 60-ft) open cofferdams were needed for construction of the main piers of the Deer Island suspension bridge in Maine, south of Bangor. Borings showed that the bedrock

FIGURE 8.11
The "tailor-made" cofferdam for the Deer Island suspension bridge, Maine, showing how it was transported to the site. (*Courtesy Merritt, Chapman and Scott Inc.*)

surfaces on which the piers were to be founded were sloping steeply away from the land, a common feature of this rocky Atlantic coastline. The contractor decided to excavate the pier areas to bedrock, place dowels in the rock to locate the cofferdams, and then frame the steel sheet-pile cofferdams, complete with bracing, on shore, with the bottoms of the piles being carefully trimmed to follow the exact contours of the rock surface. The cofferdam boxes were barged to the bridge site and then lowered into place (Fig. 8.11). Bags of dry-batch concrete were then placed by divers around the edge of the cofferdams. Concreting then proceeded normally, and the 614-m- (2,048-ft)-span bridge was soon completed.[6]

SOME UNUSUAL CASES

As with all other civil engineering works, no two bridge piers are ever the same. There is something of geological significance to be learned from every pier that is constructed. Here we will present a few unusual, yet especially useful, cases that do not fit into other sections.

The original Forth Bridge, that majestic railway structure that spans the Firth of Forth near Queensferry, Scotland, just to the north of Edinburgh (Fig. 8.12), was opened for use in 1890. Geologic conditions at the site actually made possible this great triple-cantilever design. The north main pier is founded on basalt, the bedrock underlying the Fifeshire shoreline. The south main pier at Queensferry is founded on the sandstone which underlies the thick deposit of lodgement till characteristic of this shoreline. The till extends into the Firth where it fills the lower part of a 180-m- (600-ft)-deep preglacial gorge that exists under the center of the channel, the course of an ancient river Forth. This deep gorge itself might have eliminated any possibility of such a design. Geology,

FIGURE 8.12
The railway bridge across the Firth of Forth, Scotland. (*Courtesy Scottish Tourist Board, Edinburgh.*)

however, had created Inchgarvie Island as a pinnacle of the Fifeshire Basalt and left it protruding above water level so that it provided an ideal location for the large central foundation. It was so dimensioned that the individual piers could be separated rather farther apart than those for the other two foundations, giving the central span the extra stability the designers wished to have. And, as a bonus, the basalt (''whinstone'' in Scotland) excavated from the island for the founding of the piers proved to be a satisfactory aggregate for the concrete which was needed in addition to the Aberdeen Granite facing of the piers. This 2.4-km- (1.5-mi)-long structure will always be an inspiration to civil engineers, while geologists can take satisfaction from the part that geology played in its conception, design, and construction.

Another notable railway bridge is in eastern Canada; it, too, is still in daily use, although its 1875 superstructure has been once renewed. The bridge is one of two six-span Canadian National Railway bridges over the northwest and southwest branches of the Miramichi River of New Brunswick, near Newcastle. By all appearances, both are identical in character as was assumed when simple borings suggested that bedrock would be reached at depths between 13.5 and 15.0 m (45 and 50 ft) below high water on spring tides (HWOST). Foundations were designed as timber caissons to be floated into place, filled with ''tremie'' concrete, and topped by masonry piers above water level. When work started, it was soon found that the borings were inaccurate. New borings in 1870 encountered sand and gravel, overlying ''silt'' in the northwest branch and ''clay'' in the southwest branch, bedrock being reached at a depth of 33.6 m (112 ft) in the former and 27 m (90 ft) in the latter.

The designer, Sir Sanford Fleming, was gifted with unusual intuition (soil mechanics would not be a recognized field for another 65 years!). He deduced that the southwest bridge piers would be safe. Special ''penetration tests'' were

FIGURE 8.13
Canadian National Railways Bridge over the northwest arm of the Miramichi River, near Newcastle, New Brunswick. (*Photo: R. F. Legget.*)

made on small plates at the bottom of boreholes (almost three-quarters of a century before such tests were "discovered" in modern soil studies) at the northwest bridge. Using these in-place test results so obtained, he enlarged the caissons. Knowing from his penetration tests that settlements would take place, he hit upon the idea of preloading the completed piers with loads slightly greater than the total loads to which they would be subjected, thus accelerating settlement. The piers were finished, the superstructure erected, and the bridge has performed satisfactorily to this day (Fig. 8.13). The interesting and significant difference between the soils along the two branches of the Miramichi River is wholly geological in origin, the northwest branch flowing through country underlain generally by Precambrian crystalline rock while the southwest branch runs over younger (Mississippian) sedimentary rock.[7]

The Trans Canada Highway approaches Vancouver across the Fraser River floodplain on another graceful steel arch with long approach viaducts, just upstream of the head of its great delta. Typically difficult floodplain foundation conditions were revealed by borings taken to depths of over 60 m (200 ft). Floodplain sand and silt overlay dense lodgement till at the north approach spans. In the vicinity of the river channel and especially on the north bank, 4.5 m (15 ft) of organic silt was first encountered. This was found to overlie soft organic silts and clay silts to about 12 m (40 ft). Then came a stratum of compact peat under which a clay stratum extending in places to 15 m (50 ft) below the surface was revealed. Sand strata of varying density with some silt occurred next in the subsurface profile, to a depth of about 33 m (110 ft)—and then sand and gravel for the next 3 m (10 ft). Finally, to depths of about 57 m (190 ft), the borings revealed soft-to-firm sensitive clays with some silt. To found the main piers for a 360-m- (1,200-ft)-span steel arch on such material called for the greatest ingenuity in design.

Of a number of alternative designs studied, a novel solution was adopted, probably unique in the history of bridge engineering. Holes 80 cm (32 in) in diameter were drilled from the surface to the top of the till. Into these holes were lowered 60-cm- (24-in)-diameter, steel pipe piles with heavy steel tips; the lower sections of these piles were filled with concrete to a height of 4.5 m (15 ft). This material served as a cushion for the drop hammer which then drove the piles by being actuated inside the piles. In this way, disturbance of the clay strata was avoided and sound bearing obtained. So weak was the clay that the annular ring (10 cm or 4 in wide), which was filled with mud during the pile driving, was finally refilled with cement-clay grout which displaced the mud as it was placed. The four main piers for the Port Mann Bridge were founded in this way, the engineering details of the design being a landmark in the history of geotechnical engineering.[8]

INSPECTION AND MAINTENANCE

All civil engineering structures require regular and thorough inspection. This can hardly be overemphasized, especially for bridge piers, the underwater parts of which are not seen from day to day and yet are most prone to periodic damage. Piers and abutments founded on dry land require inspection of structural integrity as well as positional and level verifications. Pier movement or material deterioration is inevitably due to some feature of the underlying geologic strata. Often foundation-bed conditions must be subjected to renewed geotechnical investigation.

An outstanding example of inspection benefit work derives from the famous Lethbridge Viaduct of the Canadian Pacific Railway, a steel-trestle structure 1,598 m (5,327 ft) long carrying a track 93.6 m (312 ft) high above the bed of the Oldman River of Alberta (Fig. 8.14). The trestle bents were founded (in 1909) on concrete pedestals supported on concrete piles driven into clay which underlies the whole site. By 1913 the deck of the bridge showed settlement of as much as 64 mm (2 in) due to settlement of one pier. This had to be underpinned by caissons carried to shale bedrock 15 m (50 ft) below ground level, and a drainage tunnel also had to be constructed. As a result of these measures, subsidence was arrested.[9]

Similar vigilance in track maintenance first detected trouble with the slender reinforced-concrete bridge at Klosters near the eastern border of Switzerland (Fig. 8.15). The bridge carries the meter-gauge Rhaetian Railway and is on a sharp curve. Its length is 75 m (254 ft) and it is characterized by its main arch of 30 m (100 ft) span. The arch rests upon mass-concrete abutments, one of which is founded on a relatively steep hillside. Soon after the bridge was placed in service, one of the track maintenance men noticed a slight rise in the track at the center of the span; this was corrected but it was soon noticed again. A detailed study revealed that the arch soffit was slowly rising. In turn, this was

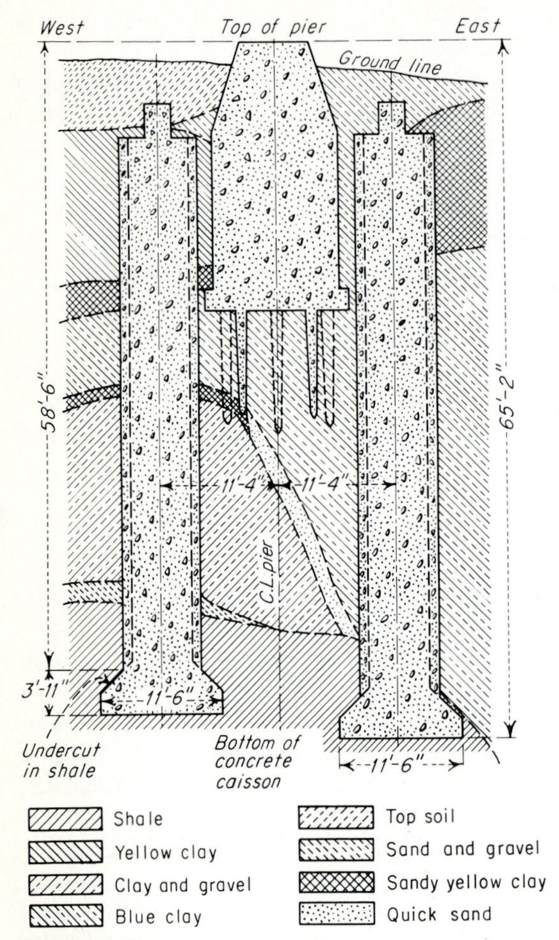

FIGURE 8.14
Section through Pier No. 59 S of the Lethbridge Viaduct, Canadian Pacific Rail, Alberta, showing the geological strata in which it was founded and the way in which the pier was strengthened. (*Courtesy Canadian Pacific Rail, Montreal.*)

found to be caused by a slow but progressive movement toward the river of the hillside abutment. Extensive borings detected widespread movement of the underlying rock talus. One possible solution was available—to brace one abutment against the other by means of a heavily reinforced concrete strut across the river at haunch level. Strain gauges are embedded in the strut, and regular observations have enabled the engineers responsible to check on the adequacy of their design. (And the residents of houses adjacent to the bridge thought that the strut was just a footbridge provided for their benefit!)

Pier inspection below normal water levels should be an extensive and regular operation, including periodic sounding of the riverbed between piers and for

FIGURE 8.15
Reinforced concrete bridge at Klosters, Switzerland, carrying the narrow-gauge Rhaetian Railway to Davos and St. Moritz, showing the special reinforced concrete strut between abutments. (*Photo: R. F. Legget.*)

some distance on either side of the bridge site. Underwater inspection should be undertaken personally by bridge engineers using diving suits. Engineers will thus be able to see personally the effect of the bridge structure on riverbed stability. This type of inspection will be a constant check on the dangerous possibility of riverbed scouring.

The importance of bridge inspection work is emphasized by two bridge failures that occurred in the United States in the late summer of 1933, both of which caused loss of life, injuries, and serious damage to stock and structures. The first occurred at the Anacostia River Bridge of the Pennsylvania Railroad near Washington, D.C.; a four-span, deck plate-girder structure, which, because of the failure of the center pier, collapsed when a train was passing over it. Subsequent investigation showed that the gravel stratum on which the piers were founded had been seriously eroded. The culmination of the erosion process was probably caused by a tropical storm which was abating when the collapse occurred. The railroad company had no record of soundings or underwater inspection from the time the bridge was built in 1904 until the date of failure; it stated that, as no settlement or cracks had developed, unsafe conditions underwater had not been indicated. At the investigation it was stated that "...it was the duty of the railroad company to keep informed concerning changes that might have any effect on the safety of the bridge."

The second accident of 1933 was on the Southern Pacific line between Hargis and Tucumcari, New Mexico. It too was due to unusual floodwaters which undermined the east abutment of the deck plate-girder structure and thus caused the collapse of the bridge as a train was crossing. The official report on the accident attributed the failure to the fact that the embankment had not been protected against erosion, and to heavy rains which had caused excessive floodwater. The position of a highway bridge (constructed 45 m, or 150 ft upstream) had increased the current's velocity and probably had diverted it against the railroad-approach embankments.[10]

These examples naturally prompt the thought that it is always easy to be wise after the event. This cannot be disputed, and yet at the same time it must be recalled that one of the surest ways of learning constructive methods is to study past mistakes and errors. Bridge inspection can provide information of general value only from records of the discovery of unsuspected features having serious consequence. These examples may still serve as a reminder of the critical importance for the design engineer to examine regularly the foundation strata in which such confidence for the support of bridge structures is placed.

GROUTING

Grouting has become one of the most widely used specialty techniques in the practice of civil engineering. It is almost always carried out to overcome geologic defects in foundation beds, or to improve upon natural geological conditions. As a result, grouting is completely dependent upon geology for its success. The uses of grouting in civil engineering work is reviewed in this section, with special emphasis given to its interrelation with geology.

Grouting involves the injection of semiviscous or slurried material into earth material, under pressure and through specially drilled holes, in order to seal open discontinuities, cavities, or other openings in the host strata. *Cementation* is a word applied to cases employing a slurry of Portland cement (either alone or mixed with sand). In addition to cement, clay, and asphalt, various chemical solutions may be used to form the necessary seal. Grouting has frequent applications as the basis for treatment of dam foundations to achieve watertightness, and is now utilized in many other types of engineered construction. Chemical grouting, in particular, is a frequent treatment for difficult foundation conditions in confined spaces, such as in urban tunnel work. As may be seen, grouting is a subject likely to be encountered by many civil engineers, and an expensive treatment that is totally dependent on geology for successful employment.

Few, if any, of the specialty processes in civil engineering have had their history so well documented as has grouting. In two classic papers published in *Geotechnique* in 1960 and 1961, Rudolph Glossop, a British foundation engineer of wide experience, presented a fascinating review of the slow development of the "injection process," as grouting is perhaps more correctly called. Glossop discloses that the process was first invented in France in 1802 for the repair of masonry. An 1837 French report was abstracted for the professional papers of the Corps of Royal Engineers in 1840 by Lieut. William Denison, an officer of the (British) Royal Engineers. Denison had also seen a simple gravity system of injection used about 1830 on the masonry of the locks of the Rideau Canal in Canada, almost within sight of the senior author's home.

First mention of grouting in United States engineering literature appears to have been by William E. Worthen, who related its 1854 use to strengthen a masonry bridge pier of the New Haven Railway line, as a result of his reading. One of the great pioneers of modern grouting was W. R. Kinipple, a British engineer, who was a grouting consultant for solidification of foundations of early dams on the river Nile. The first cement grouting to strengthen bedrock seems to have been in 1876 by Thomas Hawksley, a notable British engineer, at an earth dam at the Tunstall Reservoir. The corresponding pioneer use of grouting into rock in the United States appears to have been in 1893, at New Croton Dam, for the water supply of New York City.

Although they were slow in adopting this new aid, civil engineers today consider grouting to be part and parcel of standard practice. Frequent reports are now released, notably by committees of the American Society of Civil Engineers. Cement is now almost the universal material used for grouting, with clay and asphalt seeing limited use.

Cement is the most usual grouting material adopted. Civil engineering literature contains references to its successful application in many different types of work. Essentially, the method depends on the fact that a Portland cement slurry will, in course of time, "set up" hard and bond satisfactorily with other materials, such as rock. Originally, the grout was simply poured into open cavities and allowed to set up freely. Eventually, the idea of injecting grout under pressure into normally inaccessible cavities was developed. The origin of cement grouting is a matter of conjecture; it seems probable that the Romans were at least familiar with the idea, even though they had no means of pressure injection. It was not until the end of the nineteenth century that James Greathead invented a grouting machine for practical pressure grouting of tube railway tunnels.

FOUNDATION STRENGTHENING WITH GROUT

Grouting has become a means for shaft sinking through water-bearing strata. The pioneering work of Albert François in the early years of the century led to the development of a direct-acting pump capable of developing 210 kg/cm^2 (3,000 psi) pressure. Reliable mechanical controls permitted the use of special chemical solutions to ensure closure of all fine fractures encountered. Cemen tation is now an accepted part of mining engineering practice. Cementation programming can be accurately planned and controlled if formulated with proper geologic knowledge. Grouting has also been applied to control underground coal fires. Unmined coal below 3.2 km (2 mi) of the right-of-way (old) Midland Railway main line near Hasland in Derbyshire, England, had combusted but was extinguished by cement grouting. This unusual case serves to illustrate the flexibility of grouting techniques.

Building foundations have occasionally required the use of grouting to increase bearing capacity. The former Havana Hilton Hotel is founded on relatively porous coral bedrock, strengthened by the injection of over 1 million gal of Intrusion-Prepakt grout. This is one of several respected specialty grouting processes, consisting of injecting a proprietary grout composition into preplaced crushed rock aggregate—in effect, making concrete in situ. This scheme has long been used (e.g., in a bridge over the New Croton Reservoir before the end of the nineteenth century). The combined effect of the specialty grout and the accumulated experience in its use make the system one of great value for especially difficult foundation jobs.

Grouting may also have a direct effect on design. In West Germany a new Geislingen Autobahn bridge, 1,110 m (3,700 ft) long, was planned across the valley of the Kocher River, south of Stuttgart (Fig. 8.16). Strata in the valley sides and bottom consist of shell-limestone, which early geological reports suggested was not strong enough to carry the loads that piers for a new bridge would impose. Then a final examination of the local geology and a reevaluation of the bedrock suggested that the rock must be amenable to strengthening by cement grouting. This led to the development of a more economical box-girder design, with the bridge supported on eight free-standing piers 176 m (588 ft) high. The limestone in the valley sides and floor is of three formations; the lowest one is the strongest of the three but still in need of strengthening. Shafts were therefore sunk for three of the main piers so that they could be founded, with all the other piers, on the lowest limestone formation. All the foundation beds were then grouted with cement and a design saving of over $20 million was effected.

Cofferdams must be used in many civil engineering projects to retain water temporarily during the construction period, notably for dam foundations and the founding of bridges. Each coffer cell must have watertight strata beneath it to be effective. One of the most extensive sheet-pile cofferdam operations was carried out during the construction of Grand Coulee Dam in Washington. The Columbia River was diverted first into one-half of its original bed and then into the other by cofferdams which were about 900 m (3,000 ft) long. The piles were driven generally into lodgement till, which was difficult to penetrate, resulting in splitting and deformation of the steel piling. When excavated, the coffer cells revealed a sand horizon 10 to 25 cm (4 to 10 in) thick, exposed across the excavation at an elevation just below the steel pile bottoms.

Work in the west coffer cell went on without trouble, but in the east cofferdam, owing to some complicating factors, including till-deformed pile bottoms, a leak

FIGURE 8.16
Bridge over the Kocher River, south of Stuttgart, West Germany, on the Geislingen Autobahn, the design of which was changed to that seen here because of the use of grouting to improve foundation-bed conditions. (*Courtesy Wayss and Freitag A.G., Frankfurt.*)

through the sand developed into a flow of 132,500 lpm (35,000 gpm). A small part of the cellular dam was wrecked, and some of the steel piles were split from top to bottom. The leak was finally stopped without serious interruption of construction by placement of 10-cm (4-in) borings to locate and grout the sand layer. Cavities were found from 30 to 60 cm (1 to 2 ft) high. Special grout was required, as cement grout washed out before it could set. Four hundred batches of this grout were pumped through an 18-cm (7-in) concrete pipeline and forced into the passage under a pressure of 2.5 kg/cm^2 (35 psi).

A first hope of bridge engineers is to select a sound geologic foundation so that grouting will not be required to ensure safety and stability. Grouting is only a means of improving upon imperfections of Nature when rock or soil does not have the sound, strong properties required to serve as foundation beds carrying substantial loads. When accurate geological conditions indicate that grouting is necessary, then it can be a most useful technique for achieving satisfactory foundation conditions. Poor grouting may be worse than none at all. Its

FIGURE 8.17
Bridge over the river Danube at Bratislava, Czechoslovakia, as seen from the north bank, the famous castle being off to the right. (*Photo: Milan Matula, Bratislava, C.S.S.R.*)

application is always a job for the experienced expert who needs to know the exact subsurface conditions before applying such skills; geological investigations are an absolute prerequisite for successful grouting operations.

CONCLUSION

River scour of bridges and other structures founded in flowing water is one of the most insidious natural phenomena with which the civil engineer has to deal. Unless the greatest care is exercised in regular inspections, erosion can take place without surface evidence, and remain undetected. Considerable riverbed erosion occurred in the fine-sand stratum below the Tigris River bed at the 17th July bridge site at Northgate in the ancient city of Baghdad. This four-lane, five-span, prestressed-concrete bridge over the river Tigris, 270 m (almost 900 ft) long, is fortunately founded on deep piles for 4.0 to 6.5 m (13 to 21 ft) as scour of the riverbed had taken place within three years of construction. Site borings revealed the potentially erodable sand bed and construction operations encountered ancient artifacts from the period of Nebuchadnezzar.

As a final example to illustrate the message of this chapter, Figure 8.17 shows the new bridge across the river Danube at Bratislava in Czechoslovakia. Foundation beds under the two main supports for the bridge differ because of the ancient fault along which the river flows. On the right bank 14 m (46 ft) of alluvial deposits overlie Late Tertiary clay and sand. On the left bank, granitic bedrock is overlain by 12 to 13 m (40 to 43 ft) of miscellaneous fill. The engineering solutions to the design problems presented by these conditions show yet again the profound influence of geology on bridge design.

REFERENCES

1 D. Moran, "Sampling and Soil Tests for Bay Bridge, San Francisco," *Engineering News-Record, 111,* p. 404 (1933).

2 "Tappan Zee Bridge A Foundation Triumph," *Engineering News-Record,* 155, p. 44 (14 April 1955).

3 A. C. Stermac, M. Devata and K. C. Selby, "Unusual Movements of Abutments Supported on End Bearing Piles," *Canadian Geotechnical Journal, 5,* p. 69 (1968).

4 "Deck-Girder Railroad Bridge Has Earthquake-Resistant Features," *Engineering News-Record, 134,* p. 120 (1945).

5 Vitruvius, *The Ten Books of Architecture* (M. H. Morgan, trans.), Dover, New York, 1960; see p. 162.

6 "Tailor-Made Cofferdams," *Engineering News-Record, 121,* p. 207 (1938).

7 R. F. Legget and F. L. Peckover, "Foundation Performance of a 100-year Old Bridge," *Canadian Geotechnical Journal, 10,* p. 504 (1973).

8 H. Q. Golder and G. C. Willeumier, "Design of the Main Foundations of the Port Mann Bridge," *Engineering Journal, 47,* p. 22 (August 1964).

9 F. W. Alexander, "Maintenance of Substructure of the Lethbridge Viaduct," *Engineering Journal, 17,* p. 523 (1934).

10 "Two Railroad Bridge Failures Laid to Inadequate Inspection,"*Engineering News-Record, 111,* p. 687 (1933).

CHAPTER 9

WATER SUPPLY

Most citizens appreciate the fundamental importance of water supply. It is a prime necessity in public health, an essential to cleanliness, and a fundamental requirement of almost all modern manufacturing processes. Indeed, a satisfactory supply of pure water is one of the engineer's greatest gifts to the public and at the same time one of his or her greatest responsibilities. So vital is the maintenance of a public water supply that no possible risks can be taken in connection with the necessary engineering design. It cannot be denied that geological conditions have a profound effect on sources of water and on the means by which water is supplied for public use. Geological conditions are, however, a part of the natural order and can seldom be altered, whereas the civil engineer must design water-supply systems to accommodate particular natural conditions, which is often a task of great difficulty.

Few cities are so fortunate as to have a duplicate water-supply system backing up the function of the main supply. Unusual precautions have to be taken, therefore, in providing conservative storage capacity. Correspondingly, in no branch of civil engineering are geological considerations more important than in connection with water supply: The most careful preliminary geological investigations are always absolutely essential. The numerous engineered works required for water-supply projects include tunnels, dams, canals, reservoirs, grouting, and buildings, all subjects considered in other chapters of this text. In this chapter, only general reviews will be presented of complete water-supply systems, with special emphasis on the problems associated with groundwater supplies.

HISTORICAL NOTE

The importance of water supply was recognized by some of the earliest engineers. The inscription of the Moabite stone, dating from the tenth century B.C., contains references to water conduits and cisterns. There is evidence to suggest that reservoirs for water supply existed in Babylon as early as 4000 B.C. The Bible contains many references to water-supply systems. Some of these were of considerable magnitude, and parts of them have remained in use up to the present day. Notable among these were the waterworks for the city of Jerusalem, which were installed in Solomon's newly conquered city about 900 B.C. and showed highly developed technical skill. They were extended by King Hiskia in 700 B.C.; an inscription records that his engineers successfully bored a tunnel almost a mile long by working from the two ends. The Romans also augmented the supply, and parts of the original installation are still used today to supply the Mosque of Omar. Works in Asia Minor at Priene (about 350 B.C.), a town of only 5,000 inhabitants, are hardly excelled by those of any modern town, since every house was connected by earthenware pipes to mains supplied from a large and steady spring on the hillside above the town.

The waterworks of ancient Rome are widely known and appreciated, but the fact that the supply to the city is estimated to have exceeded 360 mld (80 mgd) is not perhaps so generally realized. The principal sources of the Roman water supply were springs in the great beds of limestone in the valley of the river Amio. Many of the springs fed into one of the 11 aqueducts that were constructed to carry the supply to the Imperial City. Monumental ruins of today testify to the engineering skill of their Roman builders. Consider the contemporary description of these works, written by Sextus Julius Frontinus, water commissioner of Rome in A.D. 97: "Will anybody compare the idle Pyramids, or those other useless though much renowned works of the Greeks, with these aqueducts, with these many indispensable structures?" In his descriptions, it will be found that Frontinus does not neglect geological features.

The Middle Ages were not devoid of their water-supply engineers. The first post-Roman artificial water supply system in England appears to have been that installed by St. Eanswide, of whom it is reported that "she haled and drew water over the hills and rocks against nature from Swecton a mile off to her Oratoria at the sea-side." This daughter of Eadbald, king of Kent (A.D. 616–640), was the first prioress of St. Peter's Priory near Folkestone.

Only a few examples can be given here but it is clear that the water engineer of today is following in a great tradition with some of our modern problems being reflected in these ancient works.

SOURCES OF WATER SUPPLY

The ultimate source of practically all water supply is the fall of rain; the use of the plural subheading implies that the natural cycle may be subdivided for classification purposes. Geological conditions govern what happens to rainfall when it reaches the ground. Part is evaporated, part will run off the surface, part will be absorbed by vegetation, and part will eventually find its way into the subsoil. All four processes depend to some extent on the geological nature of the ground on which the rain falls. That which runs off the surface to form the flow of streams and rivers is clearly of special interest to water engineers. The available runoff naturally includes also that part of the water which eventually finds its way into watercourses by seepage through local soil and rock. Geology has therefore an appreciable effect on the relation of runoff to rainfall.

The most satisfactory method of evaluating probable runoffs is by means of stream gauging and study of rainfall records. The geology of the catchment area often affects the relation of runoff to rainfall and consequently may interfere with the regularity of stream discharge. Loss to evaporation into the atmosphere and to absorption by plant life are directly written off by the water engineer. That which percolates into the ground, although representing a loss in the rainfall–runoff relationship, goes to replenish the great volumes of underground water which are of such vital importance to life on the earth's surface. Clearly, the amount that thus enters the ground is directly dependent on the geological nature of the exposed surface and on the type of vegetation that is sustained by this geologic material.

RELATION OF GEOLOGY TO RUNOFF

Calculation of rainfall losses is a vitally important part of all hydrological studies. Many empirical formulas have been evolved to suggest a general relation between rainfall and runoff for catchment areas of different sizes and different climatic conditions. Of the factors affecting rainfall losses, two are fundamentally dependent on geological conditions. Although general formulas are useful in limited application, they must be interpreted in the light of local geology if they are to be accurate.

Loss of rainfall by percolation to pervious underground strata is the first of these two factors. In all but the most exceptional watersheds (such as those used at Gibraltar), some portion of the rain will fall on pervious material and so sink below ground surface. There, in the interstices and fractures of the underground reservoir, its slow but steady movement constitutes the basis for dry-weather flow estimates of watershed losses to percolation. Determination of minimum possible dry-weather flow at any gauging station is a critical hydrological investigation, since it will often be impossible to measure this by actual gauging before a water-impounding project is started. Records at nearby completed works can usually be obtained, and these constitute a valuable guide when catchment areas are geologically similar.

D. Halton Thomson once compiled a study of more than 30 dry-weather flows for British streams, and these showed discharges varying from 0.004 cms per 1,000 ha (0.055 cfs per 1,000 acres) on the river Alwen in North Wales (drainage area 2,520 ha or 6,313 acres) to 0.14 cms (2.00 cfs) gauged on the river Avon, Scotland (drainage area 2,020 ha or 5,000 acres). The ratio of the two records from these catchment areas, almost identical topographically and in respect to climate, is 1:36. This is a surprising variation, and the explanation is wholly geological. Superficially, both areas might be regarded as impervious; the geology of this part of the Alwen Valley consists of Silurian shales and grits covered with lodgement till; that of the Avon is wholly granite. Based on this general comparison, it would seem that the dry-weather flows from the two areas should be almost equal. Further study discloses the fact that although the Alwen catchment area is watertight, as might be expected, and thus provides little or no underground storage, the granite over which the Avon flows is so jointed and decomposed that vast quantities of water are held in its interstices. This ample but shallow underground storage thus feeds the stream during persistent dry periods.

Another interesting example is afforded by flow records of two different points on the river Exe in Devonshire, the lower one just above the range of tidal influence and the other near the headwaters of the river above the junction of two important tributaries with the main stream. It was found that the unit dry-weather flow for the whole area is almost four times that of the upper part of the area alone; the reason for the variation is, again, solely geological. Only about one-eighth of the upper part of the area is formed of pervious rocks, the New Red Sandstone, and the remainder is Devonian Measures and igneous rocks; thus, practically no underground storage exists. The basins of the two tributaries that join the main river below the upper gauging station are largely formed of the pervious New Red Sandstone, and in consequence, they are both well served with underground storage to the extent that their unit dry-weather flows are so high that they raise the figure for the whole catchment area to that indicated in the table.

These examples are by no means unusual, and they illustrate vividly the effect that geological conditions have on dry-weather flows. This effect is felt to varying degrees for all stages of streamflow. Even during torrential rainfall, a pervious surface stratum will rarely be so completely saturated with water that rain falling on it will all run off, giving a flow equivalent to 100 percent of the rainfall. This is confirmed by the unique conditions that combined to cause the catastrophic floods early in 1936 in the eastern part of North America. Unexpectedly early rains fell on catchment areas that were still frozen hard, and thus impermeable. The resulting runoffs caused floods that broke all known records and in some cases exceeded the calculated "1,000 year maximum." That such floods do not normally occur is due to the regulating effect of geological formations, which affect the course of water from the time it falls as rain until the time it reaches the watercourse to join with the streamflow.

The figures quoted illustrate how geological conditions alone can explain some variation in runoff records. By means of similar studies, not only can records be compared, but the accuracy of recorded flows can be checked as possible limiting cases. Whenever runoff records have to be correlated with rainfall, the importance of watershed geology must be kept in mind if meaningful and useful correlations are to be made.

WATER QUALITY

Rainfall absorbs gases and floating solid particles from the air before it reaches the ground. Rainwater is therefore by no means pure; it is often definitely acidic, especially during thunderstorms, when nitric acid may be formed. There is currently much public concern about acidity in rainfall ("acid rain") caused by industrial pollution. Small portions of carbon dioxide are always absorbed. Consequently, as runoff rainwater flows over soil and rock of various types, there is a slight chemical effect, and the characteristics of lake and river water can therefore vary considerably. Knowledge of geological conditions causing chemical variations in water quality is a prerequisite for maintaining drinking water quality.

Many upland watersheds are covered with peat. Runoff on peat deposits inevitably takes on traces of the peat and of organic acids, becoming brackish, often slightly discolored, and usually slightly acidic. If acidity is pronounced, the water may be slightly plumbosolvent, a grave matter in England and other European countries where lead piping is still in use for plumbing. Such water may also have a serious effect on exposed concrete surfaces. Staffordshire blue brick (an engineering brick of high quality) and aluminous-cement concrete are the only normal types of open conduit lining that successfully resist the action of moorland water. Ordinary portland-cement concrete of varying consistencies is seriously affected.

Hardness, temporary and permanent, is the most widely recognized of impurities. One can gain a general appreciation of the geology from which a water supply comes, when paying a first visit to a locality, by simply washing one's hands. Hardness is entirely geological in origin since, in surface waters, it is the result of contact with rocks containing calcium or magnesium carbonate or calcium or magnesium sulfate. If a river is known to flow over limestone or dolomitic deposits, the source of the contamination can easily be found.

Sometimes the strata in the bed of the river do not suggest the cause of the salt content. The river Derwent in Derbyshire, England, for example, flows over grit beds, and yet its water is hard. The hardness is due to the fact that the

FIGURE 9.1
The distillery at Bunnahabhain, Isle of Islay, Scotland, showing the typical "granite and peat" topography that is conducive to the distilling of good Scotch whisky. (*Courtesy Scottish Tourist Board, Edinburgh.*)

streamflow is fed by tributary streams which come down from mountain limestone deposits. Study of geological features both in riverbed and over catchment area will suggest the type of hardness and the necessary treatment.

Although generally troublesome, hardness of water is not always a disadvantage, as some beer drinkers may know. It is usually the mineral content of water that gives special local flavor to pale and bitter ales; hard water is desirable for their manufacture. For example, ales brewed at Burton-on-Trent, England, are made from hard water obtained from valley gravels and Keuper Sandstone. It was once carefully calculated that in one year drinkers of Burton ales consumed 160,000 kg (350,000 lb) of solid gypsum in the process of assuaging their thirst. Soft water is required, on the contrary, for stout. Waters used in the north of Scotland and elsewhere for a similar, although stronger, purpose are also dependent on their minute mineral content for their peculiar efficiency (Fig. 9.1). Just in case one or two readers may have some interest in this rather specialized aspect of water quality, it may be recorded that one famous company which operates several distilleries once arranged for each of the distilleries to use water from one of the others, an erudite research project of some appeal. It was found that, although the resulting Scotch whisky produced was, in every case, excellent, it "was entirely different from that produced before (they) temporarily exchanged the waters."

More prosaic is the record of a study from South Africa of five pairs of streams in Natal. The streams were unpolluted, individually geologically homogeneous, and the two streams in each pair were geographically far apart. Samples were taken in both the dry and rainy seasons. This investigation clearly demonstrated the influence of the varying geological formations over which the streams ran. Total dissolved solids varied from 314 ppm (for the Dwyka Formation) to 38 ppm (for the Beaufort Formation). Breakdown of the chemical content into percentages of Ca, Mg, Na, K, CO_3, SO_4, Cl, and SiO_2 reflected the profound influence of the geology of the catchment areas.

Geology not only influences the quality of "natural waters" but also the nongeological impurities which unfortunately often reach water supplies due to geological conditions. Some years ago, for example, the Los Angeles area faced serious contamination of groundwater by the wastes from oil wells. Trouble started when sumps were excavated along the east bank of the Los Angeles Flood Control Channel for the purpose of reclaiming the oil which remained in the wastewaters from oil-well drilling. These sumps handled more than 9.5 mld (2.5 mgd) of wastewater, containing 9,000 ppm of chlorine or 13,000 ppm of sodium chloride. Adjacent wells penetrating the local sands and gravels soon showed contamination; some had to be abandoned and others deepened and cemented in to seal off the source of contamination. Tests showed conclusively that the bottom and sides of unlined ditches and sumps had not been effectively sealed by their drilling mud and oil emulsions.

This kind of problem must be faced whenever groundwater is being withdrawn from areas of industrial waste contamination. In the past, a major source of such pollution has been the wastewater from active and, especially, abandoned coal mines. Most coal-mining states have taken very strong regulatory action to eradicate the grave danger which arises from the interaction of air and water upon the sulfur-bearing minerals exposed in coal-mining operations. The problem has been aggressively tackled in the United States under the provisions of the Clean Water Act and the Surface Mining Conservation and Reclamation Act, and the results have been highly satisfactory.

This geologic contamination interdependence is well illustrated by the "Montebello incident" in California. Process wastes from formulation of the weed killer 2,4-D (2,4-dichlorophenoxyacetic acid) were discharged into the sewers of the city of Alhambra. After passing through the city's sewage-treatment plant, the wastes were discharged as a small part of the effluent into the Rio Hondo River about a mile (1.6 km) upstream from an interconnection with a subterranean groundwater basin. Within a short time, 11 wells serving 25,000 people were so seriously affected that, even after cutting off the source of contamination, large expenditures were necessary to treat the water to make it safe again for public use.[1]

WATER SUPPLY FROM RIVERFLOW

Potable water is most simply obtained from the flow of rivers or from freshwater lakes; it has been thus used since prehistoric human times. Today, some of the greatest cities in the world secure their water supplies in this way—

London, from the river Thames; Montreal, from the St. Lawrence; Chicago and other North American lake cities, from the Great Lakes. What relation does geology have to this branch of waterworks engineering? There is, first, the indirect influence that geological conditions have on the rate of riverflow. Geologic structure will affect to a considerable degree the type of structure adopted. The duplicate water-supply system at Toronto, for example, employs a tunnel in the local shale for a distance of 990 m (3,300 ft) out from the lakeshore, finishing in a vertical shaft leading up to a series of precast concrete and steel pipes laid on the lake bed in a trench, at the ends of which are located the screened intakes.

Where source rivers are bordered by porous beds of sand or gravel, advantage is taken of this geological feature, and the naturally filter-drawn water is from the porous stratum. The natural-filtration technique is also adapted in the widely used artificial, slow sand-filter beds process. Many such installations are to be found on the highly industrialized rivers of Europe. Berlin and Frankfurt in Germany and Goteborg in Sweden obtain at least part of their water supplies in this unusual way, as do some cities in the great Ruhr industrial area of Germany. Hamburg, Germany, in addition to obtaining water that has filtered through from a riverbed, uses two irrigating channels as additional "distributors"; river water is pumped into infiltration channels located to the north and south. The water supply wells of Breslau, Germany, draw water from the porous strata via well points similar to those used to dry up water-bearing construction sites. Thus, Düsseldorf, Germany, uses 30-cm-(12-in)-diameter gravel-packed suction wells spaced at 21 m (70 ft), each yielding about 110,000 lph (28,800 gph).

This interesting method is not confined to Europe. The city of Des Moines, Iowa, has long obtained its water supply in this way from the Raccoon River which flows through a valley about a mile wide consisting of impervious material superficially blanketed by water-bearing sand. Reinforced-concrete collection galleries 120 cm (4 ft) in diameter are located parallel to the river bank and about 60 m (200 ft) distant. Uniform percolation here yields a minimum supply of over 1 mld (300,000 gpd).

A somewhat more unusual installation supplies the city of Kano, an important center of northern Nigeria, with a population of over 100,000. Original supplies had been drawn from wells dug in the local laterite. The new installation was made up of five large reinforced-concrete cylinders, three with internal diameters of 2.7 m (9 ft) and two with internal diameters of 4.5 m (15 ft), sunk to depths of 10.5 to 15.6 m (35 and 52 ft). The wells rest on rock and penetrate the riverbed sand, over a stretch of about 3 km (2 mi). Intakes of 10-cm (4-in) diameter are installed in two of the wells and porous concrete blocks in three; these are located near the rock surface to draw water by gravity; the sand acts as a filter and also as a storage reservoir during dry seasons. A daily flow of over 2 million L (600,000 gal) was obtained.

To refer to a water supply from beneath a dried-out riverbed may appear to be stretching the title of this section, but the inclusion is literally correct. A somewhat similar example is to be found in Harrisonburg, Virginia. The City considered augmenting its water-supply system on account of a water shortage during the great drought of 1930. Supply was found from groundwater flow below the valley of the Dry River, from the normal streamflow of which the city already drew its main supply. An existing concrete diversion dam was extended for a distance of about 270 m (900 ft) through a relatively level part of the valley, and as far as a steep rock cliff. This necessitated a thin, reinforced-concrete wall, varying from 30 to 60 cm (12 to 24 in) in thickness and keyed into the underlying fine-grained, closely cemented Pocono Sandstone for a distance of 40 cm (16 in). The entire wall was backfilled on the upstream side with selected stone, obtained to some extent from excavation. At the lowest level of the rock, a concrete collecting gallery was constructed upstream of the dam, from which a 35-cm (14-in) supply main leads to the city.

The valley bed consists of products of weathering and erosion debris varying in size from fine sand to large boulders, found in 12 separate underground streams. These streams carried such an amount of water as to cause difficulties during excavation (Fig. 9.2). This flow averages about 3,600 lpm (700 gpm). A similar idea has been used in southern France.

WATER SUPPLY FROM IMPOUNDING RESERVOIRS

When adequate quantities of public water supplies cannot be obtained directly by river or lake abstraction, an obvious alternative is to utilize a distant source of supply such as the streams found in mountain areas. Although these aqueduct schemes are always bold engineering conceptions, they are only an extension of Roman engineering. Water is conveyed today for city use in aqueducts hundreds of kilometers long; those supplying Los Angeles, California, bring water 400 km (250 mi) to the city from the Owens Valley, 70 km (43 mi) of this through 142 tunnels. Other distinguished engineering enterprises of this type include the Hetch Hetchy scheme supplying San Francisco, the Catskill and Delaware systems for New York, the Vyrnwy scheme supplying Liverpool, the Rhyader supplying Birmingham, the Haweswater for Manchester, and those that supply other inland cities of England.

A typical water-supply scheme of this type consists of a rainfall catchment area, a storage dam, an intake structure, the main aqueduct, and a point on the city distributing system, generally a storage reservoir. The engineering problems introduced by such works are not peculiar to water-supply engineering and no unusual features are introduced by their use for water-supply purposes. The problems that they introduce can therefore be treated generally.

The Catskill system for the city of New York includes two great tunnels, one under the Hudson River 366 m (1,200 ft) below water level, and two great dams. The detailed geologic study made in connection with this and other associated works was a pioneer work in this field. The Board of Water Supply for the City of New York continued to apply geologic study to the new Delaware River system. The Delaware River aqueduct consists of 137 km (85 mi) of tunnel in

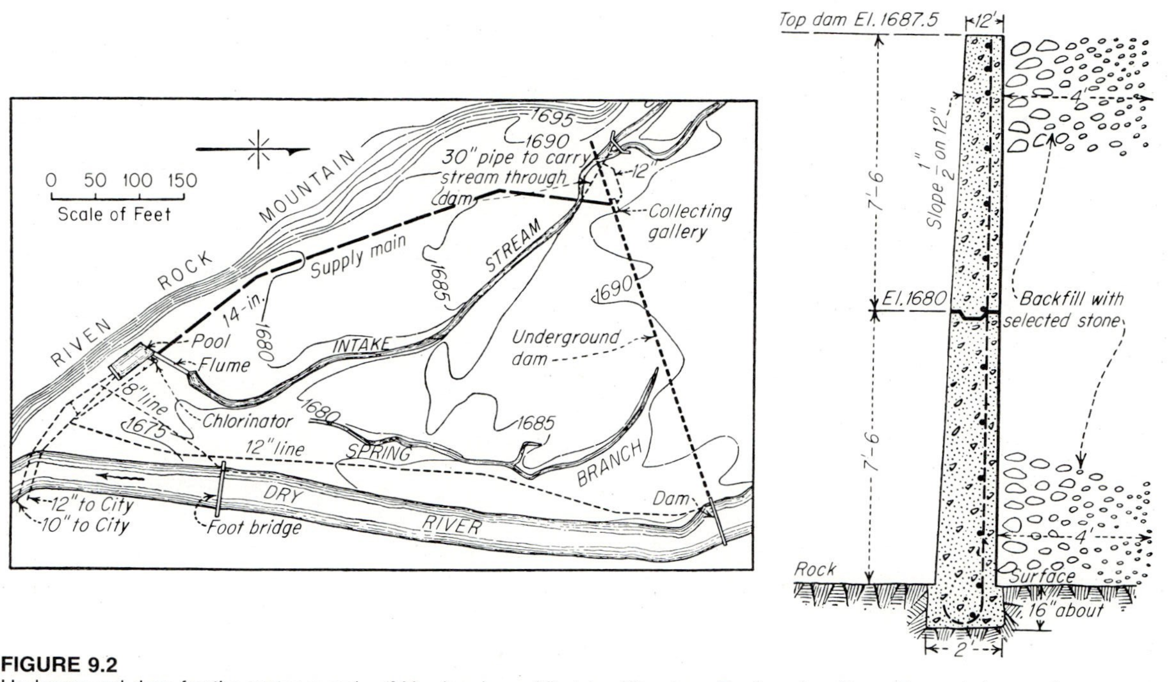

FIGURE 9.2
Underground dam for the water supply of Harrisonburg, Virginia. (*Courtesy Engineering News-Record; from reference 2.*)

six sections, the longest of which is 72 km (45 mi); the predominating rock types are shales, slates, sandstones, gneisses, and schists.

San Francisco built the great Hetch Hetchy water supply and power scheme, put into service in 1934. Water is obtained in a catchment area in the Yosemite National Park and conveyed to the city through a 250-km (155-mi) aqueduct, 130 km (82 mi) of which consists of tunnels, the remainder in pipelines. Several dams are also included; the largest is the O'Shaughnessy (impounding) Dam, 129 m (430 ft) high, founded on granite which was seen to contain many large and deep potholes when uncovered by excavation. The tunnels were driven through granite, slate, and sedimentary materials. In these last, much trouble was encountered because of "quicksand" and excessive pressure from chlorites, methane gas, and sulfureted hydrogen; an explosion of methane was the cause of a fatal incident. Despite all difficulties, the tunnels were completed; in some the granite rock was so satisfactory that the tunnels have remained in use without concrete lining.

The Colorado River Aqueduct of the Metropolitan Water District of Southern California, which conveys water for 385 km (240.5 mi) from the Colorado River to the Los Angeles area for the use of 13 cities, is an even greater project. The aqueduct was designed for a peak flow of 45.5 cms (1,605 cfs), or approximately 40 million mld (1,000 mgd). There are five pumping stations having a total lift of 485.4 m (1,618 ft). The aqueduct consists of 147 km (91.9 mi) of tunnels, 100 km (63.4 mi) of open canal, 89 km (54.7 mi) of conduit, and 47 km (28.9 mi) of siphons. Excavation in the aqueduct totaled about 19 million m^3 (25 million yd^3), and 2.7 million m^3 (3.5 million yd^3) of concrete were used. Of the total tunnel footage, 58 percent was supported by structural lining, 14 percent was coated with gunite, and the remainder was unlined.

These figures will indicate the magnitude of the project, but its unique nature will be realized fully only when it is considered that the eastern part of the aqueduct is located in a remote desert region, construction operations in which could begin only after the building of 237 km (148 mi) of surfaced highway, 750 km (471 mi) of power lines, and 290 km (180 mi) of water-supply lines. The aqueduct passes through the desert as a concrete-lined open canal, and the lining is watertight, as the surface of the ground consists generally of alluvial deposits. It crosses three active earthquake faults; flexible pressure pipelines across these locations have been considered desirable. Excavation for the impoundment at Parker Dam and the tunnel construction were affected by unusual geological conditions (see Fig. 9.3).[3]

Even in such a relatively small country as Great Britain, water-supply systems connecting different river basins have now been necessary. The Kielder project in northeastern England, an undertaking of the Northumbrian Water Authority, was conceived as an answer to steadily growing water demands—a rise from 635 mld (140 mgd) in 1961 to 910 mld (200 mgd) in 1971, and a possible demand of 1,970 mld (433 mgd) by the turn of the century. Careful studies suggested that one large storage facility would be more economical than five or six smaller reservoirs, even though the one reservoir necessitated transferring water between three river basins. Kielder Dam, to be

FIGURE 9.3
Eagle Mountain pumping station on the Colorado Aqueduct, California, which lifts water 131 m (438 ft) out of the total lift of 485 m (1,617 ft), showing typical terrain traversed by this water-supply system. (*Courtesy Metropolitan Water District of Southern California.*)

western Europe's largest earth-fill dam, was therefore designed. Its construction was obstructed far more by environmental issues than by any geological factors. The transfer tunnel, 38 km (24 mi) long and 2.8 m (9.5 ft) in internal diameter, starts at the river Tyne 24 km (15 mi) upstream of the city of Newcastle upon Tyne.

These few references to water-supply schemes involving reservoirs and connecting conduits will not be complete without including the greatest of all such projects, the California Water Plan. Operational goals of the major California water-supply projects had been met, yet further supplies were still necessary for the large urban populations in the southern part of this state. Over 70 percent of the natural streamflow occurs in the northern third of the state, and yet the great demand for water is in the south. Accordingly, in 1951 the state legislature authorized the construction of the California Water Project, by the Department of Natural Resources, to bring northern water in excess of local needs to the south. Later the Department of Water Resources was established to be responsible for this and other water concerns of the state. In 1960 a $1.75 billion bond issue was approved by the voters. The total project cost amounted to $2.3 billion. Construction occupied more than the whole decade of the sixties (Fig. 9.4).

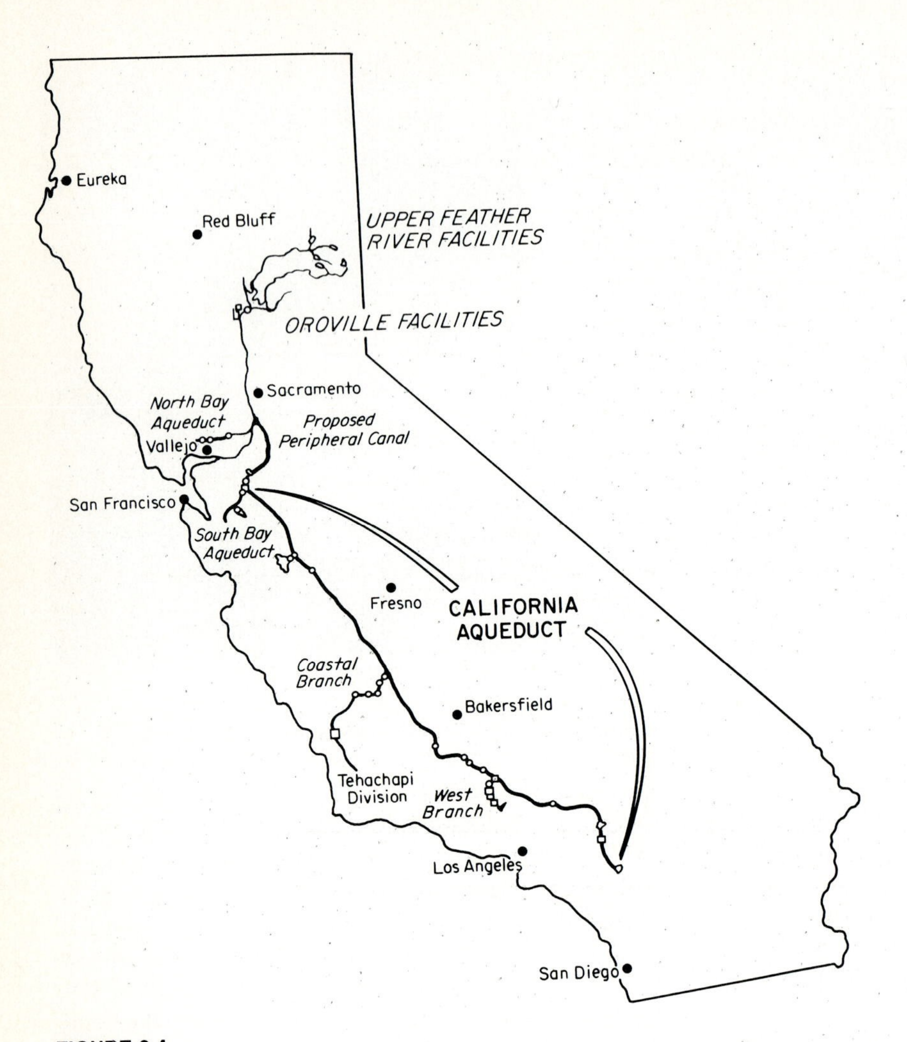

FIGURE 9.4
Outline plan of the main features of the California Water Plan. (*Courtesy Department of Water Resources, California.*)

Most parts of the vast undertaking were operational by 1973. These included 1,075 km (684.5 mi) of aqueducts, mostly in canals but with 33 km (20.5 mi) in tunnels; power plants, operated by falling water in the aqueducts, capable of generating over 5 billion kWh; pumping plants requiring for their operation almost 13 billion kWh; and no less than 21 major dams with a total volume of over 204 million m^3 (267 million yd^3) of material. Oroville Dam alone contains 61 million m^3 (80 million yd^3) of soil. Since the region in which these many works had to be constructed includes some notable faults and some active seismic provinces, design studies have added much to geotechnical knowledge in these two fields. Construction practices also presented many notable advances, one of the most unusual being the flooding of ponds in the San Joaquin Valley to induce consolidation of the unusual collapse-prone soils there, in advance of canal construction. The literature of civil engineering and of engineering geology has been enriched by the records already published of this outstanding water supply system, named the "Outstanding U.S. Civil Engineering Achievement of 1972" by the American Society of Civil Engineers.[4]

WATER SUPPLY FROM GROUNDWATER

Wells constitute the principal means of obtaining a supply of water from underground sources. Unless they pass through very stable strata, good well practice calls for casing throughout the depth of all strata, except that from which the supply is to be obtained. Occasionally, wire screens are used to protect that part of the well into which the water is being drawn. Another precaution is to fill the well annulus with specially selected coarse gravel surrounding the discharge pipe. This may decrease the yield of the well by about 5 percent, but it will prevent plugging.

In areas of plentiful labor supply wells are sometimes constructed as shafts that penetrate strata of low specific yield. Here it may also be necessary to attempt to secure a greater exposed yield surface by driving adits into the pervious strata from the well shaft. It will generally be a wet construction operation, but it has the advantage that as soon as water problems become too troublesome, the object of the adit is achieved. Roman adits are quite common in wells in the Chalk of southern England and in the Bunter Sandstone. In the Canary Islands adits alone are sometimes used as water-collecting galleries; they are driven into steeply sloping lava beds to obtain the water that seeps through this porous rock. The *karez* of northern India and the *kegriz* of Iran are primitive types of a special kind of adit of considerable interest but restricted application. The planning and building of these unusual collecting galleries appear to be a matter of intuition rather than calculation, and the work is generally hereditary. Soviet engineers working in the Caucasus have tried to construct *kegriz* scientifically but without success.

Fortunately, the usual type of well installation does not call for anything beyond ordinary scientific skill. When the excavation has tapped the water-yielding stratum, a body of relatively free water will be available which can be drawn upon as required. The level of this water may vary from time to time quite apart from the variation due to pumping. When the groundwater surface is shallow, barometric pressure may affect the level, and even temperature has been found to have some effect because the capillary properties of the strata above are affected by a change in temperature. Water levels in adjacent wells may not coincide if there is a marked change in geological conditions between them. A striking example occurred at St. Andrews, Scotland. A farm drew its supply from a shallow well 8.4 m (28 ft) deep in sands and gravel. An architect sank a deeper well 18 m (60 ft) from the borehole to supply some new cottages, but only a trace of water was found, although the well was sunk to a depth of 9 m (30 ft) in the same gravel. The poor performance was probably due to variation in level of the underlying pervious lodgement till boulder-clay.

When pumping is started in a well that has been satisfactorily completed to a free-water surface, the water level will naturally fall and continue to do so until the hydraulic gradient in the surrounding material equalizes the amount withdrawn. For uniform material, this gradient curve is theoretically a parabola. This phenomenon surrounds the well as the *cone of depression.* The circular intersection of a cone of depression with the groundwater surface will mark the range of influence of the pumping operation. The effective yield of any other well located within this range will be interfered with to some extent. Such interference is a critical matter in urban areas where groundwater resources are being taxed to the limit. In Liverpool, England, the effect of pumping has been noticed as much as 3 km (2 mi) away from the well in which pumping was taking place, in largely pervious red sandstone.

The wells so far considered have been without unusual geological features. In addition, engineers should be aware of a few special types of wells and their geologic character. Some shallow well supplies depend on *perched water,* i.e., a body of groundwater lying above and separated from the zone of saturation by an impermeable stratum. Some striking examples are found on Long Island, New York, where a thin layer of almost impermeable lodgement till separates a pervious surface deposit from the main body of pervious strata, lying as much as 75 m (250 ft) below this perched water.

Australia provides numerous examples of two other types of unusual wells. *Soak wells,* used for obtaining shallow groundwater that collects after running off exposed sloping surfaces of impervious weathered granite, are constructed in porous weathered granite at the foot of these slopes. The second type of unusual well encountered in parts of Australia (and elsewhere) is known as a *tray well* (Fig. 9.5), tapping localized groundwater perched not on a continuous impermeable stratum but on a lenticular bed of impervious material. At Challner, a saucer-shaped bed of impervious clay with some ironstone nodules about it, located halfway down through a deep stratum of sand which overlies solid granite, collects a small quantity of freshwater as it percolates through from the surface. This supply is tapped by the Challner Well; endeavors were made to increase the supply by boring further, but the bed of clay was pierced and the freshwater started to empty into the sand below. Fortunately, the well was saved by plugging the hole.

Another, 305 m away, is known as Nugent's Well, the bore of which encountered no water above the thin clay layer, so that it had to be extended to the main body of groundwater overlying the granite directly. This second well is also unusual: It cannot be extended right down to rock surface because the water immediately above the rock has a strong saline content. Only the thin layer of freshwater that tops this saltwater layer can be tapped for use.

Salt water may be present for two general reasons: (1) encroachment of

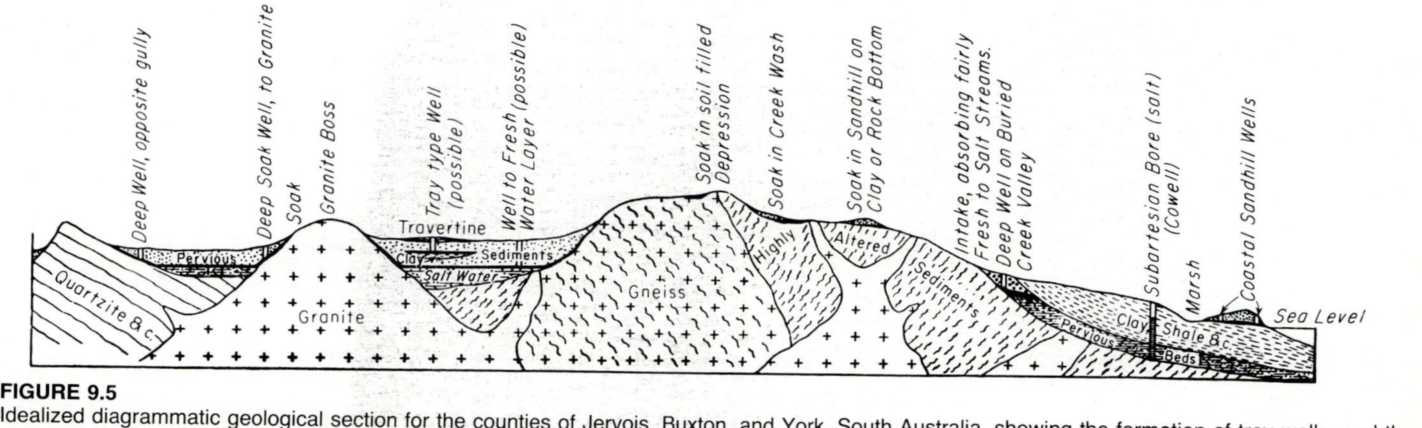

FIGURE 9.5
Idealized diagrammatic geological section for the counties of Jervois, Buxton, and York, South Australia, showing the formation of tray wells, and the dependence of these and other water-supply features upon geological structure. (*Courtesy Department of Mines, South Australia.*)

seawater into wells located near the seacoast, and (2) contamination of inland well waters with excessive mineral content. In both cases, as soon as the salt content passes about 500 mg/L, it ceases to be fit for human use. Salt contamination of inland groundwater, such as is found in parts of Australia and in the Sudan, has in the past been attributed to the leaching out of decomposition products of overlying or adjacent rocks. Consideration of the chlorine content of samples investigated in South Australia disproved this supposition and indicated insufficient drainage for diluting salts brought down by rain. Continued evaporation of water gradually concentrates these dilute salt solutions. As evidence of what rainwater may contain, it may be mentioned that in England the average chlorine content of rainwater is about 2.2 ppm, equivalent to 3.62 parts of salt. Rain falling at Land's End during strong southwest winds blowing in from the sea has been found to contain one hundred times this amount. This complex problem is further complicated by local conditions. The greatest care must always be taken to avoid driving a well into saline water, and care certainly must be taken to prevent overpumping, which introduces "man-made geological problems."

There is little need to describe the usual well field installations which supply water so economically to so many municipalities. Two unusual developments, however, will serve to illustrate the complete dependence of all underground water supplies upon local geology. The Province of Saskatchewan embarked in 1963 with the government of Canada on a study of provincial groundwater resources. An area of 259,000 km^2 (100,000 mi^2) in the southern part of Saskatchewan had been studied in detail by 1970. Geologic mapping combined with deep drilling discovered a number of buried valleys. These are the same preglacial valleys which have been frequently mentioned in other chapters, but in the prairies they are located nowhere near the surface, being found at depths such as 150 m (500 ft) below the surface. As usual, the valleys were found to be filled with glacial deposits and some till, but much sand and gravel, and this was water-bearing. These were really buried rivers which can eventually be tapped as sources of water supply. It has been estimated, for example, that the buried watercourse (sometimes known as the Ancestral Missouri) found in the Estevan Valley could supply an estimated 200 to 400 mld (50 to 100 mgd) from several wells. Water quality is such that treatment would be needed before use, but when water is scarce, this is a small penalty to have to pay. Almost 50 observation wells have been installed and Saskatchewan will have eventually a very clear picture of its deep underground water resources.

The second special case is the latest development in the water supply of London, England, one of the best-known and oldest-established systems for any major city. The city obtained its first water from the great Chalk Aquifer provided beneath it by the London Basin. When additional supplies were needed, water was abstracted from the rivers Lea and Thames, the latter now providing the major part of the metropolitan supply. One of the approaches to London's Heathrow Airport lies right over some of the impounding reservoirs. Still more water is necessary, but the Metropolitan Water Board is not allowed to take any further water from the Thames above Teddington Weir when the river flow is less than 65 mld (163 mgd), as it usually is for much of the summer. Instead of building still more reservoirs to hold water throughout the months of low flow, the Thames Conservancy (responsible for water above the weir) embarked on an experimental program of utilizing groundwater to augment low-water flows.

The Conservancy selected an area of 1,000 km^2 (386 mi^2) containing two small Berkshire streams, the Lambourne and the Winterbourne. After intensive preliminary geologic study, a three-year program was started, pumping groundwater into the streams at times of low river flow and of low rainfall. Preliminary calculations had suggested that the groundwater level would thus eventually be restored to its normal elevation through streambed loss into the high permeability underlying the chalk. It was estimated that over a ten-year period a net dry-period gain in the flow of the river Thames of up to 378.5 mld (100 mgd) could be obtained, and at an estimated cost of from one-third to one-quarter that of the conventional balancing reservoirs.

Based on the most detailed investigation of underlying geology, the experiment was quite successful. The scheme is being implemented, and a similar scheme is already in operation in the catchment area of the Great Ouse River in eastern England. When the local geology is suitable, this concept of adjusting the timing of the flow into and out of the ground to increase low flows in rivers used as a source of water supply is one that could be widely adopted, with great savings and no deleterious effects on the environment.

REPLENISHMENT OF GROUNDWATER

All over the world groundwater supplies are being obtained from springs and by pumping. And all over the world there are today all-too-many cases of "overpumping," with a variety of serious consequences, such as seawater intrusion at coastal locations and subsidence of overlying ground. Most of the water pumped out of the ground for human use is disposed of as "waste." If some of this waste could be reclaimed and returned to the ground, it could replenish the great natural resource which groundwater represents. Excess surface waters might be used in the same way. This is being done now to a steadily increasing degree in many parts of the world.

It would appear that the first attempts to replenish groundwater were made in 1881 at Northampton in England; in this case, water was fed back into the Lias Limestone. Experiments were conducted in the London Basin by the East London Water Company as early as 1890. In North America, the Denver Union Water Company may have spread water over the alluvial gravel fan of the South Platte River near the mouth of its canyon in 1889.

It was not until well into the present century, however, that the need for replenishment of groundwater became so urgent that major schemes were investigated and ultimately put into operation. Los Angeles, with its large population, has been the center of much activity in this field. Experiments were first started in the San Fernando Valley as early as 1931, and spreading grounds were developed by 1936 to affect one-third of the groundwater supply for the city of Los Angeles.

After five years' successful operation, the system took on a definite character. Basins 30 by 120 m (100 by 400 ft) were separated by dikes 1.2 m (4 ft) high and 4.5 m (15 ft) wide. Inflowing water was left to seep into the ground at the rate of between 0.9 and 3.0 m per day (3 and 10 fpd). This took between 8 and 24 hours, depending on the state of the beds, which had to be harrowed and cleaned of silt at intervals of ten days. The water reached the North Hollywood pumping station, some distance down the valley, one year later and was there pumped out for use through eighteen 50-cm (20-in) wells 120 m (400 ft) deep.

This early work led to great progress which has today made Los Angeles the leading area of groundwater recharge in the world (Fig. 9.6). Because of continuing groundwater depletion, estimated to amount to 865 million m^3 (700,000 acre-feet), a special water-replenishment district now includes 29 cities with a combined population of 2.5 million people. Costs are met by pumping assessments against water companies and large water users. Between 50 and 75 new wells have also been drilled for recharging purposes along 17.5 km (11 mi) of coastline as part of the "Barrier Project" designed to limit further encroachment of seawater.

Some progress has been made also in the Los Angeles area in the use of reclaimed sewage-plant effluent for recharging groundwater supplies; naturally this operation is under strict control. The city of Amarillo, Texas, has similarly improved its groundwater situation by recharging from the surface.

The practice is, however, not restricted to the United States. In Europe, too, falling groundwater levels have created similar needs. In Sweden, 20 municipalities obtain their water supply from underground sources supplemented by recharging. And in London, England, the Metropolitan Water Board has used old shafts and adits to put water back into the underground chalk reservoir. Importantly, some of the debt incurred by a too-profligate use of one of Nature's great bounties is at last being repaid.

It is believed that groundwater recharge is today being practiced to some degree in every state of the United States. In some cases recharge is achieved by flooding areas that are underlain by permeable soils; in others, the same result is achieved by discharging water into wells. In some cases "raw water," as from riverflow, is used for recharge; in some, storm water which is collected separately from sewage is used; and in an increasing number of cases, treated sewage is approved for use, under strict testing and with assurance that it is suitable. On a vastly greater scale, Israel operates a major wastewater reclamation scheme to add 12 percent to its total supply. Special lagoons were constructed, the treated water being allowed to percolate into the ground, being reclaimed (after an estimated stay of one year in the ground) by pumping from wells.

FIGURE 9.6
Restoring water to the land in California; the Rio Hondo spreading grounds for groundwater recharge of the Los Angeles County Flood Control District, near Pico Riviera. (*Courtesy Los Angeles County Flood Control District.*)

One of the most enlightened multipurpose developments involving groundwater recharge relates to Coors beer, well known in the western part of the United States. The Coors brewery is located to the west of Denver, in the outskirts of the municipality of Golden. Springs supply the water for making the well-known beverage. Much more water is required in their plant for industrial purposes, however, and this is obtained from Clear Creek. The stream flows over gravel 18 m (60 ft) deep in places, which was first found to be gold-bearing, and is now proving its further worth as concrete aggregate. Immediately downstream from the brewery, gravel is being mined so as to leave carefully shaped shallow pits, which are then allowed to fill with water from the creek and to become attractive lakes. The water percolates through the gravel and then is reclaimed for plant use after this period of natural filtration (Fig. 9.7). The combination of gravel mining, water conservation, and development of public recreational facilities in this area is a vivid example of what can be

FIGURE 9.7
General view of the way in which the Adolph Coors Company is using worked-out gravel pits as spreading grounds for water recharge, near Golden, Colorado. (*Courtesy Adolph Coors Company.*)

accomplished with cooperation between different responsible agencies and with farsighted policy making on the part of management, all based on a thorough knowledge of the unusual local geology.

SOME UNUSUAL SOURCES OF SUPPLY

Unusual sources of water supply have been adopted out of necessity in some parts of the world. The supply of the Rock of Gibraltar is one such example. The hard, dense Jurassic limestone is without springwater. Strategic water supplies for the fortress were limited for many years to the rainfall on the surface of the rock itself. The town also relies on a catchment area of 10 ha (24 acres) constructed of corrugated iron sheets laid on 40° sand slopes. The sheets were secured by means of timber purlins and piles creosoted under pressure. An additional catchment area of 6 ha (15 acres) was secured in unutilized steep rock slopes on the west side (Fig. 9.8). Rockfalls frequently damage the large catchment, but trouble is quickly repaired.

Additional potable water was found in the sand of the flat tableland to the north of the rock. Two wells supply about half of the total consumption.

FIGURE 9.8
The "iron-clad" catchment area on the Rock of Gibraltar. (*Photo: R. F. Legget.*)

Brackish water, used for all purposes other than drinking, is obtained from shallow wells in the same area. All drinking water is stored in underground rock reservoirs; these have a total capacity of 63.5 million L (16.8 million gal). The reservoirs are in galleries leading from two main construction adits driven from west to east at a level of roughly 112 m (375 ft) above sea level. An additional corrugated-iron catchment area of 4 ha (10 acres), together with two new reservoirs inside the rock, have been added as water supply for naval ships. This interesting example shows how unfavorable geological conditions have been adopted to give an almost perfect example of surface water supply. Gibraltar's water is used without treatment, and although rather "flat," it is quite palatable.

In Australia (notably Western Australia), outcrops of granite are remarkably solid and shaped so that small concrete walls can be constructed to channel rainfall to storage tanks, sometimes including holes excavated in granite. In areas where no solid rock is available, Australian engineers have waterproofed catchment areas of sandy soils. Such attempts were first made in 1935 in Western Australia when emulsified bitumen was used on carefully graded natural soil. The experiments proved successful, and a small catchment area located at Narrogin in Western Australia was then treated in this way. Local rainfall averages 45 cm (18 in) per year; only about 5 percent was runoff, and the remainder is retained in the previously constructed reservoir.

The city of Juneau in Alaska has been built on a small coastal plain on Douglas Sound, a section of the 1,600-km- (1,000-mi)-long inland passage that distinguishes the northwest coast of North America. Water supply for this isolated city was obtained from Gold Creek well above the town and from wells in Last Chance Basin through which the creek flows. A tunnel had been driven in 1898 by a mining company that intended in this way to drain the gold-bearing gravels of Last Chance Basin. Their efforts were defeated when the heading broke through into the gravels, the tunnel being flooded. Another company tried in 1902 to wash the gravels into the tunnel. Some success was achieved, but a flash flood in 1905 plugged the tunnel and brought operations to a halt. The tunnel was then abandoned and became clogged with debris, but it was still connected with the water-bearing gravels. Having been driven through sound basalt and gabbro, the tunnel itself was found to be quite sound when it was investigated as a possible conduit for water. The tunnel was cleaned out, lined into the water-bearing gravels, and connected to the city reservoir. The tunnel now receives and naturally warms the water slightly from its initial temperature of about 5°C (41°F) throughout the year, thus eliminating previous problems with winter freezing of well-supply water wells.[5]

SOME GEOLOGICALLY SIGNIFICANT SYSTEMS

Variable geologic conditions are routinely encountered in the development of public water-supply systems. Some of these projects have turned unusual geological conditions to their success. This short list will at least illustrate the vital role of geologic studies in water-supply engineering.

London, England

The geologic section beneath London (Fig. 9.9) has made the city fortunate in its geological setting. Wells into the chalk now supply only about 14 percent of requirements, about the same amount from the flow of the river Lea, and almost three-quarters of the total now used from the flow of the river Thames. The London Basin remains, however, a truly remarkable urban geological phenomenon; its use is strictly controlled. For London's utilization of the surface waters, many outstanding works have been necessary—tunnels, embankments for reservoirs, replenishment facilities in the Lea Valley, and the most recent development described on p. 337.

Long Island, New York

Long Island constitutes the coastal plain of the states of New York and Connecticut, separated from the mainland by Long Island Sound and isolated geologically from its water-bearing beds. The island supports a population of several million people, and in recent years a large part of its water supply has been obtained from groundwater. There are crystalline bedrock outcrops in some of the western parts of the island, and the rock surface so dips to the east that in places it is at least 600 m (2,000 ft) below the surface. Overlying the rock

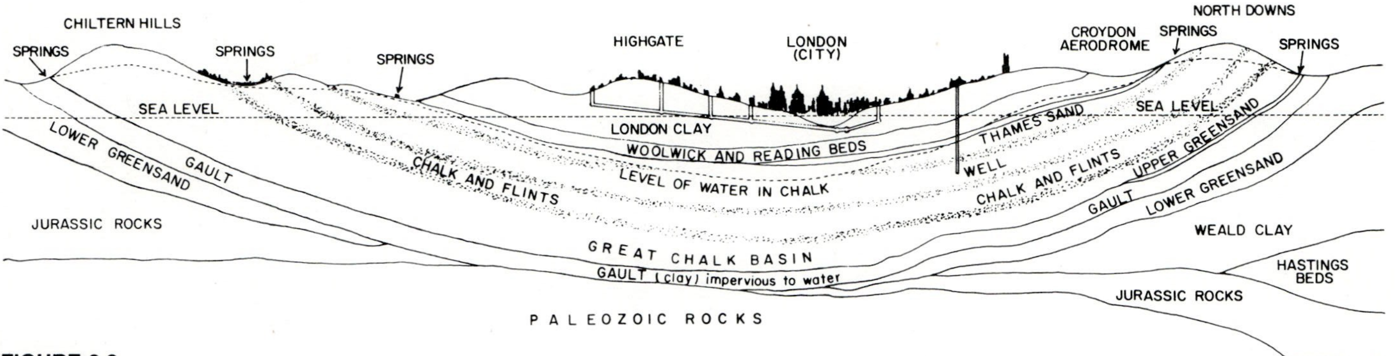

FIGURE 9.9
Simplified geological section through the great London Basin of southern England. (*Courtesy Metropolitan Water Board, London.*)

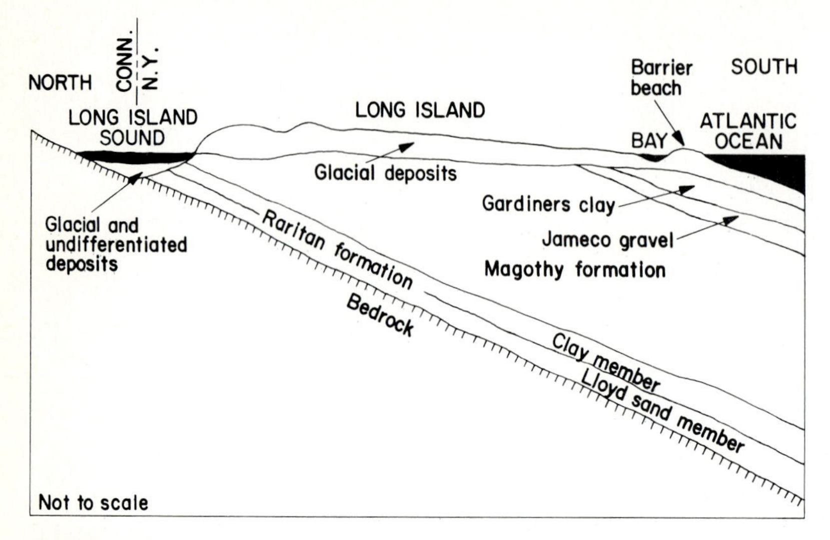

FIGURE 9.10
Simplified geological section through Long Island, New York, indicating water-bearing horizons. (*Courtesy Division of Water Power and Control, New York State Department of Conservation.*)

is a great deposit of stratified Cretaceous sands, clays, and gravels; the dip ofthe strata decreases toward the surface (Fig. 9.10). At the surface, stratified sand and gravel outwash deposits and two end moraines give rise to a fairly regular topography; the highest ground elevation is about 126 m (420 ft) above the sea.

These unconsolidated deposits constitute the great underground reservoir from which has been drawn probably the greatest concentrated groundwater supply utilized in North America. As the unconsolidated strata do not cross the sound to the mainland, the groundwater underlying the island is obtained only from the rain falling on it, and, consequently, the relation of this body of freshwater with the surrounding seawater is a delicate one. The rainfall averages about 105 cm (42 in) per year, of which possibly one-half reaches the zone of saturation. Opinions differ as to how much of this water can be obtained for use, but it is known that over 750 mld (200 mgd) are routinely recovered. Unfortunately, early pumping was concentrated in developed areas, creating serious saline contamination. This especially occurred under the older part of Brooklyn, the water of which experienced a significant lowering of surface level. By 1947 virtually all pumping of groundwater had been stopped, and the area was serviced by the New York Board of Water Supply. Even though much of the Brooklyn area is paved, natural recharge by rainfall has steadily restored original groundwater to an elevation above that which existed when some public works in Brooklyn had been built. Flooding of such works as the Newkirk Avenue subway station has required extensive remedial work.

The entire situation illustrates vividly the interdependence of groundwater, local geology, paving of surfaces, and overpumping. The *Atlas of Long Island's Water Resources* is a fine guide for all who face major problems linked to overdraft pumping of urban groundwater.[6]

South Coastal Basin of California

A little-known fact about Southern California is the singularly interesting manner in which the South Coastal Basin of California makes use of its groundwater resources. The basin comprises the Los Angeles metropolitan area and 26 incorporated cities to the east. Here the coastal plain, formed by three mountain ranges, is served by three rivers: the Los Angeles, the San Gabriel, and the Santa Ana. Within the well-defined valleys, the area is geologically complex, with 37 fairly distinct sub-basins with steep valley slopes. Junctions of alluvial fans, valleys, and the relatively flat coastal plain provide almost perfect recharge conditions. Irregular rainfall in the mountains is high and provides "flashy" replenishment to the groundwater stored in the unconsolidated coastal plain deposits. It has been calculated that its capacity in the 30-m (100-ft) layer immediately below outflow level is no less than 85 billion m^3 (7 million acre-feet), of which only about 60 percent has so far been used. The total surface area of the basins is 3,400 km^2 (1,305 mi^2), and the average specific yield of the unconsolidated strata is 8.4 percent by volume. Despite the high cost of pumping, the groundwater demand has led to a serious drop in water levels. Augmentation of this supply has brought about construction of the three great aqueducts already mentioned.

Vancouver, British Columbia

Vancouver, the western metropolis of Canada now with a population of almost a million, formed its Greater Vancouver Water District in 1926. The District has control over catchment areas of three small rivers in the adjacent Coast Range with an area of 58,500 ha (145,000 acres). High rainfall and good runoff feed reservoirs which supply a notable pressure tunnel crossing Burrard Inlet to reach the city. Since the catchment areas are underlain by igneous rocks, water quality is remarkable—hardness rarely exceeding 10 ppm, total dissolved solids only 23 ppm—so that chlorination is all the treatment that is required. The growth of Vancouver has been immeasurably aided by the good fortune of this gravity supply of pure water.

Sydney, Australia

Greater Sydney, New South Wales, now with a population of almost 4 million, has had an unbroken history of water supply since the first settlement. On January 26, 1788, "sufficient ground was cleared for the encamping of the officer's guard and the convicts who had been landed...at the head of the cove, near the run of fresh water, which stole silently along through the very thick wood...." Tank Stream, thus described, supplied water to the little settlement from 1788 to 1826. "It still flows, a dark and hidden stream, beneath the shops and offices of modern Sydney," but it had become polluted before abandonment as the main water supply in 1826. Groundwater was next extracted from "Busby's Bore" tunnel, in a corner of Hyde Park, but this supply became insufficient. In 1886 the first water was brought from the Nepean watershed to the south, collected by means of a weir across the Nepean River at Pheasant's Nest.

This was the beginning of a series of notable catchment projects. Geologic complications resulted from valuable coal seams below some of the water-supply dam sites. The N.S.W. Mines Department declared limitations on coal mining both in the vicinity of the dams and beneath their reservoirs; no leakage or settlement have been reported. The last of these major projects was the Warragamba scheme, including the largest dam in the southern hemisphere, across the Warragamba River. This great structure, 120 m (394 ft) high but only 330 m (1,080 ft) long at its crest, filling a precipitous gorge in sandstone rock, required expert geological advice during construction complications. Observations made then are an important chapter in the related records of engineering geology.

Hong Kong

The small crown colony of Hong Kong packs a rapidly increasing population of over 3 million into a total area of less than 103,000 ha (255,000 acres). Its good rainfall, annually variable, constitutes a water storage problem since the catchment supply has now been supplemented by water from China. The Chinese have reversed the flow of the Shih Ma River and pump its water through eight pumping stations, aided by six large reservoirs, to a point just north of the Hong Kong border. The colony is taking 67,500,000 L (18,000 million gal) of this water per year, giving the Chinese almost £1,000,000 ($3,000,000) in foreign exchange. Storage, however, is difficult since there are no natural large reservoir sites amid the Hong Kong hills. Every advantage has been taken for the building of small storage reservoirs, but for major storage it was finally decided to convert an inlet from the sea to hold freshwater.

The Plover Cove reservoir has converted an area of 1,110 ha (2,740 acres) of seabed that is naturally lined with 10.5 m (34.4 ft) of soft clay overlying 1.5 m (4.9 ft) of stiff clay. Dredging removed all the soft clay, the stiffer clay being left as the bottom of the reservoir to act as a seal against upward penetration of seawater through granite bedrock in various stages of weathering. The main dam is 2,070 m (6,800 ft) long and 45 m (148 ft) high, but only 16.7 m (55 ft) above water level. Geologically, it is interesting because of its extensive use of decomposed granite and granodiorite as fill, borrowed from the hillsides overlooking the famous harbor. So rapidly has the demand for water increased

FIGURE 9.11
The flowing artesian wells of the Ogden city water supply before they were submerged beneath the water of the Pine View Reservoir; a photograph probably taken about 1925. (*Courtesy City Engineer, Ogden and the U.S. Bureau of Reclamation.*)

that an even larger estuarine storage project at High Island is now in use. This water storage concept has become so well publicized among Asian water supply engineers that the freshwater-deficient Republic of Singapore is well into a comprehensive project to utilize similarly all of its estuaries.

Ogden Valley, Utah

Possibly the most unusual water-supply system of its kind supplies the city of Ogden. In 1914 the city started drilling wells in the artesian system of the lower Ogden Valley, created by the existence of silts and gravels beneath an impervious bed of 21 m (70 ft) of varved clay. By 1933 a total of 51 flowing wells had been drilled in ''Artesian Park'' (Fig. 9.11). In 1935 and 1936 the U.S. Bureau of Reclamation flooded Artesian Park behind Pine View Dam at the lower end of Ogden Valley. Before the dam was built, however, a contract was negotiated between the city and the United States government providing for the capping of the artesian wells and the construction of a collection system. Ogden reserved the fight to drain the reservoir in the event of trouble with its water system. The reservoir has been so drained, and repairs have been executed. Some damage appears to have been due to consolidation of the impervious clay stratum because of the extra load of the impounded water. In 1955, therefore, special precautions were taken before the dam was raised by 8.7 m (29 ft) to give additional storage. Flexible well connections were used to minimize the effects of future ground movement. The successful project has been described as a ''three-layer'' water system.[7]

Sheikhdom of Kuwait

Kuwait, although only 1.6 million ha (4 million acres) in extent, obtains over 2 million bbl of petroleum per day from beneath its surface. An average annual rainfall of only 8 cm (3 in) comes during the winter months of November to May, sometimes as intense local storms but more generally as light winter showers. Post-Miocene rocks underlie the generally inhospitably hot surface of the country. The Dibdibba Formation, thought to be of Pleistocene age, serves as a sand and gravel aquifer, as does also the underlying limestone. Regional dip of the strata is to the east and northeast. The original brackish groundwater supply was drawn from shallow wells. Drilling for oil started in 1936, and two major sources of brackish water supplies were discovered, the water being distilled for potable use. In 1960, fresh groundwater was surprisingly discovered during brackish water drilling on a Basra Road construction wrongly sited during a sandstorm. Geologic study revealed the freshwater to be surrounded and underlain by brackish water. The local geologic structure traps winter rain infiltration in the cemented Dibdibba Sandstone before being rendered brackish. Now, 32 mld (8.5 mgd) are obtained from this unpredictable source of groundwater beneath even the most unexpected of areas.

Honolulu, Hawaii

Eight inhabited islands comprise the main part of the state of Hawaii. World-famous for their beauty and their equitable climate, they are known throughout the world by geologists as the *type area* for island volcanism. An increasing amount of engineering work has accompanied development for the extensive production of sugar cane and pineapples, for defense, and in more recent years as almost ideal tourist resorts. The islands are formed of generally highly permeable volcanic lava, a rock that hosts some of the most unusual water-supply systems of the world. Rainfall impacts exposed rock surfaces of the rugged terrain, seeps into the ground continually, and so gives excellent local supplies. This same open-pore structure is also filled with seawater below ocean level, and so deep below the islands. The conjunction of fresh and saltwater in the Hawaiian Islands is one of the most remarkable examples of a hydrogeological phenomenon that is becoming increasingly important.

The *Ghyben-Herzberg effect,* first recognized by W. Badon Ghyben of the Netherlands in 1889 and apparently quite independently by B. Herzberg in Germany in 1901, is, in principle, very simple. Since seawater, because of its salt content, is about one-fortieth heavier than freshwater, it follows that freshwater will, if undisturbed, "float" (so to speak) on the top of the seawater (see Fig. 3.9). The depth to which the freshwater will extend below sea level will be 40 times the height of the lens of freshwater. There will be no dividing line between saltwater and fresh, but experience has shown that the actual occurrence of both types of water in a permeable formation, such as the Hawaii lava flows, does approximate this theoretical pattern when undisturbed.

Inhabitants of Oahu, relatively few in number until recent times, relied upon streamflow until the great reservoir of groundwater underlying the island was discovered in 1879. Because of impermeable surface deposits, the groundwater was then under artesian pressure in many parts of the island. By 1910, more than 400 wells had been drilled. Ten years later groundwater in the vicinity of Honolulu had dropped 6.9 m (22 ft 6 in) below its original level, which was only

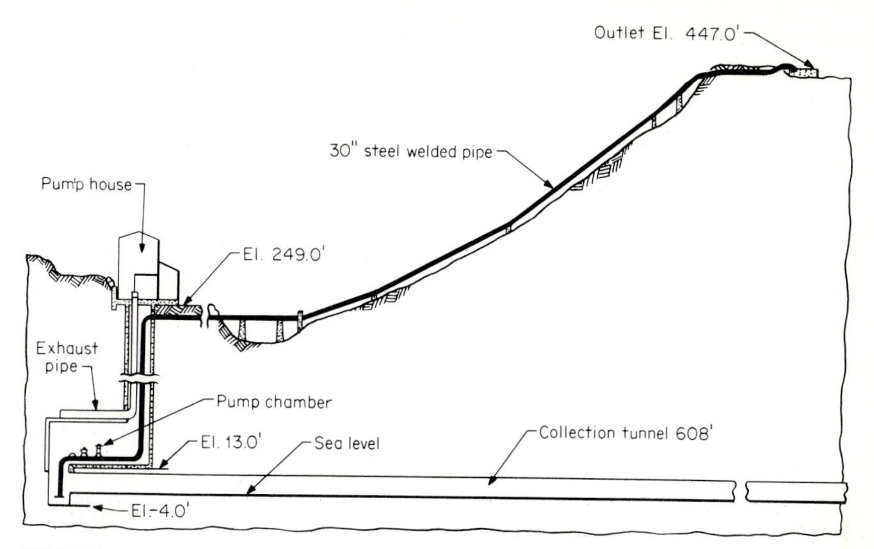

FIGURE 9.12
Simplified section of a typical Maui-type well as used in the Hawaiian Islands. (*Courtesy Board of Water Supply, Honolulu.*)

12.6 m (41 ft 4 in) above sea level. A few wells had been "salted up" by intrusion of seawater; the delicate balance of sea and freshwater beneath the island was in danger of being seriously disturbed.

In 1929 the Honolulu Board of Water Supply was established to control and manage the water supply to the city; in 1959, the board was given control over all water on the island. Conservation of this unique water-supply system began in 1936, with adoption of a type of well first used about the turn of the century on the island of Maui for the supply of irrigation water for sugar-cane plantations. *Maui wells,* as they are now widely known, consist essentially of horizontal tunnels driven in the volcanic rock at carefully selected locations, generally just below the known freshwater levels (Fig. 9.12).

By pumping at controlled rates, the accumulated supply is, so to speak, "skimmed off" the reservoir of freshwater so delicately balanced upon the underlying saltwater. The surface level of seawater is depressed beneath the island by the overlying freshwater to the extent that, for every foot of freshwater in the rocks above sea level, almost 40 ft is found below sea level. The rapidity with which saltwater rises on depletion of the freshwater reservoir will be obvious from these figures. By the use of the horizontal skimming tunnels of the Maui wells, an assured supply is now being obtained for the city of Honolulu and the other rapidly developing communities on Oahu.

It would be difficult to imagine a more vivid example than the Maui wells of the constructive use of an unusual geological situation for "the use and convenience of man" (to use the famous phrase from the 1818 London charter of the Institution of Civil Engineers). To stand in the silence of the collecting tunnels in Oahu, to see the crystal-clear water seeping from the surrounding basalt, and to think of the delicate balance between this clear freshwater and the deep expanse of seawater below one's feet, is to have the most moving experience possible for those interested in the fundamental role of geology in all the works of the civil engineer.

REFERENCES

1 "Groundwater Pollution in California Points to Industrial Waste Discharge," *Engineering News-Record, 137,* p. 785 (1946).

2 A. B. McDaniels, "Groundwater Cut-off Wall Provides New Water Supply," *Engineering News-Record, 113,* p. 757 (1934).

3 J. Hinds, "Colorado River Water for California," *Civil Engineering, 7,* p. 573 (1937).

4 H. O. Banks, "California's Water Plan," *Civil Engineering, 30,* p. 52 (December 1960); other useful papers in same journal—*36,* p. 48 (January 1966); and *42,* p. 78 (June 1972).

5 "Miner's Mistake is Modern Blessing," *Engineering News-Record, 168,* p. 37 (March 8, 1962).

6 P. Cohen, O. L. Franke, and B. L. Foxworthy, *An Atlas of Long Island's Water Resources,* New York Water Resources Commission, Bulletin No. 62 (1968).

7 R. R. Wooley, "Artesian versus Surface Supply—Ogden River Project," *Civil Engineering, 11,* p. 536 (1941).

CHAPTER **10**

DAM FOUNDATIONS

Dam construction causes more interference with natural conditions than any other civil engineering operation. Equally striking, however, is the critically important function that dams perform in storing water for domestic supply, power generation, flood control, irrigation, and recreation. Strict performance capabilities are built into each structure carrying out these functions. These projects exist in all parts of the world and in surprising numbers ranging from the smallest timber-check dam to those as large as Lloyd Barrage Dam or Hoover Dam; this is ample reason to devote particular attention to the design of their foundations. Finally, although failures of engineered works are always of serious consequence, failures of dams are more serious than others, since they generally occur during periods of abnormal weather, often without warning, and almost always with disastrous results. Defects in foundation beds are an unfortunate factor in many dam failures, another telling argument that no single feature of foundation beds can be responsibly neglected in dam design.

Constructive examples of successful completion and performance of engineered works based on geologic methods and investigations are always valuable. A slight departure from this course will be made in the case of dams, and a brief review will be presented of some failures of the past. This does not cater to the almost hysterical attention devoted to dams whenever a failure does occur, which the press would almost lead one to believe are regular occurrences. On such occasions, not a word is said of the tens of thousands of dams successfully performing their tasks, nor is the fact often mentioned that failures in many other branches of civil engineering (e.g., water purification) are equally liable to endanger the lives of people. Dam failures are news events, but they will be considered here only as useful illustrations of some of the causes and of the significance of foundation geologic investigations.

HISTORICAL NOTE

Ancient dams include one constructed by Joshua to help the Israelites cross the river Jordan, and the even older prehistoric Marduk Dam across the river Tigris. Built for river regulation, it lasted until the end of the thirteenth century A.D.The first recorded masonry dam was that built by Menes, first king of the first Egyptian dynasty, some time before 4000 B.C.Constructed about 19 km (12 mi) to the north of Memphis to divert the waters of the Nile to provide the site for that ancient city, it was 457 m (1,500 ft) long, at least 15 m (50 ft) wide, and maintained for 4,500 years. Arab literature records a great dam constructed (1700 B.C.) in Yemen, a dam of which little is known beyond its being 3.2 km (2 mi) long, 36 m or 120 ft high, and 150 m or 500 ft wide. Failure came by flood in about A.D.300.

These few examples establish the longevity of dam building as a major civil engineering endeavor. Further examples could be described, but most foundation geologic information is almost completely lacking in ancient records. Puentes Dam in Spain, however, is an early modern example. Constructed between 1785 and 1791, it was 282 m (925 ft) long, curved in plan, and 45.6 m

(152 ft) thick at the widest part of the base, with a maximum height of 50.1 m (167 ft), a perfectly safe design from the structural point of view. Constructed of rubble masonry and faced with cut stone, it was finished with such ornamentation as to make it one of the wonders of Spain. Although the dam is apparently founded in part on bedrock, a gravel pocket about 20.1 m (67 ft) wide was encountered near the center of the site. The dam was made to bridge this gap on an ingenious pile-supported grillage and protected by an upstream apron. This method of overcoming geological difficulties served satisfactorily until April 30, 1802, when that part of the dam above the grillage failed, following seepage erosion and washout of the underlying gravel pocket. An eyewitness account speaks of the disastrous collapse of this "plug" of masonry, which went out like a cork, leaving what appeared to be two massive bridge abutments with a 16.8-m (56-ft) gap, 32.4 m (108 ft) high. This failure was somewhat exceptional, since other large dams constructed in the same period functioned successfully, thus inaugurating modern dam design and construction.

FAILURES OF DAMS

The record of dam failures in succeeding years provides a useful if somewhat melancholy study. These failures indicate definitely that the main reasons have been (1) failure to provide adequate spillway capacity, and (2) defective foundation-bed conditions. Spillway capacity is determined mainly from anticipated catchment area runoff influenced considerably by geological conditions. The second cause of failure is essentially geological, although varying considerably from one case to another.

Lapworth's historic review mentions over 100 failures of dam structures between 1864 and 1876 due to undermining of water-bearing beds below the dam foundation. Justin details over 60 failures of earth dams between 1869 and 1919. In his paper, Dr. Lapworth mentions the Hauser Lake failure which bears a striking similarity to that at Puentes Dam, since a central portion 120 m (400 ft) long, founded on water-bearing gravel, was destroyed in 1908 while the remaining sections were left intact.

A relatively recent example is that of Austin Dam, constructed in 1893 to provide a water and power supply for Austin, Texas. The dam was of masonry, 327 m (1,091 ft) long, 20.4 m (68 ft) high, and 19.8 m (66 ft) wide, and rested on Cretaceous limestone, clay, and shale. The limestone strata were almost horizontal but alternating in hardness. Workers described the limestone as dissolved in places, giving rise to underground caverns, but such is not mentioned in engineering reports. The clays and shales were slippery and very broken, since the site was on a fault zone. Flood flows over the dam caused some erosion of weaker strata and at least one serious washout, which was repaired. This inevitably contributed to the sudden failure of the structure on April 7, 1900, when a central section collapsed completely. The dam was then overtopped by 3.3 m (11 ft) of water because of severe flood flow. The

FIGURE 10.1
The remains of St. Francis Dam, California, after the structure had failed. (*Courtesy World Wide Photos.*)

concentration of flow through the resulting gap moved two adjacent blocks, each about 75 m (250 ft) long, downstream for a distance of about 18 m (60 ft). The causes of failure appear to include erosion of weak rock supporting the heel of the dam, the slippery nature of at least part of the foundation bed, and the presence of percolating water under the base of the dam. The gap was closed and the dam placed in service again in 1915, but shortly thereafter another flood took out 20 crest gates and blocked the tailrace and turbine draft tubes with debris. Not until April 1940 were the dam and the associated powerhouse finally put into continuing service. A thorough job of reconstruction included extensive grouting and foundation rehabilitation that can really be called underpinning, even though applied to a dam.

The failure of the St. Francis Dam will be within the memory of some engineers. This great gravity dam, 61.5 m (205 ft) high and 210 m (700 ft) long, was completed in 1926 by the Bureau of Water Works and Supply of the City of Los Angeles to assist in water storage. Located in San Francisquito Canyon, it created a 47 million m^3 (38,000 acre-foot) reservoir. The dam was founded partially on schist and for about one-third of its length on a conglomerate with sandy and shaly layers. The contact between the two rocks is an inactive fault. Although the schist is a relatively sound rock, the conglomerate

> ...is by no means a strong rock. A test...gave a crushing strength of 500 pounds to the square inch....When wet, the rock shows a considerable change, [a sample starts] to flake and crumble when placed in a beaker of water and in about 15 minutes slumps to the bottom of the vessel as a loose gritty sediment that can be stirred about with the finger....So far as can be ascertained, no geological examination was made

FIGURE 10.2
The site of the Malpasset Dam on the Le Reyran River in southern France just after this thin concrete arched dam had failed. *(Courtesy Keystone Photos.)*

> of the dam site before construction began and no crushing or immersion tests were made of the conglomerate.[1]

Seepage was noticed through the conglomerate when the reservoir first became full (in March 1928); eventually, the dam failed on March 12, 1928, with a tragic toll of 426 lives and untold property damage (Fig. 10.1). Boards of inquiry were appointed by the State of California and by the City of Los Angeles; both agreed that the main cause of failure was the nature of the rocks under the dam. Quoting Dr. F. L. Ransome again:

> The plain lesson of the disaster is that engineers, no matter how extensive their experience in the building of dams or how skillful in the design of such structures, cannot safely dispense with the knowledge of the character and structure of the adjacent rocks, such as only an expert and thorough geological examination can provide.[1]

Just about 30 years later an almost equally tragic failure occurred; 344 people were killed when Malpasset Dam in southern France collapsed after several days of unusually severe rain (Fig. 10.2). The dam was completed in 1954; it was thin-arched and of reinforced concrete, its greatest thickness only 6.8 m (22 ft 8 in), and its height about 60 m (200 ft). Curved to a radius of 103 m (344 ft), it was built in a narrow gorge of Le Reyran River for the Department of Var to serve as both irrigation and water supply. Its sudden collapse on December 2, 1959 resulted in a catastrophic flood that carried everything before it for 12 km (7 mi) downstream. Most of the people who lost their lives lived in the town of Fréjus, which was in the path of the flood. The preliminary report of the commission of inquiry, established by the French Ministry of Agriculture, confirmed that the structural design, although daring, was sound. After reviewing all possible causes for the disaster, the commission was forced to the conclusion that the principal cause of the catastrophe was substantial shear displacement of the rock below the foundations and at the left abutment. The mica schist bedrock was reported to be sheared and jointed, with a wedge-shaped mass overlying clay-filled joints. One of the official commissions concluded as follows:

> The cause of the failure is to be sought solely in the ground below the level of the foundations...and...the most probable cause of the accident is to be attributed to the presence of a plane of sliding, or superior upstream fault,...which for a great length and at a slight distance followed almost parallel to the foundation of the arch in the upper section of the left bank. The susceptibility of the foundation material to deformation, already great, was increased locally by the presence of the sliding plane. The work was not capable of adapting to such aggravation.

Commenting upon this tragic failure two years after the collapse, Dr. Karl Terzaghi had this to say:

> A conventional site exploration, including careful examination of the rock outcrops and the recovery of cores from two-inch boreholes by a competent driller, would show—and very likely had shown—that the rock contains numerous joints, some of which are open and filled with clay. From these data an experienced and conservative engineering geologist could have drawn the conclusion that the site is a dangerous one...the most advanced means of rock exploration at our disposal ...would have added no significant information to what was known concerning the safety of the left abutment of the dam, before the failure occurred.[2]

The Malpasset Dam failure, in which some of the engineers were brought to court to answer charges of imprudence, points out that no amount of the most sophisticated testing of rock (or soil) will be effective if the basic geology of a foundation site is not first determined to be sound.

A further illustration of this principle is a dam failure within the boundaries of a great city. The dam was well designed and constructed but could not perform as designed when a foundation failure took place. The location of this dam (Los Angeles) is of significance, since downstream urban development lent unusual importance to its safety. The Baldwin Hills Reservoir was intended to store water coming to Los Angeles from the Owens River and Colorado River aqueducts. It is an earth dam 69.6 m (228 ft) high and 195 m (640 ft) long, containing 650,000 m^3 (850,000 yd^3) of soil. The site used was convenient for supplying a rapidly growing residential area in southwest Los Angeles. Excavation during the years between 1947–1951 revealed a zone of Late Pleistocene or Recent normal faults, with an 8-m (59-ft) displacement, believed to be part of the active Inglewood fault system. The gate tower was relocated to avoid this fault zone. The dam was operated with great care and with constant supervision. An inspection was made on April 3, 1963, but on Saturday afternoon, December 14 of the same year, the dam failed, releasing most of the water stored in the reservoir in a disastrous flood. Leakage that started late on Saturday morning enabled the authorities to have the Los Angeles Police Department issue warnings via motorcycle, car, and helicopter. Alerted residents were mass-evacuated from the area below the dam. Despite these efforts five persons were drowned. Property damage estimates exceeded $10 million.

In the ensuing legal actions, the City, on its own behalf and for the Los Angeles Department of Water and Power, brought suit against a group of important oil companies, contending that the extraction of petroleum, water, and natural gas from the ground at the Inglewood oil field near the reservoir caused land subsidence that triggered the collapse of the dam: Design and construction were not called into question. The total claimed was $25.8 million. The matter was settled out of court for a total amount to be paid to the City of $3.875 million. Extensive studies of the dam failure revealed—if nothing else—how complex were the geological questions raised by the accident.

The failure of the Teton Dam, Idaho, on June 5, 1976, is not only one of the most recent of major dam failures but one marked by a wealth of information made public, on all aspects of the design, construction, and failure of the dam by its builder, the U.S. Bureau of Reclamation (USBR). This frank disclosure of all pertinent information was an example of professional engineering at its very best, but such does not establish the real cause unequivocally, since the collapse destroyed the essential evidence.

The dam was 92 m (305 ft) high above the Teton River bed, 122 m (405 ft) above the lowest point of its foundation, and about 900 m (3,000 ft) long. Built to provide flood control, power generation, recreation, and (most importantly) irrigation water, it was topped out in November 1975. Filling of the reservoir behind the dam was almost complete at the time of failure. The foundation was in rhyolite, a closely and heavily jointed volcanic rock, there being even some human-sized openings found beneath the dam location. Recognizing the impact of such conditions, the USBR carried out the most extensive grouting program that they had ever undertaken, a total of 16,000 m^3 (570,000 ft^3) being injected. Despite this, the dam failed and killed 11 people, rendered 25,000 homeless, and caused about $400 million in damages.

Two days before the failure, two small seeps were noticed downstream of the toe of the dam. Early on the day of failure, an engineer noticed a small leak about one-third of the way up the dam, close to the right abutment. Attempts made to plug the leak were in vain, complete failure later the same morning taking out a large part of the dam (Fig. 10.3). Two expert commissions of inquiry concluded that the failure took place because of piping in the loess-like silt, of which impermeable material the main embankment was constructed. What caused the piping can now only be conjectured, possible causes being deformation of some sort in the relatively narrow key trench across the foundation, unsealed fractures in the bedrock, or the absence of filter zones between bedrock and core material. A large area of saturated embankment soil was found when the remaining part of the dam was carefully excavated for

FIGURE 10.3
Remains of Teton Dam in Idaho, after its failure. (*Courtesy U.S. Bureau of Reclamation.*)

study, adding still further to the complexity of determining the exact cause of failure. The overall message for the future, however, is clear—an unmistakable need to determine the essential integrity of foundation bedrock and a consideration of the interaction between bedrock and core material in all earth dams.[3]

INSPECTION AND MAINTENANCE

Ironically, in the months before the Teton Dam failure, much discussion had been focused on dam safety, concluding that regular inspection and maintenance of dams was needed. In November 1977, the President of the United States initiated a major program of dam inspection, the U.S. Army Corps of Engineers being charged with the safety inspection of 9,000 non-Federal dams. This program supplemented the dam inspection programs of individual states, which were expected to cover the remaining 40,000 dams (generally more than 7.5 m or 25 ft high) known to exist. Previous to this, in April 1977, President Carter had charged agencies of the United States government with ensuring the adequacy of their own dam safety programs.

Most developed countries have similar ongoing programs for dam inspection, quite often inaugurated after some serious dam failure in the country concerned. In Great Britain, for example, the first relevant legislation appears to have been the Reservoirs Act of 1930, passed after a number of dam failures. Annual inspections are now obligatory for all structures above a certain size. The restriction of such legislative requirements to dams was at first a matter of some surprise to many engineers. But the vital importance of dams, their frequently isolated locations, the fact that they are often seen regularly by very few people, and the tragic consequences of any failure combine to counteract these impressions.

With or without state regulations, however, regular inspection of all dam structures is an essential requirement. Examination is generally restricted to those components which can be seen—together with the inspection of galleries, tunnels, and shafts that have been left open after the completion of construction. Inspection galleries provide invaluable access. One certain indicator in inspection is leakage of water through or around the dam. Records of this must be kept regularly. As with any safety-related indicator, such as stress or strain, any increase, either sudden or gradual, should initiate prompt remedial action. Study of the sediment and chemical content of seepage waters may reveal the reason for the increase and suggest remedial measures. Even in the extreme case, in which damage is irreparable, there will at least be a warning of impending trouble. Failures should henceforth become more of a rarity, especially if safety-related associated geological conditions are always regarded as critical elements of design, construction, maintenance, and regular inspections.

Foundation geology is where dam inspection should always start. If, as it appears to the authors, this has not always been the case, the oversight must generally have been due to some administrative misunderstanding, since it is to be expected that all those engaged in this vital work will realize the vital contributions that engineering geologists can make to such work. Aerial photographic interpretation is an economical way of initially assessing dams. Aerial photographs, however, cannot take the place of detailed examination of ground conditions on foot, especially in the vicinity of dams, where occasionally quite minor details of observation may prove to be of unusual significance. Such "foot-slogging" field inspection work will be greatly enhanced if as-constructed drawings of the dam foundations are available. In some cases such drawings will contribute directly to the understanding of later observations. Their importance, especially in regard to the future safety of the dam for which they were prepared, cannot be overemphasized.

REVIEW OF DAM CONSTRUCTION

A dam is an artificial structure erected to support a waterproof barrier designed to retain water above the level that it normally occupies at the site of the dam. Suitable provision is also made for passing a calculated flow of water past the dam, through it, over it, or around it, depending on local circumstances. The barrier is generally an integral part of the dam structure, supported as (1) an earth-fill or rock-fill dam, in which the barrier is either on the upstream face or in the center as a core wall; (2) a gravity dam of masonry or mass concrete, in which the barrier is the upstream face of the dam itself; or (3) a reinforced-concrete dam of the arched type, multiple-arch, multiple-dome, or some other special design, in all of which an unbroken, reinforced-concrete skin serves as the barrier.

Dams can be generally grouped into two main divisions: (1) earth- or rock-fill dams, which depend for their stability on the natural repose of compacted soil material; and (2) concrete or masonry dams, which depend for their stability on the structural performance of the material used for construction. The type of dam to be constructed at any location must be determined mainly from geological considerations; the actual kind of dam to be constructed, once its general type has been decided, will also be dependent to some extent on geological conditions controlling material supply.

Since all dams retain water to design level, waterflow in the watercourse being regulated is seriously affected below the dam site; the flow is generally regulated to a more uniform discharge than that given by the stream itself. In addition, the underground water conditions in the valley above the dam location are completely changed; the groundwater surface is raised at least to the water level of the reservoir water line, and changes of decreasing importance occur farther up the valley. Below the dam, the groundwater level may be lowered if normal streamflow is depleted. Between the two sides of the dam, there is thus set up a considerable difference in groundwater level. Although the water retention barrier generally extends from side to side of the dam site, effectively isolating the two groundwater levels, this artificial separation of hydrostatic head will always exist while water is being retained. The dam embankment exerts unusually high unit pressures on certain parts of the underlying foundation strata. The beds will be submerged well below water level in the reservoir area and so will be subjected to appreciable hydrostatic pressure.

These are the main reasons for the assertion that a dam causes more interference with natural conditions than does any other type of civil engineering structure. In the special case of weirs, or dams founded on pervious strata, in which the barrier merely deflects groundwater flow, but does not stop it completely, the serious change in the underground conditions caused by the dam will be obvious.

Dams designed to bar completely through-flow of water, surface and underground, down the valley which they cross, present four main geological problems:

1 Determination of the soundness of the underlying foundation beds and of their ability to carry the design loading

2 Determination of the degree of watertightness of the geologic foundation and of the measures, if any, required to make these materials watertight

3 A study of the effect of exposure to water

4 An investigation of the possibility of earth movements below the dam and determination of measures to be taken in safeguard.

Geological problems also affect reservoirs, but for convenience they will be considered later in this chapter.

Dams on pervious foundations present their own special problems, associated mainly with controlling the flow of water beneath the structure. It may not be possible to make the dam foundation absolutely tight, but the slight seepage accepted on so-called "impermeable" foundation beds, although associated with uplift pressure, is generally of no other consequence. For those dams on permeable foundation beds, the exact nature of the flow of water through the underlying strata is a vital part of the design, relying on accurate knowledge of the hydraulic properties of such unconsolidated materials. Reservoir sedimentation is another associated geological problem. Scouring out of strata exposed immediately below dams because of the excessive flood velocities is also a matter of importance in the design of dams.

PRELIMINARY WORK

Proposed dam locations are generally restricted by topographical, economic, and social considerations. Areas for geologic examination as possible sites will be fairly well defined. In general, preliminary studies will follow the lines already suggested in Chapter 3. Accurate geologic sections along possible dam axes or closely jointed rock will be a major requirement. As valleys have often formed along faults, bedrock will require exploration for some depth below ground level. Therefore, determination of the top-of-rock surface across the valley will often be a first step in the preparation of the geologic section. Geophysical methods can be of great assistance in this work, when utilized in connection with strategically placed boreholes. The possibility of a buried valley, a condition that will likely compromise the function or integrity (or both) of a dam, must be investigated carefully, not only from borehole logs but also from general studies of the local geology. Valleys 21 m (70 ft) deep have been found between two boreholes spaced as closely as 15 m (50 ft).

The presence of boulders, especially in glacial debris, requires the greatest care in exploratory drilling. The almost insuperable difficulties during the construction of the Silent Valley Dam for the Belfast City and District Water Commission in Northern Ireland were due very probably to the assumption that granite boulders encountered in the original boreholes represented bedrock. Although a rock foundation bed had been anticipated at 15 m (50 ft) below ground level, the cutoff trench actually had to be carried to a depth of 54 m (180 ft), through running sand. This required the maximum working pressure of compressed air, and served to reduce the water level by only about a meter. The solution of the problems thus encountered provides a fascinating but sobering study.

The compilation of geologic sections, local topographic detail, and, more particularly, local geologic features is always important in previously glaciated valleys. Surface observations led to the deduction that the proposed Vyrnwy Dam, for the water supply of the city of Liverpool, England, was the site of an old glacial lake, once retained by a rock bar. The centerline of the dam, which was the first large dam to be built in Great Britain, was preliminarily located along the inferred position of the rock bar. Subsequent detailed investigation by boreholes and shafts proved the inference, and the dam was built as originally located. The estimated cost of a centerline deviation up or down the valley of only 200 m (656 ft) would have added £300,000 to £400,000 ($1,500,000 to $2,000,000) to the cost of the work. Even a small dam known to one of the writers was constructed without any investigation of the rock surface, resulting in a significant cost overrun because the centerline was only 15 m (50 ft) upstream from what would have been a better location.

Whenever feasible, the rock surface should be traced right across the valley being investigated. This is not always possible; eroded valley-bottom faults are sometimes filled to great depths with unconsolidated material. One of the authors has carried out investigations in northern Ontario, in a rock gorge barely 15 m (50 ft) wide, which accommodates the entire flow of a fairly large river. Solid diabase extends for a great distance on either side of the gorge, yet diamond drill holes carried 54 m (180 ft) below water level in the riverbed failed to reveal solid rock; the holes passed through boulders, gravel, and finally more than 30 m (100 ft) of compacted sand. This is typical of valley faults, which are especially common in Canada.

Forty years after the early study of this rock gorge, further site exploration was carried out and modern dam design has led to design and construction of a 51-m- (170-ft)-high earth dam, the deep gorge being bottomed out at a depth of 69 m (230 ft) below river level. Final excavation was naturally by handwork of a particularly intricate character. The narrow graywacke gorge was found to have been created by erosion along the northwesterly striking shear zone.

One of the precepts of the engineering geologist is never to take anything for granted in site studies for engineering works; this is especially so for dam sites. This was probably never better shown than in the geological study carried out on the Vlatava River in Czechoslovakia for Orlik Dam, the highest in the country when constructed in the mid-1960s and the highest of a series of dams built for the power development plan for this famous stream (Fig. 10.4). The dam is located 91 km (57 mi) upstream of Prague; it is 91 m (300 ft) high and 515 m (1,700 ft) long at the crest. Proterozoic metamorphic rocks here form part of the host rock of the great Central Bohemian Pluton, which consists of a complex mixture of granitic igneous rocks. Geologic conditions at the site appeared to be favorable with the Vlatava River flowing through a narrow valley in schist.

Careful geological reconnaissance showed, however, that the superficial geology was misleading. The narrowing of the valley at the site selected was actually due to a massive rock slide of Pleistocene age. A detailed study of the site through shafts, exploration pits and borings confirmed the geological deduction. The chosen site was abandoned and a new site selected 200 m (650 ft) upstream of the slide area. The bed of the valley here consisted of schistose, altered amphibolites, suitable for the founding of the mass-concrete gravity

FIGURE 10.4
Orlik Dam on the Vltava River, Czechoslovakia, south (i.e., upstream) of Prague. (*Photo: Q. Zaruba, Prague.*)

dam, which may now be seen at this geologically interesting location. Construction of the dam raised the normal water level for a distance of 60 km (38 mi) above the dam with the result, typical for Europe, that many buildings and structures were either flooded or disturbed. Much remedial work had to be carried out, including underpinning of Orlik Castle by the grouting of the supporting granodiorite cliff and the anchoring of its exposed face with prestressed rock bolts.

Different again were the site conditions which had to be investigated for the proposed Auburn Dam on the North Fork of the American River, planned as a part of the great Central Valley Project of California, an undertaking of the U.S. Bureau of Reclamation. The dam was initially designed as a double-curvature, concrete arch structure, with a height of 205 m (685 ft) and a crest length of 1,245 m (4,150 ft); as such, it would thus have been one of the largest arch-dam structures of the world. Amphibolite bedrock, with minor metasediments, was generally of a competent nature. These rocks are, however, interlayered with considerably weaker zones of talc schist, chlorite schist, and occasionally talcose serpentinite. The importance of the proposed dam and the character of the foundation materials led the Bureau to initiate one of the most extensive site studies ever undertaken for any such structure. Designed to give complete information for the dam site and the area within 150 m (500 ft) of its axis, the investigation involved approximately 8 km (5 mi) of surface trenches, 1,065 m (3,550 ft) of exploratory tunnels (six of them), 11 shafts totaling 669 m (2,230 ft) driven from selected points within the tunnels, five raises totaling 308 m (1,028 ft), and a shaft of 45 m (150 ft) leading into one of the tunnels. NX drilling was carried out from 306 holes and totaled 26,400 m (88,000 ft).

To facilitate this work and the detailed examination of all the exposed bedrock in the foundation area, the entire site was stripped and cleaned down to bedrock. Ten years and $280 million were spent on site investigations and design work. Occurrence of an earthquake in August 1975, 80 km (50 mi) from the dam site, raised questions about the seismicity of the proposed site and the safety of the proposed arch dam. This led to independent contract studies and much public controversy with the result that construction was postponed, the future of the project remaining undecided as this volume goes to press. No matter what the final decision may be, nothing can detract from the remarkable and exhaustive site investigation, an example on the largest scale of what should be the procedure for all proposed dam sites.

Preliminary site studies will usually enable those in charge of the engineering work to decide on the general type of dam to be used and to begin their economic and design analyses. If, for example, top-of-rock floor cannot be reached by ordinary drilling methods, an earth- or rock-fill dam or a "floating" concrete structure may be necessary. If sound rock is found at reasonable depths, economic considerations will guide the designer in a choice of type. Once the choice of dam sections has been made, geological work continues, especially adapted to serve the particular requirements of the kind of dam selected and carried as far as needed to ensure that the dam can be completed within the estimated cost.

EXPLORATORY WORK DURING CONSTRUCTION

Geologic investigations at the dam site do not cease when construction begins; on the contrary, geological study should be extended, exposed rock surfaces examined, and design assumptions verified. For all major dams, the services of a geological expert will be advisable, and the expert's work should be continued until all excavation has been completed and construction begun. Final inspection of the foundation excavation will be the most critical part of the investigation work. As active construction operations proceed, additional equipment and power supply will continue to aid exploration, beyond that possible in ordinary preliminary work. Extensive test pitting is a first possibility, followed by excavation of shaft exploration and tunnels. The purpose of such shafts and tunnels is to permit visual inspection of the actual rock structure to complete the geological sections across the site, and on rock properties in relation to design loading. In many deep gorges, where borings cannot easily be placed in the riverbed, underwater tunnels drifted from shafts sunk in the gorge sides can prove invaluable. Exploratory excavation work of this type can often perform a double purpose by providing grouting access to render the foundation strata watertight. Such internal grouting can be made thoroughly effective and more extensive in its influence than is possible from surface operations.

Exploratory tunnels were widely used for the main dam of Le Sautet water power development in the south of France on Le Drac, one of the headwaters of the Rhône (Fig. 10.5). Located in a magnificent gorge 180 m (600 ft) deep and 0.8 km (0.5 mi) long, the dam is 124 m (414 ft) high and yet has a crest length of only 79 m (263 ft). The dam proper is arched, to bear on the walls of the gorge, but it is backed with lean concrete designed to buttress-retain the two

FIGURE 10.5
Le Sautet Dam, powerhouse, and gorge, on the Drac River, France; the dam is 124 m (414 ft) high. (*Courtesy Société Forcés Motrices Bonne et Drac, Grenoble.*)

sides of the gorge; the resulting cross section approximates a gravity dam. The gorge rock is limestone, which called for ten years of most careful exploratory work. Nearly 500 m (1,650 ft) of exploratory tunnels were driven, resulting in determination that the rock was as sound as could be desired, without faults or other imperfections. Watertightness of the gorge sides was assured by an extensive grouting program. About 6,000 m (20,000 ft) of grout holes were drilled and grouted, absorbing about 2.7 million kg (6 million lb) of cement under pressures as high as 35 kg/m^2 (500 psi). Much of the grouting was done from the tunnels, some of which have been left open so that future grouting can be done if inspection shows this to be necessary.

The program of continued geologic study of construction of Hoover Dam provides an unusually good example of the invariable requirements of dam construction work. In 1928, a board of engineers and geologists recommended after much study that the dam be located at the Black Canyon site on the Colorado River. The canyon walls were formidable, defying intimate inspec-

FIGURE 10.6
General view of the site of Hoover Dam, as finally excavated, showing the canyon walls and the "inner gorge"; this photo was taken as the second bucket of concrete was being placed on June 6, 1933. (*Courtesy U.S. Bureau of Reclamation.*)

tion. The gorge floor was buried deep below the swift-flowing water of the river, heavy boulders, and river sediment. Geologic predictions were made, however; explorations were renewed, and active construction soon started. By 1933 the canyon floor had been cleaned up (the river flow was diverted through tunnels), and site geology was available for direct inspection (Fig. 10.6).

Dr. Charles P. Berkey, a member of the original board, made a final inspection at this time. It revealed that every major contention and assumption made by the board five years before was confirmed by close examination of the dry gorge. Close inspection was made of all rock surfaces as they were stripped and a final and thorough inspection made before concreting started. All the gorge rock was found thoroughly capable of bearing the load and resisting the thrusts of the dam. Stability and general watertightness of the walls and floor, as well as of the four great tunnels in the canyon walls, exceeded all expectations, although grouting had to be carried out to control occasional leakage and otherwise guard against the great pressure of the water to be retained in this unusually deep reservoir. It was also learned that the foundation rock did not soften under prolonged submergence. The gorge did not follow a fault zone, but a most interesting feature revealed by excavation was the existence of an "inner gorge," forming a narrow and tortuous channel roughly along the center of the main gorge, but at an additional depth of 23 to 24 m (75 to 80 ft) below the rock benches on either side. The side rock benches were generally smooth and uniform (although showing some erosion potholes); but the inner gorge was pitted and fluted, being generally very uneven in form and depth. This form suggested that the whole of the previously bedload mass of sediment had moved in great flood "tides," being subject to scour, whereas in the center there had been a considerable whirling action which set up "pothole erosion" at greater depth.

Small dams have their foundation problems, also. Nothing can ever be taken for granted in the founding of a dam, nor in its performance, ceaseless vigilance being essential. And in the case of every dam, no matter what problems are encountered in its founding, *as-constructed drawings showing every detail of the rock or soil on which the dam is founded must be prepared as the final contribution of the engineering geologist to the construction program. This record must show all the revealed geologic conditions as well as all variations from design assumptions. Such records are absolutely essential as an aid to continued maintenance of the dam. And they can only be prepared once, i.e., before the first concrete is placed or the first load of soil deposited for compaction.*

SOUNDNESS OF BEDROCK

The essential soundness of bedrock below the dam is a prime requirement for investigation. Civil engineers desire answers to two questions well before construction operations are initiated. These concern the probable nature of the bedrock surface and the possible presence of weathered or altered rock, both of which contribute to increased cost. Exploratory pits and core drilling may give an indication (although certain types of weak rock may yield surprisingly good cores), showing the bedrock's deleterious nature only when exposed to the atmosphere. Thus, a careful geologic study is necessary to determine whether or not disintegration is to be expected. A geologist may be able to deduce this from knowledge of the rock type encountered and from a detailed study of cores.

In Washington, D.C., granite has been found weathered to saprolitic soil to a depth of 24 m (80 ft), and as such could be removed with pick and shovel. In Georgia, limestone has been found weathered to soil to a depth of 60 m (200 ft); in Brazil, shales have been found disintegrated to a depth of 118 m (394 ft) below surface level. Actual examples from construction practice will indicate how serious this problem can be. Before the start of construction of the Assouan (Aswân) Dam on the river Nile in Egypt in 1898, it was assumed that all highly weathered granite would be eroded in the deeper channels of the riverbed. When the site was unwatered, it was found repeatedly that decomposed granite lay at the bottom of most of the channels; the removal of this material added greatly to the cost of the work. For the complete structure, excavation was 100 percent in excess of that calculated. At Sennar Dam on the

Blue Nile of Sudan, the reverse experience was encountered; the excavation necessary was less than originally calculated, and there was a consequent saving in cost. In a North American example, a cutoff wall resting on andesite had to be carried 12 m (40 ft) lower than the level originally suggested by core borings in order to reach sound rock.

An added difficulty is determination of exactly what is meant by "sound rock"; the term may have quite different meanings for engineer and for geologist. The engineer associates structural soundness with the strength of the rock and its permeability, whereas the geologist may tend to consider the matter from the mineralogical standpoint. The subject is complex, and its consideration requires experience and a close cooperation between engineer and geologist. A final note of warning must be stressed with regard to rock-mass quality near fault zones.

Although it is probable that no recorded dam failure has resulted from a direct failure of foundation rock in compression, this in no way minimizes the importance of the subject. Laboratory compressive tests usually deal with dry rock specimens, whereas at least part of the rock supporting a dam will be exposed continuously to water. The effect of prolonged exposure to water should be included in testing. Bedrock as such is not only exposed to water, but to water under appreciable pressure, which will quickly aggravate any rock failure. If weathered rock is displaced as it weakens (as by seeping water), we have the example of St. Francis Dam as the tragic result.

Preliminary testing can remove most doubts related to rock bearing strength. Tests made for Madden Dam, at Alhajuela, Panama Canal Zone, were conducted by the National Bureau of Standards. Among those tested was a bluish-gray, fine-grained sandstone of the Gatun Formation, found to have a compressive strength of 245 kg/mc^2 (3,500 psi) when dry but only 60 kg/m^2 (850 psi) when wet. Similar figures for a light-gray, medium-grained sandstone of the Caimito Formation were 160 and 38 kg/m^2 (2,300 and 550 psi), respectively. The values are the extremes of a range. Petrographic examination disclosed a clay mineral occurring as a film coating feldspar and other grains in the sandstones; abundant glauconite grains also contributed to the weakness. This example serves not only to show the importance of wet rock testing but it illustrates the troubles encountered because of the unwanted presence of clay minerals.

Gravity dams must also be analyzed against the possibility of sliding failure. The critical sliding plane is usually dependent on geologic structure. Various values have been suggested for the coefficient of friction between either concrete or masonry and mortar and rock surfaces. Tests reveal that, provided a rock floor is properly cleaned and prepared, the bond obtained between concrete and rock can be so efficient that failure will take place only by a shearing fracture of solid rock mass or of the concrete, or along rock discontinuities below the dam. Special preparation of glaciated rock surfaces may be necessary to overcome their typically smooth surfaces. Perfect bond may be obtained between concrete and the rock surface, but the rock mass itself may slide forward under the influence of the pressure transmitted to bedding planes or low-angle joints or shear planes. This especially occurs when normal resistance to movement has been in some way removed, as by hydrostatic uplift. The potential for such movement is greatest on bedding planes that are horizontal or that dip downstream from the dam. This situation can be detected by adequate preliminary geologic study. Austin Dam, already mentioned, provides a case in point.

When foundation rock is finally seen after dewatering and riverbed excavation, it is not uncommon for unexpected geological features to be revealed. Again, constant attention to all geologic details is imperative. Few dams have probably undergone the extreme difficulties with exposed foundation rock as has Bort Dam, key structure of the Dordogne power system in southwestern France. A gravity structure, curved in plan, its exact location and shape were determined by the necessity of utilizing two types of foundation rock, a sound gneiss capable of sustaining 73,000 to 97,500 kg/m^2 (15,000 to 20,000 psf) and a weaker mica schist with a strength of only 2,449 kg/m^2 (5,000 psf). A fault zone of crushed rock separated the two formations and crossed the gorge immediately under the dam axis. Great difficulties were experienced during construction, notably with rock falls in the mica schist when it was exposed in the side of the gorge. A series of reinforced-concrete "rockslide protection works" had to be constructed to retain great masses of mica schist until the main mass of the dam reached and constrained them.

Warragamba Dam was the largest dam in the southern hemisphere when completed in 1962; it is a major element in the water supply system for Sydney, New South Wales, Australia. It is located on the Warragamba River about 64 km (40 mi) from the seacoast, in the foothills of the Blue Mountains. The project began as a modest 15-m- (50-ft)-high weir, constructed in 1940 as a temporary water storage measure, at what appeared to be the obvious site for the ultimate high dam. Detailed geologic studies were completed as World War II came to a close. Geologic fieldwork and drilling included calyx holes 1.2 m (4 ft) in diameter, enabling engineering geologists to examine the local sandstone rock in situ. The record of this investigation is an excellent example of the employment of calyx exploration.

As a result, a new and superior site was found 1,200 m (4,000 ft) upstream of the "obvious" site and was adopted for the mass-concrete Warragamba Dam. The dam is 120 m (394 ft) high, and yet only 350 m (1,150 ft) long at its crest, dimensions which demonstrate what an ideal site was used. The gorge had been eroded in an uplifted plateau, generally of Triassic-aged Hawkesbury Sandstones and associated shales. A fault zone was discovered under the dam site and was carefully excavated and backfilled with concrete. Tests on the bedrock showed strengths when wet of 60 to 80 percent of corresponding dry strengths. The most unusual feature encountered was evidence of bedrock displacement, as exhibited in calyx holes and smaller drill holes, a lateral movement of as much as 12.5 cm (5 in) being observed. This was attributed to

FIGURE 10.7
Morris Dam for the water supply of Pasadena, California, located over a fault in the San Gabriel Canyon. (*Courtesy Water Department, City of Pasadena.*)

the unloading of the superincumbent rock. Arrangements were made for the measurement of rock stresses and strains in the bedrock adjacent to the dam, following its completion.

POSSIBILITY OF GROUND MOVEMENT

Bedrock displacement of dam foundation is of vital importance. The rigid nature of many dams is such that any movement of the foundation beds may lead to serious structural damage in seismic regions, unless due allowance has been made in design.

Allowance for earthquake forces must always be made, even for dams on the most solid rock foundations. A usual allowance for small dams is to consider an equivalent static horizontal acceleration force of a fraction of that of gravity; vertical acceleration is usually neglected. For large dams, dynamic time-histories of horizontal and vertical acceleration are employed. Morris Dam, in San Gabriel Canyon near Azusa, California, was built to augment the municipal water supply of Pasadena, California. The natural period of vibration of the dam (98.4 m or 328 ft high) was estimated at 0.16 seconds or less (Fig. 10.7). It was further deduced that the volume of material in this dam had to be increased by 15 percent to increase the damping thickening necessary to resist seismic forces. Special designs to withstand seismic forces have been evolved to be proof against even severe earthquake shocks.

Many older dams have failed as a result of earthquake shock. It is surprising that dams built at right angles across active faults have not failed under earthquake displacement. Two good examples are provided by earth-fill, puddle-core dams built across the San Andreas fault in California in 1870 and

1877, respectively. San Andreas Dam is 28.5 m (95 ft) high above the stream bed and 39 m (130 ft) above the bottom of its cutoff trench; and the Upper Crystal Springs Dam is 25.5 m (85 ft) high above stream bed and 57 m (190 ft) above the bottom of its cutoff trench. Both were constructed on clay foundation beds with the fault passing almost at right angles across their axes. The disastrous San Francisco earthquake of 1906 reached a maximum intensity of 10 on the Rossi-Forel scale. At the two dams, permanent displacement up to a maximum of 3.6 m (12 ft) took place; and although outlet tunnels around the dams were badly fractured, the dam structures themselves remained stable. Despite this appreciable movement, they remained watertight and so stand today. A large mass-concrete gravity dam (the Crystal Springs Dam, constructed in 1877 and enlarged in 1888 and 1890) located only 0.4 km (¼ mi) from the fault was undamaged.

Earth dams in Japan have similarly withstood severe earthquake shocks without serious damage. The antiseismic design for a water supply dam at Rangoon, Burma, also in a seismic region, was based on the Morris Dam example. In the absence of any clay suitable for use as a puddled core, the dam was designed with a flexible concrete core wall. The dam is 37.8 m (126 ft) high. The concrete core wall was built as a series of panels connected by means of ingenious grooved joints filled with asphalt.

The presence of active geologic faults—certain signs of past earth movement—will make it increasingly difficult to find sites for dams in many seismic regions. At Morris Dam, California, for example, a minor fault intersects the dam foundations near the base of the right abutment in a direction almost normal to the axis of the dam. An age determination study of old stream-bed strata revealed that no appreciable movement along this and other fault planes had taken place for a period of approximately 10,000 years. The fault was accommodated by special design; an open joint with vertical slip planes was built into the dam, over the fault trace. The joint ran from top to bottom of the dam between two of the blocks into which it was divided for construction purposes. The four sliding planes are at an angle of 45° with the dam axis and so lie along the strike of past movement. Planes of contact are separated by a bituminous filler which can yield slightly if motion is not as anticipated. A somewhat similar type of sliding joint, but horizontal, was incorporated in the design of the 63-m (210-ft) Lake Loveland concrete arch dam near San Diego, California. The joint was installed above a shelf in the left abutment in an attempt to eliminate cracking of the concrete at this location when the dam is under load.

PERMEABILITY OF BEDROCK

The second essential requirement of dam foundations in rock is that the entire geologic structure underlying the site of the dam, in addition to being strong enough to carry the design loads, will be a watertight barrier to the water impounded by the dam. One example only need be mentioned; it has been

FIGURE 10.8
Hales Bar Dam on the Tennessee River, while remedial work to overcome leakage beneath the dam was being undertaken by the TVA. (*Courtesy Tennessee Valley Authority.*)

termed "the greatest object lesson that the history of engineering foundations had to offer." The Hales Bar Dam on the Tennessee River was founded on a soluble limestone, well known for its cavernous nature (Fig. 10.8). Cavities were encountered to such an extent during the excavation of the dam site that completion was delayed several years, and the cost was increased far beyond the original estimate. Several hundred thousand barrels of cement were injected into the fissures and openings in the underlying rock. About ten years after the dam had been completed (in 1926), leakage through this limestone had become quite serious, and many unusual methods were used in attempts to provide an effective seal, including the use of various kinds of mattresses and the dumping of rocks, gravel, clay, and bales of hay. Success was achieved only by drilling a large number of holes to an average depth of about 27 m (90 ft) and injecting into these hot liquid asphalt, the flow of which was assisted by ingenious devices; over 11,000 bbl of asphalt were used. The process was later repeated, and leakage was eventually stopped. By 1941, however, leakage had again increased, amounting then to 48 cms (1,700 cfs), so that further remedial action had to be taken by the new owner, the Tennessee Valley Authority.

The TVA geological and engineering staffs made careful studies, and many rehabilitation schemes were evolved. It was finally decided to form a continuous curtain of concrete along the face line of the dam by means of continuously connected calyx drill holes, 45 cm (18 in) in diameter. More than 200 holes were drilled to depths up to 18 m (60 ft); the total length of drilling exceeded 3,600 m (12,000 ft). The holes were lined with cement-asbestos pipe and then filled with concrete. Study of the Bangor Limestone cores showed a rate of dissolution much greater than would be expected for most limestones. Finally, the TVA decided in 1963 to give up the unequal battle and constructed Nickajack Dam 10 km (6.4 mi) below Hales Bar Dam, which submerged the Hales Bar structure. Over $10 million had been spent in attempting to correct the flawed geologic setting of the original dam site.[4]

The first reason for founding dams on watertight rock obviously is to make sure that no water escapes. Not only is this necessary for water conservation, but it is also essential because of the erosion potential of flowing water. Limestone and all weak rock types are suspect until proved to be sound. Although limestone is notorious as the most troublesome of foundation rock, it was a shale-sandstone combination that presented one of the most unusual examples of an undesirably permeable dam foundation. Santa Felicia Dam, of the United Water Conservation District of California, is a 60-m (200-ft) rolled-earth dam located due north of Los Angeles on Piru Creek. It lies over sandstones and shales of Miocene age, much folded and known to be oil-bearing. One oil well was actually located in the reservoir and further drilling was contemplated. The district therefore had to pay for the abandonment of this part of the oil field and for the capping of producing wells in the reservoir area; this cost $200,000. Oil at the dam site was therefore suspected, and oil was encountered in every one of the borings and in the 70 grout holes drilled into the old stream bed. Gas under pressure was encountered in some holes. All holes had to be sealed off, and the many oil seeps had to be capped prior to placement of embankment fill. Anticipated hydrostatic reservoir pressure would more than counteract the oil and gas pressure when the reservoir was filled, but this unusual foundation "permeability" construction made for an uncommon operation.

In addition to the possibility of major leaks, dams have been compromised by minor leaks. Springs are included in this category; if preliminary investigations suggest that such may form along discontinuities after construction is complete (or if noted during construction), careful measures must be taken to suitably connect them with the drainage system. This last possibility is probably the most difficult to counter, since even slight leakage between the rock surface and the dam structure will give rise to what is generally known as uplift pressure. The need to counter uplift pressure has been known at least since the design of Vyrnwy Dam (in 1896):

> Although no visible springs of water issued from the beds of rock thus exposed, it was by no means certain that, when the reservoir was formed and the head on one side became 144 feet, springs subject to that pressure would not occur...and the Author agreed with him [the late Mr. T. Hawksley] in thinking it desirable to provide relief drains, which so far as he is aware, had not been done in connection with any former masonry dam.

The geologic implications of uplift pressure beneath dams have been admirably reviewed by W. H. Stuart. "Geological conditions such as the direction, dip and spacing of joints, fractures and bedding, and the location of impervious boundaries were used in designing the grout and drainage systems'' of the six dams dealt with, which varied in height up to 113.7 m (379 ft). Foundation-bed conditions varied from sandstone over shale to granite, but the essentials of the drainage systems used were similar, all having 7.5-cm (3-in) drain holes spaced at from 3- to 6-m (10- to 20-ft) centers, with depths up to 18 m (60 ft) (apart only from the 111-m (370-ft) Detroit Dam, where they were drilled to 36 m (120 ft). Measured uplift pressures were all less than the conventional conservative assumption (full head at the toe of a masonry dam to zero at the heel). Drainage systems, as Stuart points out, are similarly dependent upon geologic conditions, even the smallest detail of geological structure being of possible significance. When dams have been completed and drainage systems are operating satisfactorily, regular inspection and maintenance of these facilities is essential, especially against mineralization plugging.[5]

It will be clear that the presence of water in rock foundation beds for dams and the possibility of flow through such bedrock is a complex geologic matter. The use of carefully controlled and observed water-pressure tests in boreholes provides a fairly reliable confirmation of underground structural soundness. If the holes hold water under the maximum test pressures, it is clear that the strata penetrated by the test holes are watertight, at least in the neighborhood of the holes. Drill-hole water-pressure tests are not an infallible guide; when considered in conjunction with ascertained geological structure, however, such tests can be a reasonably reliable indication of watertightness. During geological investigation of the Madden Dam site in the Panama Canal Zone, water was forced under pressure into flowing wells to detect possible hydraulic connection to other wells or any other possible means of escape for the test water. Bright-color dyes especially are of value. Complementary laboratory tests were also carried out during the Madden investigation on the permeability of core samples taken from boreholes. The specimens were packed into sections of steel pipe with lead wool and steam packing and subjected to a head of 5.6 kg/cm^2 (80 psi). This was an early example.

Today, water-pressure testing in exploration drill holes is a standard procedure in studies of dam foundations, although there is never any standardization of the test conditions. Only rarely will bedrock be intrinsically permeable; rather, flow will usually occur along joints, bedding planes, and along shear zones, thus pointing again to the vital necessity of geologic exploration.

DAMS ON PERMEABLE FOUNDATION BEDS

Discussions up to this point have been confined to the common dam situation offering watertightness on either impermeable foundation rock or by installation of a cutoff wall from the dam structure to a solid-rock floor. Many potential dam sites remain at which impermeable rock exists only at great depth. Sites on unconsolidated deposits present many unusual problems of design and construction. ''Floating'' dams, placed over sands and gravels, depend for stability on a predetermined safe rate of underflow. Emphasis must be placed on preliminary geologic studies of such dam sites, to learn all that is required to utilize the unconsolidated deposits as constituent parts of the structure. Here, detailed testing is performed in material that otherwise would be excavated or penetrated by cutoff trenches or sheeting. Sampling of undisturbed soil therefore becomes a major feature of preliminary work. The soil characteristics to be investigated are common to other uses of unconsolidated materials, yet there is no other branch of civil engineering work in which close correlation of geology and soil studies is so absolutely essential.

Fort Randall Dam, one of the group of major river-control structures on the Missouri River, is sited on a permeable foundation (Fig. 10.9). Located in southeastern South Dakota, this dam controls the runoff from one-twelfth of the entire area of the United States as part of the water conservation and utilization of the Missouri River basin. The spillway outlet works and associated powerhouse (40,000 kW units under a head of 33.6 m or 112 ft) are located at the northeast end of the dam, all founded on Niobrara Chalk which underlies the entire site. The main part of the dam, however, is an embankment constructed of 20 million m^3 (26 million yd^3) of the glacial drift overburden which had to be removed elsewhere on the site. Fourteen and a half million m^3 (19 million yd^3) of chalk and shale bedrock were placed as the 450-m-(1,500-ft)-long downstream berm. The embankment was constructed over the natural alluvial soils in the valley floor which consist of pervious interbeds of silty clays, sands, gravelly sands, and clayey gravels.

Dam design here is based on a need to dissipate underflow head by seepage travel. An impervious blanket was therefore placed in 30-cm (12-in) lifts for 450 m (1,500 ft) upstream from the embankment, varying in thickness from 3 m (10 ft) at the upstream end to 6 m (20 ft) where it joins the embankment proper. Embankment soil was rolled, at about optimum moisture content, in 20-cm (8-in) layers. Figure 10.9 illustrates the section similar to that used for innumerable smaller dams on permeable foundations. Essential to this design is the row of relief drains in the downstream berm.[6]

CONSTRUCTION CONSIDERATIONS

A litany of geologically related problems can be anticipated during dam construction. Success in defeating such problems begins with thorough geologic investigations from the feasibility stage up to and beyond the final approval of the cleaned and prepared foundation rock surface. A selection of cases included from around the world illustrates the infinite variety of problems that can arise when a major dam is under construction.

Of universal consideration is the removal of all rotten and disintegrated rock, while keeping accurate records of the type and quantity of such spoil. When finally the approved foundation bed is ready for construction, a detailed geologic map of the bed must be made; this is one of the most critical duties of the resident geologist. Although record surveys of this kind may seem to be

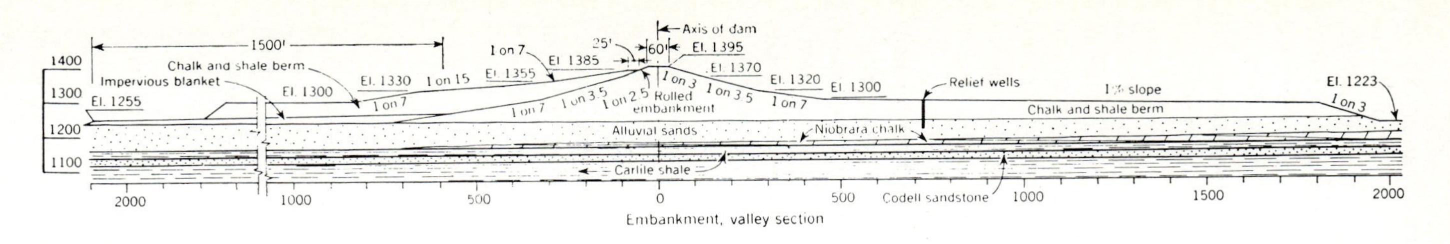

FIGURE 10.9
Cross section at right angles to the axis of Fort Randall Dam on the Missouri River, showing foundation-bed conditions. (*Courtesy American Society of Civil Engineers.*)

unimportant at the time, they are of inestimable benefit in all cases of dealing with operational problems such as seepage, deformation, or instability. Once buried beneath the dam, these essential details can never again be obtained.

Matters of detail begin with preparation of the rock foundation bed. The essential construction objective is to secure a good bond between concrete and rock. To obtain this, the rock surface must be clean and free from all loose particles and standing pools of water (however small) when concreting operations start. Smooth glaciated rock surfaces should at least be roughened by "pop shots," to assist in developing a structural bond. Failure of the Gleno Dam in Italy has been attributed to sliding of the structure on a glaciated rock surface. The exact location and regular discharge of all springs encountered must be most carefully recorded and brought to the attention of the geological adviser for examination and evaluation. Detailed inspection of all unusual features, in addition to general supervision of excavation operations, is just as essential a part of the duties of the geological adviser as are the preliminary geological investigations.

Only during construction will the full extent of unusual foundation-bed features be revealed. During the construction of the dam for the Merwin Hydro Electric Project on the Lewis River in Washington State (downstream from and associated with Swift Dam), a buried preglacial gorge immediately below the location of the powerhouse in the 375-m (1,250-ft) dam was detected. The preliminary studies found it to be 21 m (70 ft) wide and 36 m (120 ft) deep. In a bold solution, a massive concrete arch was completed in 1931 to span this inner gorge. Noxon Rapids Dam in Montana revealed similar glacial action upon the riverbed, which was discovered only after dewatering. Figure 10.10 shows only one of the potholes extending 24 m (80 ft) below riverbed level which required 17,000 m^3 (22,000 yd^3) of concrete to fill. The site is on indurated argillite but lies in a wide valley that was once the bottom of glacial Lake Missoula. Failure of an ice dam in glacial times about 40 km (25 mi) above the dam site created strange and awesome rock erosional features. These features necessitated design changes employing a mass-concrete spillway and powerhouse section, flanked by two large earth-fill wing dams.

Geological features revealed during construction can sometimes affect progress schedules. During construction of the Beechwood Dam and powerhouse for the New Brunswick Power Commission on the St. John River 160 km (100 mi) north of Fredericton, a relatively small fault in the argillite that made up the entire riverbed was discovered only during the second construction season, despite careful preliminary drilling. A modification of plans to deal adequately with the fault zone was readily arranged, which, in turn, delayed placement of earth fill in the adjacent wing dam. Facing the forthcoming Canadian winter, the contractor ingeniously employed an asphalt-mixing plant to heat the soil before placement, and the work was finished almost as originally scheduled.

Many years before, difficulties were encountered during the construction of one of the major rock-fill dams of the western part of the United States. The

FIGURE 10.10
Unusual pothole detected in preliminary investigations and uncovered during excavation for the foundation of Noxon Rapids Dam, Montana; workers indicate the scale. (*Courtesy Washington River Power Company.*)

Los Angeles County Flood Control District was set up in 1915 to carry out flood-protection and water-conservation work for Los Angeles County. In 1924, $25 million of the funds raised by bond issue was allocated for construction of a concrete gravity dam of unprecedented size in San Gabriel Canyon. Work was begun under contract in 1929, and completion of the dam was expected within six years.

> While excavation was still under way a serious earth slip at the west abutment resulted in holding up the work for a further study of foundations. Before the end of 1929 the contract for constructing the dam was cancelled. Surveys for a new dam site developed what was thought to be a safe and satisfactory location just above high water level of the reservoir of the Morris Dam, constructed by the city of Pasadena lower down the San Gabriel Canyon. At this site it was proposed to build a rockfill dam... to have a height of 300 feet above streambed, a crest length of 1,670 feet, and a base thickness of 900 feet at streambed level....

After further conferences and reports, work on the new contract was temporarily held up, and the dam design was revised again to take into consideration the rock available from the quarry. On this basis, the work was satisfactorily completed in July 1937. Over 7.5 million m^3 (10 million yd^3) of material was

FIGURE 10.11
San Gabriel Flood Control Dam No. 1, under construction in August 1934; concrete toe wall in the foreground and Quarry No. 10 in the background. (*Courtesy Los Angeles County Flood Control District.*)

finally placed in the dam; about 750,000 m^3 (1 million yd^3) was a sand-clay mixture used as an impervious blanket near the upstream face, and all other material was quarried rock. All loose rock fill was placed with the aid of sluicing; shrinkage approximated 6 percent. The dam is functioning satisfactorily, having already been tested by severe floods (Fig. 10.11).[7]

Dez Dam, on the Dez River in the Khuzistan area of southwestern Iran, designed to provide irrigation water, flood control, and power generation, is located in a narrow gorge identified in the course of photogeologic surveys in 1956. Gorge walls rise about 420 m (1,400 ft) from river level to the surrounding plateau. A 35-km (22-mi) access road was required just to reach the gorge. Access to the gorge itself had to be arranged by tunneling 6 km (4 mi) with a maximum downgrade of 8 percent. A remarkable system of spiral tunnels was needed to keep the tunnel (7.5 m, 25 ft, in diameter) close to the side of the gorge so that short adits could be provided at intervals, through which excavation was disposed of economically by being dumped into the gorge below. The rock is a competent fused limestone conglomerate with no faults or cavities, in great contrast to other limestone formations which have been mentioned.

Two faults, normal to each other, were found to intersect beneath the center of Seminoe Dam of the Kendrick Project of the U.S. Bureau of Reclamation. The dam is an arched concrete structure, 78 m (260 ft) high, and located in a narrow gorge. The two faults were uncovered only when excavation was

FIGURE 10.12
Folsom Dam, California; a close-up view of the excavated fault zone, showing some of the concrete filling. *(Courtesy U.S. Army Corps of Engineers, Sacramento District.)*

possible after dewatering the site of the dam. Engineers decided to clean out the two fault seams and backfill them with concrete through shafts formed in the concrete base of the dam. Despite the hazard of the work, 2,600 m^3 (3,500 yd^3) of fault breccia and gauge were removed and replaced with concrete within six weeks of completion of the concrete foundation.

A similar problem involving construction discovery of a significant fault was faced at Folsom Dam on the American River 32 km (20 mi) east of Sacramento, California. This is a combined mass-concrete and earth-fill dam with a total embankment length of 7 km (4.4 mi), built as a joint project of the U.S. Army Corps of Engineers (dam) and the U.S. Bureau of Reclamation (powerhouse). The mass-concrete section is founded on sound granite rock, but the earth-fill part of the dam and eight of the small saddle dams are founded on weathered granite, requiring cutoff walls. The fault was uncovered in sound granite beneath the concrete part of the dam, and was traced back under some of the existing concrete. A tunnel was therefore excavated along the plane of the fault (Fig. 10.12) to its termination. About 37,000 m^3 (50,000 yd^3) of additional concrete was required.

Cutoff walls nearly always provide the contractor with a geologic challenge. Mount Morris Dam on the Genesee River in northern New York was built on a limestone and shale site. The mass-concrete flood control wall is 75 m (250 ft) high, but it was necessary to carry the cutoff wall well into both abutments. The

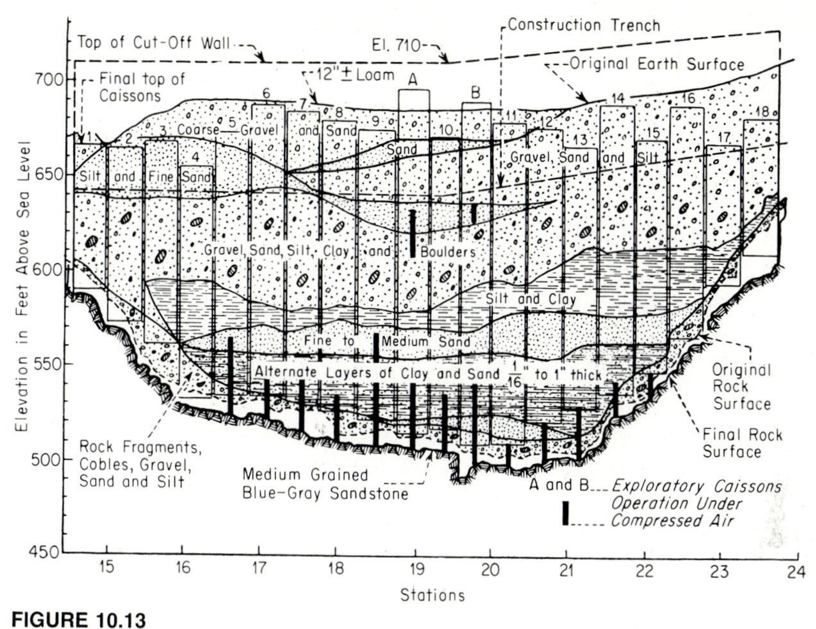

FIGURE 10.13
Diagrammatic section through the soil strata encountered by the concrete caissons forming the cutoff of Merriman Dam of the Delaware Aqueduct works for the water supply of New York. (*Courtesy Engineering News-Record.*)

shale walls of the cutoff excavation proved unstable and the contractor resorted to tunneling through competent rock and then stopping (upward) into the shale for a distance of 10.6 m (35 ft), while mucking out through the access tunnels. The process was then repeated after each individual segment of the excavation had been infilled with concrete.

At Merriman Dam, part of the Delaware Aqueduct water supply of the city of New York, a complex assembly of glacial fluvial and lacustrine sands, gravel, silty clays, boulders, and lodgement till had to be penetrated to cutoff depths up to 54 m (180 ft) for the 60-m- (200-ft)-high earth-fill dam. It was decided that the best way to achieve this would be by caissons so designed and constructed that compressed air could be applied when and if necessary. Twenty reinforced-concrete caissons for the compressed-air environments measured about 3.6 by 13.5 m (12 by 45 ft), with heights varying with depth to bedrock. Figure 10.13 diagrammatically illustrates the heterogeneous glacial profile penetrated by this unique cutoff wall.

Many shales are overconsolidated and therefore troublesome during construction. Contractors should expect difficulties and should consider a method such as that followed on the Oahe Dam on the Missouri River. Shale exposed at final grade was excavated in relatively small areas and immediately covered with a bituminous coating. Concrete was placed within 48 hours and swelling and slaking was avoided. When this initial concrete slab had set, anchor-bolt holes were drilled through it and swelling tendency was further restricted. Morrison Formation shale at the Green Mountain Dam project of the Blue River was also protected from swelling with an asphalt-bitumen emulsion applied within half an hour of excavation.

GROUTING AS A SOLUTION TO GEOLOGICAL PROBLEMS

An impressive job was completed in 1961 in the coastal mountains of British Columbia. Mission Dam looks today like a normal earth rock-fill dam, but it was plagued by a continually unfolding train of problems during construction. A grout curtain 150 m (500 ft) deep had to be installed in water-bearing alluvial materials, down to the underlying bedrock, and this for a dam which itself is only 60 m (200 ft) high. Intrepid diving work, often in mud, was necessary at an early stage. Later, 2 acres (.5 ha) of polyvinyl-chloride sheeting had to be installed to assist in the consolidation drainage of the clay used in construction. Dr. Karl Terzaghi was the special consultant on this dam; he regarded it as one of the most challenging projects he had ever faced. The dam was renamed Terzaghi Dam (rightly so) at the Sixth International Soil Mechanics and Foundation Engineering Conference held in Montreal in 1965.[8]

Aqueous emulsions of bitumen have been used as permeability reduction grouts. Asphalt has been so used at Hales Bar Dam, as has been noted; it also formed the cutoff wall at Claytor Dam, on the New River near Radford, Virginia. Founded on stratified gray dolomite with shale and chert present in seams, this dam was located over some large and threatening dissolution channels. An asphalt-sand mixture was chosen as prepared in a standard construction premix plant, and dumped into place. Its consistency was well formulated as it penetrated into fine seams and acted as a satisfactory grout. It is believed that asphalt has also been used in grouting operations carried out in the U.S.S.R. In sealing rock at the Tennessee Valley Authority, Great Falls development, asphalt grout was employed in association with a cement-grouted curtain (Fig. 10.14). The Authority has used similar grouting at the Timms Ford Dam, though not in the vicinity of actual structures, where stress redistribution might lead to unwanted deformation.

One of the first dam foundations to be effectively grout sealed was that of the Kinder Embankment of the Stockport Corporation, England. During the construction of this work (from 1903 to 1905), the bedrock was found to be so fissured and faulted that it could not support the originally intended masonry dam; a satisfactory shale bed was eventually found at a depth of 54 m (180 ft) below ground level. In the construction of the cutoff trench to this depth, much trouble was encountered with groundwater. The drainpipes installed to tap the water as concreting advanced were carried up until water ceased to flow from them and were then grouted up under pressure, a practice now frequently followed.

A pioneer dam application in the United States was the pressure grouting of Estacada Dam constructed in Oregon in 1912. No less than 10,200 m (34,000 ft)

FIGURE 10.14
Limestone abutment of Great Falls Dam on the Collins River, Tennessee, just below its confluence with the Caney Fork River; as seen (top) before, and (bottom) after an extensive grouting program was carried out in 1946 with asphalt and cement grout. (*Courtesy Tennessee Valley Authority.*)

of grout holes were used for the operation, including diamond and Calyx drilling and a Canniff air-stirring grouting machine, then in common American use. The dam was built on very permeable volcanic breccia as described in this extract: "The general idea provided for drilling a double row of holes of an average depth of 50 ft under the heel of the dam across the entire valley to and under the shore abutments and the subsequent forcing into each of these holes of grout of such consistency as to percolate through the entire substructure."

High-pressure grouting was carried out for a very different purpose in the construction of the Abitibi Canyon hydroelectric project in northern Ontario. This large dam, located on a river flowing into Hudson Bay, is founded on gabbro which preliminary borings indicated to be free from discontinuities. When the canyon dam site was unwatered, it was found that along the stream bed lay a fracturing and alteration zone about 12 m (40 ft) wide. Since it was not known whether or not this zone would be impervious to water, an exploratory shaft was sunk to a depth of over 30 m (100 ft) below river level at the upstream face of the dam. No water was encountered, and the rock appeared to be sound; but it was decided to carry on exploratory drilling from the shaft bottom. Whenever porous zones were found, cement grout was injected at pressures not less than the full reservoir static head. After grouting, the shaft was filled with concrete so that dam construction could continue. A smaller shaft was completed at 3 by 3.6 m (10 by 12 ft), with an enlargement at the bottom (Fig. 10.15). From this shaft, grout holes were drilled, and injection was carried out at 3-m (10-ft) vertical stages, using pressures up to 17.5 kg/cm^2 (250 psi). In addition, grouting was carried out in old exploratory drill holes in the canyon walls and from inspection galleries left in the dam. A total of 1,527 bags of cement were used with satisfactory results.

The foregoing examples have been cited because of their major geological interest. It has been, however, during the second quarter of this century, that construction of large dams in the United States has led to the greatest progress in techniques of cement grouting. Notable have been the practices developed by the Tennessee Valley Authority, which have been born of necessity because of the poor foundation conditions (usually in limestone) for several of their large dams. Foremost among the TVA examples is Norris Dam, a gravity structure on the Clinch River about 130 km (80 mi) above its confluence with the Tennessee River; it is 85 m (285 ft) high and floods an area of 14,000 ha (34,200 acres) at normal pool level. The entire valley above and below the dam site is underlain by Knox dolomite; in many locations there are underground caverns and dissolution-riddled beds. Dolomite at the site was described in official reports as "a comparatively hard and substantial rock... quite massive, compact and sound. Some jointing has occurred... [with] occasional seams, of importance in regard to possible leakage." Complete exploration of the dam site confirmed previously expressed opinions of consultative boards to the effect that the foundation rock itself is characterized by massive, thick strata of excellent rock interrupted at definite intervals by horizontal bedding planes which are partly open, partly clay filled, and for a portion of the area, closed.

FIGURE 10.15
Abitibi Canyon water power project in northern Ontario (now a part of Ontario Hydro): a general view of excavation for the dam, looking upstream and showing the main cofferdam and deep shaft excavation for unsound rock. (*Courtesy Ontario Hydro.*)

A comprehensive grouting program was designed to consolidate the entire base area of the dam, spillway apron, and powerhouse. The grouting treatment had two elements: shallow, low-pressure grouting to cover the entire area of the foundation; and deep, high-pressure grouting to form an impermeable curtain under the heel of the structure. On the reservoir rim, the work involved determining the location and grouting of these portions needing treatment to prevent excessive leakage.

New limestone grouting experience has been gained at almost every one of the major TVA structures. Chickamauga Dam, for example, founded on a jointed and cavernous limestone, exhibits an overburden of 12 to 15 m (40 to 50 ft) of good, sandy, clayey silt to mixed gravel, then finally limestone slabs and clay. Subsurface conditions were found to be so variable that most of the original grout holes on 30-m (100-ft) centers missed the areas of highest grout take. One large cavern was found as much as 27 m (90 ft) below normal river level. Larger caverns were filled with a grout mixture of sand, bentonitic clay, and cement. Published records give full details of this invaluable experience.[9]

Grouting records also exist for dams of the U.S. Bureau of Reclamation.

Hoover Dam was one of the great pioneer grouting projects, specifying in detail materials, pressures, and procedures. Wherever concrete was to be placedagainst rock, advance provision was made for forcing grout into rock joints and bedding planes, if any existed. While concreting was in progress, special pipe connections were installed through which grout could later be forced when the dam was carrying full water load, in order to seal effectively the concrete-rock contact. A grout cutoff curtain was formed in the bedrock across the dam site, from grout holes drilled from the outlet tunnels and extended as deep as 45 m (150 ft). After the construction of the dam had advanced appreciably, grouting was also carried out from the inspection galleries left in the body of the dam. Quality assurance review of the grouting was made by coring of grouted rock, which showed satisfactory results. Injection pressures used varied up to 70 kg/cm^2 (1,000 psi); the average quantity of neat cement used throughout was 0.8 sack per foot of hole for all the longer holes.

Many modern European dams, especially the daring arched dams that now distinguish many of the spectacular gorges in high mountains, are assisted in their functioning by hidden grout curtains carefully placed in the surrounding rock. St. Pierrett-Cognet Dam, an important unit in the Drac development in France, is typical. Although the dam is 75 m (246 ft) high, the gorge in which it is located is so narrow that the entire structure contains only 38,000 m^3 (29,000 yd^3) of concrete. The gorge is in a highly jointed Bajocian Lias, reasonably watertight but surface-loosened by weathering. The gorge walls absorbed 300 tonnes of cement at the rate of 30 kg/m^3 (55 kg/yd^3) of finished curtain. Figure 10.16 illustrates clearly the extent of the grouting necessary to seal weak rock such as marl and jointed rock such as limestone at Foum el Gherza Dam, Algeria.

From relatively insignificant beginnings pressure grouting, or cementation, has developed into an important feature of almost all major dam construction operations. In practice, grouting operations are generally managed by engineering geologists with planning and modifications based on accurate knowledge of underlying geologic stratigraphy and structure. Grouting must therefore always be a cooperative endeavor of engineer and geologist, and the successes of the foregoing examples have been dependent to a large extent on such joint action.

PROBLEMS WITH DAMS IN SERVICE

Operational problems at dams usually involve unacceptable leakage. Although trouble has been experienced because of concrete deterioration, the usual cause is some inadequacy of the underlying geology. Aside from the critical need for regular inspection, it may be helpful to note particular actions that have been taken.

Mulholland Dam, located in Weid Canyon above Hollywood, California, is a mass-gravity structure containing 130,000 m^3 (175,000 yd^3) of concrete. It was built in 1924–1925 as part of the water-supply system of Los Angeles. After the failure of St. Francis Dam, local residents sued the owner, the City of Los

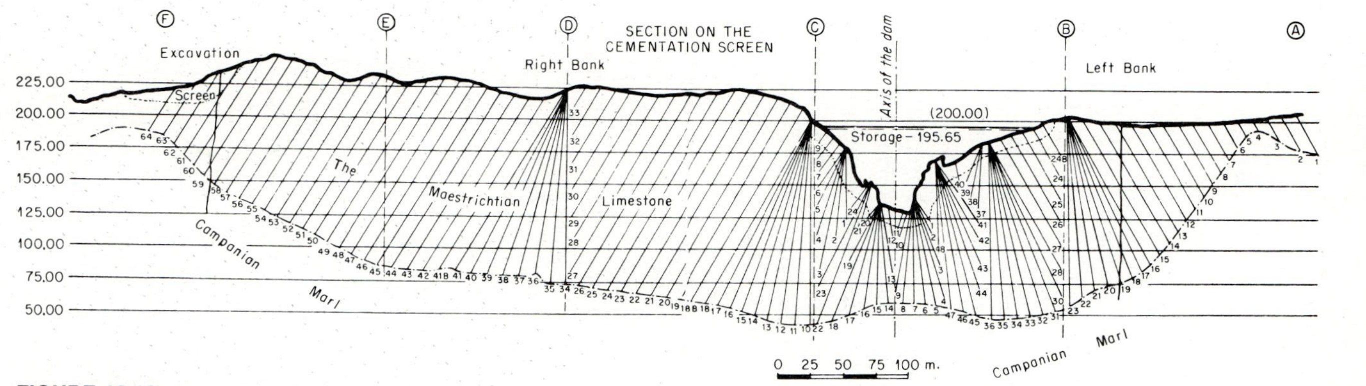

FIGURE 10.16
Cross section showing the extensive grout curtain formed at the site of the Foum el Gherza Dam in the Sahara zone of Algeria, to seal the marl and limestone strata shown; sulfate-resisting cement was used for the structure. *(Courtesy La Société Nationale Electricité et Gaz d'Algerie and the editor, Travaux, Paris.)*

Angeles, in August 1928 to force the abandonment of the reservoir formed by the dam on the ground that it constituted a menace to life and property. Several favorable inspections were made by recognized geologists and their reports were almost unanimously favorable. The lawsuit was dismissed, yet Los Angeles yielded to public opinion and in 1934 placed 200,000 m^3 (300,000 yd^3) of earth-fill buttress against the downstream face and camouflage-landscaped the fill. The dam stands today, still effectively hidden, its stability justifying the geologic opinions of half a century ago.

Very different were conditions at Wolf Creek Dam on the Cumberland River in south-central Kentucky, built by the U.S. Army Corps of Engineers between 1941 and 1951. The mass-concrete dam is 539 m (1796 ft) long, carrying ten Tainter spillway gates and abutting directly to the left wall of the valley. A rolled-earth embankment, 1,182 m (3,940 ft) long, extends to the right abutment in the valley wall. The dam has a maximum height of 77 m (258 ft) and can impound 7.5 billion m^3 (6.1 million acre-feet) of water. Mississippian to Ordovician bedrock includes limestone and shale of different formations. The shale is at a shallow depth beneath the spillway section of the dam and passes beneath the Leipers Limestone foundation bed. Groundwater has dissolved Leipers Formation joints and bedding planes, resulting in a karstic structure. This was well recognized when the dam was built and a deep cutoff trench was carried into bedrock from which grouting was carried out.

After satisfactory performance for 16 years, the first sign of possible trouble came in October 1967 when a muddy flow was noticed in the powerhouse tailrace. Within a few months a sinkhole was observed in the downstream embankment. An extensive exploration program found high piezometric pressures under the dam, indicating leakage through the limestone. Water level in the reservoir was lowered 6 m (20 ft) and an extensive program of cement grouting completed in June 1970. On the basis of embankment deformation monitoring a board of consultants concluded that a positive cutoff wall extending well into the bedrock was necessary. Final remediation included a concrete cutoff wall starting at the mass-concrete section of the dam and extending along the axis of the embankment at about its center point, with an additional wall adjacent to the nearby switchyard (Fig. 10.17). The slurry-trench method was chosen, using circular clamshell holes varying from 1.3 to 1.2 m (52 to 47 in), with bentonitic clay slurry filling the holes until concrete was placed. These primary holes were 1.3 m (4 ft 6 in) apart, the space between them being excavated and filled with secondary concrete elements once the primary elements were set. The resulting permanent concrete wall was completed in 1979, in 60 m (200 ft) of embankment soil and extended 24 m (80 ft) into the underlying limestone.[10]

GEOLOGIC INFLUENCES ON RESERVOIRS

Dams are generally built to store water above the elevation at which it normally stands. This impounding may vary from control of water levels within a few

FIGURE 10.17
Wolf Creek Dam on the Cumberland River, Kentucky; showing construction operations in progress for the creation of the cutoff wall. *(Courtesy U.S. Army Corps of Engineers, Nashville District.)*

feet of normal elevation by a regulating weir to the creation of a large lake. Apart from specially restricted cases such as low-head regulating weirs, reservoirs serve as artificial storage basins. Usually the dam structure forms but a small part of the periphery of the reservoir; its bottom and sides consist of in-place natural earth material. A superficial assumption is that as long as the dam site is watertight, all will be well. However, unknown or unheeded geologic conditions can defeat the entire function of a reservoir. The presence of a dam can create a number of problems due to the hydrostatic head of retained water. All material in the bed of the reservoir and especially close to the dam will be subjected to considerable hydraulic pressure, all flooded areas are subject to the action of water, and, finally, the groundwater level up the reservoir valley will be directly affected by the rise in water level, generally for a considerable distance into the reservoir, from the shoreline. In consequence, there will be a tendency for the impounded water to find some means of escape through any available geologic pathway. Many earth materials are affected by exposure to water and may subsequently lose their stability; landslides are only one possible threatening result. Wells within the area of influence of the reservoir will be correspondingly affected. All these matters depend fundamentally on the geologic structure and stratigraphy of the reservoir basin. Reservoir geology is therefore an essential complement of geologic investigations at any dam site.

Geology also affects reservoir capacity. For reservoirs floored with impervious strata, this calculation is a simple matter. However, if attention is given to the geological material underlying the basin, it may be found that large masses of pervious rocks are in contact with the impounded water and are so arranged that they can drain additional storage as the reservoir level is lowered, with consequent economic benefit.

RESERVOIR LEAKAGE

Leakage from reservoirs is an obvious source of trouble, though attention has not always been given to such a possibility. Of many unfortunate examples, there needs to be mentioned only the abandonment of Jerome Reservoir in Idaho and of Hondo Reservoir in New Mexico as a result of excessive and uncontrollable leakage. That structures of such magnitude should be abandoned, not because of defects of design or construction but because of undetermined site geology, is telling evidence of the need for adequate study of every feature connected with the geology of a reservoir site.

One of the most singular cases of what can result from the neglect of geology must be described anonymously, but the lesson it tells does not suffer from this lack of identification. There is standing today in the northeastern part of North America a fine buttress-type, reinforced-concrete dam about 12 m (40 ft) high and 540 m (⅓ mi) long, constructed in 1910 and still in good condition, even though it has never retained water. It was built by an owner who spurned professional advice. Cursory reading of a then-available geologic report would have shown that one buttress and one complete side of the intended reservoir consisted of a glacial moraine made up of highly pervious, small-boulder ablation till. As could have been foretold, the reservoir leaked like a sieve as soon as impoundment commenced. Not willing to be beaten by mere geological circumstance, the owner had a vast area paved with an asphalt-coated, reinforced-concrete slab; later a cutoff wall was placed to a maximum depth of 24 m (80 ft) below ground level, carried up the valley from the dam. All to no avail; the reservoir still leaked to such an extent that all hope of power generation had to be given up, and penstock and powerhouse were dismantled. The dam stands today in mute testimony to what the neglect of geology can do. Beavers who inhabited the stream built their own series of dams which did retain water; one of the beaver dams was carefully constructed around a sinkhole, the leakage through which was thus stopped by Nature's own engineers.

Failure by excessive leakage may take place because of geologic defects such as faults, close or open joints, disintegration of rock under exposure to water, downdip drainage in pervious rock strata, or the absence of adequate impervious barriers at critical points along the reservoir perimeter. Geological defects and water-degradable rocks are the greatest potential danger in the vicinity of the dam, where the impounded head will be greatest. The most doubtful rocks are those which are soluble in water. Rock salt is naturally the most dangerous of all, but as it does not occur frequently, it is not of great significance. Gypsum, though less soluble than rock salt, is seriously affected by exposure to water. Of unusual significance in all reservoir investigations is the presence of limestone. Although it is the least permeable of the rocks mentioned, its relatively wide distribution renders it comparatively well known.

All limestones are at least partially soluble in ordinary water. If there is continuous flow of fresh water, dissolution will continue. Special precautions must be taken at all limestone sites subject to movement of groundwater to make sure that no potentially serious dissolution zones or caverns exist. In most cases, aerial photographs and medium-scale topographic maps (larger than 1:25,000) will exhibit telltale geomorphic evidence of the presence of karstic features. Sometimes these are the result of the action of flowing streams, which for a part of their courses flow underground, as in parts of Derbyshire, England. The troubles at the Hales Bar Dam were due to cavernous limestone. In the Panama Canal Zone, pioneer work on clay grouting at Madden Reservoir was ordered on the basis of careful studies of the underlying limestone strata which had been made in the preliminary geologic investigations. Malad Reservoir, near Malad City, Idaho, represents another example. Construction was started in 1917 on a dam 132 m (440 ft) long and 21 m (70 ft) high. Work was stopped when the crest reached a height of only 15 m (50 ft); although the impounded water had risen only 3 m (10 ft), leakage was occurring at a relatively large volume. Investigation suggested that the leakage path was through a bed of soluble limestone 150 m (500 ft) long and 7.5 m (25 ft) wide, but hidden from view by a valley-side talus slope.

A final example is Lone Pine Reservoir in Arizona, the abandoned 30-m-(101-ft)-high earth-fill dam completed in 1934–1936. Built as a WPA project without geologic advice, the reservoir lies over salt beds and is near obvious sinkholes. The main rock types are basalt, limestone, and sandstone, with jointing which tends to increase the probability of leakage. Water level reached a high mark of 22 m (74 ft) on April 6, 1936, but receded to the 13-m (43-ft) mark by May 8; further filling proved impossible.

The prime cause of reservoir leakage is flow through a pervious stratum dipping downstream (Fig. 10.18). The inclined sequence of pervious and impervious strata will naturally have one result—any water in contact with the pervious beds in valley A (as, for example, water impounded in the valley by a dam with crest level about X) will flow underground until it is discharged into valley B. Discharge might take the form of springs, but there are many variations responsible for leakage. For example, the lower pervious stratum might not outcrop in valley B but may remain underground with no natural outlet, if the surface stratum is impervious. Artesian pressure would thus be set up, and a potential recharge water would occur.

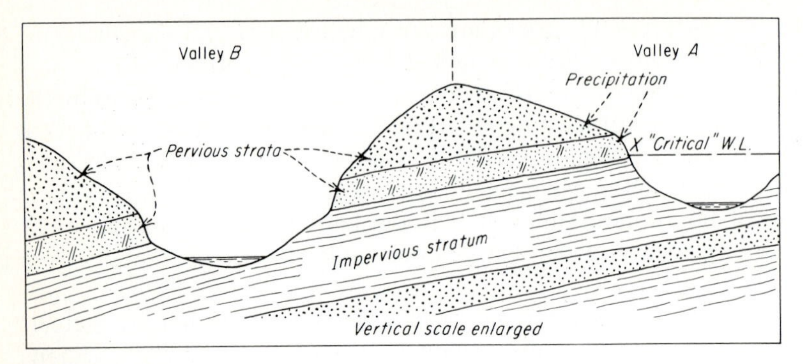

FIGURE 10.18
Simplified geological section illustrating possible leakage from a reservoir.

Leakage may also take place through some permeable stratum at a low point around the reservoir sides. Any depressions should be geologically investigated, since they are departures from regular topography and such is always an indication that some geologic process has been at work at some time in the past. Glacial action is frequently responsible for saddle depressions between adjacent valleys. A section through a simple saddle in which the rock level has been worn down well below surface level, pervious deposits making up the observed level of the ridge at this location, is shown in Figure 10.19. If water is impounded above "Critical WL" and the broken line extending through "Pervious material" is the necessary underground hydraulic gradient for flow through that material, it is clear that leakage will occur from valley A into valley B. In addition to causing loss of water from the reservoir, this underground flow of water may cause pore pressure trouble in valley B, through reduction of shear strength.

All too many cases of reservoir leakage are due to this cause. A remarkable example is that of Cedar Reservoir in Washington State. In 1914 the City of Seattle constructed a concrete dam in the Cedar River gorge, a tributary of the Snoqualmie River, with a crest length of 239 m (795 ft) and a total height of 65 m (217 ft). Before the dam was built, Seattle was advised by its consultants of the danger of severe leakage around the northeast abutment, but this advice was neglected. Filling of the reservoir started in 1915, and leakage soon reached 110 mld (30 mgd). Despite this, filling was slowly continued. Leakage increased and finally washed away the small town of Moncton, whose 200 inhabitants had to be relocated. A new board of consultants employed exploratory drilling. The findings of this board were just the same as those of the earlier consultants, namely that the northeast abutment of the dam was part of a large glacial morainal deposit of open-textured gravel, infilling a deep preglacial valley.

Much effort and money were expended in attempts to seal the reservoir. Toward the end of 1918, water was again allowed to rise, resulting in increasing leakage and a great earth movement on the morainal embankment approxi-

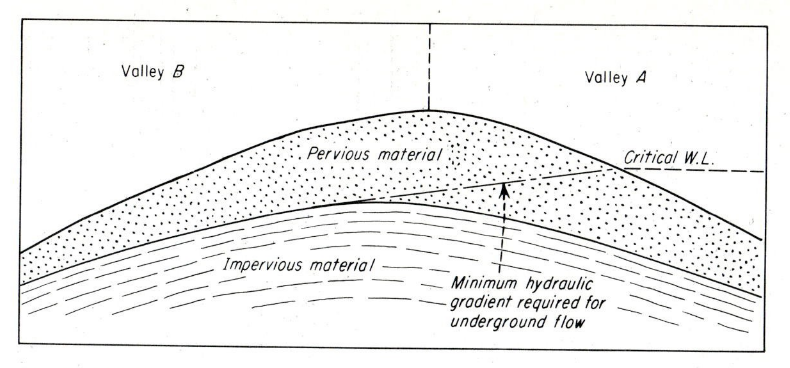

FIGURE 10.19
Simplified geological diagram illustrating another possible cause of reservoir leakage.

mately 1,800 m (6,000 ft) from the dam. Well over 764 thousand m^3 (1 million yd^3) of material was then carried away by the flood, estimated at between 85 and 570 cms (3,000 and 20,000 cfs). The flood rushed down to the Snoqualmie River, taking out the small town of Edgewick, as well as sawmills and other property, and destroying some track of the Milwaukee Railroad. The dam still stands but no attempt has again been made to use it to impound water. J. H. Mackin's paper on this failure provides a rational explanation of the cause of the failure.

Reservoir leakage is generally visually obvious and often substantiated by water balance computation. Sudden increases in streamflow below a dam site without a corresponding increase of rainfall will be a direct indication of some escape of water; and the appearance of springs, boils, or other surface indications of underground water in locations previously dry and stable also indicates unsatisfactory conditions. Minor leakage is more difficult to trace, especially when it flows along limestone dissolution channels. Small quantities of dye can be added to groundwater at wells or sinkholes and samples taken at breakout points such as springs. The use of bacteria has been suggested, but for practically all special work a chemical indicator is used. Many chemicals are thus employed, among them fluorescein and its sodium salt, uranin, along with permanganate of potash, or ordinary bluing. Characteristic colors can be detected even when dyes are diluted to contents of less than 10 ppb.

Radioisotopes are now used to detect groundwater flow. Potential difficulties that have stood in the way of their widespread adoption are now solved by employing postactivation irradiation of nonradioactive isotopes that are not otherwise found in groundwater. One of the early applications of isotope tracers was in the Libyan Desert of Egypt, to determine watertightness of a proposed reservoir site where groundwater was shown to be static, rather than the reverse. Rubidium chloride was specially prepared for this application at Harwell, the British Atomic Energy Research Establishment, flown out from London to Cairo, and put to use little more than a day later. Similar applications have been made in several countries. Radioactive hexacyanocobaltate isotope was used to trace the subsurface loss of Yarkon River water into the limestone aquifer at the Lydia well field in Israel. Naturally, background for freshly pumped water has to be determined before the radioactive material is introduced. The single-well pulse technique has been found to be reasonably effective and economical, even though results obtained have to be interpreted with care if the aquifer being studied is extensive. Expert assistance must still be utilized for all such uses of radioactive materials, but they do hold promise of extending subsurface water investigations greatly.

Most of the potential possible sources of reservoir leakage should normally be discovered in the course of preliminary geologic investigations. The expected nature of potential leakage will be some indication of whether or not it can be prevented by available remedial actions. Where dipping strata cross a valley at any angle and extend into adjacent areas, little, if anything, can be done to interfere with the natural seepage direction. When saddles occur with low spots in the bedrock elevation, as indicated in Figure 10.19, cutoff dams can be constructed as barriers to underground flow. Simple structures will suffice for this purpose, provided a cutoff wall of some type is carried to the rock floor at elevations above the reservoir floor. In other cases grouting may effect a satisfactory seal, especially where jointing, bedding, or shearing is localized and if the grouting is based on geologic study.

Grouting operations may also be employed for remedial work if leakage is discovered after a dam has been completed. Seldom can reservoirs be emptied completely, or even partially; remedial work to eliminate leakage from a full or partially filled reservoir is often an extremely difficult matter. Another method of reducing and possibly eliminating leakage is to employ natural sealing material, such as a prepared clay mineral or even the silt content of the watercourse feeding the reservoir.

SECONDARY EFFECTS OF RESERVOIR FLOODING

As a reservoir is initially filled, areas of dry land will be flooded as the water rises, and the flooding will often entail extensive civil engineering construction of what may be termed a secondary nature. Buildings, settlements, and transportation routes may require relocation. Many of the potential relocation sites will be located on hillsides, and the possibility of landslides should always be given special attention. Early in the 1960s the Ottawa River was spanned by the mass-concrete Carillon Dam of Hydro Quebec, at a point 96 km (60 mi) below the city of Ottawa. A major bridge had to be reconstructed and much rehabilitation work carried out along both banks for many kilometers above the dam. Small landslides occurred, attributed to the slight rise in the water level, 80 m (50 mi) above the new barrier. This is typical when any dam is built in developed areas.

Landslides are a natural development as reservoir levels are raised or lowered. Changes in groundwater level and consequent changes in pore pressure in soils are secondary effects of dam construction which must receive careful design attention. Pumped storage projects often bring about an aggravated slope stability problem due to their frequent water level changes.

The vital necessity for this study of potential slides unfortunately has been confirmed by the most tragic of all failures in modern times, the overtopping of Vaiont Dam in Italy. A massive rock slide on October 9, 1967 cast several hundred million cubic meters of soil and rock into the reservoir, creating a wave estimated at 90 m (300 ft) high. This overtopped the 257-m- (858-ft)-high arched dam and released a flood destroying several villages and almost 3,000 lives. The engineers responsible were arraigned in the Italian courts, adding further to the tragic aspects of this dreadful disaster. The dam still stands, since it was only slightly damaged as a result of the slide and ensuing wave. It is a monument to the skill of its arched-dam designers, but its distinction as the world's highest arched dam now has long since been forgotten. Located in a very narrow gorge of the Vaiont River, a tributary of the Piave River in northeastern Italy, the dam was completed in 1960. It is keyed into the competent rock of the sides of the gorge. It is only 3.4 m (11 ft) thick at its crest and 22.7 m (75 ft) thick at its base. It was carefully instrumented so that its performance could be checked. It stands today retaining mainly rock and soil. The area in which the dam was built is characterized by a thick section of sedimentary rock, predominantly limestone, with interbedded marl and clay beds. The entire area had been subjected to geologically recent movements. Slow creep movement of part of the reservoir wall had been noted but the final movement was sudden.

Perhaps the most extensive of all stability investigations was carried on in the mountains of British Columbia at the ancient Downie Slide, the toe of which has been flooded by the Revelstoke hydroelectric project. This slowly creeping mass contains about 1.5 billion m^3 (1.9 billion yd^3) of sheared and weathered metamorphic rock and alluvium; the slide is thought to have occurred between 9,000 and 10,000 years ago. It is on the west bank of the Columbia River about 65 km (40 mi) north of the town of Revelstoke. Disturbed material is evident for about 2.5 km (1.5 mi) along the river. The slope of the slide surface is about 18°, except near river level. The construction license was issued with the specification that the reservoir level will be restricted to 9 m (30 ft) below full design pool level until stabilization measures have been completed and are shown to be effective.

RESERVOIR SILTATION

In all but exceptional cases, dam or weir construction will raise the water level upstream and thus increase the cross-sectional area of the stream and decrease the average velocity, causing a part of the suspended streamload to fall to the bottom of the reservoir. A considerable number of large dams have been rendered useless by unscheduled and rapid siltation. Three of these are the so-called New Lake Austin on the Colorado River in Texas, which lost 95.6

percent of its capacity in 13 years; La Grange Reservoir on the Tuolumne River in California, which lost 83 percent in 36 years; and Habra Reservoir on the Habra River in Algeria, which lost 58 percent in 22 years. An unusual degree of siltation occurred at Lake McMillan Reservoir of the Pecos Irrigation District in the United States, which lost 55.5 percent of its storage capacity between 1894 and 1920. Since 1920 no further accumulation of silt has taken place because an extensive growth of tamarisk (salt cedar) in the flatlands above the reservoir has so reduced the stream velocity that the main part of the silt load is deposited before the reservoir is reached.

Preliminary engineering for all dams must consequently include an investigation of sediment transport. Often the quantity is found negligible. Detailed studies include a careful correlation of information of sediment yield of each exposed geologic unit. This information is necessary in view of two groups of related engineering problems: (1) methods of removal of the transported material, either before it is deposited in the reservoir or after it has come to rest behind the dam; and (2) the effect of unavoidable deposits of such material on the stability and utility of the project. The high silt content of the Rio Grande and Colorado rivers is well known. Elephant Butte Dam is on the Rio Grande, a short distance north of the junction of Mexico and the states of Texas and New Mexico. It was completed in 1916 and has a drainage basin of 67,000 km^2 (25,923 mi^2). Continuous measurements have been made for many years. Estimates suggest that the useful life of Elephant Butte Reservoir will be about 150 years.

On the Colorado River, Hoover Dam has created one of the largest of artificial lakes, Lake Mead, which receives all the silt load of the river. Studies of sedimentation have been actively pursued ever since the dam first began to retain water. Estimates of the useful life of the reservoir have naturally varied, but one of the most detailed suggests that the reservoir will not be filled with sediment to its crest level until some 445 years after infilling began on February 1, 1935. Though a sobering figure, the estimate is in keeping with the suggestion that the useful life of the reservoirs constructed in North America in the second quarter of this century averages about 200 years.

Cleaning of filled reservoirs is a normal part of operation of the usually dry flood-control dams, such as those of the Los Angeles County Flood Control District (Fig. 10.20). Reservoirs for water supply, however, experience the same sort of problem, although usually at a slower rate. The San Fernando Reservoir of the Los Angeles Department of Water and Power, for example, lost 24 percent of its capacity in about 20 years. Arrangements were made for dewatering the reservoir for a period of 38 days, during which a major sluicing operation was undertaken. About 15,000 m^3 (20,000 yd^3) of sediment were sluiced out of the reservoir into a specially constructed debris basin; it was found that about 69 m^3 (90 yd^3) could be moved in this way per worker-day. Philadelphia faced a similar problem with its Queen Lane sedimentation basin in which 450,000 m^3 (500,000 yd^3) of silt had accumulated during its first 50 years of operation. The large accumulation of silt is accounted for by the fact

FIGURE 10.20
Sluicing operations in progress upstream of Santa Anita Dam, removing flood debris retained by this arched concrete structure. (*Courtesy Los Angeles County Flood Control District.*)

that the high silt content of the Schuylkill River is removed at the Queen Lane plant by chemical coagulation.

Sediment deposition in reservoirs is so obvious a result of dam construction that there remains little more to be said about it. Silt contents of streams and rivers can be determined at the time of design. Predictions can be made of the probable continuation of this amount of silt reaching the dam after its construction. The effect of the dam upon the riverbed downstream also can be estimated and the possible effects of the degradation of the stream bed investigated. Finally, the economics of the entire project can be evaluated in making the final decision regarding building. A key factor in such a study would be the probable finite useful length of life for the reservoir.

Behind all such mundane proceedings, however, is the tragic significance of the origin of most of the sediment that will be trapped in the reservoir. To some extent, the sediment may be derived by normal riverflow from the beds of the streams and rivers that bring it to the dam site. In almost all cases, however, most of the silt load will represent topsoil that has been washed off the surface of the land in the watershed above the dam. No civil engineer should be able to view a silt-laden stream without being disturbed by the critical problem of soil erosion. Much progress has been made in the agricultural and construction control of soil erosion but much still remains to be done. Engineering geologists can provide

FIGURE 10.21
Long Lac Diversion Dam of Ontario Hydro, which raises water level sufficiently to divert some flow that would normally reach Arctic waters into Lake Superior and the St. Lawrence system. *(Courtesy Ontario Hydro.)*

the basis for soil-loss estimation through their mapping of engineering soil and weak rock units of the watershed. All such work demands the attentive interest of civil engineers and engineering geologists, since upon its success depends the long-term usefulness of many dams and other structures.

RIVER DIVERSIONS

Diversions of water between watersheds are infrequent, but those that do exist are all of unusual interest. Sometimes water is diverted from one side of the continental divide to the other to augment the water supply of burgeoning cities such as Denver and Quito. These major engineering projects are routed and designed with a firm basis of geologic mapping, and always involve a variety of works, dams, canals, tunnels (usually), and reservoirs. At least 19 interbasin river diversions have been carried out in Canada to date, all made feasible because of sound geologic investigations. Numerous glacial outwash channels

FIGURE 10.22
A beautiful dam in an ideal setting, typifying the significance of the local geology in the founding of all dams; Escales Dam on the Ribagorzano River in northern Spain. (*Courtesy Empressa Nacional Hidro-Electrica del Ribagorzano, Spain.*)

were created in the northward recession of the last ice sheet from Canada. Now covered with vegetation, these still can be detected by careful geomorphologic study. These Arctic waters have been diverted through old glacial outwash channels north of Lake Superior, between the great river basins of northern Canada and the relatively narrow area of land draining into the north shore of Lake Superior. Additional power was generated as the diverted water flowed into the lake and thereafter as it augmented the flow of the St. Lawrence on its way to the sea (Fig. 10.21). There are relatively few locations where geology can provide such a convenience, but their high value warrants mention so that civil engineers may be more likely to avail themselves of geologic advice.

CONCLUSION

Few civil engineering works attract as much public attention as do dams, large or small. Few other works so often seem to be a cooperative effort between humans and Nature. To see a thin-arched concrete dam in a narrow river gorge or a massive concrete dam in a great river valley spanning between exposed rock walls is to experience a sense of aesthetic satisfaction not always aroused by the works of the engineer. Figure 10.22 illustrates a striking example of this "aesthetics of dams." This impounding dam for the Escales hydroelectric project in northeastern Spain is a gravity structure, 125 m (410 ft) high and 200 m (650 ft) long. Impounded water flows through tunnels to a large underground power station in which three 12,500-kW units operate under a head of 117.5 m (385 ft). The dam is located on the Ribagorzano River which has its headwaters in the Pyrenees. Flanked by supporting alternating strata of Cretaceous marls and limestone which dip almost vertically, this dam—as do so many others in similar locations—shows a sense of fitness with its surroundings that makes it a fine symbol of the interrelation of geology with the design and construction of major civil engineering works.

REFERENCES

1 F. L. Ransome, "Geology of the St. Francis Dam Site," *Economic Geology, 23,* p. 553 (1928).

2 K. Terzaghi, "Failure of the Malpasset Dam," *Engineering News-Record, 168,* p. 58 (February 15, 1962).

3 "Teton Dam Failure," *Civil Engineering, 47,* p. 56 (August 1977).

4 "T.V.A. Gives up on Hales Bar Dam," *Engineering News-Record, 170,* p. 26 (April 18, 1963). Contains an excellent list of other papers on this dam.

5 W. H. Stuart, "Influence of Geological Conditions on Uplift," *Proceedings of the American Society of Civil Engineers, 87,* SM6, Paper 3008, p. 1 (1961).

6 H. J. Hoeffer, "Fort Randall Dam to Provide More Storage on Missouri River," *Civil Engineering, 22,* p. 474 (1952).

7 "San Gabriel River Flood Control," *Engineering News-Record, 114,* p. 113 (1935). Contains useful bibliography on p. 116.

8 K. Terzaghi and Y. Lacroix, "The Mission Dam," *Geotechnique, 14,* p. 13 (1964).

9 "Corewall Grouting at Chickamauga Dam," *Engineering News-Record, 122,* p. 551 (1939).

10 A. L. R. diCervia, "Seepage Cut-Off Wall Installed Through Dam is Construction First," *Civil Engineering, 49,* p. 62 (January 1979).

CHAPTER 11

TRANSPORTATION FACILITIES

Construction of transportation facilities has been an historic preoccupation of civil engineering—from the first efforts of primitive peoples to bridge a stream with a fallen tree trunk to the conception and construction of modern expressways with all of their associated works. Beaten tracks to drinking places were probably the first artificially constructed pathways, followed by communication ways between settlements. Many references to practical road building are to be found in early historical records. Roadways of the Roman Empire are still in use today, but many centuries had to pass before road construction regained the eminence it possessed in Roman times.

Road, rail, and now air transport have become so familiar that it is easy to forget that before the beginning of the industrial revolution, transport by water was by far the most important means of movement. Beginning with the primitive local use of rivers, the record of water transport is continuous throughout the whole range of human history. Associated marine works are of critical importance in receiving and shipping of goods. Rivers as transportation routes were joined by canals to link natural waterways or to give access from the sea to inland points. Canals played an important part in the growth of modern transportation, and although the building of new canals is now rare, the canals in use today are vital to the human economy and are built and maintained with due geological influence.

Canals

HISTORICAL NOTE

Master builders throughout the ages have always recognized the need to extend inland waterways and to penetrate narrow land barriers between seas by means of canals. A canal connecting the Nile and the Red Sea is said to have been begun in the fourteenth century B.C.; work on the project was abandoned by Necho about 610 B.C. after 120,000 lives were lost in the excavation work. According to Strabo, the canal was finished by Ptolemy II, who is said to have constructed locks with movable gates. Xerxes, in his war against the Greeks, constructed a canal across the isthmus of Mount Athos; trouble was encountered because the sides (up to 15 m or 50 ft deep) were excavated at too steep a slope. The Isthmus of Corinth was crossed by canal only as recently as 1893. This crossing was conceived by Periander of Corinth, one of the Seven Sages of Greece, who built the best alternative to a canal, what might be termed the first marine railway. This *dioclos*, a paved roadway upon which vessels could be portaged from the Gulf of Corinth to the Saronic Gulf, 6.4 km (4 mi) away, was rediscovered in 1957 by Greek archaeologists. It was 3.9 m (13 ft) wide and built of heavy rectangular blocks of the local limestone. Nero, who personally inaugurated work on the canal in A.D. 67, cut the first sod with a golden spade. With the aid of 6,000

FIGURE 11.1
The Corinth Canal, Greece. (*Courtesy The Embassy of Canada, Athens, Greece.*)

slaves (sent by Vespasian from Judea) he completed 2,000 m (2,200 yd) on one side of the isthmus and 1,450 m (1,600 yd) on the other, to be stopped because of a revolt. The modern canal is 6.4 km (4 mi) long, 21 m (70 ft) wide, and 8 m (26 ft) deep and is superimposed in the ancient cuts (Fig. 11.1). It was seriously damaged during World War II but was soon back in service.

THE CANAL AGE

Modern civil engineering surely got its start in canal building. This work began with improvement of navigation on rivers in Europe, and especially in France and Great Britain, starting as early as the seventeenth century. The first modern canals were conceived as interconnections between rivers, one of the first being the Midi Canal in France, still in use. In England, the first was St. Helen's Canal, but it was the Duke of Bridgewater's Canal, opened in 1757, that marked the start of the canal era.

It was in the building of canals that the first notable British and French engineers gained their experience. Canal fever soon passed to the United States and to Canada, and many truly remarkable works were carried out early in the nineteenth century. The railroad put an end to the canal age about 1840. Many of those canals are still in use, while some are being restored to use. In fact, canals are still being built today to bring highland water to the burgeoning cities of the South American Andes.

THE PANAMA CANAL

Superficially, the excavation of a water channel appears to be a simple operation, but when the details of design and construction are considered the true complexity of the work becomes evident. Canal bed and canal banks should be impermeable, canal banks must be stable, and bridge foundations must be sound and secure. Geology has a fundamental control over all of these requirements. Geological complexities and problems of canal construction are nowhere better illustrated than in construction of the canal across the Isthmus of Panama. The French began this immense project at the instigation of Ferdinand de Lesseps after his success with the Suez Canal, which opened in 1869 without the impact of adverse geology. It was, however, geology (combined with tropical disease) that defeated the French at Panama.

The Panama Canal, now one of the most important waterways of the world, crosses the Isthmus where the great continental divide dips to its lowest elevation, about 94 m (312 ft) above sea level. The route of the canal follows existing river valleys and crosses the divide in the great Gaillard Cut; it is 64.5 km (40.27 mi) long. Construction was started in 1882 by the French company which operated until 1889; a reorganized company started work in 1894 but was eventually bought by the United States of America in 1902, under whose auspices the canal was completed in 1914. The canal was officially opened on July 12, 1920.

Of geological interest are the following excavation volumes, largely in weak rock and soil, carried out to June 30, 1959.

	M^3	Yd^3
Excavated by French	22,800,000	28,908,000
Excavated by Americans	449,000,000	588,181,200
Excavated in Gaillard Cut	144,500,000	189,832,700
Reexcavated in Gaillard Cut (attributed to slides)	58,300,000	76,228,000

Geology underlay the difficulties encountered in the excavation of the great Gaillard Cut, 14 km (8.75 mi) long. Host rock here is shale, sandstone, basaltic dikes, volcanic agglomerates, and volcanic tuff, complicated in distribution by numerous faults. Giving the greatest trouble was the synclinal trough of the Culebra section, about 1.6 km (1 mi) wide at canal level. The trough is filled

FIGURE 11.2
A landslide on the Panama Canal, September 18 and 19, 1915, showing an island formed overnight in 10 m (30 ft) of water. (*Courtesy The Governor, Panama Canal Zone.*)

with a fine-grained sandy clay of the Cucaracha Formation, structurally weak, and responsible for repeated failure of most of the slides that interfered so seriously with canal construction. Slides began as early as 1884 and have continued as maintenance problems. One slide of over 380,000 m^3 (500,000 yd^3) happened overnight; others were of varied types and of considerable duration, described thus by Engineer Colonel Gaillard:

> Most of the slides of the past year 1911–1912 were breaks resulting from the failure of an underlying layer of rock of poor quality due to the pressure of the enormous weight which crushes the underlying layer, forces it laterally and causes it to rise up and heave in the bottom of the cut. The heaving at times is 30 feet.

An accompanying photograph (Fig. 11.2) shows such an upheaval in the form of an island which appeared overnight out of a depth of 10 m (30 ft) of water. From a study of this and by recognizing the structural instability of the *cucaracha*, one can appreciate the origin of the slides, especially intensified in some places by the extra loading of spoil dumped too near the crown walls of the cut. The unsatisfactory nature of the *cucaracha* was reported on in 1898 by the French geologists Bertrand and Zurcher, whose opinions were probably not utilized.

Careful maintenance has been a continuing necessity of canal operation. A small fracture was noted in 1938, at the top of Contractors' Hill; regularly observed by 1954, it had increased to such an extent that a drainage tunnel, 1.5 by 2.1 m (5 by 7 ft), was quickly excavated to drain porewater. Removal of about 1.9 million m^3 (2.5 million yd^3) of the Pedro Miguel Agglomerate forming the hill was carried out to relieve the pressure on the underlying *cucaracha*. Work was successfully completed before the end of 1955; the finished appearance of the cut is shown in Figure 11.3.[1]

FIGURE 11.3
Excavation on Contractors' Hill, Panama Canal, required to stabilize the Culebra Cut, as described in the text. (*Courtesy The Governor, Panama Canal Zone.*)

THE ST. LAWRENCE SEAWAY

The other major North American ship canal, built jointly by Canada and the United States half a century after the Panama Canal, is the St. Lawrence Seaway. The full Seaway includes improved river channels, canalized sections of the St. Lawrence, crossings of the Great Lakes with much local dredging, and two canals. Sod was turned on August 10, 1954; less than five years later, the Seaway was officially opened. At one time more than 22,000 workers and 500 professional engineers labored to create one of the greatest of all civil engineering projects. An 8.1-m- (27-ft)-wide waterway, capable of handling oceangoing vessels, extends from the head of the navigable St. Lawrence River at Montreal to Lake Ontario, and thence into the Upper Lakes. This involved new locks, lift locks, a guard lock, and a canal, with some deepening and much dredging of existing channels, and construction of a large number of ancillary works, some of which (such as the Long Sault Dam) are major structures.

The two lift locks in the international section of the river were located on the United States side. All the earlier, smaller locks had been built on the Canadian side, where geological conditions are favorable. The Eisenhower and Snell Locks are notable structures—24 m (80 ft) wide, 258 m (860 ft) from pintle to pintle, and a minimum of 10 m (30 ft) of water over sills. All but the western portion was a major new cut, with a bottom width of 133 m (442 ft) and almost 15 million m^3 (20,000,000 yd^3) of excavation. Extensive boring, soil sampling and testing, and a demonstration excavation were carried out. Even so, serious difficulties were encountered with this excavation because of the properties of the lodgement till and marine clay involved as described in Chapter 4.

The Welland Canal is the fourth canal to bypass Niagara Falls, the first having been opened in 1829. Major improvements to the relatively simple original canal resulted in the second (opened in 1845) and then the third canal (opened in 1887). The fourth Welland Canal was started in 1912 but, because of World War I, it was not officially opened until 1932. It followed a new route for part of its 42-km (26-mi) length, its total drop of 99.5 m (326.5 ft) being achieved in only seven locks (compared with 40 locks in the first canal). Each of these 258-m- (860-ft)-long by 24-m- (80-ft)-wide locks were much bigger than all other Canadian locks, possibly having been influenced by the example of the Panama Canal. Ten distinct Ordovician and Silurian formations, including dolomites, limestones, sandstones, and shales outcrop along its course. Of these, the Queenston Shale gave unusual trouble during excavation, disintegrating on drying. Therefore, excavation left a thin cover in place until just before concreting began. Trouble had been experienced here a quarter of a century before, with difficult excavation in dense lodgement till.

The Welland Canal, however, was only one part of the billion-dollar St. Lawrence Seaway. Dry excavation amounted to 126 million m^3 (166 million yd^3) with 27 million m^3 (35 million yd^3) of dredging in addition. Over 19 million m^3 (25 million yd^3) of spoil were used in the dikes forming the navigation canal between Montreal and Beauharnois. Corresponding figures for the international power project are 62 million m^3 (82 million yd^3) of dry excavation, 9 million m^3 (12 million yd^3) of dredging, and 13.5 million m^3 (18 million yd^3) placed in dikes. In the Canadian section alone, 1,700 exploratory borings were put down, totaling 9,600 m (32,000 ft) in soil and 12,000 m (40,000 ft) in bedrock. Along with supplementary test pits and auger holes, these special investigations resulted in equally impressive progress with construction, all major works being completed within less than five years.

Fortunately, much of the geological experience has been published. Especially notable was the use of lodgement till as the impermeable core of miles of dikes; no serious problems were experienced with this most economical use of excavation waste. Problems were encountered with groundwater, especially in the sandstone excavation at the upper Beauharnois lock. These substantial inflows into excavated areas were overcome, however, with no serious delays in construction, as based on careful advanced geological investigations.

CANAL LOCKS

Lock foundations are of particular interest because of the varying hydrostatic conditions to which the completed structures will be subjected. When locks are founded on solid rock, few problems usually arise. Some locks, however, must be founded on weak rock or soil, and this necessitates special care in the design and construction. An example is the main sea lock of the Ijmuiden Canal, Holland (Fig. 11.4). As local freshwater supply is obtained from groundwater trapped beneath the top clay stratum, there was concern that construction operations might interfere with this underground reservoir and lead to ingress of seawater. As a result of thorough subsurface exploration, the conditions shown in Figure 11.4 were determined, and this permitted the complete sheet

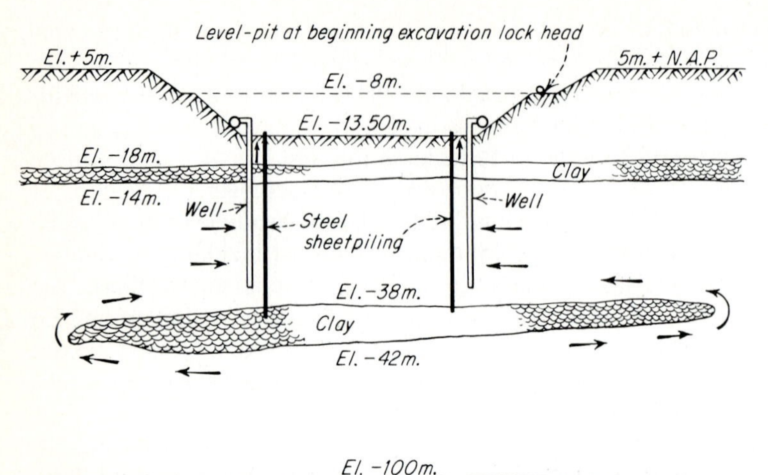

FIGURE 11.4
Subsurface conditions at the site of the Ijmuiden Lock of the North Sea Canal, the Netherlands. (*Courtesy Engineering News-Record.*)

piling enclosure of the lock site—394 m (1,315 ft) long by 48 m (161 ft) wide. Successfully completed, this is an outstanding example of the adaptation of local geological features to the scheme of construction.

Entirely different were the geological conditions beneath one of the very few lock walls to fail in North America. Wheeler Dam and Lock, built on the Tennessee River near Athens, Alabama, in 1936 by the U.S. Army Corps of Engineers, were preceded by the usual careful subsurface investigations. The installation was later transferred to the Tennessee Valley Authority and the lock performed well until June 1961. Excavation was then in progress for a new and larger lock when 131 m (436 ft) of one wall of the original lock suddenly moved outwards. The TVA had also conducted detailed subsurface studies for the new foundations and had found nothing amiss. Critical study later found sliding to have occurred on a thin (1- to 6-mm- or 1⁄16- to 3⁄8-in-thick) seam of clay in a 15-cm (6-in) stratum of Fort Payne Shale underlying the site. Most of the remainder of host rock is argillaceous to crystalline, fine- to somewhat coarse-grained limestone, all practically horizontal. Careful exploratory drilling had failed to detect the thin clay seam during both the subsurface investigations by two prestigious engineering firms. Constructive open discussion of the failure included a number of suggestions for improving borehole observations, now reflected in modern practice.

Maintenance work on the Panama Canal in 1958 resulted from inspection that had shown some trouble with the functioning of underlock culverts in the Pedro Miguel Locks. The lock floors were found to be perforated with many 4-in-deep holes, but no as-constructed record could be found to indicate what

FIGURE 11.5
Major maintenance work in progress on the west side of Pedro Miguel Lock on the Panama Canal, February 1965. (*Courtesy Panama Canal Commission.*)

they were for. It was said that they had been installed to relieve hydrostatic pressure in the bedrock because notable rock-floor uplift had been experienced during the long construction period. This uplift was in reality almost certainly due to rebound in the bedrock following the deep excavation required for the lock chamber, but this was an unrecognized phenomenon at that time. Bedrock of the weak-rock Culebra Formation contained shear failures when the rock was uncovered during rehabilitation work. Planar separations thus created permitted water to circulate, and this had washed out weathered material, resulting in voids which had in turn affected the culverts (Fig. 11.5).

SOME EUROPEAN CANALS

Extensive use of inland water navigation in Europe that began on the Rhine and other major rivers has extended to all manner of interriver waterways. Major canal construction has thus been a feature of European civil engineering for

FIGURE 11.6
The cutting at Lanaye on the Albert Canal, Belgium, during construction (above) and as completed and in use (below). (*Courtesy La Ministère des Travaux Publics et de la Reconstruction, Belgium, Brussels.*)

many years. The Albert Canal, a masterpiece of Belgian engineering, was completed just before World War II and became a vital factor in some of the war's most crucial fighting. This dual-purpose canal served as a defense barrier and as a means of conveying 2,600-tonne barges from the sea to the important industrial area around Liège. Its 130-km (80-mi) length employs only six locks to drop 55 m (184 ft) from the river Meuse to the Antwerp docks. Excavation of 68 million m^3 (89 million yd^3) of soil and weak rock was governed by geologic factors. Near the Liège end, an open cut ranging from 19.5 to 63.6 m (65 to 212 ft) in depth was carried through tufa and chalk, both of which harden on exposure to the atmosphere. Steep slopes, as shown in Figure 11.6, were therefore possible because of this geologic characteristic. Steam shovels were used to remove both clay and gravel overburden and also the rock. Pneumatic tools were used for dressing both the finished slopes and berms, 0.9 to 1.8 m (3 to 6 ft) wide, and at intervals of 10 m (33 ft). Another major cut is near Briegden, some kilometers "downstream" from Lanaye, where preliminary exploration had revealed waterlogged sand and clay at a depth of 39 m (130 ft). To have avoided these strata would have necessitated two extra locks to raise the canal level above them. The bold decision was made to stay with the original concept of a deep cut, using special drainage installation construction methods throughout. The great cut, over 45 m (150 ft) deep, has continued to perform well.[2]

Roads

HISTORICAL NOTE

Modern concepts of roads, their construction, and utilization are rooted deeply in the grand roadways of the Roman Empire. Roman road building was always carried out with intuitive geological appreciation. At its zenith, the Roman legions had built and used over 80,000 km (50,000 mi) of first-class highways. The greatest of all early single roads, however, was on the South American continent. Built by the Incas, it was 6,400 km (4,000 mi) long and stretched from Quito in Ecuador to Tucumán, Argentina. Much of its 7.5-m- (25-ft)-wide surface was paved with bitumen. A second road followed almost 3,200 km (2,000 mi) of coast. Much of these great roads, constructed in the 350 years prior to the Spanish conquest, can be seen today, and some portions are still in use.

With the dawn of the industrial era road building assumed a new importance; Britain and France led the way in the early branch of modern civil engineering. John Metcalfe ("Blind Jock" of Knaresborough), probably the first of the pioneers, overcame his blindness even to the extent of appreciating geology in his road-building work. Sixty years later Thomas Telford and James Loudon McAdam started their notable road work, the results of which can still be seen in Great Britain. The meager writings of these great men show clearly that geology (even though not so called) was a vital part of their thinking about roads.

North America had its later road pioneers, also. As early as 1906, Dr. C. M. Strahan, then county engineer of Clark County, Georgia, started experiments on the properties of soils for road building, his first published statement appearing in 1914. Just a year after the start of this work, the Wisconsin Geological Survey was granted $10,000 by the state legislature for investigations and experimental work in road building. By 1927 the Illinois Geological Survey organized its own areal and engineering geology division. Progress in other states followed. In Kansas, the State Highway Commission appointed a chief geologist in 1937, a position later occupied by S. E. Horner, who made notable contributions to the applications of geology to highway engineering.

Today, there is probably no branch of civil engineering in which the potential contributions of geology are more generally appreciated than in the field of road design and construction. Long-established annual meetings are devoted to this one subject, and annual symposia on geology applied to highway engineering have been held in the United States since 1950. Preeminent are the annual meetings of the Transportation Research Board of the U.S. National Academy of Sciences. Ultimate recognition is given by the American Society for Testing and Materials' (ASTM) standard test methods for investigating and sampling highway subgrade soils, of real assistance to those who plan and direct such exploratory work.

In this branch of civil engineering only, the wheel may be said to have turned full circle since there are now available highway geologic maps such as those of Ontario. A number of regional handbooks now explain and interpret the geology of the land adjacent to main highways for the information and enjoyment of travelers. One of the earliest handbooks is the 1958 West Texas Geological Society guidebook to the geology of U.S. Highways 90 and 80 from Del Rio to El Paso.

ROUTE LOCATION

In land transportation, there are two main problems to be faced in the design and construction of both roads and railways: (1) route selection, and (2) choice of road section and bearing surface of reasonably permanent stability, capable of carrying intended traffic. The final choice in both categories depends largely on economic aspects, but geology generally has considerable bearing on the final solutions. In very early days, road location was a matter of adapting older pathways that had been formed by primitive peoples. Roman road construction marked a significant break from this primitive practice; Roman engineers laid their roads as straight as possible, a practice that continued in France until the eighteenth century. When changes in direction had to be made, this was generally done on high ground or at station houses—curves were used only infrequently. The Fosse Way in England is about 320 km (200 mi) long (joining Lincoln and Axminster), but the greatest departure from a straight line between the two end points is only 10 km (6 mi).

Only with the coming of the railway was attention directed to the basic problems of economical route selection, of severely limiting grades, and of keeping curvature to a minimum. In flat country, selection was largely a matter of convenience and of land availability. The route chosen was selected to give the shortest possible convenient link between the centers to be joined, although deviations were made, if necessary, to obtain improved river crossings. In hilly country, however, many alternatives could be considered. The ridge route up a valley, for example, might be chosen in preference to the valley route; the climb to the ridge would more than compensate in convenience for the numerous river crossings usually necessary if a valley route was used. Such selections were inevitably influenced by the underlying geology, since it determines the local topography.

Glacial landforms in general had a profound influence on the location of early roads. Older towns in Ontario and in similar glaciated areas frequently will be found to have a "Ridge Road" as a well-established older highway, the name clearly indicating the presence of a prominent glacial landform such as a moraine.

The building of superhighways has resulted in many similar vivid examples of the effect of local geology upon road location and upon the costs of construction within the limits thus determined. The similarity of modern road location is nowhere better illustrated than by the construction of the Pennsylvania Turnpike, much of the route of which utilized an abandoned railway right-of-way. This pioneer modern toll highway was opened to traffic on October 1, 1940, after a remarkably short construction period of only two years. The Pennsylvania Bureau of Geology and Topography published a notable guidebook to the geology of the entire highway soon after the end of World War II. This 54-page description gives a good general account of route geology and a succinct description of the subsurface investigations and of engineering features affected by geology, especially in the old railroad tunnels. The well-illustrated guidebook concludes with a "geologic itinerary" from one end of the turnpike to the other.

New roads and railways will always be built. What techniques are now available for evaluation of final route selection? Current techniques contrast greatly with the laborious work of field location practiced until the 1930s. Today, office study of topographic and geologic maps is a first requirement, and almost always a possibility, even though the scales of maps available for more isolated areas are still small. Concurrently, photogeologic interpretation of the proposed routes is a well-established and invariable aid.

Preliminary work for the Ohio Turnpike Project No. 1 is an early example of the application of geology to modern highway route selection and design. This modern dual highway is 390 km (241 mi) long and cost about $1 million per mile. Almost 30 million m^3 (40 million yd^3) of excavation was involved, with a corresponding volume of fill, and over 5.8 million m^2 (7 million yd^2) of concrete paving. Consecutive aerial photographs to a scale of 1:10,000, covering a corridor 2.4 km (1.5 mi) wide were first studied. A band 600 m (2,000 ft) wide was then selected for detailed study. All available soil survey and geologic maps and reports were used to portray all main soil types and soil boundaries, geologic contacts, and drainage patterns. Unpublished material on abandoned glacial shorelines was provided by the Ohio Geological Survey. Preliminary correlation of pedological soil characteristics with engineering soil properties was verified by the Ohio State Highway Testing and Research Laboratory. This work saved about 80 percent of route geologic mapping, as well as adding immeasurably to the accuracy of such work. Aerial photographs also provided preliminary correlation of all existing information. Before field studies commenced, 34 candidate stream crossings and 114 road and railway crossings were studied in detail in the office. Twenty-two exploration contracts were awarded to 11 contractors, for 18,013 m (60,405 ft) of earth borings and 2,644 mi (8,812 ft) of rock borings, all at a cost of only two-tenths of one percent of the total project cost. All rock cores and soil samples were passed to the State Geologist for safekeeping and public use, a good example of the reciprocity that now characterizes cooperative geological–engineering investigations.

With railways and roads now so commonplace in developed countries, it should not be forgotten that surveys for many of the pioneer routes of the New World provided the first information about the local geology. Many papers give invaluable geological information as by-products of such preliminary engineering work. A singularly useful paper of this sort was published as early as 1888; one of the most recent examples is the record prepared of the geology along the Alaska Highway, a wartime emergency route that penetrated the relatively unknown Canadian Northwest. Its construction was a remarkable wartime achievement; many geological problems were encountered then and have had to be dealt with since its maintenance and improvement as a fine civilian highway.

CLIMATE

Roads always involve disturbance of natural ground surfaces and, as such, they are unusually susceptible to the vagaries of the weather. Even in North America, there are areas in which a working season of only a few months is available each year. Slow progress on the Trans Canada Highway through the Rocky Mountains, for example, was attributable to this short working season. During the building of the Detroit Industrial Expressway, large, wide cuts in exposed glacial clay were exposed to freezing weather, its laminated character leading to its rapid surface disintegration and subsequent sloughing during the spring rains. More unusual was the effect of heavy rain on the sub-base rock of Florida's Sunshine State Parkway. Soon after its opening in 1957, trouble was experienced between Stuart and Fort Pierce at the northern end of the road because of "alligatoring" of the pavement. This was traced to the fact that, during paving operations, heavy and unseasonable rainfall (up to 52.5 cm or 21 in) was experienced. This had the effect of seriously wetting the Belle Glade, or Okeechobee, rock. This weak limestone became "greasy" when wet and under load, leading to the pavement failures. Tests of this rock in its dry state had been quite satisfactory. This salutary reminder raises again the importance of the combined importance of climate and geology.

DRAINAGE

Drainage is one of supreme importance in all highway work; waterlogging ruins subgrades. Drainage methods vary but they must inevitably be suited to the nature of the materials to be drained, the physical properties of which—in the case of unconsolidated materials—can be tested in the same way as other soils. The underlying geologic principle is to keep the groundwater level sufficiently deep so as to promote gravity flow through porous roadbed materials by suitable artificial drainage channels or by the use of a layer of sand or gravel in "self-draining" impermeable materials. The reverse of a drainage process is an infrequent requirement, to be sure, but a concrete roadway in Ellis County, Kansas, was found to be settling unevenly and causing displacements between adjacent slabs. Investigations showed that the differential movement was due to changes in the moisture content of the subgrade soil occurring at slab joints and along cracks. This was successfully corrected, and the concrete slabs were releveled by introducing water into the subgrade material through an installation of well points jacked into place about 25 cm (10 in) below the slab from the roadway shoulder.

Successful drainage is possible if the design is carried out with knowledge of the geologic strata beneath the roadbed and in the area surrounding the road. Drainage facilities added during construction represent only a very small part of total road cost. If initial drainage provisions turn out to be inadequate, remedial drainage facilities can lead to astronomical costs. Though good drainage practice is now well recognized, as recently as 1979 a leading North American engineer had this to say, inter alia:

> Pavement designed without good internal drainage are doomed to early failure the day they are completed, because failure mechanisms are built right in.... On a national basis it would be impossible to place an exact monetary value on the total loss caused by the undrained pavement practice, but it is clearly rising to near catastrophic levels...[and using the writer's estimates] the potential loss [in the United States] that can be blamed on poor drainage is about $200 billion for those 15 years [1976 to 1990] or a little over $15 billion annually.[3]

Harry Cedergren, in this arresting statement, reminds his readers of the statement of John McAdam, made as long ago as 1820: "The erroneous opinion...that a road may be sufficiently strong, artificially, to carry heavy carriages though the subsoil be in a wet state...has produced most of the defects of roads in Great Britain." The following statement was made with remarkable insight in 1841:

> Drainage is an affair of primary importance in roadmaking, and requires much skill to execute in a proper manner...it may even be necessary to carry the process of draining far beyond the area of the road.... In this country, the roads sustain much injury from heaving up by frost. This would be, in great measure, prevented by adopting a better system of drainage....

These words come from a 42-page pamphlet published in Toronto by a remarkable civil engineer and geologist (as was William Smith), named Thomas Roy. Little is known of his private life although he served as the equivalent of the city engineer of the fledgling city of Toronto during his final years (he died in 1842), but his pamphlet on road building can be read with profit today.

Even with such warnings, probably more highway maintenance problems have been due to inadequate drainage facilities than from any other cause. Even so fine a modern highway as the New York Thruway has experienced serious trouble, apparently because of inadequate drainage, on the Berkshire Extension very soon after it was opened. The troubled area is underlain by glacial till overlying thin horizontal strata of clayey silt and fine sand. The roadway consists of a 23-cm- (9-in) -thick reinforced-concrete pavement on a basecourse of select granular material. Cracking and "pumping" did serious damage to the pavement, parts of which had to be removed. Material below the basecourse was then removed, some to a depth of 3 m (10 ft), and then replaced before the roadway was rebuilt. This case is regrettably typical of widespread damage due to inadequate drainage, even on modern highways. Successful highway drainage systems are designed with full appreciation of route geology. Drainage facilities are essentially simple, a variety of plastic, concrete, and galvanized steel pipe now being available to meet almost any combination of ground conditions.

Water is often the prime cause of trouble when the route lies over terrain that is so unsatisfactory that a road cannot be built directly upon it. Peat bogs, swamps, marshes, muskeg, and tidal flats are examples of such a condition; all are relatively new geological deposits which are underconsolidated and which contain a high percentage of water. Geologic mapping aided by an adequate program of borings will usually reveal the extent of these deposits, since they are not the "limitless sinkholes" sometimes popularly imagined. Early civil engineers generally dumped solid fill along the route until it finally stopped settling or, alternatively, laid mattresses of brushwood or similar material to serve as artificial foundation beds for fill. The crossing of Chat Moss in 1827 by the original Liverpool and Manchester Railway in England, with George Stephenson as the engineer, is an early example of this treatment. Today, with accurate geologic information, jet removal of peat at pressures up to 2,000 kg/cm^2 (100 psi) is possible underneath dry fill, which is allowed to settle as the disturbed peat is displaced laterally.

Displacement of unstable deposits by blasting is even more widely employed beneath dry fill which has been dumped in place or well down in the soft material in front of the fill deposit. Such work, on the Baltimore to Havre de Grace highway, Maryland, was necessitated at swamp crossings. The surface vegetable mat was first removed by blasting with 50 percent nitroglycerine, and the fill was then advanced and blasted into position; a 40 percent gelatin explosive which has a slow heaving effect was used. This type of construction operation has undergone considerable development in North American in more recent years.

At the other extreme from the road-building liability problem posed by swamp conditions is the problem of routing roads through loess, from which water must be kept away if its stability is to be maintained. Ancient Chinese roads are to be found worn deep into the dry, friable loessal soil by the passage of wagons through the centuries; the wind has blown away the dust formed by the steady disintegration of the loess by wagon wheels. Pumpelly reported such roads were worn 15 m (50 ft) below the general level of the country traversed. Widely distributed throughout the world, from New Zealand in the south to the

FIGURE 11.7
Vertical cut in loess, just east of Vicksburg, Mississippi, showing how well it will stand if protected from water. (*Courtesy U.S. Geological Survey.*)

northern parts of the United States and the U.S.S.R., loess has been carefully characterized in relation to its occurrence on highway routes. Splendid exposures of this interesting soil are found in Iowa. During the relocation of U.S. Highway 127 near Magnolia, the Iowa State Highway Commission experimented with a new type of terraced design for deep cuts in the loess, cuts which involved the movement of 610,000 m^3 (800,000 yd^3); some cuts were 24 m (80 ft) deep. Terraces 4.6 m (15 ft) wide were cut between vertical faces each 4.6 m (15 ft) high; the terraces were graded slightly toward their inner edges so that precipitation could be quickly drained before causing collapse of the metastable open structure with minimal interlocking of the grains (Fig. 11.7). The loess here is typical; it is composed of angular to subangular quartzose and feldspathic silt and fine sand with montmorillonite as the textural binder.

MATERIALS

Road-making materials are almost all used in their natural state (cement being the main exception), and so geologic influences are important. Nobody recognized this more acutely than the Romans, who used a basecourse of sand upon which four separate layers of masonry were laid with the addition of a simple kind of concrete. The Romans always selected the hardest stone they could readily procure in each district, especially for the top layer (the *summa crusta*), just as did the Incas in their almost incredible Andean road construction. Among the regular Roman road materials was iron forge waste, especially the cinders from the forge fires. Roads so constructed were known as *viae ferriae,* a name that is said to be the origin of the French term *chemins ferrés* and almost certainly is the origin of the English expression crushed stone *road metal.* Thomas Telford and James Loudon McAdam ("macadamized" roadways) were the two great British pioneers who appreciated the importance of geology in road planning and in the selection of natural road materials.

Nineteenth-century city streets had to resist the abrasion of steel-rimmed wheels yet give traction for horses. These needs led to the development of special pavements using both granite setts or paving brick. Granite setts, usually 10-cm (4-in) cubes, used to be the main product of all granite quarries, and may still be seen in use in the older parts of industrial cities, even in North America. Naturally shaped hexagonal blocks of columnar-jointed basalt were sometimes employed directly, a practice that may still be seen in use today for the revetments along the banks of the Rhine. Some of these hexagonal blocks so used come from the widely utilized lava fields of Czechoslovakia.

With the twentieth century also came the beginnings of great changes in road and street traffic, especially in cities, and a corresponding turn from the empirical to definite design and construction procedures for roadways. Investigations initiated in 1906 were among the earliest attempts to apply some of the facts concerning soil behavior to road work. At the first conference of state highway-testing engineers in 1917, the matter received the considered attention that has grown to today's well-established discipline of highway geotechnique.

Soil, concrete, and asphalt are the three major materials of road work. Economics usually dictate using material as close as possible to the road location, and, obviously, study of the local geology is an essential prerequisite to deciding what local materials can be used. In some cases, unusual materials have been put to good use after necessary laboratory studies of their properties. In Liberty County, Texas, for example, seashells obtained from reefs in Galveston Bay have been successfully used, when mixed with sand, to form a satisfactory subbase for main road construction. More unusual but of potentially wider application is the practice of in-place calcining of local clay. Apparently this idea was first tried out in Australia where local adobe clay, as found in parts of New South Wales and Queensland, was successfully treated in this manner. The resulting road is made stable in wet weather and at relatively low cost. The same technique has been used with the black cotton soils of the Sudan and Nigeria, in specially built kilns adjacent to airfield-construction jobs.

With the continued increase in road construction in the developing countries, there is much activity in the utilization of residual soils for road construction. At the Road Research Laboratory in the United Kingdom, work has been done on which this progress report was made:

> The volcanic soils have a varied clay mineralogy reflecting environmental factors and derivation from a wide range of volcanic rocks. Common clay minerals which may be

present are halloysite, disordered kaolinite, goethite, montmorillonite, hematite and vermiculite. These greatly influence the properties of soils such as the East African red clays in which aggregation of clay results from flocculation by ionized iron and cementation by iron oxides with resulting decrease in plasticity.

Accurate foreknowledge of road material properties will not only assist in accurate cost estimating by engineers, but also in the preparation of bids by contractors, as based on carefully planned construction methods; expenditure of money on careful preliminary work always pays good dividends. Displacement of material directly from cuts into fills is especially economical earth moving.

CONSTRUCTION

Construction methods should incorporate geological information, a practice which is always helpful, sometimes imperative, and occasionally critical. Fill placement, for example, requires rigorous compaction control. Geologic study can assist in suggesting the uniformity that is to be expected in borrow deposits, in excavation of cuts, and as continued observation during excavation to ensure that further excavation will be in materials as expected—or to forewarn of unsuspected changes. And when troubles develop with a finished road or as a highway approaches completion, geology may be able to provide the explanation and suggest the solution.

Placing fill by pumping is an old method that is still applicable under certain geologic conditions. Construction of Interstate Route 10 about 50 km (30 mi) southeast of Baton Rouge had to contend with many low-lying areas containing swamp deposits and other unstable surface soils. The Louisiana Department of Highways completed subsurface exploration of these conditions, even though tidal waters often covered the ground. About 6.6 m (22 ft) of unsuitable material were removed and replaced with suitable load-carrying soil at reasonable cost, over a stratum of dense clay for a distance of 32 km (20 mi). An ideal granular material was found in the bed of the Mississippi River, at distances of 16 to 24 km (10 to 15 mi) away. Three-stage construction involved (1) clearing, (2) excavation of an initial access channel by a dragline dipper dredge, (3) followed by a large hydraulic dredge which did the actual mucking. Muck-water slurry was deposited on the right-of-way perimeter and in other disposal areas and the channel was excavated to depths of about 3.3 m (11 ft) below the surface. Granular fill was pumped from the bed of the Mississippi by a 75-cm (30-in) hydraulic dredge at a solid-to-waste ratio of between 14 and 18 percent. The method was judged economical for distances up to 40 km (25 mi), and provided a saving of about 44 percent against the cost of a structure-supported design.[4]

As an example of troublesome hard rock sites, preliminary study of geologic maps indicated subsurface cavities on a 14.3-km (8.9-mi) section of Britain's M4 motorway in Glamorgan, South Wales, from Pencoed to Stormy Down. Bedrock was known to be a limestone conglomerate overlying shale, sandstone, and a thin coal seam. Air-percussion drilling was carried to depths of 20 m on a 20-m (65- by 65-ft) grid pattern. In areas of surface depression, the grid was reduced 5 m (16 ft). Inconsistent results prompted use of borehole TV cameras, which showed that the voids between the limestone blocks were generally clay-filled, being open only where percolating water had eroded the clay. Special unit-price provisions were used to modify the contract documents so that the contractor could be suitably paid for special work required in filling individual exposed cavities with aggregate.

Other problems may not be readily identified even during preliminary investigations. An extreme example is the "solid rock" excavation undertaken by the California Division of Highways as a part of the Southern Freeway network serving central San Francisco, through Potrero Hill. The route lay above an existing 1906 tunnel of the Southern Pacific Railroad, one-half km (1,600 ft) long and lined with unreinforced-concrete walls and invert. In September 1967, the general contractor completed his excavation with an arched and rock-bolted retaining wall up to 16.5 m (54 ft) high constructed along the east side of the excavation. Highly jointed strata included folded shales with sandstone boulders of the Franciscan Formation. Serpentinite made up the host rock of both the tunnel and the cut. The maximum depth of road excavation was about 18 m (59 ft), and the minimum clearance over the tunnel was 8 m (26 ft). Even before excavation was complete, owners of buildings adjacent to the cut had noticed cracks in walls and roofs; cracks appeared next in sidewalks, and then water and gas mains began to rupture.

By the time the excavation was finished, the tracks in the railroad tunnel were found to be slightly out of alignment, and movement was detected in the concrete walls and invert of the tunnel and in the brick roof arch. Inclinometers were installed to measure the movement that was obviously taking place in the tunnel, the only Southern Pacific access to San Francisco. The entire western side of the tunnel moved inward at a rate of 0.6 mm (0.02 in) per day. Possible solutions were quickly reviewed; the method adopted is shown in Figure 11.8. Repairs were carried out without interference with rail traffic and with only minimal interference with the highway.

Three large truck-mounted drill rigs were used to advance 60 1.07-m- (42-in)-diameter holes—half being inclined to depths of 27.4 m (90 ft). Into the holes were inserted 0.9-m- (30-in)-wide flange steel beams weighing 342 kg/m (230 lb/ft). These were immediately secured with pumped, high-early-strength concrete. Steel struts were then wedged into specially excavated trenches between the pairs of piles before being embedded in weak concrete, which then constituted a good foundation for road surfacing. Movement decreased after only nine frames had been installed. One month after the completion of this unusual rehabilitation job, movement had virtually stopped. No serious maintenance problems with the tunnel have since been experienced. The strut design was chosen to transfer the load that was actuating the movement of the west wall of the tunnel. The transfer mechanism was a strong framework bridging into the continuous body of rock on the east side of the tunnel. Rock

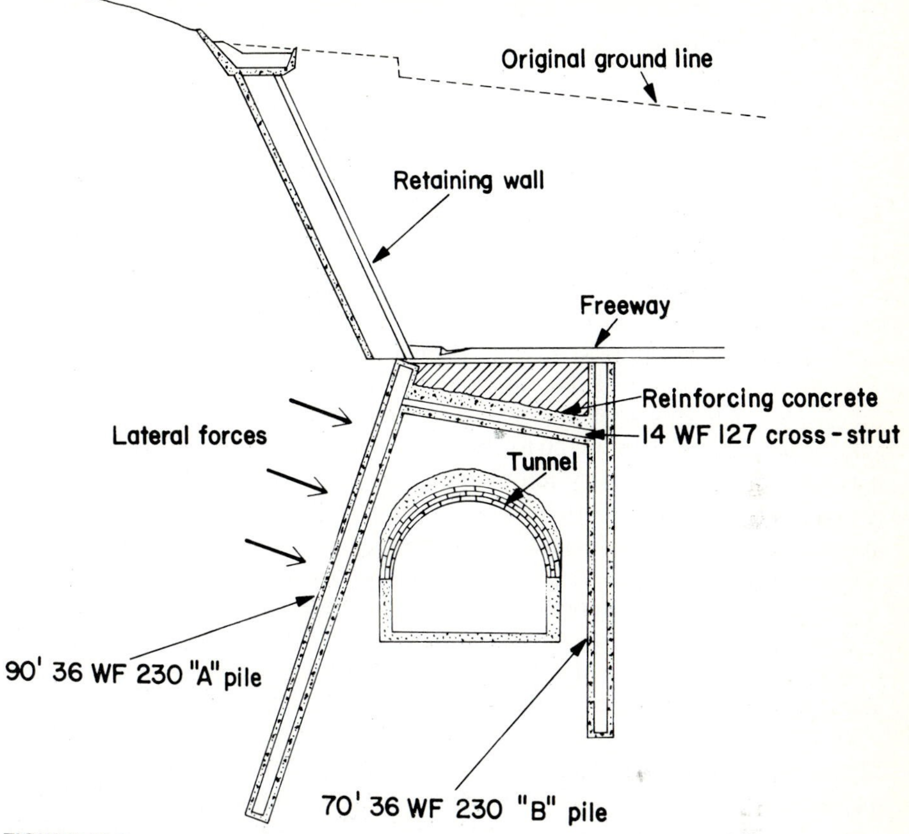

FIGURE 11.8
Simplified cross section through new freeway cut and the Southern Pacific Railroad tunnel, San Francisco, California, showing the remedial measures described in the text. (*Courtesy American Society of Civil Engineers.*)

movement appeared to have resulted from release of internal rock stresses by the deep road excavation, a salutary reminder of the state of stress which always exists below the ground surface.[5]

One environmental aspect of highway construction has become increasingly important in recent years—the erosion of soil into streams and other bodies of clear water from areas disturbed by highway excavation. Rainfall running off newly placed fill or discharging from drainage channels during excavation will naturally carry off soil particles. Knowledge of local rainfall patterns and erodability of each geologic unit is really essential prior to the start of highway construction. This knowledge will indicate what arrangements should be made for controlling or reducing runoff from the work areas in progress. Typical field experiments were conducted by the Pennsylvania Department of Transportation and the U.S. Geological Survey during the construction of Interstate

Highway 81. Four adjoining small watersheds were observed for predominantly loose clay-sized particles, even though the clay content of the construction area soils was only 20 percent. Sediment control technology is now available to control runoff from construction jobs. This technology features essential factors of climatic information and knowledge of local geology.

SOME GEOLOGICALLY SIGNIFICANT ROADS

Geologically influenced road design and construction is common nearly everywhere today. The state of California contains many notable examples, such as the large cut on U.S. Highway 101 near Dyerville Bridge, over the south fork of the Eel River, in Humboldt County. Careful preliminary investigation showed that the cut would be in the Yager Formation of Upper Cretaceous age; sandstone and shale with numerous thin interbeds of conglomerate. The upper 12 to 15 m (40 to 50 ft) were weathered and closely jointed. On the basis of the preliminary information, the cut was designed at a 1:1 slope with 10-m- (30-ft)-wide horizontal berms at 18-m (60-ft) intervals (Fig. 11.9). Major cuts are often geological display cases even though the original geological exposure may become concealed in time by weathering or vegetation. Those who use the Carquinez crossing at San Francisco today probably never realize that the great cut through which they drive, 105 m (350 ft) deep, involved the excavation of 6.5 million m^3 (8.5 million yd^3) of soil and rock, about half the excavation quantity of the Panama Canal.

In many cases, the geological interest of a highway is beneath the surface. There are few better examples of this aspect of geology in highway design than on Ontario Highway No. 11 between Fort Frances and Atikokan. It crosses Rainy Lake on what is known as the Rainy Lake Causeway, 4.5 km (2.8 mi) long, leapfrogging from island to island, with water depths up to 15 m (50 ft) deep. Local Precambrian Shield geology was sound bedrock, even at considerable depths, but overlain in the channels by varved clays which created their own construction problems.

Spain has built some notable roads, one of the most recent being that between Bilbao and San Sebastian on its northern rocky coast. Its 117-km (73-mi) length was engineering-geologically mapped in advance of construction to a scale of 1:1,000. Bedrock consists of thinly bedded limestone, marl, sandstone, and shale, rocks that often cause problems. By skillful design, the length of the road carried in tunnels and over bridges was reduced to about 12 percent of the total, but this involved cuts up to 69 m (230 ft) deep and fills up to 79 m (262 ft) high. The four-lane toll highway is now a geological exhibit in its own right.

The "Highway of the Sun" in Italy, especially its section from Florence to Bologna running through the complex Apennines geology, has much exposed weak rock susceptible to weathering. In view of this and the extremes of local climate, the design was the reverse of that adopted for the Spanish road, tunnels and bridges being used in preference to big cuts and fills. Seventeen km

FIGURE 11.9
Major cut on U.S. Highway 101 in Humboldt County, California, adjacent to the south fork of the Eel River; top of cut 144 m (480 ft) above the road; scale can be judged by automobile on road. (*Courtesy Division of Highways, Department of Public Works, California.*)

(11 mi) of the total length of 82 km (51 mi) consists of 45 bridges and viaducts, totaling 11 km (7 mi) in length, and 25 tunnels, 6 km (4 mi) in length. Four of the tunnels are artificial structures, erected to protect the road from possible landslides, a real hazard because of the character of the sedimentary rocks. For the same reason, such elaborate protective devices as massive retaining walls are to be seen all along the highway.

Landslides are one of the most widespread of all hazards to be encountered in road design. The new seacoast highway in the New Territories of Hong Kong had to be designed and constructed in the face of unusually complex geology. Weathered and disintegrated granite harbored seven bad slides, usually on old and weathered joint planes, occurring even during harbor construction. The rugged terrain and complex geology necessitated cuts up to 59 m (198 ft) deep and viaducts up to 39 m (130 ft) above ground surface. ''Miserable geology'' is the only appropriate term to describe the local environment of this highway!

FIGURE 11.10
The Alaska Highway crossing a large alluvial fan, the result of erosion; British Columbia near Mucho Lake, mileage 467. (*Photo: H. W. Richardson.*)

In another part of the world, one of the most unusual examples of highway-geology interaction is illustrated in Figure 11.10, which shows the Alaska Highway crossing a major alluvial fan.

Railways

Railways still play a vital part in land transportation, even though the average citizen may have lost sight of this because of the all-pervading attachment to the automobile for personal travel. The volume of freight handled by railways is at an all-time high, and even passenger travel is making a slow comeback in many countries of the Western World. In developing countries, new railways are being built, and this aspect of ''development'' is likely to continue for some time. The recurring critical energy situation lends added importance to the fuel economies of travel and transport by rail. The automobile must soon take second place in urban travel behind light rail systems and subways in all large urban areas. In the major cities of the world, subway construction may be expected to provide some of the greatest construction challenges for many years to come. A realistic appraisal of the future of railways indicates that new main lines may be constructed only in some more isolated parts of the world, but that existing lines will continue to be consolidated, maintained, and upgraded to carry a mounting volume of traffic.

FIGURE 11.11
The Melide Causeway across Lake Lugano, built on top of a submerged glacial moraine, carrying the Gotthard line of Swiss Federal Railways on its way to Italy. (*Courtesy Schweizische Bundesbahnen, Zurich.*)

RAILWAY LOCATION

Railway route location methods are almost identical with those adopted for road location, with recognition of significant geomorphic features, especially in glaciated regions, being a first and important step. Figure 11.11 shows a classical case of geological determinants, the Lake Lugano crossing of the Gotthard line of the Swiss Federal Railways. This 270-m- (900-ft)-deep lake bars the otherwise direct route to Como. Swiss engineers found the solution to their routing problem by noticing that Melide was located on a rather odd-looking promontory in the lake. This suggested that it might be part of a submerged glacial moraine. Soundings in the lake delineated the ridge as running across the lake at quite a shallow depth, clearly the remainder of the suspected moraine. An embankment was constructed on the moraine, and made wide enough to carry out not only the railway but also a main highway.

Aerial photo interpretation has now developed to such an extent that it has its own voluminous literature and there are few applications where it has more value than in route selection for highways and railways. Field surveys are still necessary, but these can now be restricted to the final candidate route chosen, through photo interpretation. Field survey work can thus be speeded up and carried out with thoroughness and a high level of certainty. Geologic surveys along the route and in adjacent areas to locate necessary fill or roadbed material are generally extensions of preliminary information gained from a study of aerial photographs. Most target areas can be identified closely by use of available geological and soil survey maps, along with aerial photographs.

Illustrative of route location time savings is the record of the preliminary surveys conducted for the extension of the Nigerian Railway into Bornu Province in central Africa. Aerial photographic coverage was flown for the total length of 980 km (611 mi) to include possible alternative routes; the work involved just over 50 hours of flying time and was done in 11 working days. Geologic and topographic mapping of the route selected by means of the aerial photographs, a distance of 710 km (443 mi), was completed in 15 months. The only unusual geological feature detected was *black cotton soil,* one of the most difficult of all tropical soils in engineering work. Despite the isolated tropical location of the work, the total cost of aerial photography and ground mapping was approximately $865 per mile of final alignment.

RAILWAY CONSTRUCTION

Railways have probably provided more vivid examples of the interrelation of geology and civil engineering than have most other types of construction. All the elements of highway construction are present in railway work, and there are the added imperatives of achieving strictly limited grades and curvature. Only with the most modern type of superhighway is road building approaching the same level of excellence that railway construction has had for more than a century. It is interesting to reflect that even the practice in modern highway work of balancing cut volume with fill requirements was a common feature of early railway construction work, at least once with quite surprising results. George Stephenson, the great early British engineer, designed and supervised construction of the Clay Cross Tunnel (north of Nottingham) for the North Midland line. During tunnel excavation, first-class coal seams were encountered. Stephenson quickly formed the Clay Cross Colliery Company, one of the most successful of such British enterprises, which was in continuous operation for over a century.

The stone houses and railway buildings that distinguish the railway town of Swindon were all constructed of limestone excavated from the famous Box Tunnel of the former Great Western Railway. The Lockwood Viaduct on the (old) Lancashire and Yorkshire Railway's Huddersfield-to-Penistone line was built with 23,000 m^3 (30,000 yd^3) of stone excavated from the Berry Brow Station cut and then dressed by masons at the adjacent site of the viaduct. Possibly even more remarkable is the fact that the clay excavated from the long Penge Tunnel of the Southern Railway in the southern outskirts of London was formed into bricks, fired in a brickmaking plant specially constructed just outside the tunnel, and then used for local houses and railway buildings. These early British railway engineers knew their local natural building materials and also their engineering economics. Their example may be useful today in developing countries, where labor is available to transform such spoil into valuable construction material.

Most major railway lines display striking exposures of local bedrock and the ingenuity of location engineers in circumventing natural geological obstacles. It

FIGURE 11.12
Royal Gorge on the Arkansas River, Colorado, through which runs the single-track line of the Denver and Rio Grande Western Railroad, here seen suspended by a steel structure from the side of the gorge. (*Courtesy Denver and Rio Grande Western Railroad.*)

is still possible to construct and maintain a main-line railway more economically than a first-class highway, and many railroad lines penetrate country where there are still no roads. The most remarkable example is the stretch of the Denver and Rio Grande Western Railway through the Royal Gorge of the Arkansas River in south-central Colorado. The railroad enters this 300-m-(1,000-ft)-deep gorge, starting just above the river level. Precipitous walls slash through upturned Precambrian strata. Tunneling was basically out of the question, as topography would have dictated a completely buried alignment. The narrow confines of the gorge made bench excavation also impracticable. The responsible engineers took the unique course of suspending part of the line over the river by means of special steel structures. Figure 11.12 illustrates the result of this bold piece of engineering, long one of the scenic attractions of travel through Colorado.

At the other extreme is the use of a natural tunnel by the Bristol-Appalachia section of the Southern Railway System (now Norfolk and Southern) through Powell Mountain in Virginia. The tunnel exists in a 236-m- (788-ft)-long natural erosion cavern in jointed limestone. When cleaned out, it was found that its walls were already smoothed by past water action, and the cross section, 36.5 m (120 ft) wide and 27.4 m (90 ft) high at the east portal, was more than enough to provide easy alignment for the projected rail line, merely by the use of an easy reverse curve within the tunnel. This unique tunnel is still in use, though for freight traffic only.

The Dore and Chinley line of the old Midland Railway (now British Railways) was an important rail link in the Peak district of north-central England. It is unusual in that one-quarter of its entire length is underground, partly in the Totley and Cowburn tunnels. These two long tunnels, both of them through limestone, provided a great construction contrast—the Cowburn Tunnel was practically dry, but the Totley Tunnel workings were always wet.

Modern excavation, tunnels, bridges large and small—all are regular components of new railway construction and are treated elsewhere in this book. One variant example is the new main line, eventually to link Amsterdam with Leiden, that has been constructed by Netherlands Railways to link Schiphol Airport and Amsterdam while augmenting commuter services. The main railway line is cut-and-cover tunneled beneath the airport in 6 km (3.7 mi) of slurry-trench and steel sheet-pile walls with chemical consolidation of soil. Soil conditions include the usual high groundwater level, surface soil being clayey sand overlying sandy clay, with a 45-cm (18-in) layer of peat about 6 m (20 ft) below a hard sand surface (Fig. 11.13). Prestressed tension piles were necessary to resist hydrostatic uplift. The entire project, based on sound knowledge of the local geology, is a masterpiece of modern civil engineering.

Even so specialized a technique as soil freezing has been used in modern railway construction. When Central Station in Montreal was formed by a major enlargement of the old Tunnel Station, it was necessary to rearrange the southern end of the Mount Royal Tunnel which terminates at the new station. A central track dividing wall had to be replaced by a single concrete arch over both tracks. The 52.5-m (175-ft) tunnel is in limestone bedrock, here 9 m (29.5 ft) below ground surface, the overlying soil being lodgement till overlain by a stratum of silt and then by stiff brown clay topped by an exceptionally busy street with the usual underground maze of utility conduits and multistorey buildings. The silt was known to be troublesome and construction was to be taken beneath the groundwater level.

After reviewing every possible construction expedient, it was decided to freeze and remove the silt and clay, and to place the concrete arch while the ground was still frozen (Fig. 11.14). A freezing-pipe network was driven vertically into the till, and connected at the surface by a pipe network fed from a central refrigeration plant. This winter work brought freezing of 1,528 m^3 (2,000 yd^3) of clay and silt in 30 days. Excavation of the frozen soil was accomplished by drilling and blasting as rail traffic continued to use the tracks below. When excavation was complete, the old arches were used as a working

FIGURE 11.13
Schiphol Airport, near Amsterdam, the Netherlands, showing new rail line under construction. (*Courtesy Nederlandse Sporrwegen, Utrecht.*)

platform for the erection of the new arch and were removed only in the final stage of this unique operation.

"MAINTENANCE OF WAY"

This old term is still used on many railways and the "engineer of maintenance of way" is responsible for the safe functioning of all manner of rail facilities. Detection of needed repairs or replacement is generally based on the long experience of a senior track foreman who is able to discern the first sign of trouble long before an accident results. Railways and highways directly affect public safety if the routes cannot be kept clear and intact. Geologic influences on track maintenance are tied directly to the nature of the earth materials used in rail embankment construction, as foundations for rail structures, and as landforms adjacent to the right-of-way. Deterioration of rail ballast rock, consolidation or slippage of embankment soil, and stability of adjacent hillslopes are of prime importance.

An extreme case is the 1894 statement by the chief engineer of the (old) London and North Western Railway of England that "there was neither a bank nor cutting between Euston (London) and Rugby that had not slipped at one time or other"; this is only the first 160 km (100 mi) of one of the main railway lines of England. Railways have historically suffered from landslides, both from nearby hillsides and in cuts and fills constructed to carry the roadbed.

FIGURE 11.14
Freezing shallow ground above the south portal of the Mount Royal Tunnel of Canadian National Railways, Montreal, showing proximity of buildings. (*Photo: F. L. Peckover.*)

Probably no railway has suffered more than the Hill Section of the former Assam-Bengal Railway in India. This 183-km (114¼-mi) section was opened to traffic in February 1904. Its construction was undertaken largely for military reasons, and it has been described by a viceroy of India as a "millstone round the neck of the Indian Finance Department."

From the very beginning of operation, slips and washouts gave trouble. yearly maintenance costs were significant, and the line was closed by failures on several occasions. In 1915, an excessive rainfall of 66 cm (26 in) in 48 hours closed the route for two years while remedial work was carried out at the then outrageous cost of over £225,000. Engineers charged with maintenance of this line dealt with vagaries of local geology—tertiary rock of alternating beds of carbonaceous shales and sandstones so affected by past landsliding that they possess little durability and break up into small fragments on exposure to the atmosphere. The shales vary from rock as hard as slate to material with the consistency of clay, and the sandstones vary from friable to first-class building stone. The following account illustrates a notable case of slope readjustment in this project—first by a slide and subsequently by remedial measures.

Figure 11.15a represents a cross section at a point on this line near the south portal of the Chamartalla Tunnel. Figure 11.15b shows how this appeared after the slope failure of 1913. The rock cut was excavated at a steeper slope, but after many falls of rock and soil had occurred, the slope was cut back, as shown. Newly exposed shale was protected from exposure to the atmosphere by revetments. The retaining wall at river level was discovered to be the cause of much trouble. When it had been rebuilt in 1899, its foundation was on hard shale at riverbed level and no erosion was anticipated, but in 1902 and again in 1908 the river flood flow had undercut this wall to a depth of 3.6 m (12 ft) and eroded the hard shale completely; the wall had to be rebuilt again. In the slope failure of 1913, a great mass of material slipped from the hillside onto the tracks. The load was so great that the retaining wall was forced into the river and the whole railbed was carried away for a length of 45 m (150 ft). Traffic was restored temporarily within 15 days. Permanent reconstruction introduced many problems, but the final solution is shown in Figure 11.15b; the design of the covered way was so prepared that it would offer a minimum obstruction to further slides, which were intended to pass over it.

Another example of slides is in Alaska, on the pioneer Alaska Railroad. Probably the most troublesome part of this railroad is where it crosses the Alaska Range (miles 322 to 385). The line has to deal with yearly winter heaving in thawed summer frost-susceptible soils beneath the track. Continual slumps and earth flows along the Nenana Gorge are caused by slow movement of rock debris on the underlying steep slopes to depths of as much as a few hundred meters. These displacements move the entire railroad down toward the river at rates varying from a few meters per year to a few meters per hour. Near Mile 351, an ancient landslide, long ago perennially frozen to stability, is slowly thawing and becoming active again because of the inevitable disturbance of local surface conditions. The central Alaskan earthquake of October 1947 triggered a serious landslide that disrupted traffic for several days and led to a request to the U.S. Geological Survey for assistance in defining geologic sources of troubles. The north end of a tunnel near Garner crosses the lower part of a broad talus cone of large schist blocks slowly creeping toward the adjacent river under the increasing load of debris falling from the cliffs above. The largest of individual chutes supplying detritus directly from the cliffs at the time of the USGS study was 30 m (100 ft) wide and was creeping at the rate of 60 cm (2 ft) per year toward the edge of a break-in-slope over which it will eventually fall onto the track below. Many geologists have recommended that the only recourse is to realign the track permanently.

Most railway maintenance work may appear to be pedestrian—constant and regular inspection of track, especially after heavy rains, with keen observation needed to detect the slightest sign of deformation. Incipient landslides can usually be anticipated by observations such as "pumping" under track, as an indication of failed drainage systems. Even replacement of ballast has a definite

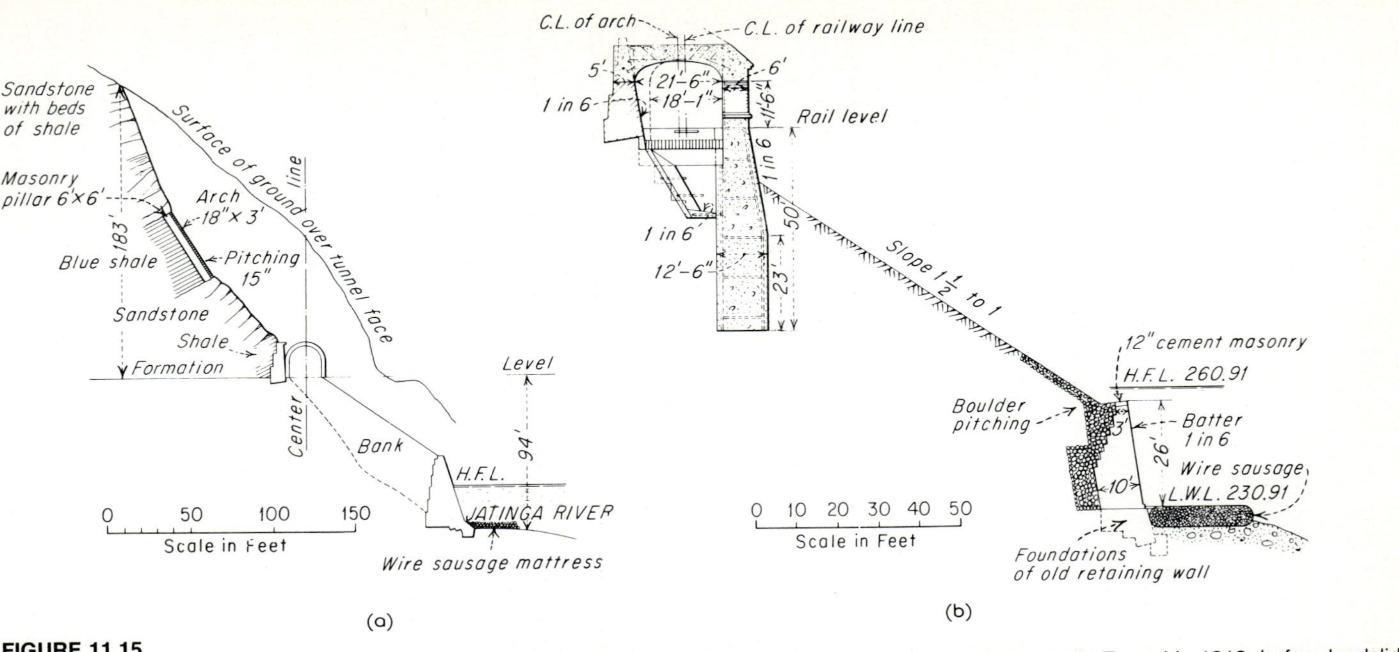

FIGURE 11.15
Landslide remedial work on the Hill section of the Assam-Bengal Railway, India; (a) cross section near Chamartalla Tunnel in 1913, before landslide; and (b) covered way structure, erected following the 1913 slip and undamaged by the 1915 slides. (*Courtesy The Institution of Civil Engineers.*)

geological input, since it is not just a case of using the most economical crushed rock available, but crushed rock that is geologically suitable for the special use that service as ballast involves.

ROCKFALLS

Rockfalls adjacent to steep rock faces are a special hazard in track maintenance. Although such faces and cuts may be perfectly safe and stable when completed, exposure to weathering influences eventually leads to loosening of rock along joint and bedding planes, with ultimate dislodgement of rock fragments, some of which may be large enough to block rail traffic. Peckover found that western Canada exhibits a definite correlation between rainfall and the incidence of rockfalls. Regular and careful inspections, coupled with a detailed and continuing study of weather records, remain a first requirement.[6]

Many railways have stretches of line that require special maintenance procedures, such as multiple tracks of the (old) Pennsylvania Railroad (now Conrail) along the Ohio and Monongahela rivers in the vicinity of Pittsburgh. The alternating strata of sandstones and shales exposed above these tracks generate continuous spillrock and rockfalls; as a result a team of trackmen patrol the most critical 9.6 km (6 mi) of track day and night. On many lines where rail traffic is light, mechanical or electrical sensor devices have been installed. One of the first of these was on the Callander and Oban line of the (old) Caledonian Railway in the Highlands of Scotland (an area subject to frequent rockfalls), along the north side of the Pass of Brander. As long ago as 1881 a special 2.7-m- (9-ft)-high wire fence was erected alongside the line and connected to 14 signals located throughout the critical length of 5 km (3¼ mi). The fence was so arranged that if a falling rock bounded clear of the fence, it would be deflected from the track, whereas if the rock lodged against a wire, one or more signals would activate at the danger location. Broken wires would also cause an electric bell to alert local track-maintenance personnel. In situations such as this, we can see how geologic mapping techniques can be used to delimit areas of susceptibility to rockfalls.

Some deep cuts are also susceptible to rockfalls. The main line of the Western Australian Government Railways between Perth and Northam was rerouted in 1944 to a deep cut in jointed granite, some 450 m (1,500 ft) long and prone to rockfalls upon weathering. Steel fences, 2.5 m (8.5 ft) high, were erected on both sides of the cut, with wires forming continuous energized circuits. Automatic track-circuit signaling activates a danger indicator at the Swan Valley signal box to show the operator there that a rockfall (or other failure) has occurred.

The former Great Northern Railroad system (now Burlington Northern) maintains 36,000 m (120,000 ft) of protective fencing of this general kind. Peculiar to this system, however, is the fact that this total includes about 12,000 m (40,000 ft) of "mud-slide fence," which consists of two wires stretched near the ground so that they will detect the characteristically thin geologic hazards.

Nearly all this fence is installed between Seattle and Vancouver, British Columbia, on the coast line.

Equally as important as these warning devices are measures that can be taken to prevent rockfalls from occurring. This requires careful (and sometimes dangerous) examination of suspect rock faces. Rock bolting usually follows scaling to remove potentially loose rock and to divert runoff from entering the slope mass. In some cases, rock faces that might cause trouble can be underpinned by rock masonry or concrete buttresses.

SPECIAL HAZARDS

There are other railway hazards of a geologic nature that must be reviewed, even though they are not so general in nature as are rockfalls and landslides. One very serious problem in mountains and in regions of weak rock is the possibility of mudflows—streams of mud and rock debris washed down from the lower slopes of bare mountainsides by the runoff that follows intermittent, cloudburst-type torrential rains. In general, these are experienced widely in tropical countries and are especially severe in sections of South America and in India. The extreme climatic conditions promote rapid disintegration of exposures of relatively weak rock, and the concentrated rainfall acts as an efficient transporting agent. If mudflows reach small-discharge bridge culverts, flow blockage may occur, followed by washouts. Several sections of the Bolivian Railway have been relocated because of this. In one place a town adjacent to the railway had to be completely relocated because the original site was buried deep under the unusual detrital cone formed by blockage of a large mudflow at a bridge opening. This bridge, originally 3 m (10 ft) above stream-bed level, is now approached by way of a slot cut into the mudflow deposit.

Protection against floods is largely a hydrologic matter, depending on the accuracy with which preliminary flood predictions can be made for the design of necessary waterways beneath roads or railways. Protection from bridge scour and bridge destruction from mudflows *(lahars)* relies on geologic input. The latter case is of special geological significance and caused one of the most tragic accidents in the history of railroading (Fig. 11.16). Late on Christmas Eve, 1953, the main part of the Wellington-to-Auckland Express in New Zealand was swept away when it plunged into the swirling waters of Whangaehu River at the site of what had been the Tangiwai Bridge. In this accident, 151 persons were killed; 20 bodies were never found. The main finding of the official board of inquiry was that:

> The accident was caused by the sudden release from the Crater Lake on Mount Ruapehu through an outlet cave beneath the Whangaehu Glacier of a huge mass of water which was channeled down the Whangaehu River carrying with it a high content of ash from the 1945 eruption and blocks of ice due to the collapse of large volumes of the glacier. This flood, which can properly be called a "lahar," proceeded down the mountain as a wave, uplifting huge quantities of sand, silt and

FIGURE 11.16
Tangiwai Bridge over the Whangaehu River, North Island, New Zealand, (a) carrying a fast freight train on the day before the disaster; and (b) the scene of the disaster on Christmas Eve 1953 while carrying a crowded passenger train of New Zealand Government Railways. (*Courtesy (a) Derek Cross, Ayrshire, Scotland; and (b) editor, The Dominion, Wellington.*)

> boulders. It was most violent and turbulent and of great destructive effect. It destroyed portions of the railway bridge at Tangiwai before the arrival of train No. 626, which was engulfed when proceeding across the bridge.

No blame was attached to the design engineers or to the train personnel involved; the tragedy was one of those that can only be described, with reverence, as an "act of God." The complex combination of an active volcano and a nearby glacier, the discharge from which led directly down a relatively short river crossed by the railway, is unusual, but it would appear that similar damage can originate from other geologic combinations, always linked by the presence of a massive volume of stored water subject to release, or from torrential rainfall. The board of inquiry in its definitive report suggested that the geological term *lahar* should be better known by civil engineers.

Lahar potential in all volcanic areas is an added hazard to be carefully considered by transportation route designers. *Lahars* have a first cousin in the form of *jökulhlaups,* the catastrophic slurry flows that may be released from glaciers. One such event took place on the main line of the Canadian Pacific Railway on September 5, 1978, just as a freight train was leaving the first of the famous spiral tunnels. A slurried mass of soil and water swept down from the mountains above, derailing the locomotive and damaging both trains and track. Fortunately no lives were lost. An estimated 175,000 m^3 (230,000 yd^3) of debris were washed onto the track by the outflow of possibly 1 million L (250,000 gal) of water from the base of Cathedral Glacier.

RELOCATION OF RAILWAYS

Although few new main railway lines are now being constructed in North America, there are many necessary railway relocations due to geologic events, to assist in maintenance, to accommodate new reservoirs, or as part of urban mass transit systems. As an example of maintenance-based relocation concerns, more than a mile of the Great Northern (now Burlington Northern) is the double-track main line north of Seattle, which is plagued by continuous rockfalls. A subsurface investigation identified prime routing to be seaward of the old line. The new embankment is, on the average, only 30 m (100 ft) away from the old, the maximum separation being 50 m (165 ft). Specially selected dense volcanic rock was brought from a quarry in the Cascades 100 km (65 mi) away, and dumped from a temporary timber and steel trestle. Seven 1.8-m (72-in) concrete culverts were installed in the rock fill to allow for tidal flow in and out from the lagoon formed between the old and the new tracks. The fact that the contract was for a sum of $1.25 million (in 1958) shows the monetary cost of the trouble caused by such simple recurring natural hazards as rockfalls.

Many urban railway relocations adapt outlying railway yards and freight lines for the passenger service that was once so conveniently provided to a downtown terminal or through station. A few are more elaborate operations than this. An extensive rearrangement of railway lines was necessary to bypass the Welland

FIGURE 11.17
The Townline Road-Rail Tunnel beneath the Welland Canal Bypass, Niagara, Canada, opened for use in 1972–1973. (*Courtesy St. Lawrence Seaway Authority.*)

Canal at the eastern end of the Niagara peninsula in Ontario. This 13.6-km- (8.5-mi)-long new ship canal enhanced operations on the Welland Canal, partially by eliminating road and rail crossings, necessitating relocation of almost 160 km (100 mi) of railway. All roads and rail lines were combined in a master plan and placed in two new tunnels, both of which were built in the dry before the newly excavated canal prism had been completed and flooded (Fig. 11.17). Railway facilities were placed in one combined tunnel designed as a reinforced-concrete box section 32 m (106.5 ft) wide and 9 m (30 ft) deep. With 9 m (30 ft) of water in the new canal, this meant that the invert of the new tunnel was 22.5 m (75 ft) below ground level. Because of the limiting grade for the three railway lines approaching and leaving the tunnel, the necessary approaches had to be each 4 km (2.5 mi) long, representing 12.5 million m^3 (16.4 million yd^3) of excavation. Favorable geologic conditions included dolomitic limestone overlain by up to 24 m (80 ft) of soil, usually a thin stratum of lodgement till followed by glacial lake clay, the upper portions of which are generally weathered, with the usual 60 cm (2 ft) of topsoil and organic matter. Construction procedures, although massive, were not unusual except for the extensive groundwater control system, since the jointed bedrock was under subartesian pressure, with the piezometric surface often close to the excavation bottom grade.[7]

SUBWAYS

Planning and construction of rapid-transit subways, necessary for moving large numbers of people efficiently and with a minimum of pollution, are now under way in many of the major cities of the world. Such railways must be close to the ground surface for convenience and economy, which dictates a certain degree of acceptance of subsurface conditions encountered (rather than chosen, as would be the case with deep tunnels). In most cases, excavation will be in surficial deposits, possibly also in rock, as a combined operation designed to minimize mixed-face conditions and related construction difficulties. A further problem is that locations close to ground surface will sometimes show the cut-and-cover method of construction to be the most desirable and economical, if permission to disrupt traffic temporarily can be gained. Despite sociological problems and the high capital costs involved, subway construction may be expected to continue, probably at an accelerated rate, as traffic problems become more acute and the energy situation more critical.

London and Subway Construction

London is preeminent in its use of subways, not only for passenger traffic but also for such special purposes as the conveyance of mail from the central main post office to railway terminals, the "post office tube" being a subway in its own right. A fascinating account of the first half-century of the London tubes depicts how fortunate London has been that its underlying geology is so suitable for tube railway construction. The London Clay is a stiff, fissured, and overconsolidated Eocene deposit. Its Eocene age shows that these clays are much older than those normally encountered in the glaciated parts of the world. When handwork is necessary during excavation of tunnels in the clay, the clay is cut off in large slices that have the consistency of a medium cheese. The sight of these large pieces of this material being thrown so easily into mucking cars belies other problems when structures within the clay are disturbed.

During the construction of the great Shell Building on the south shore of the Thames in the center of London, for example, the presence of tube railways under the site added to normal construction difficulties. This complex of buildings is one of the largest aggregations of London buildings, with an area of 3 ha (7.4 acres). One basement underlies the complete site! Beneath one of the main buildings and lying very close to the 20-storey tower were the twin tubes of the Bakerloo Tube, one above the other, the top one having only 1.5 m (5 ft) of clearance below the basement slab for the large building. Although the building load could have been carried on the clay, that part overlying the two Bakerloo tunnels was founded on belled caissons, excavated in the clay to a depth well below and isolating the tubes, and then filled with reinforced concrete. There was concern that the tubes might be deformed upward by the excavation-induced rebound of the clay. As a precaution, excavation over the tubes was carried out in strips each 9 m (30 ft) long. The concrete foundation slab was placed in each strip as soon as it had been excavated and before the adjacent strips were excavated. Each tube was instrumented throughout its excavation length and kept under constant surveillance so that the job was completed without any disturbance to traffic in the tubes and with only negligible total movement.

Although much of London's subway excavation has been in "free air," compressed air has had to be used occasionally in saturated soil below groundwater level. Special precautions and extra cost indicate this to be a measure to be avoided if at all possible. It was evident from a detailed geologic study of the ground around the proposed Victoria line in London that this technique would have to be used here. An associated and unsuspected danger occurred during construction when compressed air was used in the Woolwich and Reading water-bearing sands that lie beneath the London Clay. This air pressure forced long-stagnant pore gases into a neighboring tunnel, also under construction. This "air" was deficient in oxygen and nearly led to fatalities.

Freezing of soil is always an expensive construction technique, but when geological conditions are so bad that any other solution is questionable, its use may become economical, especially when time is considered. This technique has even been used on the London Underground, where host ground is usually the easily handled London Clay. Subsurface exploration for the modern Victoria lines however, showed that the upper half of the inclined tunnel for the access escalator at Tottenham Hale Station was to lie in dense, water-bearing cobbly sandy gravel. Few worse combinations can be imagined for excavation in confined quarters. Chemical consolidation of the gravel (a proven method) was first considered, but borings revealed lenses of silt within the gravel, material that would only questionably respond to chemical grouting. Freezing by liquid nitrogen was adopted to consolidate the ground for excavation. This relatively new approach (as compared with the usual cold brine method) makes possible an increased rate of freezing against the warming effect of groundwater flow through gravel, and thus allowed for working a full face in the inclined tunnel, more than compensating for increased costs.

Some Other Subways

The effect of nearby foundations on subway construction is not peculiar to London. Significant problems have to be faced also with underpinning of existing structures on nearly all new routes. Brussels, now well into the building of a 70-km (44-mi) double-track subway system, has had to tunnel under a famous 68-year-old ceremonial masonry arch (weighing 23,600 tonnes; 26,000 tons) resting on poor foundations, as well as pass under two large brick conduits holding the flow of the river Seine and storm water from the city's sewers. The reverse of this happened in Paris in 1972, when the foundation for Europe's tallest office building, rising 197 m (656 ft) above ground level, had to be constructed around and under an operating subway of the Paris metro. In São Paulo, Brazil, the presence of many multistorey buildings flush with the street lines adjacent to the city's subway made mandatory the use of a tunneling shield.

Even more remarkable was the construction of the tunnels in which the Stephanplatz Station of the Vienna subway would be located. As usual, these tunnels were located beneath the center of a main thoroughfare, in this case that which runs in front of the famous St. Stephen's Cathedral, justly regarded as the city's greatest cultural monument. The Virgin Chapel, dating back to

1367, is here located flush with the street line and hence subjects the ground beneath to unusually large loads. Foundations of the cathedral structure vary but it was known that some rest on loessal soil. An intensive study of the site geology revealed from ground level down: loess (clay and silt), fragmented slate, clayey silt and gravel, and fragmented slate and gravel, with groundwater about 15 m (50 ft) below the surface.

Construction methods were selected which would avoid damaging the cathedral. Methods adopted included special grouting, the use of compressed air for shield excavation, and low-vibration bored piles to support station structures. Careful deformation measurements were taken throughout the entire construction period, which was successfully completed without damage to the historic structure.

Few subway lines have encountered geological conditions that had a determining effect upon their major design features and overall construction planning, the geology of the routes usually having to be accepted without question. Design and construction of one major subway, however, was entirely controlled by local hydrology. The new city center in Rotterdam, a leading city of the Netherlands, is one of the finest examples of enlightened city planning. This entire area has been rebuilt since the end of World War II, following its bombing devastation just after the outbreak of hostilities. The main part of a rapid-transit subway system is now in use, including a second notable tunnel under the Nieuwe Maas that gives ready access to the business area while not adding any vehicular traffic to this already crowded central section of the city.

The Rotterdam subsurface is dominated by a stratum of gravelly medium to coarse sand, a Pleistocene deposit brought down by the rivers Rhine and Meuse. This sand stratum lies at a depth of from 16 to 18 m (52 to 59 ft) below the local datum, equal to average sea level. It extends to a depth of more than 40 m (130 ft) in the eastern part of the city, diminishing to about 30 m (100 ft) in the western area. All major structures in Rotterdam are pile-founded. Above the sand is a stratum of alluvial clay about 1 m (3.25 ft) thick, followed by a layer of organic peat, and a final stratum of marine clay with a total thickness of about 2 m (6.5 ft). From there to the surface is a variety of shallow clay and peat deposits, with pockets of fill.

Much of outlying Rotterdam is of very recent development, typical of many of the world's major cities. Here, however, roadways are being reconstructed to offset the slow settlement of the organic peat beneath them. Sand drains and preloading are now used to accelerate consolidation. Because the general level of the city is below sea level, as with so much of the Netherlands, the Dutch dikes against the sea are justly world-famous. Groundwater throughout the city is close to ground level, never more than 2 m (6.5 ft) below the surface. Fortunately, it is reasonably constant in elevation, and so no difficulties are experienced with deterioration of the timber piles so widely used all over the city. But any deep excavations necessarily interfere very seriously with groundwater conditions. Herein lay the great problem to be faced in constructing the subway system.

FIGURE 11.18
Rotterdam subway under construction, looking toward the main railway station and subway terminal, the view giving a good idea of the scale of the works by comparison with automobiles on adjacent streets. (*Courtesy G. Plantema, Gemeentewerken, Rotterdam.*)

All the rapid-transit lines on the north side of the Nieuwe Maas, i.e., in the central part of the city, had to be constructed as tunnels. Once the line emerges to the south from beneath the Maas, it becomes elevated, then a surface line extending into the suburbs. Of the initial 7.6 km (4.7 mi), 3.1 km (1.9 mi) had to be in tunnel with the remaining 4.5 km (2.8 mi) on a prestressed reinforced-concrete superstructure. With an invert depth of about 11 m (36 ft) below ground level, the combined twin tunnels required a structural width of about 10 m (33 ft). To construct a tunnel of these dimensions right through the central city area, close to many large buildings founded on piles, and with its terminal adjacent to the new Central Railroad Station, would have been difficult and hazardous, especially with high groundwater. Cut-and-cover methods would have involved a massive pumping operation for each section of trench for a five-year period and great disturbance of groundwater conditions around each trench, even with the most expert use of well points and controlled pumping. To avoid pumping, the decision was therefore made to build even these sections of the subway tunnel as precast concrete tubes and to float them into place right in the center of the city. This bold operation was successfully completed and the first section of the subway was opened in 1968 (Fig. 11.18).[8]

Modern subway building is in the great traditions of both railway and tunnel construction. Great efforts are being made to reduce the cost of tunneling, one reason for this concern being the knowledge that more subways are going to be needed if the traffic problems of major cities are to be solved. There is much written and said today about the innovations needed in tunneling, one complaint being that tunneling practice in New York (where 50 percent of all urban tunnels in the United States are said to be located) is conservative, and yet is slavishly followed elsewhere in the United States. Innovations are needed, as always; tunnel costs should be minimized to the maximum extent possible. But the fact that the local geology always determines the final result is a fact of life that is mentioned all too rarely in these current discussions, and sometimes not actively considered. In all railway and subway construction, adequate knowledge of route geology is the first requirement. Unless route geology is known with reasonable certainty, then all other considerations may be "of nothing worth." As the subways of the world proliferate—as they must do—urban geology will present challenges such as not yet seen. Answers to these challenges will depend on success of the railways yet to be built.

Airfields

Civil aviation all over the world has experienced a phenomenal growth in the last half century, especially since the end of World War II. There is probably no branch of civil engineering work about which the public is so impatient and responsible officials so frustrated as the construction of airfields—the building of new fields where necessary and the enlargement and improvement of existing facilities. Singapore had to build a complete new airport within 20 years of the opening of the former one. Cities such as Chicago have had to construct duplicate fields. And all too many fine landing facilities, built apparently for the foreseeable future only a few years ago, are inadequate and proving incapable of expansion. Modern airfields require large, reasonably level areas without serious flight impediments. Desirable flat areas owe their flatness to their geology and are generally under intense competition for a variety of other uses. Typical examples are the floors of old lakes or seas, which are usually clays, and alluvial plains often underlain by coarse materials. Airfields must be capable of being thoroughly drained at all times, and above all, they must be as conveniently situated to the cities they serve as possible.

In view of these requirements, site geology almost always has to be accepted; convenience, access, and economics of reasonably level ground are the basic determinants. Nevertheless, within these large areas, geology will govern optimal placement of individual facilities. One case of site rejection on the basis of unfavorable geological conditions involved the presence of an extremely sensitive marine clay which gave promise of unacceptable long-term settlements. Some airfields have had to use filled ground; examples include La Guardia Airport, New York, and the Baltimore-Washington International Airport. In such cases, serious design problems can be met and solved utilizing soil mechanics applied to site characteristics and three-dimensional distribution developed by application of geology.

It is imperative that geology be applied to airports in screening the entire construction area and selecting optimal locations for structures. Geology is also vital in determining drainage influences and the amounts of cut and suitable fill material available. Knowledge of these factors will be necessary to prepare estimated costs, tentative time schedules, and ultimately contract documents for actual construction. Correspondingly, accurate subsurface information is essential for the proper design of pavements (especially because of today's large wheel loads) and of foundations for the structures that will serve the airport. Finally, and perhaps of greatest importance, adequate drainage facilities can be properly designed and installed only if the soils to be encountered, their drainage characteristics, and the local groundwater conditions are known with accuracy. While not "exciting" applications of geology, they are vital and call for the same degree of careful attention as does the most spectacular service provided by geology for tunnels, dam foundations, or landslide correction.

Landing-field areas have to be graded to a given level, then drained and provided with suitable runways. Design of the latter is comparable to highway design; similar materials are used with similar design requirements. Drainage is the essential and often the most difficult part of airport work, especially since the field has to be practically level. An engineering soil survey is essential to choose the type of drainage system. If porous materials underlie the site, the system can be simple; if clay or similar material is found, an elaborate system of field drains and main drains may be necessary. Installation of drains must be undertaken with great care. Backfilling (select porous material) up to within 15 cm (6 in) of the surface is essential, to avoid development of landing-field defects, indicated as responsible for one accident out of every eight in United States aviation.

All these work items are common to other branches of civil engineering. There is nothing of special note in the design and construction of airfields beyond the vast area always involved, the great costs inevitably associated with installations of such size, and the necessity for the closest possible attention to all details of design, such as drainage facilities and pavement cross section. These factors are important, not only because of the extensive use that has to be made of the basic designs once they have been established, but also because of the imperative need of absolute safety in the performance of airport runways. A summary account of the site studies for the Mirabel Airport near Montreal stands as an illustration of the extent of areal studies for major airports. This is one of the most recent entirely new major airports to be constructed in North America.

Selection of this Montreal site involved preliminary terrain studies. Following site selection, more detailed study commenced with the aid of all available aerial photographs and all available information on the local geology and hydrogeology. An area of 35,500 ha (88,000 acres) was expropriated so that control could be exercised on all environmental concerns relating to the new facility. The site consists generally of rolling country, flattening towards the east. Most of the area is a plain at about 75 m (250 ft) above seal level, underlain by the Leda (Champlain) Marine Clay so characteristic of the St. Lawrence Valley. The clay has a desiccated crust of from 1.5 to 2.4 m (5 to 8 ft) and is underlain by lodgement till resting on bedrock. Local depressions in the till were found to be filled with uniform fine sand and organic material. Groundwater generally lies between 1.5 and 3.0 m (5 and 10 ft) below the surface.

With the surface geology of the estimated area known in general terms, the first step in the detailed study was an intensive program of 50- to 75-cm (20- to 30-in) auger borings, yielding disturbed samples for soil identification. Concurrent seismic refraction surveying was carried out over the site to determine the depths to lodgement till and to bedrock. All field observations were fed into a computer through which a bedrock contour map was prepared. Based on this and the results of the shallow borings, a program of 200 deep soil borings was carried out at an average spacing one to each 2.5 km (1 mi^2), irregularly spaced in accordance with presumed data needs. Soil samples from every hole were subjected to a full set of laboratory tests. Final refinement of the resulting information formed the basis for the engineering design of runways, taxiways, and other features of the airport and for the layout of drainage facilities. Construction started in 1973 and Mirabel Airport became operational in 1975.

DRAINAGE

Drainage is one feature of airport design, the importance of which cannot be overstressed. The high percentage of pavement cover makes such a change to the local hydrogeology that the drainage system for an airport is a major part of the overall design. Sound knowledge of the geology underlying the site of a proposed airfield can significantly affect drainage design.

During World War II, a small Army airfield was constructed at Bowling Green, Kentucky. With an area of only 107 ha (265 acres), providing four 1,170-m (3,900-ft) runways, this relatively small project was designed with the usual care exercised of the U.S. Army Corps of Engineers. Recognition that the site was located in "sinkhole country" gave designers information that the cavernous limestone bedrock (the upper stratum known locally as "cathedral rock") could contain subsurface dissolution cavities. Existing sinkholes were located and other potential downward drainage channels were surveyed. The entire field was graded to drain directly to 12 manholes, themselves discharging directly into the underlying limestone. Some 600 m (2,000 ft) of French drains were used in flatter areas to conduct drainage to the nearest manholes. The airport has now been in use for over four decades, and the unusual drainage system has worked quite satisfactorily. It illustrates what can be done by constructive use of a local geological condition that is ordinarily regarded as undesirable (Fig. 11.19).

FIGURE 11.19
Bowling Green Airport, Kentucky, showing sinkholes (right front), some of which were used for drainage of runways, roads, and hardstands. (*Courtesy GRW Engineers Inc., Bowling Green.*)

Airports serving coastal cities are frequently located on low-lying land not suitable for other purposes. Subsoils here are sometimes of poor quality, with high water content. In such cases, not only is surface drainage important but also subsurface drainage, in order to induce and control settlement of such soils, especially under runways and taxiways. Sand drains have proved their great utility for this purpose in a number of cases. Three-quarters of the original La Guardia Airport was located over thick marine organic silt. Six meters (20 ft) of ash and debris till had been placed over this in 1938, but differential settlement of the early runways had taken place. Major rehabilitation incorporated sand drains, which were installed (at 4.5-m or 15-ft centers) to speed and finalize this settlement. Conditions at Newark Airport were similar; up to 7.5 m (25 ft) of soft and highly compressible peat and organic silt overlies sand and varved clay, bedrock being at about 20 m (65 ft) below the surface. Sand drains spaced at 3.0- to 4.2-m (10- to 14-ft) centers were used to consolidate the compressible organic materials.

CUT AND FILL

Estimation of cut and fill quantities reveals the necessity of adequate geological study more than anything else in airfield work. Much of what was said about open excavation in Chapter 4 applies directly to airport construction since airport excavation does not usually extend to great depths. One airport example usefully illustrates an extreme to which airfield excavation may have to proceed. One of the many smaller post-World War airfields was built to serve the metropolitan district around Charleston, capital of West Virginia. Kanawha Airport is located in rugged, hilly country northeast of Charleston. The severity of the local terrain necessitated moving slightly more than 7.4 million m^3 (9.7 million yd^3) of material to produce the level area necessary to accommodate the main runway, which is 1,800 m (6,000 ft) long, and subsidiary runways of 1,740 and 1,560 m (5,800 and 5,200 ft), respectively. The maximum difference in elevation between the highest point in a cut section and the toe of the deepest fill was 135 m (450 ft); one fill alone extended 69 m (230 ft) from its toe to runway level. Alternating layers of shale and sandstone were encountered; the absence of groundwater eliminated any real problems with excavation. Benches filled with rock from 0.8 to 6.1 m^3 (1 to 8 yd^3) in size were used to support all slopes steeper than 1 in 3. About one-half of all excavation had to be drilled and blasted, but good fragmentation was obtained; drop weights were used for reducing large rock fragments to manageable size for handling.

Frequently, airport construction projects have given rise to unusual experiences with geological conditions not normally encountered in the general run of civil engineering work. One of the many World War II challenges of the U.S. Navy Seabees (Construction Battalions) was the unusual airfield construction on Iwo Jima. This small Pacific island, measuring 9 by 5 km (5 by 3 mi), is of Quaternary volcanic origin. An active volcano (Mount Suribachi) is the main outcrop of solid rock on the island. The general plateau constituting the main part of the island, at an elevation of 102 m (340 ft) above the sea, consists of volcanic ash in two main forms—a loose black cinder commonly called "black sand," and a consolidated buff-colored ash known as "sandrock." The black sand was easy to handle as stable fill material. The buff sandrock was similarly well suited for use as a "stabilized" surfacing for airstrips and roadways. This latter material was found to have an in-situ moisture content greater than that required for optimum compaction, at field densities of only 1,150 kg/m^3 (72 lb/ft^3).

REFERENCES

1 D. McCulloch, *The Path between the Seas,* Simon & Schuster, New York, 698 pp. (1977). A splendid account of the whole Panama project.

2 "War Christens Belgium's Albert Canal," *Engineering News-Record, 124,* p. 729 (1940).

3 H. R. Cedergren, "Poor Pavement Drainage Could Cost $15 Billion Yearly," *Engineering News-Record, 200,* p. 21 (June 8, 1976).

4 J. W. Starring, "Sand Fill Pumped 15 Miles for Interstate Construction," *Civil Engineering, 41,* p. 44 (February 1971).
5 J. P. Nicoletti and J. M. Keith, "External 'Shell' Stops Soil Movement and Saves Tunnel," *Civil Engineering, 39,* p. 72 (April 1969).
6 F. L. Peckover and J. W. G. Kerr, "Treatment and Maintenance of Rock Slopes on Transportation Routes," *Canadian Geotechnical Journal, 14,* p. 487 (1977).
7 "Wet Site Turned into Dry Home for Massive Tunnel," *Engineering News-Record, 187,* p. 22 (November 4, 1971).
8 G. Plantema, "Rotterdam's Rapid Transit Tunnel Built by Sunken Tube Method," *Civil Engineering, 35,* p. 34 (August 1965).

CHAPTER **12**

MARINE WORKS

Almost three-quarters of the surface of the globe is covered by the seas, surrounding all inhabited lands. Large percentages of people on all continents live within 100 km (61 m) of the sea. Residents of many islands frequently see the sea every day of their lives. It was along seacoasts that some of the first human settlements developed. The first long-distance travel was by sea. For many centuries, ocean and coastal travel was the only means of mass transport. The Mediterranean Sea, in particular, may be regarded in many ways as one of the cradles of human history. Along its shores some of the earliest harbors were constructed. Some have long since disappeared, while others provide vivid reminders that the construction of marine works is always a battle between humans and Nature, where human success has always required a singularly careful appreciation of the natural forces that have to be contended with and knowledge as accurate as possible of site conditions and local geology.

THE TIDE, WAVES, AND CURRENTS

Ocean dynamics are more important to the civil engineer than the specific properties of seawater. Never does one see the ocean at rest. In motion, as at the height of a great storm, it presents one of the most majestic of all natural phenomena. The wind is the chief factor in the formation of waves and currents, but the regular movement of the tides is perhaps the greater determinant in the design and construction of the marine works of the civil engineer. The civil engineer has to tame these "great forces of Nature" as they break upon and physically impact marine works founded on and supported by geologic formations. Local currents near harbors and in estuaries affect the siting and layout of marine works and must be accepted as basic conditions and integrated into design. Wave action, however, is of far more serious consequence. In general, waves are of two types: *swell waves* and *storm waves*. Both are generated by the action of the wind on the surface of the sea, storm waves more directly than swell waves. Of major importance is the maximum force that is to be expected from wave action at the location of any proposed works.

Those who have never seen the sea, particularly at the height of a great storm, find it difficult to imagine the force that the sea can exert. Even well-authenticated figures tend to seem unrealistic. As an indication of what waves can be and do, it may be noted that, although ocean waves do not normally exceed 7.5 m (25 ft) in height in midocean, there is a well-established record of a storm wave in the Pacific with a height of 33.6 m (112 ft). Atlantic storm waves over 15 m (50 ft) high are not unusual. Approaching a coastline, wave size will naturally decrease, but the potential force of breaking waves is remarkable. One of the stormiest parts of all the oceans is the Pentland Firth, separating the northern coast of Scotland from the Orkney Islands. Close to the eastern end lies the small harbor of Wick, famous in the annals of civil engineering. When its breakwater was destroyed in 1872, there was clear proof that blocks of masonry weighing up to 1,350 (long) tons had been moved intact by the sea. A new pier was constructed, but in 1877 it was also destroyed by a great storm; accurate observations showed that a block of masonry weighing 2,600 (long) tons had been carried away from its original location. Dunnet Head lighthouse at the western end of the Firth frequently loses windows atop its 90-m (300-ft) cliff when they are broken by storm-tossed stone. Tall (but true) tales of the force of the sea make it obvious that the design and construction of marine works are supreme examples of the art of the civil engineer.

Tidal movement is a natural phenomenon of an entirely different but equally majestic character, involving the simultaneous movement of all the water in the oceans. Under the combined attraction of the moon and the sun, but chiefly the moon, the oceans are pulled away from the earth in a dual cycle of about 24 hours and 50 minutes. The extra 50 minutes makes for a significant scheduling problem in marine construction operations. As the moon waxes and wanes each lunar month, so do the tides vary from neap (or low) tides twice a month, when the pull of the sun and the moon are opposed, to spring (or high) tides, also twice a month, when the two bodies are in line and so pulling together on the waters of the sea. While tidal movement is simply explained, it is surprising to find so great a variation in tidal ranges along the coasts of the world. Remarkable variations occur along the coastline of northeastern North America. At Nantucket Island, near Boston, the range is less than 30 cm (1 ft). As one goes north into Canada, the range steadily increases until, in the Bay of Fundy between New Brunswick and Nova Scotia, some of the highest tidal ranges of the world are experienced, spring tides reach a height of 16.2 m (54 ft) at Minas Basin, only 640 km (400 mi) from Nantucket. Tides even higher than the famous Fundy tides have been observed in the Canadian Arctic. Other strange variations are that many coastlines have almost negligible tides, that some locations have only one tide each day, and that other locations have a tidal period of an even 12 hours, explained only after meticulous oceanographic study. Accurate knowledge of local tidal conditions is obtained directly from published tide tables for locations listed in such tables and by interpolation from them for other points, always verified by readings from a local tide gauge.

Civil engineering works on large lakes are properly called "marine works," even though they are in freshwater. There are no tides in the ordinary sense on such lakes, but certain large, shallow lakes experience *seiche* movements of water that may be more troublesome than the tide. Rapid changes in water level may be caused by changes in barometric pressure or, more commonly, by changes in the wind. The free water surface is subject to wind shear; the transfer of energy results in water displacement, the magnitude of which depends on the depth of the lake, the wind speed, and the fetch over which it blows. Lake Erie, the shallowest of the Great Lakes, is renowned for its seiches. Differences of 4 m (13.5 ft) have been observed between the levels of the two ends of the lake at Buffalo and Toledo, with wind "setups" (as they have come to be known) of as much as 2.5 m (8.4 ft) in the harbor of Buffalo. The effects upon harbor structures can well be imagined. One indirect result is the development of currents at intermediate points along the lake; the mouth of

Conneaut harbor experiences additional current velocities as high as 30 cm (1 ft) per second in response to a major seiche.

THE EARTH BENEATH THE SEA

Some features of ocean bed geology are of importance to civil engineers in the prosecution of their marine works. There does exist "solid ground" (for want of a better term) beneath the surface of the sea, exposed so starkly up the wide stretches of low-tide beach in the Bay of Fundy. The crust of the earth is continuous over the entire globe, even though large parts of it are shielded from view by the waters of the sea.

The first deep-sea sounding, to a depth of 4,430 m (2,425 fathoms), was obtained by Sir James Clark Ross in 1840 during one of the early voyages of exploration to the Antarctic. He used a weight on the end of a manila line, and this crude method had to suffice until 1870 when Lord Kelvin used piano wire for the same purpose and so extended appreciably the depths that could be plumbed. Our real knowledge of bathymetry has come mainly from geophysical echo-sounding devices.

Geologic explanation of the formation of the continental shelf, in terms of its composition and its surface characteristics, is a matter of real importance for the engineer of marine works. Seabed sediments are predominantly sand, some mud, and a small percentage of gravel, whereas the percentage of mud is much increased in the deposits on the continental slope. Immense mountain ranges, correspondingly great "deeps," and gigantic submarine canyons, the dimensions of which surpass any to be seen on dry land, have now been located. The irregularity of the seabed is well recognized, but perhaps the most unusual underwater physiographic feature of all is the wide extent of the continental shelf.

This remarkable area has been defined officially as the "zone around the continents, extending from low-water line to the depth at which there is a marked increase of slope to greater depth"; this outer slope is called the *continental slope*. The average depth to the shelf is 130 m (72 fathoms), the greatest depth about 365 m (200 fathoms). The widest parts of the shelf are those bordering glaciated lands; the Arctic coasts have the broadest shelf of all, the average width being about 67 km (42 mi), the average slope about 1.9 m to the kilometer (10 ft to the mile). Vividly demonstrated on all large-scale ocean charts, the continental shelf is clearly of great practical significance to fishermen, to mariners, to those concerned with cable laying, and in relatively recent years, to those in search of petroleum. Civil engineering structures for both oil drilling and defense purposes now stand upon the shelf at selected locations off the American coast. Some North American lightships have been replaced by tower-type structures founded on the shelf.

This submarine shelf is directly related to Pleistocene geological phenomena. Sea level has risen from about minus 78 m (260 ft) to possibly 6 m (20 ft) below the present level between 17,000 and 6,000 years ago. Changes in sea level in the last 6,000 years are not so well defined, but it is in the character of a rise, about 10 cm (4 in) along the Atlantic coast of the United States between 1930 and 1948. Many examples of "raised beaches" in all parts of the world are vivid evidence of a much higher sea level in the past. The reverse process appears closely related to the existence of the continental shelf that is today such an important feature of submarine geology. Undersea core boring, although fraught with many mechanical difficulties, has steadily progressed since World War II, and continuous cores can now be obtained from great depths. Study of bottom sediments is opening up new phases of geological study.

Visual inspections of geologic bottom conditions at shallow depths are of great practical importance. The standard diving suit provides access to depths of about 15 m (50 ft). The value to civil engineers of personal inspection of the geology of the seabed at the site of the works for which they are responsible is inestimable. On one job, the dredging contractor was being hampered by the presence of "boulders as big as cottages." It required only a short survey in a diving suit by one of the authors to see that the expression "cottages" was a poetic exaggeration; the correctness of the dredging was similarly demonstrated by this personal inspection. Skin diving also presents great possibilities for use in connection with civil marine work.

TYPICAL COASTAL PROBLEMS

Wide is the range of civil engineering problems that come within the general subject of marine works. The following summary gives some of the problems experienced on one short length of seacoast. The British Isles shoreline is never more than 112 km (70 mi) from any inland point. The best starting point for a quick survey of this famous coastline is the third (and final) report of the Royal Commission on Coast Erosion. Although published in 1911, this masterly document is still of great utility. One of its many historical lessons is that, without the most careful planning and design, engineering works on one stretch of coast may cause serious trouble on adjacent sections.

Morecambe Bay is prominent on the coastline of northwestern England; at low tide much of the area of the bay is dry sand. About 405 ha (1,000 acres) were reclaimed from the bay when the Furness Railway was built along its northern shore, embankments across the many little estuaries serving a double purpose. The sand of the bay is typical of coastal deposits in this part of England. Sand dunes around the estuary of the river Ribble have had to be stabilized by vegetation. Some of the dune grasses growing as natural protection of these dunes were not plants native to Great Britain. It was found that they had grown from seeds swept out of ships in the neighboring port of Liverpool after cargoes of grain feed for poultry had been brought in from the United States during World War I. Westward, along the coast of north Wales, lies the dubiously sited (but necessary) main road and main railway line in close juxtaposition to the exposed rock shoreline. Then there is the striking coastal

FIGURE 12.1
Aerial view of Dungeness on the south coast of England, showing the trend of the shingle ridges. (*Courtesy Committee for aerial photography, Cambridge University; photo by J. K. St. Joseph; Crown Copyright, British Air Ministry.*)

scenery of the island of Anglesea, which leaves no doubt that this engineering location was dictated by geology, even though it has required ceaseless maintenance work along most of the coast. In still another example, Harlech Castle, on the west coast of Wales, built right on the coast in 1286 by King Edward I, now lies more than 800 m (0.5 mi) from the sea.

The Bristol Channel features some fine man-made ports. The small port of Saundersfoot, now disused, has the remains of a fine shingle beach, the result of dumping ballast from across the sea when the port was in use. During the 1928 excavation of Port Talbot, an upright stone bearing the date 1626 was found 6 m (20 ft) below the surface, clear evidence of sand movement along this coast. Even more remarkable is the history of the church at Penard. Built about 1270, the church and neighboring castle were inundated by advancing sand dunes in 1528 and remained covered until rediscovered in 1861. Major landslides along the south coast of England, such as at Lyme Regis on Christmas night, 1839 (precipitating 8 million tonnes of rock into the sea), are geologic constraints to reckon with. Chesil Beach, one of the most remarkable beaches of the world, leads to the Isle of Portland, for centuries the source of world-famous building stone. The beach is 29 km (18 mi) long, shingle composed and consistently graded from pea size at its northwestern end to coarse gravel (up to 7.5 cm or 3 in in diameter) at the Portland end. Dungeness, farther to the east, is the largest shingle foreland in Great Britain. Figure 12.1 shows clearly the stages in its growth, probably since Neolithic times.

Ninety-eight percent of the shingle here is flint (from the neighboring chalk). Its seaward growth has averaged about 4.8 m (16 ft) a year throughout the last four centuries, at least.

Shoreline erosion along the east coast of England was a recognized problem as early as 1391, when a jetty had to be built at Cromer to protect fishing boats; the work was not successful. The twin ports of Yarmouth and Lowestoft give good evidence of the effects of engineering works on neighboring coasts. The great expanse of the Wash is a "museum" of efforts to control coastal erosion, remains of Roman works having been found as reminders of a long engineering tradition. Immediately north of the Humber between Spurn Head and Flamborough Head, there is a world-class example of an eroding coastline. The lodgement till boulder clay cliffs reach 30 m (100 ft) in height, yet the entire stretch of coast south of the resort town of Bridlington has been eroding steadily since Roman times at almost 1.8 m (6 ft) per year. A strip of land 4 km (2 mi) wide along this entire coast is the average loss to the sea since the Romans left England.

The Lost Towns of the Yorkshire Coast is the melancholy title of a volume that presents a detailed review of this extreme case of coastal erosion. Displaced erosion debris accumulates at Spurn Head, where it provides the inverse problem of steady extension of this headland at a rate of about 12 m (40 ft) per year; here necessitating many relocations of the entrance lighthouse to the Humber. Coastal instability here prompted late nineteenth-century attempts to construct a new dock at Sutton Bridge to serve, in part, the Great Northern Railway. The dock was opened with due ceremony on May 14, 1881, but it was discovered on the very next day that leakage was taking place from the entrance lock. A thorough inspection of the site was made, but the consensus of expert advice was that the site was so unsuitable that reconstruction was impracticable. It is said that the only ship which entered the dock on its one day of operation brought a load of timber from Norway and sailed the next day with a cargo of coal. The entire project was abandoned. It is not unreasonable to imagine that consultation with a Roman engineer who worked at this location many centuries before might have forestalled one of the few really complete failures in the history of modern civil engineering.

DOCKS AND HARBORS

Dock and harbor engineering dates from the very earliest of all civil engineering works. Because of the inevitable interference with natural processes caused by dock construction, many ancient harbors, although initially successful, eventually proved unsuccessful for reasons that can only be classed as geological. One of the most famous of all ancient ports was that at Tyre, often mentioned in the Bible, and second port of the Phoenician empire; it was founded on an island and included two harbors protected by rock-fill breakwaters. In 332 B.C., Alexander the Great destroyed the city after building a causeway access to the island. The causeway trapped the local coastal sand, and the channel soon filled up; the site of Tyre is now on a peninsula. The great Roman harbor of Ostia, having marine structures which even today command respect, was finally silted up by sediment brought down by the Tiber, despite the ingenuity of the Roman engineers in designing countermeasures. Ostia's site is now 2.4 km (1.5 mi) from the sea. Throughout the intervening 2,000 years, engineers have given ceaseless effort to overcoming similar difficulties, but today rely heavily on dredging equipment and control devices to deal with the great volume of sediment carried to sea by rivers or moved along the coast by the sea.

Modern dock and harbor structures have included some of the most notable civil engineering works of recent years. The extensive dock system of Liverpool, England; the steady development of the Port of New York; and the many and varied engineering works in the gulf and tidal portion of the River St. Lawrence—these and many similar developments testify to the magnitude of dock and harbor works. Geological features affect all construction of this kind to some degree, although usually in one or more of the ways described in the other chapters of this book. A notable underwater geophysical survey was carried out at Algiers harbor, to obtain information on the position of bedrock underlying superficial deposits. Construction of the rock-fill breakwater at the new harbor of Haifa, Israel, required a large quarry location—a striking reminder of the similar construction methods adopted at the adjacent harbors of Tyre and Alexandria over 2,000 years before.

Ports on both sides of the English Channel cater to the busiest international exchanges of traffic to be found anywhere. The Southern Railway Company of England decided to proceed with a train-ferry scheme connecting a watertight dock at Dover with the French side of the channel. Seabed geology at the dock site consists of the Lower Chalk Measures. Geologic mapping, borings, and the nearby portals of the proposed Channel tunnel, which had been practically dry for 50 years, all suggested that the chalk would be of a solid and homogeneous nature. It was therefore proposed that dock construction should be carried out in the dry within a cofferdam constructed of steel piling driven into the chalk. This did not prove immediately possible, as chalk hardness limited the penetration of the piles, but a slight modification of the design permitted the work to go ahead. When pumping of the enclosed area was started, it was "found that, although a head of from 10 to 20 feet could be sustained, the difference in pressure...caused an inflow of water through the seabed in the immediate vicinity of the works greater than the pumps could discharge." Usual methods of sealing were tried, but failed. Small bags filled with permanganate of potash were placed by divers in fissures in the seabed outside the cofferdam, and the color showed up all over the enclosed area. Further consideration showed that increased pumping might enlarge the fissures and make matters worse. Eventually, large quantities of underwater concrete were placed as a seal.

An interesting explanation of the unusual chalk characteristics held that the chalk encountered consisted of ancient rockfall debris from adjacent chalk cliffs, consolidated by the upright long-term wells at the littoral drift of gravel for which the adjacent coast is noted. Another explanation, based on wider experience with the surface of the Chalk of England when exposed in excavation, is that near-surface shattering of the rock was the result of fossil permafrost conditions which probably existed when much of Great Britain was under Pleistocene glacial ice.

Civil engineering works on major lakes of the world, particularly on the Great Lakes, are marine works in every sense except that the water they are built in is fresh and not salt. Almost at the other extreme, from the geological point of view, was the building of a port on the Great Lakes to serve a new taconite (low-grade iron ore) mining development. Taconite Harbor is located on the northern shore of Lake Superior, 128 km (80 mi) northeast of Duluth and 120 km (75 mi) due east of the taconite plant at Hoyt Lake. Two natural islands were conveniently present at this point, about 450 m (1,500 ft) offshore. When the islands were connected and extended, it was possible to construct a breakwater large enough to protect loading facilities. Local bedrock is a fairly sound vesicular basalt, laid down in successive surface flows, with a strike parallel to the shoreline and a dip of about 25° toward the lake. This favorable geological orientation was incorporated into design and rock was excavated to shape the finished wharf. Excavated rock was covered with a relatively thin wall of reinforced concrete after anchoring the rock with long steel dowels against possible movement on the dip toward the lake. Figure 12.2 shows the finished cross section; the rock excavation was carried out with the area enclosed by a cellular cofferdam. Almost 750,000 m^3 (1 million yd^3) of rock was removed, excavation starting with line drilling at 30 cm (1 ft) centers. Some evidence of open lava flow surfaces was noticed as rock was removed from the vertical face. Some of these were water-bearing, and some contained loose fragments. Extensive grouting was carried out both before and after excavation, and all anchor rods, designed on geologic details, were grouted in place.[1]

Only infrequently can local rock conditions be adapted for harbor design so effectively as at Taconite Harbor, but harbors seldom have to be developed from scratch, so to speak, without modifying natural features to form the desired harbor. Most of the major ports of the world have developed from initial use of a natural harbor. Some of the most famous—Sydney, Australia, and St. John's, Newfoundland, the shelter provided by San Francisco Bay, and the dock facilities of London and New York—are supreme examples. These and many other harbors can quite aptly be described as "shelters provided by Nature." This old expression is really another way of saying "shelter provided by local geological conditions." Most splendid harbors have been formed by the rise of the sea, forming such natural sea-filled depressions. A study of the geological factors responsible for major American harbors was published by the U.S. Geological Survey as early as 1893. There is probably only one harbor in the world that engineers formed by imitating Nature in this way, i.e., by flooding a natural depression that could thus be connected to the sea. This unique example is in Italy, on the north coast of the little island of Ischia, lying immediately to the west of Naples. Port d'Ishia until the 1850s was a

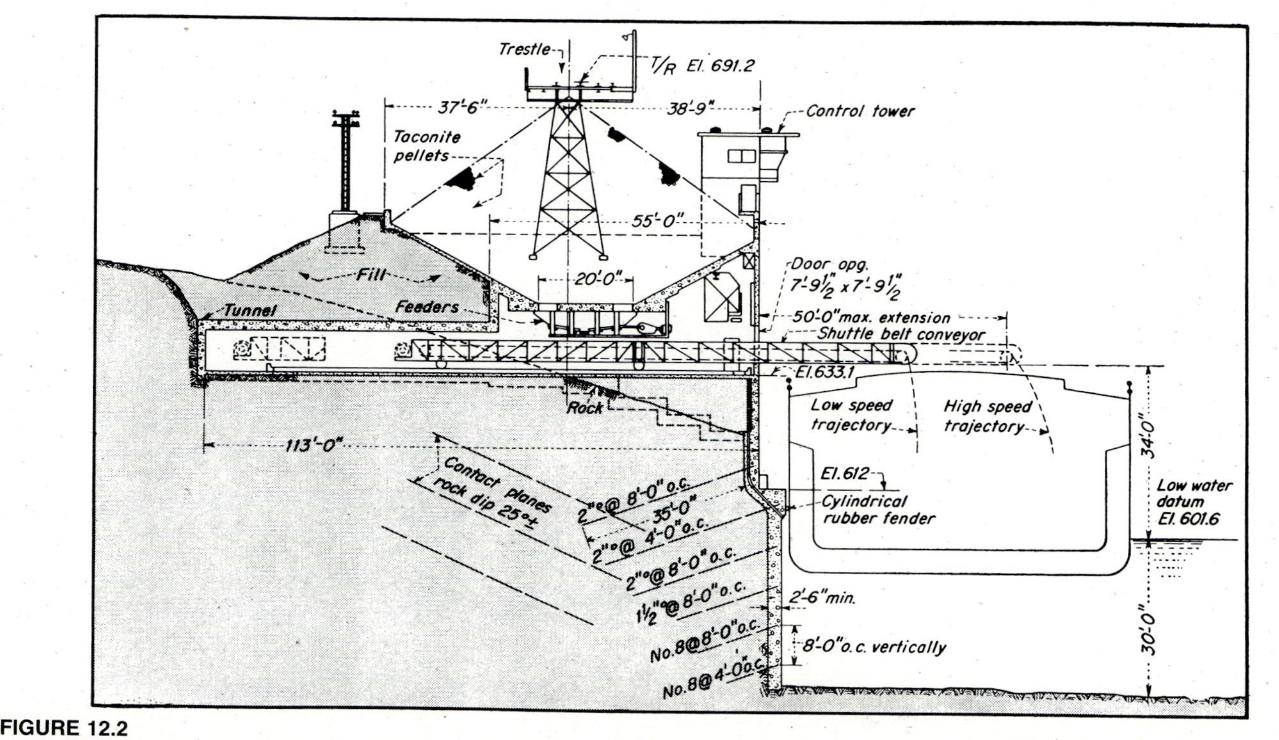

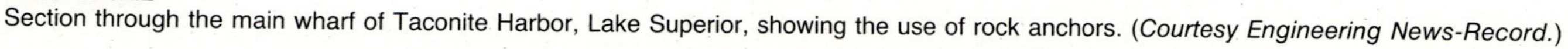

FIGURE 12.2
Section through the main wharf of Taconite Harbor, Lake Superior, showing the use of rock anchors. (*Courtesy Engineering News-Record.*)

landlocked volcanic crater lake, separated from the sea by a narrow neck of land. Italian engineers spent two years excavating an artificial opening through this neck. In this case, relatively little work converted a natural feature into a perfect harbor, but usually major civil engineering work is necessary to form a new harbor. Possibly the greatest challenge to the civil engineer engaged in dock and harbor engineering comes when natural conditions do not provide all the shelter needed and protective works, such as breakwaters, have to be constructed, almost in defiance of the elements.

BREAKWATERS

From the very earliest times attempts have been made to develop or improve harbors of refuge by constructing protective breakwaters at locations of inadequate natural shelter for shipping. One of the earliest examples of harbor dangers in this branch of civil engineering is that of the Gulf of Ephesus. Strabo recorded this history, going back for many centuries B.C. Under King Attalus Philadelphius of Pergamum, "engineers" constructed a breakwater across the end of the gulf in an attempt to provide a protected harbor, but the breakwater interfered with the natural channel so greatly that the main objective of the work was defeated by accessory siltation. Many years later, Roman engineers attempted to rectify the mistake, but it was too late and the natural harbor was infilled and ruined. Today, it is difficult to imagine that the sea once extended 24 km (15 mi) into this Greek gulf.

Similar examples punctuate the intervening 3,000 years, but all would demonstrate that breakwater construction demands great care in investigation of all possible interference with natural geological conditions *before* construction is started. As recently as 1939, a small breakwater was constructed at Redondo Beach, California, only to reveal itself soon as a serious impediment to natural littoral drift. Expensive remedial measures have been necessary since that time. This is one of many examples that give solid support to the suggestion of J. B. Schijf, a leading Netherlands coastal engineer, that in all coastal engineering "[you should] be sure to put off to tomorrow what you do not absolutely have to do today." This a cryptic way of pointing out that design and construction of breakwaters *must* follow unusually careful preliminary study of all phases of local geology, static and dynamic, before design is finalized and construction begins. Not only must the seabed be characterized for stability, bearing capacity, and ease of removal if dredging is involved, but the adjacent stretches of coast must be quantified in terms of local currents, littoral drifts, and any features that may in any way be affected by the proposed structure. This is a procedure that can take years, even for a small harbor such as that at Forestville in the Gulf of St. Lawrence. The investigation was for an initial harbor development on a small scale (now greatly enlarged), yet the preliminary field studies required more than two years. Methods of investigation are only slightly modified from those treated generally in Chapter 3; the added difficulty of working in tidal waters is often the main variation.

Seabed character at the site of a proposed breakwater does not ordinarily present unusual problems; the action of the sea itself generally results in satisfactory bed conditions, at least in relatively shallow water, although these conditions must always be carefully investigated. The main problem of design is to ensure structural stability against the anticipated action of wave impact. Modern marine hydraulics utilizes physical and digital model studies as a guide to design, and the abundance of meteorological records now on hand for most coastal regions is today more determinate than it was in the early years of modern civil engineering practice. Many traditional empirical design rules still are useful, however.

Breakwaters are most economically constructed of large blocks of suitable rock. Frequently, such quarrying operations are large in scale, calling for careful preliminary geological investigation. Where no suitable rock is economically available, expensive structures of massive concrete may have to be fabricated, leading to critical structural design requirements. Alternatively, artificial concrete "rocks" may be used, such as *dolos* or "tetrapods," developed by French engineers for use in harbors of North Africa, but now employed at such widely spaced locations as Kahului Harbor on the Hawaiian island of Maui, at Crescent City, California, and at another Pacific coast location shortly to be mentioned.

Still another alternative for locations where rocks of only limited size can be obtained is to construct the breakwater to the usual mound form with grout pipes embedded in it. Through these, special cement grout can be forced as soon as the rock is stabilized, thus reversing the usual procedure of fragmenting large masses of rock into smaller pieces for handling by recementing small pieces into a solid mass of artificial conglomerate large enough to withstand all anticipated seas. This procedure was successfully followed at the small Quebec harbor of Forestville already mentioned, and at the adjacent port of Baie Comeau.

Geological processes have to be kept ceaselessly under observation throughout construction and long afterward in all breakwater work. One of the largest and most important "artificial" harbors formed by large breakwaters is that at Valparaiso in Chile, South America. The base for one of the main breakwater extensions here was formed by depositing dredge spoil sand through as much as 48 m (160 ft) onto a seabed of black clayey silt. Deformation of the entire seabed in the vicinity of the work caused slides of over 50 percent of the material deposited. It was found, during deposition of the sand, that everything of a light character—"shells, mud and things of that sort"—was washed out of it; however, sand that reached the bottom was so dense that an anchor fluke would not enter it. This provides an interesting example of the principles of sedimentation in operation on a major scale.

COASTAL EROSION

Most engineers know of stretches of coast on oceans, lakes, or inland seas that are being eroded at a noticeable rate. Although sandspits and bars give some evidence of corresponding accretion, the balance always appears to be against the land in this constant battle with the sea. Rockbound coasts and cliffs are similarly affected, although their increased resistance as compared with that of unconsolidated beach deposits usually renders erosion on rocky shorelines of small immediate consequence. This relative unimportance is emphasized by the fact that coastlines distinguished by continuous rock outcrops are not so favorably placed for development, either for pleasure or for commerce, as are coastlines formed of unconsolidated material. Although erosion is similar for all types of shore and although the erosion of rock cliffs often forms unusual geological features, attention here will be devoted more particularly to the erosion of low-lying lands.

Despite the fact that coastline erosion is part of the natural geological cycle, shoreline protection has become a worldwide problem of economic importance. Steady increases in land values, development of large industrial estates, the desire to own homes near the water—these and other factors ensure that this problem will rise to steadily increasing prominence each year. Great Britain is committed to substantial coastal-protection works such as are found on many North American coasts, while the concern of the United States is indicated by the formation of the Beach Erosion Board as a special agency of the U.S. Army Corps of Engineers. In Japan, with its 27,000-km (16,214-mi) coastline for a land area of only 370,000 km^2 (142,338 mi^2), the problem is equally serious. From Australia, South America, India, and many other lands come reports of engineering studies of how to protect valuable coastal lands from the erosive force of the waters of the sea or of great lakes.

Even the crudest estimate of the total cost of coastal-protection works would be so astronomical as to defy belief. Typical figures are a round average of about $1,000,000 per lineal km; $26 million for works on the coast of Long Island, and $1.5 million to protect merely about 600 m (2,000 ft) of the tracks of the Pere Marquette Railroad along the shore of Lake Michigan. Such figures could be duplicated hundreds of times over and would still fail to indicate the full magnitude of this branch of civil engineering work which depends so completely upon a full appreciation of the geological conditions of the coast that has to be protected.

It may be surprising that coastal erosion is also a serious problem on lakeshores. The natural forces that have to be countered are just the same as those along the seacoast. In the case of the Great Lakes, storm damage can equal that on many of the seacoasts of the world. Intensive industrial and residential development around the shores of lakes Ontario, Erie, Huron, and Michigan has made coastal erosion as serious today as anywhere in the world. Intensive studies in progress by U.S. and Canadian agencies are designed to limit the damage to the billions of dollars' worth of property fronting on these lakes. It has been estimated that the loss of land and property on the Lake Erie shore in Ohio alone, during a period of merely 20 years prior to the end of World War II, amounted to almost $9 million. Study of long-term records

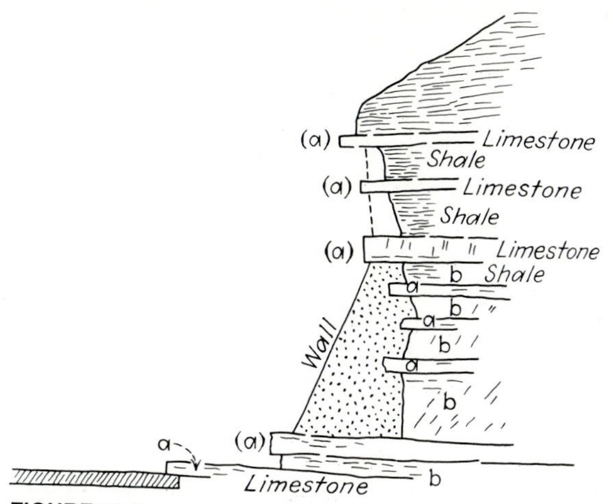

FIGURE 12.3
Simplified geological section of the sea cliff at Lyme Regis on the south coast of England, showing protection wall to halt erosion. (*Courtesy John Murray, London; from Sixty Three years of Engineering by Francis Fox.*)

shows a clear correlation between rapid shore erosion and years of high lake-water levels. These, in turn, are related to cyclical variations in precipitation.

All that is involved in this major branch of remedial civil engineering work is one of the simplest of geological processes—water crashing against geological materials—but this process is exercised on a grand scale. The erosive action of pounding waves and swirling water is a combination of the direct hydrodynamic forces exerted by waves and the erosive effect of eddies. When the nature of a beach is such that waves carry with them sand or gravel in suspension, the result is intensified erosion. Wave action is only one factor in erosion; the determining factor is the geological character of the exposed coastline. On rock-bound coasts characterized by regular strata of appreciable thickness, erosion will generally be uniform. When strata of varying types having differing resistances to abrasion and dissolution are exposed, differential erosion will result, often with somewhat fantastic effects such as the formation of pinnacles and natural arches. When erosion threatens development right up to the cliff edge, the civil engineer's skill and ingenuity are taxed. The cause of erosion will usually be obvious after even a preliminary survey of coastline geology.

A section similar to that shown in Figure 12.3 might be found in many places. The section is through a small cliff close to the famous old church of Lyme Regis in Dorset, England. The alternating layers of softer shale were easily eroded by the sea waves, and the alternate limestone beds were left overhanging. When the overhang became too great, a fracture of the limestone bed resulted, and renewed erosion of the shale took place. Other forces at work in completing the disintegration were the action of heavy seas, which lifted the projecting layers of limestone and so loosened them from their beds, and mass wastage of the upper layers of unconsolidated materials due to seepage of groundwater along the face of the receding cliff. The remedy successfully adopted was the construction of a continuous wall protecting the underlying layers of shale from further erosion, thus stabilizing the entire cliff face.

Shores consisting of unconsolidated material can rarely be regarded as stable; even prevailing winds sometimes affect their stability. Any development adjacent to the high-water line on such beaches should consider general beach conditions. Protection walls will then be a frequent recourse. Foundation conditions must be given unusual attention to resist scour, and allowance must be made for every change in beach conditions that can be visualized. Cutoff walls of steel sheet piling below reinforced-concrete superstructures provide a convenient and effective method of construction ensuring scour protection, provided the wall structure does not encroach far below high-water mark. Bulkhead walls constructed within the tidal range on the beach or below the low-water line must always consider possible consequences of construction.

Geologic aspects of beach formation are largely empirical, since local conditions are of unusual significance. In general, there are two main types of coastlines: the *shoreline of emergence* and the *shoreline of submergence*. The first is the result of a gradual uplift of the seabed; the floor of the sea emerges to form a flat coastal plain consisting of unconsolidated deposits. Wave action then erodes the edge of the emergent landmass, carrying this away from the waterline to form an offshore barrier beach which will gradually work nearer to the original shoreline until ultimately the two will merge.

Shorelines of submergence are the result of a gradual depression of a coastline; the original coastal area is submerged beneath the sea, which will thus stretch far inland and be in contact with an eroded land surface. The initial characteristic is extreme irregularity of outline; features such as fjords and deep bays are indicative of submerged valleys. Erosion by the sea will continue, and its first effect will be a tendency to straighten the shoreline, a process in which harder strata offer greater resistance and consequently give rise to special features such as stacks, arches, and small offshore islands. The products of erosion will be rapidly washed into deep water in initial stages of development, but eventually a small beach will form at the foot of the cliffs. Where littoral currents exist, they will sweep erosion debris along the coast past headlands to be deposited in the deeper water of bays. With continued action, these strands of debris reach water level and become spits or hooks; in some cases they almost completely block a whole bay. Eventually, shorelines lose their irregularities, acquire a beach formation and tend toward equilibrium identical with that of a shoreline of emergence, with regularity.

Currents and waves carry the finer material into deeper water while the coarser material will form the main beach deposit. Beach composition varies considerably from place to place, but gravels are generally found within 1.6 km

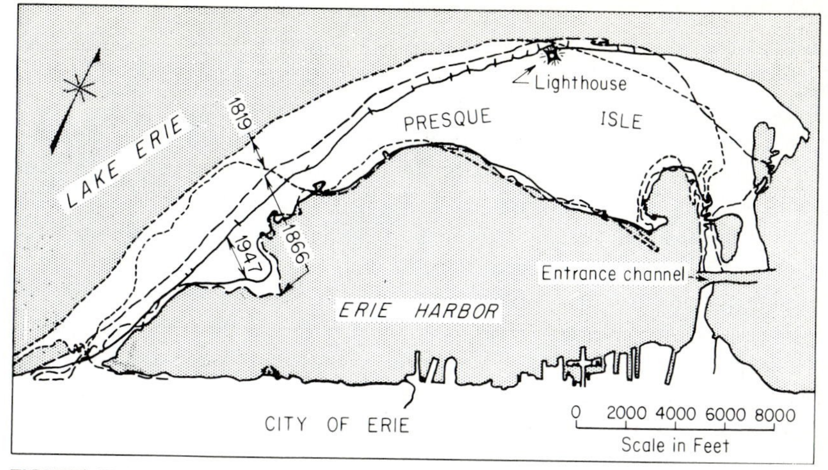

FIGURE 12.4
The harbor of Erie, Pennsylvania, in Lake Erie, showing a "classical" example of a compound recurved spit, formed by littoral drift. (*Courtesy Engineering News-Record.*)

(1 mi) of shore in depths of not more than 9 m (30 ft) of water; sand rarely extends beyond a depth of 90 m (300 ft). Particle size of sand grains or gravel is directly related to the slope of a gradient beach.

Two of the most notable Great Lakes coastal features are the compound recurved spits that form the harbors of Toronto, Ontario, and of Erie, Pennsylvania; the latter is shown in outline form in Figure 12.4. These two major projections from the normal lakeshore have been caused by the interaction of alongshore movement (but in opposite directions in the two cases) and the outflow from a river, deflecting this material away from its mouth. Nature has thus provided two of the finest harbors on the Great Lakes, although much work is required to maintain them. At Toronto, the eastern entrance was created by an early twentieth-century storm. Once formed, it was kept open by dredging. Knowledge of the drift coming from the Scarborough Cliffs to the east was put to good effect by a contractor for one of the great coal docks, which required a large quantity of sand fill. The low bidder for the work anticipated dredging the sand from a continually replenished point just outside the east entrance, necessitating little if any moving of his dredge. His geological knowledge paid off; once his dredge had made its initial cut, it did not have to be moved again.

Erie Harbor was the first such project undertaken by the U.S. Army Corps of Engineers following the original act of March 3, 1823, authorizing its improvement. There is a continuous record of the civil engineering works that have been necessary to keep the narrow connecting neck of land unbroken as a convenient access to the great park that has been developed on the spit proper. Surveys indicated that natural replenishment of beach erosion was insufficiently protected by the groins and other protective works (at an expenditure of several million dollars). In 1955 tenders were received for dredging of over 3 million m^3 (4 million yd^3) of sand to form an artificial beach as a final erosion barrier.

When shoreline facilities are damaged by natural hazards, geologic implications of reconstruction may be appreciable. Hurricane destruction of more than one-half of U.S. Highway 90 between Pass Christian and Biloxi, Mississippi, in 1915, led to the then-astronomical damage of $13 million. After the highway had been replaced, the local county authority constructed a seawall 38 km (24 mi) long adjacent to the road at an additional cost of $3.4 million. The beach disappeared completely in a very short time; fill behind the wall escaped, and the highway was again threatened. The Corps of Engineers was asked to help in the restoration work, but before it could provide its expert assistance, another hurricane occurred and did more damage. Repair work was eventually carried out at about one-third of the original total cost of the wall. The beach was replenished by pumping. It is estimated that 25,000 m^3 (32,500 yd^3) is lost from the beach every year, but in view of the large amount of pumped fill placed in 1951, routine maintenance has been sufficient to keep it in good condition.

LITTORAL DRIFT

One of the most substantial of coastal processes is the extensive movement of sand in a regular and steady manner along many stretches of coasts. This is generally known as littoral drift. Somewhere along the path of drift, accretion of erosion products will be obvious, especially at river mouths, where drift and estuarine conditions interact. *Barachois* (a French word meaning sand and gravel bars) become a prominent feature, especially along parts of the Atlantic coast of Canada. Rockaway Point, Long Island, a sandspit near New York, advanced at an average rate of more than 60 m (200 ft) per year from 1835 to 1927. When the progression of the point was halted by a rock-mound jetty, 230,000 m^3 (300,000 yd^3) of sand was found to have accumulated behind it in the course of little more than a year.

Where artificial structures project into the sea below mean tide level, littoral drift may become a difficult problem when sand or other drift accumulates on one side of the structure and a corresponding, although not necessarily equal, erosion of material occurs on the other side. If the accumulation of drift is of any magnitude, nearby harbor entrances may become obstructed. On the other hand, trapped drifting sand may actually be used for nearby bench replenishment to build up a good beach section. The coasts and major harbors of India have provided some of the most valuable information yet available on managing littoral drift. At Madras on the east coast of India, approximately 1 million tonnes of sand pass any given spot on the coast each year.

Much accumulated experience was applied to harbor design at Vizagapatam between Madras and Calcutta on the east coast of India in the late 1920s. Initial studies suggested that no drift trouble would be encountered, but the start of dredging operations at the harbor entrance soon demonstrated conditions similar to those at Madras. Eventually, it was found that (1) the volume of sand traveling along the coast amounted to 1 million tonnes per year; (2) the travel took place largely in shallow water, and was generally confined to a zone 180 m (600 ft) wide; and (3) the travel was primarily the result of wave action, although sometimes it was accentuated by shore currents. Eventually, a scale model of varying degree was used to define wave action on the coastal section and to indicate the final solution, a breakwater of two scuttled ships loaded with rubble. Drift so deflected falls into the dredged sand trap on the inner side of the entrance channel; from here the sand is excavated by a dredge which discharges the sand onto the other side of the entrance channel, whence it is moved away by drift action.

An appropriate quotation of 1912, dealing with the Madras case, is worthy of repeated consideration by all concerned with coastal engineering work.

> The chief lesson to be learned from a study of sand-travel at Madras appears to be that it is absolutely necessary, when an engineer is called on to advise about a work situation, on a coast where there is any suspicion of traveling sand, mud, or shingle, that he should be allowed adequate time to make observations, conducted with due precautions, of the directions and causes of travel; and that if he should arrive at the conclusion that such travel is likely to affect the proposed works in the near or distant future, he ought at least to acquaint his employers with what he conceives to be the broad general facts of the case and their probable financial effects, whether on the proposed works or on other works or interests.[2]

Hydrodynamic aspects of littoral drift are well understood. Wind had been shown to be a dominant influence. This is well illustrated on stretches of coast where the direction of drift is found to change, generally at a well-defined location. On the north shore of Lake Ontario (in the vicinity of Whitby), this location can be linked directly with the direction of the prevailing wind on the lake. Along the west shore of Lake Michigan, the change appears to occur in the vicinity of 55th Street, Chicago. The source of drift material is always a geologic question. Where exposed cliffs are obviously being eroded in the vicinity of the area of accretion and in the direction of drift, the source is obvious. Not so obvious, but probably more important, is the sediment brought down by rivers and streams discharged nearby. For example, stream and river discharge into the Pacific Ocean between Point Conception and the Ventura River amounts to almost 1.5 million m^3 (2 million yd^3) of sediment annually, and much of this moves steadily along the coast. If the supply from such sources—cliffs or streams—experiences interference by a misplaced engineering structure built on the beach, erosion is quick to develop.

Progressive erosion of many stretches of sand beach can be protected with groins. Groins are wall structures generally built at right angles to the shoreline and extending below the low-water line to interfere with drift travel and trap sand until a stable beach is formed. The shores of many parts of the British Isles have been so protected, as has the low-lying eastern seaboard of the United States. Erosion of the New Jersey coast has been particularly severe: about 60 cm (2 ft) per year (45.9 m or 153 ft between 1840 and 1920); and figures for Florida are similar. U.S. Army Beach Erosion Board recommendations note that groins need not extend higher than the level of uprush at high waters, construction should be at right angles to the coastline, and a spacing of one to four times the length from the bulkhead or shoreline was tentatively suggested, longer spacing being successful for large volumes of drift, and vice versa.

Older groins have typically solid walls, providing a complete barrier to material coming into contact with them. Modern practice tends toward groins with "windows," short sections of wall constructed to a lesser height, with the intention of allowing some of the drift to pass the groin while retaining enough to maintain the beach. Permeable groins are another method of achieving the same objective. A notable installation of this type of groin is at Sheridan Park, Wisconsin, on the west shore of Lake Michigan, where erosion at 60 cm (2 ft) per year was arrested in 1933.

Solutions to each individual coastal problem will be found only after evaluation of exhaustive field studies at the site and by comparing that situation with the experience of similar problems. No two coastal-protection problems will be exactly similar; local variations will spring from site geologic conditions. The impressive publications of the Beach Erosion Board should be constantly consulted. British engineers have likewise documented experience of well over a century's works along a coastline of the most varied geologic conditions. Figure 12.5 shows a typical example of British coastal-protection work.

MAINTENANCE OF TIDAL ESTUARIES

The formation and persistence of bars at the entrance to many tidal river estuaries has been fought for centuries; the Abbot Castelli included the subject in his studies in the years around 1660, which were followed by other early Italian and French engineers. Two general classes of these bars exist: (1) those less commonly of hard material not affected by the scouring action of normal currents passing over them, and (2) those more commonly of sand, shingle, or other unconsolidated material which, while retaining their general features, are constantly shaped by winds, waves, and varying currents. In addition, temporary bars of sand or shingle, such as the famous Chesil Bank, are occasionally heaped up by storm waves and afterward displaced by currents.

An example of the first type of bar existed at the entrance to Aberdeen Harbor, Scotland. It originally consisted of lodgement till and when removed by dredging was reformed from beach sand from the neighboring shore. Continued dredging finally produced a permanent channel by the eventual removal of the sand source. This left exposed only the underlying till, at a depth not susceptible to further erosion by tidal currents.

Common examples of the second type of bar appear to form primarily by interaction of tidal and other shore currents, and secondarily through local

FIGURE 12.5
Coastal protection works at Sherringham, Norfolk, England, showing clearly the action of the groins. (*Courtesy J. Duvivier, London; photo by Photoflight.*)

topographic detail, and rarely by the influence of river water discharge into these estuaries. Once formed, a sandbar will always exist at the entrance to any new channel, even though the exact location of a river discharge channel may change appreciably. One of the most remarkable impediments to navigation was the large sand bar at the entrance to the river Mersey at the port of Liverpool. This small river, only 90 km (56 mi) long and draining only 4,500 km^2 (1,724 mi^2), has a high tidal range (up to 9.6 m or 32 ft). Lower reaches of the river are swept by large volumes of water on every tide, yet leaving well-defined sandbanks exposed at low water. Through these, the river maintains a channel with a depth of between 9 to 15 m (30 and 50 ft) below low-water level migrating across a 4.8-km (3-mi) band, yet always forming a large sandbar. Before dredging, each bar lay in 15.9 m (53 ft) of water at low tide, with the depth at the bar itself being only 2.1 and 5.1 m (7 and 17 ft), thus limiting ship clearance. It was only in 1890 that this obstruction to navigation was dredged experimentally. For some years, extensive and systematic maintenance dredging was practiced until the site of the bar was stabilized as a channel by the construction of training walls and rock revetment. Training walls are an attempt to modify a cause of silting or bar formation by forcing a river into a permanent course in which its own velocity counters bar formation by scouring action. The use of training works is expensive and, if not successful, is liable to make matters worse.

Development of the entrance to New York Harbor required dredging for four long bars obstructing the main ship channel and Gedney Channel. Between 1884 and 1890, the channel was dredged to 9 m (30 ft), and since that time a new channel 600 m (2,000 ft) wide and 13.5 m (45 ft) deep has been dredged from the main entrance to New York Harbor. Practically no maintenance is now required on this channel owing to the normal scour action of the lower section of the Hudson River.

Many rivers and river entrance channels have been successfully developed and deepened by the judicious use of training works. An example is the mouth of the Mississippi River, which was improved from 1875 to 1879 by Gen. James B. Eads (at his own expense until governmental requirements were satisfied). The method Eads adopted concentrated current scour in one of the "passes" of the river mouth by two parallel jetties of fascine (woven baskets of saplings filled with soil and broken rock) work. The Fraser River of British Columbia is one of the great rivers of the Pacific coast of North America. It has a length of 1,250 km (790 mi) and drains an area of 238,000 km^2 (91,700 mi^2); the average yearly flood discharge is 10,700 cms (380,000 cfs). The river passes through ablation till and other glacial deposits and is heavily charged with sediment. This has formed an unusual delta, as river currents dominate tidal currents and wave action is not great. Tides and tidal currents have given an unusual form to the seaward part of the delta and have led to "training" the estuary. Geologic study and borings disclosed that at least 240 m (800 ft) of deltaic materials and calcareous sandstone has actually formed in relatively recent years in the delta. Local conditions affecting tidal estuaries are so variable that detailed geologic study of each cause is usually required. In recent years, specially constructed scale models have proved of assistance.

DREDGING

Dredging applies to any open excavation under water, generally with floating equipment, and often utilizing full-sized oceangoing vessels. The simplest form of equipment consists of barge-mounted mobile cranes, followed by a great variety of craft ranging up to the largest of suction dredges. These usually discharge into barges, but some dredges are self-contained, equipped with their own holds for carrying dredged material to disposal areas. A thorough knowledge of the material to be dredged is always a prerequisite for any successful dredging operation.

Of unusual geological interest was the combined flood-protection and navigation project for Florida waterways, initiated in 1930 to prevent a repetition of the disastrous floods of 1926 and 1928. Included in the project was the construction of 106 km (66 mi) of levees around Lake Okeechobee. Levee material was to be obtained from the bed of the lake. Much of Florida is geologically new country; bed materials of the lake vary widely, and uniform deposits are exceptional. Under the recent deposits are marl and limestone strata of varying thickness. Recent deposits often contain considerable quantities of seashells, sand, and other material so finely ground as to be almost colloidal, not very favorable material for dredging operations. Almost all of the contractors engaged started with draglines for excavation, but a powerful hydraulic dredge proved so successful that other dredgers were brought in. Special cutters had to be used at the end of the suction pipes, and the marl and limestone layers had to be broken initially by drilling and blasting.

This example emphasizes that dredging is really only open excavation carried out under water with the economy of floating equipment. Adequate preliminary investigations are even more important for this kind of work than for excavation work in the dry, since the material to be removed cannot be seen in advance. Removal of unconsolidated material usually revolves around the selection and the availability of suitable floating equipment for the dredging work. Suction dredgers fitted with cutter heads are the most common type in wide use; ladder dredgers are excellent for special purposes, and large bucketwheel dredgers are suitable for shallow and heavy work. Suction dredgers should always be fitted with a rock catcher in the suction line. This is a simple trap device for catching boulders and rock fragments, which may be sucked up the discharge line, before they get into the pumping unit and possibly cause damage to the impeller.

Initial dredging is often plagued with the need for underwater rock excavation. Two methods are in general use, one employing underwater drilling and subsequent blasting, the other using a floating rock breaker. Both methods require accurate knowledge of the nature of the rock, along with bedding and jointing. This knowledge will be of great value, not only in planning construction methods but also in estimating construction schedules and specifying the amount of acceptable overbreak. An early (and remarkable) example of underwater rock work was the removal of Flood Rock, a rocky ledge in Long Island Sound, New York: 61,000 m^3 (80,000 yd^3) was removed in this operation in 1885. The work had to be continued in later years, and a modern channel was completed only in 1920. Another interesting rock work project was carried out in New York in 1922 when a subaqueous reef in the East River was removed to improve navigation. The work had to be carried out under tidal waters in a swift current, to a final depth only 4.5 m (15 ft) above the roof of a busy subway tunnel. So accurate was the drilling and blasting of the crystalline Manhattan Schist that the project was successfully completed without damage to the tunnel or interference with subway traffic. Seismographs were used to measure and control the extent of blast vibration.

As in most aspects of civil engineering, wartime exigencies can bring rapid and ingenious equipment development, generally on a vastly increased scale.

The U.S. Seabees were faced with the job of removing 17,000 m^3 (22,000 yd^3) of coral, ranging from soft to hard, in depths of water up to 5.4 m (18 ft) at a harbor in the Solomon Islands. Standard Navy pontoons were lashed together to form a barge large enough to mount three standard 10-tonne cranes with 10.5-m (35-ft) swinging leads hung from 12-m (40-ft) booms. Pile-driving equipment was mounted on the leads and used to drive steel pipes into the coral, in which dynamite charges were then set. The barge was moved 150 m (500 ft) away for the firing and then returned so that the blasted rock could be removed by a 1-yd^3 clamshell bucket, operated by a 25-tonne crawler crane sitting on the bow of the work barge. In blasting through coral reefs in Samar, Philippine Islands, the Seabees adopted the simpler technique of detonating patterns of surface blasts that could be varied according to the type of reef and the character of the coral. Although uneconomical in the use of explosive (requiring just over 2.4 kg/m^3 or 4 lb/yd^3 of dynamite as compared with an average working figure of 0.6 kg/m^3 or 1 lb/yd^3), the method was quick and was carried out without any really suitable equipment.

Dredging projects generally represent potential negative environmental impacts, particularly of an ecological nature. Geologic studies are generally essential to determine means of mitigating or avoiding such impacts so that reasonable and necessary developments may be undertaken.

SOME SUBMARINE PROJECTS

Specialty marine structures are sometimes required to be placed in relatively shallow water. Geologists are then called upon to characterize seabed sites. Mining of the Cape Breton coalfield of Nova Scotia has been conducted beneath the seabed for many years. Tunnels beneath the sea or under tidal rivers, though not numerous, now are not the novelty they were when the Brunels, father and son, succeeded in implementing the first underwater tunnel under the river Thames at Wapping in 1843. A more recent underwater tunneling operation placed the explosive charge necessary to remove Ripple Rock in the coastal waters of British Columbia.

Ripple Rock was the name given to the double-humped summit of a basalt "hogsback" near the south end of Seymour Narrows in the inland passage between Vancouver Island and the Canadian mainland. Long an impediment to navigation, the rock became a hazard of increasing danger. Its peaks reached to within 2.7 and 6.0 m (9 and 20 ft) of low water in a navigable channel marked by strong tidal currents—a channel difficult to navigate even without the added problem of avoiding Ripple Rock. Elaborate exploratory drilling showed that the host rock was generally fresh basalt, with some andesite and a few bands of volcanic breccia. Tunneling was extended from an adjacent island and then filled with a large enough charge to blast off the entire twin peaks. First, an access shaft, 2.1 by 5.4 m (7 by 18 ft) and 171 m (570 ft) deep, was sunk on Maud Island, leading to a 2.1- by 2.4-m (7- by 8-ft) tunnel below the channel to a point under the center of the rock mass. Two 2.1- by 4.5-m (7- by 15-ft) raises,

FIGURE 12.6
The destruction of Ripple Rock, Inner Channel, British Columbia, on April 5, 1958. *(Courtesy National Film Board, Ottawa.)*

each about 90 m (300 ft) long, were then driven up to within 12 m (40 ft) of each summit. Finally, about 966 m (3,220 ft) of "coyote" workings were driven under the rock, 1.5 by 1.2 m (5 by 4 ft) in section. A total charge of 1,240,000 kg (2,736,324 lb) of Nitramex 2H was then placed throughout the workings and detonated on the morning of April 5, 1958; the charge represented 4.1 lb of explosive per pound of rock and water to be removed (Fig. 12.6). The operation was entirely successful, and the hazard to navigation was completely removed.

Aside from these carefully planned and premeditated displacements beneath sea level, other submarine earth movements are completely unexpected and may wreak serious consequences on engineered structures. Underwater slides are common to Dutch coastal engineering work. Koopejan reported 224 major slides in a 65-year period involving the mass movement of about 2.3 million m^3 (3 million yd^3). Dr. Terzaghi reported on an underwater slide in a clean sand and gravel delta in Howe Sound, British Columbia, which took out part of a wharf and warehouse. More serious, however, are the major submarine slides in some western Norwegian fjords, even though the physical damage they have done is relatively small. Damage of a very serious nature occurred on November 18, 1929, when more than 20 breaks developed in the transatlantic cables south of Newfoundland on the edge of and down the continental slope from the Grand Banks. The breaks occurred very shortly after a serious earthquake. The times and locations of the breaks followed an orderly sequence down the continental slope, suggesting a speed of about 90 km/h (60 mph), gradually decreasing along the 560 km (350 mi) over which the breaks occurred, but giving an overall average velocity of about 37.5 km/h (25 mph).

Kuenan has suggested that the breaks were caused by turbidity currents, but Terzaghi suggests rather that they were the result of temporary liquefaction and displacement of sediments forming the seabed.

Geologic influences on cable routes will be obvious. Undersea geophysical techniques and underwater photography are useful tools. The significance of seabed movements was overwhelmingly demonstrated by the undersea slides at the mouth of the Magdalena River, Columbia, in August 1935. The crown of the slide carried away 480 m (1,600 ft) of the western end of a breakwater and most of the rivershore; several hours later, tension breaks occurred in the submarine cable lying 24 km (15 mi) offshore in 1,280 m (700 fathoms) of water. When the cable was brought up for repair, grass of the type that grew near the jetties was brought up with it. Clearly, seabed geology remains an overall parameter of vital concern in engineering marine works. This concern took on entirely new proportions in the 1960s and 1970s as the worldwide search for oil extended beneath the surface of the sea, often far from land.

OFFSHORE STRUCTURES

Lighthouses come immediately to mind as the oldest type of offshore structure, sometimes standing in isolation on a hidden shoal or rocky islet, their stability dependent on the geologic structure of their location. The Pharos of Alexandria, built by Sostratus of Cnidus in the reign of Ptolemy II between 283 and 247 B.C., was probably the most famous of all early lighthouses. Built on a small island near the mouth of the Nile, it served until destroyed by a thirteenth-century earthquake. Of lighthouses in use today, that at Dordonan, near the mouth of the Gironde River in the Bay of Biscay, is the oldest. The first light at this location was constructed as early as 805 A.D., the second by the Black Prince; the recent structure was built between 1584 and 1611. With more recent additions it is now 60 m (207 ft) high.

The most famous of lighthouses guards the Eddystone rocks, 22.5 km (14 mi) into the English Channel near Plymouth, England. The first light on the rocks was a timber structure built between 1695 and 1698 by Henry Winstanley but tragically washed away by a great storm in 1705. It was replaced by "Rudyard's Tower," built of oak but destroyed by fire in 1755. Then John Smeaton constructed the first masonry Eddystone lighthouse, one of the truly pioneer civil structures. Not only did Smeaton study site geology and that of its stone guarry, but he also applied two elements of Roman civil engineering technology. Wooden wedges were to strengthen the stone dovetails by which he interconnected each course of masonry with the assistance of pozzolanic cement. Smeaton imported the Italian ingredients of the cement, a material proven by the Romans. Started in 1756, the tower was first lit in 1759; it served continually until 1882 when it in turn was replaced by the structure to be seen today (Fig. 12.7). This was built by J. N. Douglas and is located 36 m (120 ft) south-southeast of Smeaton's tower; one reason for constructing the new tower was that Smeaton's had been undermined somewhat by the sea.

FIGURE 12.7
The Eddystone Lighthouse in the English Channel, off Plymouth; the "stump" of John Smeaton's pioneer structure of 1759 may be seen on the right. (*Courtesy Kemsley Picture service, London.*)

The most famous of lighthouses founded on sand shoals instead of on bedrock guards the entrance to the river Weser in Germany. Built between 1881 and 1885 on sand, it employs a concrete mattress.

Pioneering deep foundation designs developed for the petroleum industry for offshore steel tower platforms have already enabled a number of lightships to be replaced by permanent offshore lighthouses. Small, near-shore trestle structures for drilling had been used for many years; more permanent steel structures are now common, with increased size and greater depths of water capabilities. This modern era started in 1947 when the first steel oil-drilling structure out of sight of land was constructed off the shore of Louisiana, in 6 m (20 ft) of water. Support piles were driven as deeply as possible into the seabed with the largest pile hammers then available. New techniques in submarine geotechnical engineering now add to a general knowledge of seabed geology needed to justify all selected locations.

In 1964, the U.S. Coast Guard began replacing Atlantic coast lightships with offshore steel towers. Eight such towers have satisfactorily served since the late 1960s. Tower sites were given a detailed survey of the surrounding seabed, to a radius of 1.6 km (1 mi). Borings were advanced to a depth of at least 48 m (160 ft) unless bedrock was encountered in the interval. Towers are used only when the sand and/or gravel are predominant. Typical is the Chesapeake light station, now standing in 19.2 m (64 ft) of water, 21 km (14 mi) east of Cape

FIGURE 12.8
The Ekofisk oil-drilling structure being fitted out before being towed to its location in the North Sea. (*Courtesy Phillips Petroleum Co., Norway; S. Martinsen, photographer.*)

Henry, Virginia. Soils in which the 97.5-cm-(33-in)-diameter pipe piles were driven consist mainly of sand, with some sand-clay and clay with shell fragments, with only one layer of cobbles present.

The most remarkable offshore platform developments have taken place for oil drilling, initially in the Gulf of Mexico but today in locations all around the world. Installations in water as deep as 61 m (200 ft) were made in the 1960s. This was remarkable at the time, but in 1976 the Hondo platform was installed in the Santa Barbara Channel off the coast of California in 259 m (850 ft) of water. Twelve structures, designed to withstand massive ice storms, were erected as early as 1964 in Cook Strait off the coast of Alaska. It has been in the North Sea, however, that some of the most remarkable developments have taken place, pride of place probably going to the Ekofisk complex, a 93-m-(304-ft)-diameter, reinforced-concrete structure weighing 444,000 tonnes (489,000 tons) and floated into place after being built in Norway (Fig. 12.8). It rests safely in 70 m (230 ft) of water.[3]

These engineering achievements all depend upon seabed knowledge for their safe performance. The geology of any one part of the bed of the ocean gradually becomes known as the proprietary information gathered by individual operators is released for public use. Most of the geologic formations encountered are of Pleistocene age. Unusual geologic features have included explosively charged, gas-laden sediments beneath the slope of the continental shelf and the soft-grained calcareous sand in Bass Strait between Victoria and Tasmania, Australia.

FIGURE 12.9
Rincon Island, constructed for the support of oil-exploration installations, off the coast of California. (*Courtesy Richfield Oil Corporation Inc.*)

The Pacific coast of the United States between Santa Barbara and Ventura now harbors one of the few artificially created open-ocean islands of magnitude. Rincon Island was constructed for the Richfield Oil Corporation (now ARCO) as an alternative to an open-tower platform (Fig. 12.9). The island permits the owner to explore, with "slant" drilling techniques, a large proportion of an offshore oil lease granted to it by the State of California. At the island site, water varies from 12.3 to 14.4 m (41 to 48 ft) in depth. Extensive site investigation revealed that the bottom consists of a silty sand grading into sandy silt, increasing in thickness with increasing depth of water. Recent shale or siltstone underlies the sediment. Sand movement on the sea floor at the site was found to be negligible; thus, the construction of a solid island structure was a feasible proposition. The 2.4-ha (6.3-acre) island has a working area of 0.45 ha (1.1 acres) at an elevation of 16 m (52 ft). Rock-faced rock embankments of Eocene sandstone of the Cold Water Formation enclose the sand fill. Rock faces are protected from the action of the sea by 1,130 tetrapods having a total weight of 35,000 tonnes. Rock was obtained from a quarry on Santa Cruz

Island, previously used for the construction of the Santa Barbara breakwater. Quarry rock proved to be variable in quality, requiring visual acceptance screening before shipment.

The sea's awesome power dominates all marine design. There are few better ways of demonstrating this than by recounting the tragedy of Hallsands, a small fishing village on the south coast of Devonshire in England. It was located above a fine shingle beach and backed by a steep rock cliff. Records showed that the shingle had been stable for at least a century, until the great dockyard at nearby Devonport had to be enlarged in 1897. Large quantities of material were required for the construction work. Permission was given to the responsible contractor to dredge shingle from below the beach at Hallsands. The villagers, 126 people living in Hallsands' 37 houses, protested, but it was not until 1902 and after the removal of 660,000 tonnes that dredging was stopped. But the damage had been done. Despite construction of a protection wall, the beach level had dropped by 5.7 m (18 ft); without natural protection, heavy waves washed right over the wall. Some residents of the village moved away, but most remained in their homes until, between January 25 and 27, 1917, there occurred such a combination of strong northeast winds and high spring tides that 24 of the remaining 26 houses disappeared into the sea. The village of Hallsands disappeared, not from natural causes but because of construction work carried out without full appreciation of the power of the sea.[4]

RIVER TRAINING WORKS

For more than 3,000 years we have tried to "tame" the great rivers. Today, the training of rivers is an important and diverse branch of civil engineering, widespread in its practice, and specialized for each individual river. One underlying fact is supreme: meandering of rivers, which river-training works are designed to control, is a part of the natural order, a phenomenon of physical geology. Design is largely empirical; no two rivers are alike in the geologic parameters controlling their flow and channel modification. Local application of the results of experience elsewhere is often fraught with uncertainty and is liable to misinterpretation, especially when geology is neglected in design evaluations.

River-control work aims to keep riverflow within a certain well-defined course, usually that which the stream normally occupies. Deviations from this course are expected during periods of floods and control work deals with regulating these waters. Prevention of bank erosion during normal flow is another important concern. Finally, channels are improved in connection with civic or industrial development, such as straightening of a riverbed in connection with town planning and canalization to facilitate navigation. In undertaking work of this nature, an engineer is dealing with a fundamental geological process, and all designs must be based on an acceptance of the natural characteristics of riverflow. Most river-training work is conducted on a relatively small scale. Much can be learned, however, from experience with the larger rivers of the world.

Four of the major river systems of the world have well-developed floodplains long occupied by human settlements and river-training work. They are the Mississippi, the Nile, and two Chinese rivers, the Yangtze and the Yellow (or Hwang Ho). Other rivers of comparable size are the Amazon, the Congo, the Lena, and the Brahmaputra, but only the last of these has seen development comparable to that which has been carried out on the two great rivers of China for 30 centuries. The Yangtze River, 5,100 km (3,200 mi) in length, is navigable for a longer distance than any other river in the world. Its flow on July 19, 1915 was recorded as 71,600 cms (2,531,692 cfs), the greatest flow ever recorded for one river channel, much in excess of the maximum flood flow on the Mississippi. In 1871, the level of the Yangtze rose to the phenomenal height of 82.5 m (275 ft) above normal in the Wind Box Gorge, 1,750 km (1,100 mi) from its mouth. About 180 million people live in its watershed. Its floods have caused untold tragedies in loss of life down through the ages. Dikes have been heroically built in attempts to control its channel; 50,000 people were at work on one dike alone in repairing flood damage after an inundation in April 1927. Today, the Yangtze is one of the most completely diked rivers of the world.

Even more remarkable is the Yellow River, "China's Sorrow," 4,000 km (2,500 mi) long, with at least 100 million people living in its basin. Its name derives from the high content of loess that it carries out into the sea, sediment being visible far from its mouth. At one time the river ran north by Tientsin and at another it emptied in central China into the Yangtze River, near Chekiang, several hundred km to the south. For 700 years it flowed eastward into the Yellow Sea, but in 1852 it broke its banks at a point more than 480 km (300 mi) from its mouth, formed a new channel across the province of Shantung, and finally arrived at its position today in the Gulf of Chihli, almost 480 km (300 mi) from its old mouth. When flow velocity, which averages 2.4 mps (8 fps), drops below 0.9 mps (3 fps), silt is deposited rapidly, the waters overtop the existing dikes, and a new page of sorrow starts. It was an old Chinese tradition that restless dragons stirred the mud in the rivers, causing great calamities.

Much Chinese effort, coupled with expert engineering advice, controlled the Yellow River when it broke its banks again, for 2.4 km (1.5 mi), in 1935 near the towns of Tung Chuang and Lin Pao Chi. The ancient Chinese contraction method was used to close the gap. The ends of the breach were reinforced and new dike structures were gradually extended from both ends, with the final closure being made with the aid of immense, stone-filled wire sausages and willow-fascine mattresses. As many as 25,000 men were employed to place and tamp the soil; ingenious stone "flappers" lifted by eight men with attached ropes suggested that the Chinese had a good appreciation of the necessity of soil compaction long before it became a well-established technique in modern soil mechanics.

India's share of flood sorrows is typified along the Brahmaputra. This great river left its normal course spectacularly toward the end of the eighteenth century, seriously affecting the ancient city of Mymensingh. With the development of substantial British irrigation works in the nineteenth century, river-training works became an established part of Indian engineering. It was in 1887 that J. H. Bell advanced new proposals for training the Chenab River at Sher Shag where a large new bridge was proposed. He suggested concentrating the low riverflow in a relatively deep channel by means of low embankments, soon known far beyond India as "Bell's bunds."

THE MISSISSIPPI RIVER

Nowhere, perhaps, has the river-training precept been better illustrated than in control of the "Father of Waters," America's greatest river and its lusty tributary, the Missouri, commonly known as "Big Muddy." The Mississippi is fed by a drainage area of 3,220,000 km^2 (1,242,000 mi^2) and has a total length about 3,840 km (2,400 mi). Its average discharge at Vicksburg is about 14,000 cms (500,000 cfs); flood discharge is as high as 56,000 cms (2 million cfs). In general profile, the river is flat and slow-moving for the greater part of its length. The valley of the Mississippi follows the general form for such river valleys. Floodwaters, before confinement by levees was effected, covered a much greater area than the normal stream bed, at times as much as 77,500 km^2 (30,000 mi^2). Early nineteenth-century attempts to confine the river during flood periods employed raised natural levees, thus increasing successive flood depths and promoting channel deepening. In consequence, no local addition to a levee could really be made without all levees being raised similarly. The first levee, built at New Orleans in 1717, was 1.2 m (4 ft) high. In 1932, the average height was more than 3.9 m (13 ft), and 4,000 km (2,500 mi) of artificial levees were being maintained on the river. All this work, up to 1932, was based on the ideas of maintaining the existing channel of the river.

In June 1932, a new policy was inaugurated under the direction of Gen. H. B. Ferguson, president of the Mississippi River Commission. Another diagnostic characteristic of the river was adopted for use. The natural process whereby the river makes cutoffs at narrow necks between bends over long intervals of time was expedited by making dredged cutoffs. These cutoffs have materially reduced the length of the river and have lowered its flood stages. Supplementary dredging is done in the reaches between cutoffs to assist the river in restoring the steepened gradients to normal. A famous example of a natural cutoff is at Vicksburg, Mississippi, where in 1863 during the Civil War, General Grant tried to isolate the city from the river by cutting a canal through a narrow neck of land. This attempt was not successful, but in 1876, the river made the change itself. Fortunately, Vicksburg was able to maintain contact with the main river through diversion of another stream; it is now well known to engineers as the location of the large-scale river models at the Waterways Experiment Station (of the Army Corps of Engineers).

River shortening began at 12 sharp bends (Fig. 12.10), removing nearly 185 km (116 mi) in a stretch of 530 km (330 mi). Associated relief floodways were

FIGURE 12.10
The Sarah Island cutoff on the Mississippi River in initial operation; some of the river flow is still following the old river course, although this had started to silt up when the photograph was taken. (*Courtesy Mississippi River Commission, U.S. Army Corps of Engineers.*)

provided, designed to allow floodwaters to spill over the natural banks but only under control, and in specially selected locations. Reservoirs have been created on tributary streams, so that discharge in the main channel has been further reduced. Finally, continued bank-protection work has led to ingenious mattress designs.

This great program of river training had just got well under way when the remarkable flood of 1936 and 1937 occurred, the greatest ever to flow down the Mississippi to the sea without any break in the levees. With the engineering organization already in the field, a detailed study characterized the entire flood, directing attention to gaps in available knowledge, and a major program of investigation was started shortly afterward. Investigation concentrated on geologic and geomorphic features of the valley related to the great river. The work is one of the most comprehensive engineering-geological studies ever made, taking in the entire valley not only by means of surface observations but with the aid of the most complete collection of boring records that could be assembled. At least 16,000 boring records were made available by engineering organizations operating in the valley; the results of the extensive exploration drilling for oil near the Mississippi delta were also added.

The results were originally summarized in a number of papers and in a comprehensive volume by H. N. Fisk, who was directly responsible for much of the geological part of this notable undertaking. As with all such studies, the results have proved of mutual benefit—to engineering, in providing the essential background for the continuing improvement of the great system of training works for the Mississippi, and to geology, in revealing hitherto unsuspected aspects of the Pleistocene history of the valley sediments. It has been shown, for example, that there has been considerable subsidence in the delta area under the increasing weight of the deposited sediment; this fact accounts for the limited extension of the delta deposits within recent decades. Jetties constructed at the mouths of the Mississippi in 1875, for example, have not had to be extended since then. Corresponding elevation of adjacent areas on dry land appears to have occurred. Over 40 years later (1985), a second, more detailed engineering-geological assessment of the Lower Mississippi River Valley was begun by the Waterways Experiment Station; this product will concentrate on development of absolute means of prediction of geotechnical conditions based on geologic site observation and soil boring logs.

Of the many aspects of the geology of this interesting area, one is the most dangerous of all. For some time the future of the lower 480 km (300 mi) of the Mississippi River has been threatened by the possibility of the "capture" of its flow by the Atchafalaya River through what has long been known as the "Old River diversion." The river, though only minor in comparison with the Mississippi, it still the third largest American river that flows into the sea. The Old River connection is 9.6 km (6 mi) long; its location is shown in Figure 12.11. In 1956, the Old River took 23.5 percent of the flow of the main river, diverting it into the Atchafalaya; it was estimated that in the absence of any control measures, this diversion could reach 40 percent by 1975. At that stage, lower flow would promote the silting up of the main channel, with probably disastrous effects upon the major port of New Orleans. A master control plan was prepared by the Army Corps of Engineers, and the first stages of it are already complete. Two control structures were built upstream from Old River, a navigation lock connecting the two rivers, with the corresponding connecting channels. Additional levee work was involved, and the Old River itself is now sealed off. This vast project was necessitated merely to control the further development of an unusually large case of natural "river capture."[5]

DELTAS AND ESTUARIES

The Nile valley has provided a vivid annual example of the silt-bearing function of rivers, providing the natural mineralized fertility to support one of the cradles of civilization. Records at the famous Roda Gauge near Cairo (maintained for many hundreds of years) show average sediment deposition of about 10 cm (4 in) per century. Sediment formation of deltas at the mouths of several of the major rivers of the world, for example the Ganges and the Mississippi, in addition to the Nile, is one of the most obvious demonstrations of the transporting power of rivers. Entrance bars to tidal estuaries below even small rivers is another reminder of the great transporting power of the sea. Delta growth provides space for development, but such land is always characterized

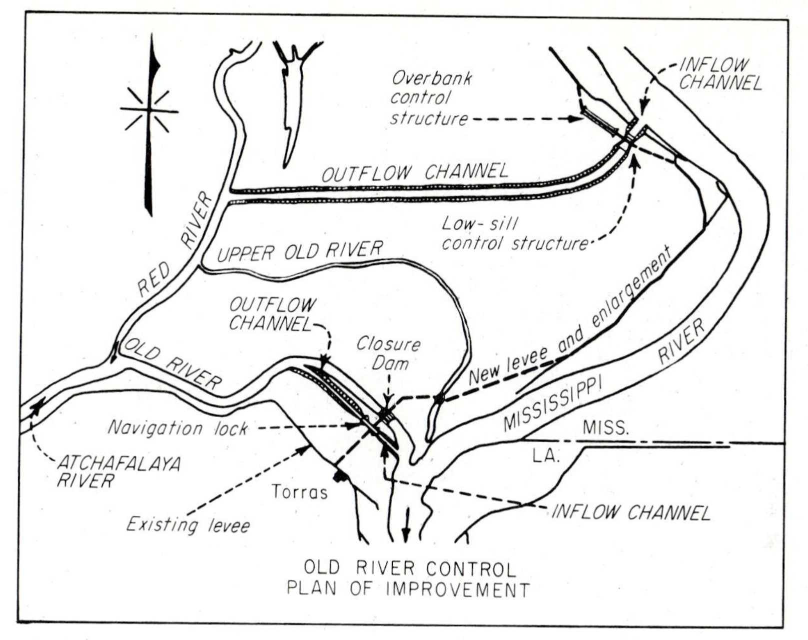

FIGURE 12.11
Sketch plan of the Old River works on the lower Mississippi River, constructed for control purposes. (*Courtesy Engineering News-Record.*)

by complex geology and structural foundation problems. The Rhone delta in France (observed since 400 B.C.) has been growing laterally at 11 m (36 ft) yearly; the corresponding rate for both the Danube and the Nile is 3 m (10 ft) per year.

Physiographic conditions at the mouth of a river are a significant control over deltaic features. Rivers such as the Severn in England and the Amazon, although they carry large quantities of silt, are without broad deltas; the funnel-shaped estuaries promote self-maintenance and sediment flushing of the main channels. One of the largest deltas in the world is at the unusual mouth of the Mackenzie River in Arctic Canada (Fig. 12.12). It is 240 km (150 mi) long and up to 80 km (50 mi) wide, underlain by permafrost and enclosed by highlands to either side; the main load of sediment is deposited in a great underwater delta in the Beaufort Sea.

Geological conditions affecting the formation of deltas are well recognized; suspended and bedload capacities are reduced when the streamflow merges with the quiet water of the sea or lake. Part of the remaining suspended load may be deposited by saltwater flocculation of colloidal material. These regular beds constitute some of the newest sedimentary deposits. Surprising depths were detected by a boring put down near Venice on the Po River delta, Italy. A depth of 150 m (500 ft) was reached without penetrating the bottom of the

FIGURE 12.12
Typical view in the delta of the Mackenzie River on the Arctic coast of the Northwest Territories of Canada, all the ground to be seen being perennially frozen (permafrost). (*Courtesy National Research Council of Canada; photo, G. H. Johnston.*)

deltaic beds. Although deltas are popularly perceived as above-water features, they are essentially underwater geological structures significant for their engineering influence. Deltas have complex geologic substrata, formed of lenses, pockets, and truncated channels of clay, silt, and sand. Most of the internal geologic structure is formed by breakages of the natural levees above the river mouth.

In estuaries without deltas, sedimentation is a particular problem. At Baltimore, Maryland, the harbor was designated a port by the General Assembly of Maryland in 1706. This great American port lies in the estuary of the Petapsco River and its tributaries. In the course of 200 years, the head of navigation has been pushed steadily seaward for a distance of 11 km (7 mi); depth at low water at one well-defined location decreased from 5.1 m (17 ft) in 1845 to about 15 cm (6 in) in 1924. It is not surprising to find that dredging began here (by the Federal government) as early as 1836. There is no doubt about the origin of the sediment deposited in the harbor and then removed at such expense by dredging. Soil erosion in the Petapsco watershed is recognized in the correspondence of the river silt load to the amount of sediment known to be deposited in the harbor. For this economic "bottom-line" reason, civil engineers must appreciate the menace of soil erosion.

RIVER DYNAMICS

Most river-related civil engineering work is on a considerably smaller scale than what is required for major rivers. An overriding fact is that *any* engineering work on *any* river, large or small, must consider the river as a dynamic element in the hydrologic cycle. Construction at a river bend, for example, must consider possible effects both downstream and upstream, incorporating records of annual flow variations. Inaccuracies in estimating the twelve-month cyclical performance of a river raises the greatest risk that training or diversion works at any one location may prove to be inadequate and therefore subject to destruction. River dynamics is, therefore, a term of real meaning. When looking at a section of smooth-flowing water, one may all too easily forget that this is but a part of an intricate system, fed by runoff or through groundwater seepage. In both cases, geology is the basic determinant. Once the initial stream has formed, its entire development depends on the geology of its course; rock exposures creating rapids, rapid downcutting along fault zones, deflection from lateral movement against harder rock, erodible soils adding to the load of sediment carried, and flow over limestones even leading to sporadic disappearance into underground channels. Streamflow can be one of the most potent of the natural forces now at work on the surface of the earth: an integral part of the geological cycle; denudation—deposition—earth movement—denudation. It is always salutary in river work to remember that the carrying (erosive) power of moving water varies as the sixth power of its velocity.

The study of river systems is a well-established branch of geomorphology. Some of the basic concepts were first advanced by a notable hydraulic engineer, Robert E. Horton, followed in more recent years by Luna Leopold, for many years chief hydrologist of the U.S. Geological Survey. A basic theory of river hydraulics is the concept of the *four orders of streams*. A stream that has no tributaries, such as one in the headwaters of a river system, is considered to be of the first order. Two first-order streams unite as a second-order stream. Third-order streams are formed by the union of two second-order streams, and the fourth order, the final classification, is the designation given to the watercourse formed of two third-order streams. For typical river systems, both the number of tributaries and their total length bear a linear relationship with their orders of magnitude.

When river systems are considered in a regional context, the different types of rivers are recognized on the basis of geologic age. Relatively young rivers begin, ideally, as streams flowing over level plateaus and eroding their courses into the material of which the plateau is composed. This will lead to very rapid development and spectacular scenery, the Grand Canyon being a superb example. When the landscape reaches its ultimate stage of development, rivers enter a mature stage, irregularities in their beds being smoothed out. Old age arrives when the coalescence of flat and gently sloping surfaces of adjacent rivers begins to take effect, working up from coastal regions at the expense of eroding bluffs and divides.

These broad concepts come into focus when river-training work considers the complex motion of water flowing around a bend. As a stream starts flowing around a bend, the surface water, since it is faster than that at the bottom of the stream, will move toward the concave bank. It will tend to erode the bank and as it does so it will be deflected downward. Water at the bottom must move upward in a compensating movement. The combination induces in the flowing water of sort of helix movement of individual water particles. Viewed in cross section, water will tend to flow across the stream bed away from the concave and eroding bank toward the other side of the stream, depositing some of its load from the erosion as it does so. Just as concave banks tend to be eroded, so convex banks tend to accumulate sediment and grow outward, in keeping with the erosion of the facing bank. On a large scale and over a number of years, it is this water action that explains the meandering of rivers when flowing in flat plains.

In ordinary practice the civil engineer encounters this natural phenomenon when unexpected flood flow down a stream or river has caused serious erosion at a critical bend. Basic hydraulics governs part of the design of remedial works carried out. Erosive forces will have increased because of the enlargement of the bend. Variations in the type of distribution of soil and rock will control the distribution of hydraulic forces and the rates of their erosion and deposition. Protective works must, therefore, take all of this into account. Once the immediate danger has been overcome, the overall situation must be carefully reviewed before any permanent works are decided upon, all against a background of the geology and the fundamentals of river hydraulics .

SMALL-SCALE WORKS

The simplest form of river-training work is channel restoration after sudden flooding. Such experiences will always reveal the weak spots in stream routing, such as at bends where significant erosion may take place. There is an immediate temptation to eliminate such trouble spots, but the best possible action to follow is to try to reproduce earlier natural conditions as closely as possible. Stream beds must be cleaned out and all debris moved by the flood removed; eroded ground must be replaced, if possible, and bank-protection work installed to prevent as far as possible any further erosion at the same bends. All too often bank-protection works are placed at needless expense on convex banks where, in all normal cases, things can take care of themselves. The U.S. Soil Conservation Service has accumulated much experience in small-scale stream training work. The works by E.A. Keller and the President's Council on Environmental Quality (U.S.) apply to stream maintenance as well as to the environment and ecology.

Phenomenal rainfalls cannot be prevented, so floods are a part of the natural order. But with increasing deforestation and urban development, the incidence and magnitude of floods have been increasing as a direct reflection of interference with the flood protection and river training that Nature provides if people do not interfere. Vegetation is one of the best of all runoff controls; if it is removed, the rate, and so the volume, of runoff will increase, enhancing the possibility of floods.

The Garfield smelter is located 35 km (22 mi) west of Salt Lake City, close to the mouth of a small canyon. Although a number of protective works were installed along this steep stream bed, a flood of unprecedented magnitude swept down the canyon in July 1927, damaging not only the protective works but the plant itself. An earthen barrier dam was then constructed right across the mouth of the canyon, almost 1.6 km (1 mi) in length and founded on alluvial soil. The dam is as much as 30 m (100 ft) high at the stream invert and is equipped with a spillway of canyon-debris boulder masonry. The pool this created so reduced the water gradient that sand and gravel brought down by small floods naturally accumulated so that by 1937 the fill had risen in places to the crest of the dam. Despite this, the spillway safely passed as a 20-minute flash flood estimated at 425 cms (15,000 cfs). These small control works are still in effective operation

BANK PROTECTION

Bank protection is an ordinary element of river-control work. Methods of protection range from the construction of permanent retaining walls in front of natural soil slopes for preventing the initial formation of soil gullies, to the use of expedients such as mattresses of brushwood, riprap, dumped rock, and even sandbags in case of emergency. The nature of the river action is the key to successful bank protection. At many river bends, the use of a wall of deep-driven interlocking steel sheet piling has often proved successful where other expedients have failed through riverbed scour. In the case of levee protection, many ingenious designs of flexible mattresses built of different materials, including reinforced concrete, take up initial variations of the riverbed and also any alteration caused by further minor erosion at the edge of the mattress. Flexibility is also the keynote of the "wire-sausage" type of protection adopted so successfully in India and China; cages of wire are filled with loose stones and so placed that in case of the failure of one sausage the superimposed cages will automatically move down to take its place. Another idea has been the planting of low-growing willow trees at or about normal waterline, so that the foliage may impede the flow at flood periods and therefore reduce its eroding power. Throughout all this work, however, as in all satisfactory river-training work, the basic concept has been to cooperate as closely as possible with the natural process of riverflow, to control it within necessary limits without attempting to usurp the place of Nature.

"Wire sausages" *(gabions)* are now commercially available as ready-made wire-netting baskets. The idea is an adaptation of techniques developed by the British Corps of Royal Engineers well over a century ago. The ease of installation and general utility of such rock-filled wire boxes has now led to their widespread use (Fig. 12.13). For large-scale bank protection, interlocking steel sheet piling effectively resists further scouring. Experienced operators are required to drive this type of piling, and installation in lodgement till or coarse

FIGURE 12.13
An imaginative use of gabions (rock-filled wire-net boxes filled with large stones) for control of drainage flows along Interstate 88, adjacent to the Schenevus exit in Otsego County, New York. (*Courtesy New York Department of Transportation.*)

alluvial material may be defeated by bending and deformation of individual piles on impact with boulders. Alternatively, large-scale riprap (suitable resistant rock) can be used, if carefully placed.

CANALIZATION OF RIVERS

So thoroughly used are the world's rivers that it is easy to forget that many have been canalized as a form of river training. Efficient operation of impounding dams, spillways, and associated navigation locks and powerhouses often requires complete control of inflowing rivers. Many storage reservoirs on subsidiary streams have been developed primarily to control inflow to rivers that are developed as transportation canals.

There are examples of such river-training works all over the world, ranging from one of the earliest in North America—the complete canalization of the small Rideau and Cataraqui rivers in Ontario to form the Rideau Canal, constructed in 1826–1831—to the mighty works required for the canalization of the upper reaches of the Mississippi River, now extending to the center of Minneapolis. The upper reaches of the Rhine, between France and West Germany, provide another quite spectacular example, and even some of the tributaries of the Rhine have now been similarly controlled.

Every river-training structure or those serving navigation or power facilities must meet geologic problems. The Moselle River of West Germany, used for

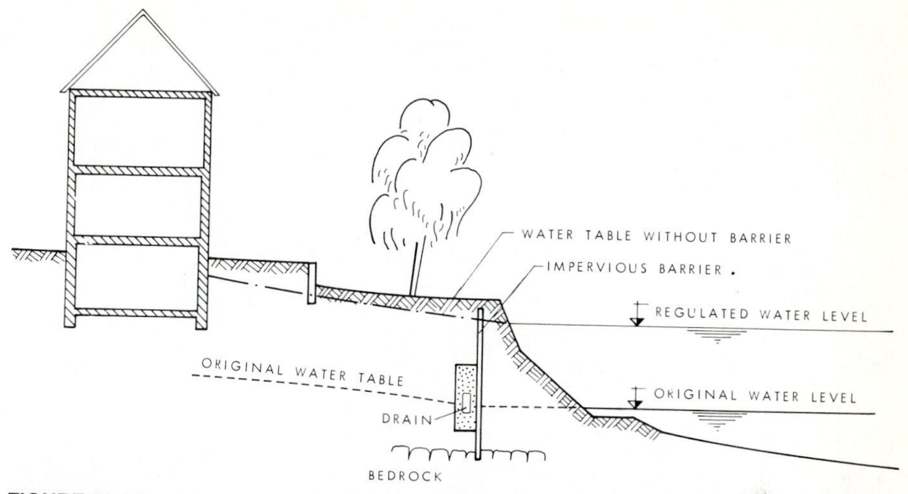

FIGURE 12.14
Simplified cross section showing the effect on river-bank buildings of the raising of water levels in the Moselle River, and the method adopted for protecting them. (*Courtesy Internationale Mosel-Gesellschaft mbH, Trier, West Germany.*)

navigation since Roman times 2,000 years ago, on May 20, 1964 inaugurated a new chapter in its long history. On that day it was opened as a fully navigable waterway, as a result of the Moselle Treaty of 1956 between France, Luxembourg, and West Germany. The three nations share the 200-MW river-generated power locks, permitting the movement of 3,500-tonne-capacity barge tugs. Eleven dams and associated works were necessary. In the relatively short length of river (only 271 km or 170 mi) involved in the canalization project, the water level drops 90 m (296 ft).

The presence of individual dams raised the river water close to or even over the top of previous river banks. Many ancillary works were necessary for protecting riverside towns and villages from flooding, for adjusting water supply and sewerage systems, and for providing necessary clearances under bridges. Figure 12.14 shows one of the attendant groundwater problems. The precanalization groundwater level sloped down and away from the river bank, dependent upon flow and the permeability of the adjacent ground. This condition had existed for centuries and so river-bank buildings, such as the warehouse indicated, had foundations designed to suit. The basement of this particular warehouse was used for the storage of Moselle wines. With the rise in water level, following construction of the next downstream dam, the groundwater would have risen to flood such storage. All along the Moselle, therefore, special provisions had to be made in such locations to maintain groundwater conditions in their previous state. One solution is illustrated, the construction of an impermeable slurry-trench barrier (and allied methods) and then the provision of a drainage facility behind this, located so that it would

FIGURE 12.15
St. Anthony Falls on the upper Mississippi River at Minneapolis, Minnesota, showing (upper left) the lock and control works mentioned in text. (*Courtesy U.S. Army Corps of Engineers, St. Paul District.*)

maintain groundwater at or very close to its original level. In some cases (without wine storage), fill had to be placed to raise ground levels; in a few cases villages and buildings had to be moved. All this work was due merely to changes in groundwater level.

Canalization of the upper Mississippi required completion of two locks adjacent to St. Anthony Falls in the harbor of Minneapolis (Fig. 12.15). Complications were provided by the existence of a stratum of resistant Platteville Limestone overlying the St. Peter Sandstone, found throughout the subsurface of the Twin Cities. The limestone bed thins to zero just above St. Anthony Falls. A thin bed of impervious siltstone within the St. Peter Formation further complicated the foundation designs. Upon excavation, the St. Peter rocks were found with open horizontal bedding planes which supplied flow under dewatering pumpage. Eventually a well-point system had to be installed because of the inflow. Installation of the well points was itself critical, so as not to penetrate the impervious silt stratum.

Works such as these are a far cry from the "dredging over sandbars and removal of snags" with which the U.S. Army Corps of Engineers commenced its work of controlling the Ohio River a century and a half ago. Today's river-control works involve some of the most expensive and complex of construction operations. Every such project is unique in itself and may present previously unforeseen problems or hazards, as the experience at the Wheeler Lock (Chap. 5) so clearly demonstrated (p. 408). The most accurate possible knowledge of the subsurface throughout the entire site is the first and essential requirement, knowledge of the hydrology of the river and its banks being equally important preliminary information.

LAND RECLAMATION

Coastal centers of population and commerce often become so crowded that expansion into the sea of tidal areas offers at least an avenue of consideration for the reclamation as developed land. Some protective works in the constant battle against seacoast erosion can be secondarily utilized as safe areas in which fill can be placed to form new land. The concept of "winning land from the sea" requires accurate information on the subsurface conditions and natural processes, such as littoral drift and local currents. Examples of land reclamation from the sea are to be found in many countries. The Back Bay reclamation scheme of Bombay, India, the airports at Hong Kong and Singapore, and many such facilities in the United States and Canada have been extended over made land that was recently under water. Many dock and harbor projects have included the reclamation of land for port service areas, often utilizing material obtained from dredging approach channels to the dock facilities. Interesting and significant as are all such works, they pale into insignificance against the protection and reclamation works extension of the Netherlands.

Zuider Zee Reclamation

Forty percent of the Netherlands, known throughout history as part of the Low Countries, lies below mean sea level. Some 1,930 km (1,200 mi) of dikes stand against inroads of the sea. Steady subsidence of the land, combined with a corresponding rise in the level of the North Sea, has led to a continuing battle against the encroachment of the sea into the fertile low-lying plains of Holland. Roman records mention a small freshwater predecessor lake to the Zuider Zee. Historical records go back for at least 1,000 years of dike building. *Poldering,* or the reclaiming of land from the sea, started as early as A.D. 1200. Windmill pumping of reclaimed lands came into use about 1600. But the battle had been a losing one until very recent years. It is estimated that from 1200 until quite recently, the sea claimed 567,000 ha (1.4 million acres), whereas coastal reclamation pumping won back less than 526,000 ha (1.3 million acres). In 1421 alone, a dreadful storm destroyed 65 villages through inundation and took the lives of 10,000 people.

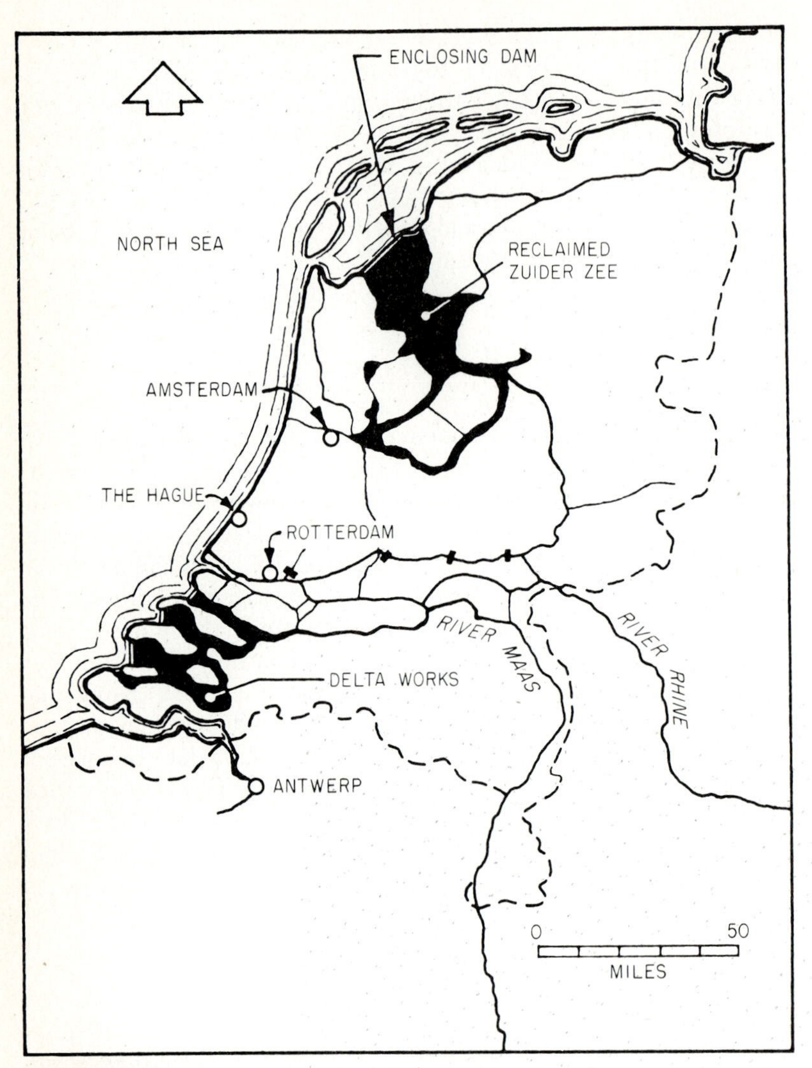

FIGURE 12.16
Map of the Netherlands showing generally the location of the Zuider Zee project and the areas to be enclosed by the delta works.

In 1918 the Dutch began the greatest land-reclamation project that the world has yet seen—transformation of 222,000 ha (550,000 acres) in the Zuider Zee to a freshwater lake with an area of almost 400,000 ha (1 million acres) (Fig. 12.16). Completion before the end of the present century will increase arable land in Holland by 10 percent. The task is being carried out in five stages. The key to the entire scheme was the construction of the great enclosing dam across the northern end of the Zuider Zee, 32 km (20 mi) long and 90 m (300 ft) wide

FIGURE 12.17
Reclamation of the Zuider Zee, the Netherlands; the circular dike of Eastern Flevoland, before the polder was dried out. (*Courtesy KLM Aerocarto n.v., the Netherlands.*)

at its crest, constructed between 1927 and 1932 in the open sea (Fig.12.17). The enclosing dam had the effect of pushing the sea back 85 km (53 mi) from the southern end of the old Zuider Zee; the shoreline of the country was thus reduced by 300 km (186 mi).

Some Smaller Examples

Extensive land reclamation schemes frequently convert land that was otherwise unusable, such as swampland, into usable land. Interference with wetlands can only be done after proper ecological studies, but when approved and carried out with care, such interference can be socially valuable. A combined dredging and filling operation was used to create a new central market in San Juan Bay, Puerto Rico. The 40-ha (100-acre) site adjoins the mouth of the Puerto Nuevo River. When the river mouth was to be dredged and rerouted to improve flood discharge, some of the area was filled with dredge spoil. The site was an old mangrove swamp; soft organic silt extended from the surface to firm residual clay 7.5 to 12.0 m (25 to 40 ft) below. Clearing was followed by placement of a 1.2-m (4-ft) sand-blanket working area, in which 33,500 sand drains, 45 cm (18 in) in diameter, were installed to predrain the site. Areas required for specific and immediate building projects were overloaded with an excess amount of spoil, which was later removed when the additional weight had caused the necessary consolidation of the silt.

Close to the teeming metropolis of New York lies Hackensack Meadows, a large area of low-lying and little-used land between Newark and the diabase sill of the New Jersey Palisades. All travelers going south from New York by road, rail, or air see this vast area of swampland:

> ...once a glacial lake, the Meadow is an inhospitable mix of sand, silt and clay topped with a layer of black organic mud, green reeds and swamp grass. For generations the area has been used as a refuse dump for the New York–New Jersey metropolitan area, so a thick layer of garbage blankets much of the marsh.

Exploratory borings have detected two pre-Pleistocene buried valleys cut more than 30 m (100 ft) below present sea level into the underlying Triassic sandstone, one valley extending under the city of Newark, the other underlying the eastern margin of the Meadows. Light buildings have been erected in the northern section of the Meadows. Subsurface conditions typically include 1.8 and 2.1 m (6 to 7 ft) of highly compressible surface tidal deposit, "meadow-mat," or fibrous peat, overlying a thin stratum of still varved clay, below which is a deep (7.5- to 22.5-m or 25- to 75-ft) deposit of soft-to-firm varved clay. Dense lodgement till then overlies the sandstone bedrock. One-storey buildings have been built and have performed well here, being sited on carefully placed select fill resting directly on the stiff varved clay, after removal of the fibrous material. This shows what can be done, based on careful study of the properties of the materials of the foundation beds, with a minimum disturbance of natural conditions, a tribute to the work of the Hackensack Meadowlands Development Commission.[6]

The problem of managing excavation waste *(spoil)* is steadily increasing in complexity as cities grow larger and available sites for disposal diminish in number and size. It is a problem governed largely by the high cost of trucking. In earlier days, a simple solution was to find an unused area of land, an old quarry or even an unused ravine, and merely end dump the rock or soil into it from trucks. Uncontrolled dumping of this kind led to much trouble with potential settlement under later buildings, premium costs for deep foundations to penetrate the loose spoil, and slides on steep slopes. Many cities lost the use of lovely ravines which had been filled with spoil. Today, uncontrolled spoil disposal is illegal in many industrialized nations. End dumping on private ground is still widely practiced, however; owners see an easy and cheap method of obtaining what they consider to be good building lots as depressions are filled in and level areas developed. Unless fill is properly placed, without appreciable organic contents, in layers and compacted, there may be continued settlement of the ground surface due to the slow, natural consolidation of the fill.

Many cities have considered accepting the responsibility for disposal of excavated material since it can often be desirable material and conveniently disposed to create high-value land adjacent to downtown areas. Generation of

FIGURE 12.18
Toronto's new breakwater (foreground) constructed entirely of material excavated in the main city area, at almost no cost. (*Courtesy Toronto Harbour Commission.*)

9.5 ha (23.4 acres) of new land in downtown Manhattan was created by controlled placement of excavation wastes from the World Trade Center. This is not a new and revolutionary idea; when the first railway into Ottawa was built from the St. Lawrence River, all the fill from a big cut at the Prescott end, to obtain an area of 3.2 ha (8 acres) of new land in the St. Lawrence River, was enclosed by about 300 m (1,000 ft) of wooden piling. This 1852 concept has been used on a large scale only in recent years. Milwaukee has reclaimed 15 ha (37 acres) from Lake Michigan, to be used for harbor and marina developments. Excepting the special case of the Netherlands, Japan probably leads the world in the creation of new land for cities. A national program of reclaiming no less than 47,000 ha (183 m^2) around its major ports was announced in 1970. Much of the fill will be pumped from adjacent seabeds, but Japan for many years has used select urban waste (excavation spoil, ash, and slag) for fill. Two areas in Tokyo harbor were filled between 1957 and 1963 with 24 million m^3 (31.5 million yd^3) of soil, most of it hydraulically placed, with the upper 1 m (3 ft) thickness consisting of granular wastes.

As Toronto harbor (Fig. 12.18) is incapable of further extension within its natural Toronto Bay, plans have placed new harbor facilities in the open water of Lake Ontario, protected by a new 4,920-m-(16,000-ft)-long breakwater. The new breakwater is being constructed slowly but economically of excavation fill from the Central Business District. Locally excavated glacial soils and shale bedrock are suitable for fill, and the bottom of Lake Ontario, formed of the same materials, provides an excellent foundation bed. Blocks of concrete and rock are broken and crushed in a special installation which issues a steady supply of by-product surfacing for harbor roads. Waterfront or shoreline fills generally do not have the natural advantage of lateral confinement, as do valley fills. Subsurface investigations are a must in order to provide stable foundation conditions.

Management of domestic and industrial waste is steadily becoming a major municipal problem throughout the developed world. Garbage dumps have been replaced since the mid 1960s and early 1970s by major engineering works for incineration and sanitary landfilling. U.S. Public Health Service estimates hold that the total annual amount of solid wastes in the United States in the late sixties was more than 250 million tonnes, to which must be added more than 550 million tonnes of agricultural wastes and more than 1.1 billion tonnes of wastes from the minerals industry. The cost of removing past waste has been estimated to be $4.5 billion annually (80 percent of the total being trucking costs). Geology is now an absolutely crucial element in selection of disposal sites.

New ecologically and environmentally acceptable sites are now selected only after detailed geological investigation. In some parts of Great Britain, municipal officials have opted to consolidate wastes for regional landfilling. An area of 164 ha (405 acres) of the south-coast city of Portsmouth, known as Paulsgrove Lake, but harbor-situated between the mainland and Hornsea Island, has now been enclosed from the sea by the construction of relatively simple embankments made of chalk excavated from a nearby hill. Refuse is now being compacted in controlled lifts for a total of more then 2.25 million tonnes. An incinerator plant will be in operation when the area is about half-full (Fig. 12.19), producing about a 90-percent reduction in waste volume, so that its ash residue will extend the life of the landfill from the original estimate of 20 years to perhaps 50 years. Advance predictions of the characteristics were based on geologic mapping and verified by subsurface exploration of 28 borings and four hand auger probes. Holocene marine deposits lay at the surface, overlying earlier alluvium and then the Upper Cretaceous chalk.

CONCLUSION

Land reclamation, begun as an earnest effort of winning land from the sea, has spread to encompass a variety of reclamation options, some with social overtones. Although most of these operations are simple in character, they will be successful only when planned and designed on the basis of accurate knowledge of the subsurface and of the characteristics of the material to be used for fill. The Bombay Back Bay land-reclamation scheme mentioned earlier was initiated in the mid-1920s, and was an imaginative and simple project for giving the great city of Bombay some much-needed new land. The area to be reclaimed was surrounded by a rock-fill dike, and a large dredge was commissioned to fill this area with material pumped from the adjacent seabed. Few residents of the area now realize that they are living on reclaimed land.

Before completion, a great deal of trouble was experienced because of the

FIGURE 12.19
New disposal area for garbage and waste, in Portsmouth Harbour, England; embankments enclose areas to be reclaimed; material (chalk) from new road constructed in the background is being transported by belt conveyor to form foundation material for new building site. (*Courtesy City Engineer, Portsmouth.*)

neglect of one apparently minor factor, the settling characteristics of the material obtained from the seabed. This was unusually fine-grained and so did not settle quickly, with the result that much material was lost by seepage through the open structure of the surrounding rock-fill dike. This early modern engineering lesson should not be forgotten.

Marine works, whether they are constructed to resist coastal waves, to train rivers, or to reclaim shorelines, all serve in active resistance to the forces of standing or flowing water. Some such forces are dynamic, as with waves, while others are passive, as with pore pressures of water held within the banks and foundation soils of marine works. Marine works are designed to hold land against the sea or rivers, yet they must also provide suitable foundations for many uses. Two main geologic concerns are always present: (1) finding suitable earth materials for construction or infilling, and (2) locating suitable foundation conditions to support the structures without unacceptable settlement or lateral movement. Both concerns are generally difficult to meet.

Most shoreline earth materials, by virtue of the winnowing action of the sea, or from the decreased particle-carrying ability of seacoast rivers, are fine-grained. Many of these silty or clayey soils are additionally laced with the organic content of sluggish waters; all of which call for special fill placement and internal drainage provisions. When the two main geologic concerns are not met adequately, and placement and drainage are not optimal, then marine works are hardly able to perform against the waters that they are designed to resist or train.

REFERENCES

1 A. de F. Quinn, "Great Lakes Port for Shipping Taconite Built by Ore Industry," *Engineering News-Record, 156,* p. 38 (October 18, 1956).

2 Francis Spring, "Coastal Sand Travel Near Madras Harbour," *Minutes of Proceedings of the Institution of Civil Engineers, 194,* p. 153 (1913).

3 B. C. Gerwick and E. Hognstad, "Concrete Oil Storage Tank Placed on North Sea Floor," *Civil Engineering, 43,* p. 81 (August 1973).

4 Allison Wilson, "The Lesson of Hallsands," *New Scientist, 45,* p. 311 (February 12, 1970).

5 "New Control Dams Remove Threat to Ol' Mississippi," *Engineering News-Record, 164,* p. 32 (January 7, 1960).

6 "The Hackensack Meadowlands Project," *Civil Engineering, 47,* p. 52 (May 1977).

CHAPTER **13**

NATURAL HAZARDS AND ENVIRONMENTAL CONCERNS

Sooner or later civil engineers come to learn that the crust of the earth is not the rigid, unyielding stratum that is often popularly imagined. Earth materials have properties similar to those of all other solid (and liquid) materials. They react to stress and exhibit strain, and they are not immovable, as many natural phenomena clearly show. In this chapter, some of the special features of the earth's surface crust and its materials are reviewed as they impact on works of the engineer.

Vertical displacements of the ground are a first consideration. An infinite variety of movements include both rising and lowering ground surfaces as well as the entire spectrum from instantaneous to long-term movement. Such movements sometimes follow the occurrence of earthquakes that result in ground rupture or severe ground shaking. After one New Zealand event the town of Karamea, located on deltaic soils, sank 60 cm (2 ft). Some ground subsidence, however, is so gradual that it has only been detected by precise leveling. A subsidence of 37.5 cm (15 in) in the vicinity of Kosmo, Utah, was determined in this way; it followed an earthquake of March 12, 1934, but the sinking was not evident to the eye.

More significant to the engineer, although from a long-term point of view, are regional movements affecting large areas of land, especially in Scandinavia and in northern Canada. Corresponding changes in the respective levels of sea and land have inundated some coastal areas while leaving some parts with uncomfortably shallow water.

Archaeological sites along the Italian coasts bear acute testimony to crustal movement. West of Naples at the ancient city of Pozzuoli stand the ruins of the Temple of Serapis, its floor now beneath the waters of the sea. Clam borings in the columns rise to a height of 6 m (20 ft) above present sea level. Fossils of similar marble-boring clams are found in adjacent bluffs to a height of 7 m (23 ft) above present sea level (Fig. 13.1). The only explanation that fits these facts is that the ground on which the temple was built must have subsided after construction, and to such a depth as would have permitted the clams to do their boring to the height on the columns. Thereafter the level of the land relative to the sea must have risen again but not to its original elevation. Some evidence shows that at least part of this uplift took place about 1500 A.D. The fact that similar evidence of changing relative levels is not found on all other coasts shows that this must have been a local phenomenon, but movements are found at various locations in the Mediterranean where ancient cities have been found beneath the sea. The Temple of Serapis is of special interest since it was first described by Sir Charles Lyell in his famous *Principles of Geology*, published in 1828, the first great textbook on geology.

On the average, sea level has risen about 100 m (330 ft) since the end of the last glacial period. This additional volume of water had been locked up in the mammoth glacial ice cover in North America, perhaps 3,000 m (10,000 ft) thick at its maximum, which extended from the Far North to as far south as St. Louis. The release of its melting water had this profound effect upon the sea. A major secondary effect has also been the *isostatic* uplift of the land itself as the tremendous load of ice was removed from it. This rebound process is still going on.

It has been estimated that melting of all the ice in Antarctica and in Greenland would force the sea level to rise at least an additional 70 m (230 ft). On the other hand, high-order leveling surveys show that the crust of the earth is still rising very slowly in those areas that carried the greatest loads of Late Pleistocene ice, some 11,000 years ago. This chapter therefore summarizes the main natural hazards which develop from normal "earth processes" with which civil engineers have to deal in connection with remedial work and with which they must be familiar.

FIGURE 13.1
Ruins of the Temple of Serapis at Pozzuoli, Italy, showing marks of marine clam borings on the columns. (*Courtesy Ente Nazionale Italiano per il Turismo, Rome.*)

VOLCANOES AND EARTHQUAKES

Volcanism and seismicity demand respect from the civil engineer as the most dynamic and far-ranging of the geologic processes. Volcanic action is clearly related to the existence of hot springs and zones of high heat flow, demonstrating the great temperatures that exist well below the earth's surface. Earthquakes are now widely understood to be related to the geologically slow movement of the massive plates of the earth's crust. Most active volcanism is found at plate boundaries or along major splits *(rifts)* in those plates. *Plate tectonics* theory has been one of the most exciting advances in the history of geology. Continental drift was first suggested over a century ago, and seriously promoted at the start of the present century by Taylor and Wegener. The concept, which holds that the continents in the southern hemisphere can be fitted together to form the protocontinent of Gondwanaland, although ridiculed for many years, is now so well regarded that there is little doubt in geological

FIGURE 13.2
It can happen here! The eruption of Mount St. Helens as seen on May 18, 1980. (*Courtesy U.S. Geological Survey; Photograph No. 80,53137.*)

circles that "the continents are drifting," and that the great plates are indeed moving and creating earthquakes and triggering volcanic eruptions.

Earthquakes and volcanism are often catastrophes, all too often leaving behind death, injury, and severe property damage. Engineering works are inevitably affected, and the civil engineer is frequently a leader in rescue and repair operations. These catastrophes may have to be accepted as part of the natural order, but geology should be used through prediction and forewarnings to produce measures to prevent harm. Appropriate planning must avoid locating structures over active faults. Active volcanoes are well known and under constant observation, but even within recent years *new* volcanoes have appeared in Mexico and on Iceland, while the largely unexpected 1980 eruption of Mount St. Helens in Washington State (Fig. 13.2), tells us emphatically that new and reactivated eruptions are to be expected in known volcanic regions. Correspondingly, the world's active seismic regions are now well delimited, and universal international efforts are at the threshold of accurate earthquake prediction.

North Americans naturally think first of the west coast of their continent as an area of known seismic activity, but there is also a serious long-recurrence earthquake potential in several midwest and east coast areas. Both Iceland and New Zealand are distinguished by hot springs, active faulting, volcanoes, and recurrent seismicity, with those of the former being associated in some way with the spreading of the mid-Atlantic ocean ridge. New Zealand, fascinating for its geological "newness," finds its capital of Wellington located on the site of a severe 1855 earthquake. A "world watch" to chronicle these catastrophic events was established in January 1968 as the Center for Short-Lived Phenomena by the Smithsonian Institution. About 3,000 registered correspondents in more than 150 countries report each event to the Center immediately upon occurrence, then observe each phenomenon as long as is necessary to establish its character and probable cause. The Center now issues premonitory warnings when indicated; typical geologic events reported on are major earthquakes, volcanic eruptions, and tsunami. It also has an annual report.

Volcanoes

Geophysical determinations suggest that the earth's central core has a density of 10.72 g/cm^3 and that it is surrounded by a mantle density of only 4.53 g/cm^3, although this is higher than the thin outer-crust density of just less that 3 g/cm^3.

History contains many records of devastating volcanic eruptions. The destruction of the ancient city of Pompeii and its port of Herculaneum by the 79 A.D. eruption of Mount Vesuvius is now well known, following the discovery of the ruins of Herculaneum in 1748 and the wide publicity given to the remarkable excavations of the buildings and streets of the old cities. The town of St. Pierre was completely destroyed in the violent eruption in 1902 of Mont Pelée, Martinique. Another tremendous natural explosion of recent times was the eruption of Krakatoa in the Strait of Sunda, Indonesia, in 1883. The conical volcano Paricutín, of Mexico, formed in 1943 from level farmland, whereas the new Icelandic volcano, Surtsey, has grown by underwater emission of lava along a sea-floor fault. The Hawaiian Islands are the eroded tops of volcanoes rising from the depth of the Pacific Ocean and are well known for their general mild activity.

Observatories now exist on or near a number of active volcanoes close to developed areas; warnings are issued when renewed activity is identified. Contingency plans can be put into operation by responsible local authorities for minimizing damage and loss of life. Mount Rainier in Washington State is such an example. Its last significant eruption was between 125 and 155 years ago, with some minor activity at the end of the last century. One of the remarkable "miscellaneous" geological maps published by the U.S. Geological Survey vividly estimates the potential hazards created by future eruptions. It is accompanied by useful explanatory notes, and by a small map showing the extent of tephra deposits more than 2.5 cm (1 in) thick from eruptions of the last 7,000 years, and the probable extent of mudflows and lava flows for this same period. The main map shows high-, moderate-, and low-risk areas for mudflows and lava flows and for tephra (airborne deposits). A subsequent study has defined probable impacts on public service delivery systems in the Puyallup Valley, identified by the USGS as a hazard area. Preparation and publication of such vitally important information is as far as the geologist can go; further action is in the hands of elected and appointed public officials.

Earthquakes

Earthquakes, as the most terrible and devastating of all natural phenomena, have real relevance to engineering. Although not frequent in occurrence, earthquakes can wreak such havoc that it is surprising to find that their scientific investigation in the Western World is of relatively recent date. In China, on the other hand, 3,000 years of earthquake records support earthquake prediction as one of the highest priorities in Chinese science. The Lisbon earthquake of 1775 was the first for which any scientific description exists in the West. The first seismograph was developed in the 1880s, following the pioneer seismic studies of Robert Mallet (published in 1862) describing the great Neapolitan earthquake of 1857. Mallet's papers include many excellent accounts of damage to buildings. These and other studies of the engineering effects of earthquakes have been a stimulus to the more fundamental study of the science of *seismology*.

The 3,000-year-long Chinese record chronicles some 10,000 earthquakes, 530 of them causing major disasters. The worst one was in Kansu Province in 1556 when more than 820,000 people were killed, all victims recorded by name. The earthquake that devastated Tokyo on September 1, 1923 cost 140,000 lives and caused damage estimated at $3 billion. The 1935 earthquake that devastated Quetta, India, took 35,000 lives in seconds. In many instances, permanent deformation of the earth's crust results, as with the San Francisco earthquake of April 18, 1906, which caused a maximum horizontal movement of 6.3 m (21 ft) along 435 km (270 mi) of the San Andreas rift. No vertical movement occurred in this case, but the earlier Yakutat Bay (Alaska) earthquake of 1899 caused a permanent vertical set of 15 m (50 ft) over a large area, completely altering the local topography and creating new waterfalls on local watercourses.

Less perfectly understood active seismic areas are to be found along the St. Lawrence Valley, in Ontario and Quebec, and in adjoining parts of New England. The long record of earthquake activity in these areas is without serious quakes in very recent times. On February 28, 1925, however, an earthquake of relatively low intensity affected a surprisingly wide area centered about Quebec City, and brought serious damage to heavy reinforced-concrete columns supporting a large grain elevator. No lives were lost in the main event, but several deaths were due to an aftershock. On May 2, 1944 a small earthquake occurred west of Montreal, its epicenter not far from the seaway city of Cornwall, Ontario. Damage amounted to about $1 million, but no lives

FIGURE 13.3
Anchorage, Alaska; damage in the Turnagain Heights area due to major sliding caused by the Good Friday, 1964 earthquake. (*Courtesy U.S. Geological Survey.*)

were lost because of the late hour of the disaster. Had it been earlier in the day, the collapse of many chimneys and parapets would almost certainly have caused a number of deaths, many of them among schoolchildren.

The same fortuitous timing led to a relatively small loss of life during the most recent of major North American earthquakes, which took place at 5:36 P.M. on Good Friday, March 27, 1964, at Anchorage, Alaska. Twice as much energy as the 1906 San Francisco earthquake affected an area of 130 million ha (320 million acres). It left 114 people dead or missing, but had it taken place on an ordinary working day, the possible loss of life is disturbing even to consider. A billion dollars of damaged public and private property followed uplift or subsidence of at least 9 million ha (22 million acres), as separated across a well-defined "hinge" line (Fig. 13.3). To the east of the hinge, land levels rose as much as 2.25 m (7.33 ft); to the west, subsidence of as much as 1.62 m (5.33 ft) took place. In the country around, thousands of avalanches were triggered, as were massive rockslides. One rockslide, 217 km (134 mi) distant from the epicenter, unfortunately killed a Coast Guardsman on duty at the Cape St. Elias lighthouse at the southern tip of Kayak Island. The Alaska Railroad lost most of its bridges along 300 km (185 mi) of right-of-way. Deep-seated coastal landslides took out much of the Valdez and Seward waterfronts. Homer, 256 km (160 mi) southeast of the epicenter, lost its small harbor into a "funnel-shaped pool." Incredible though it may seem, the water level in wells in which automatic level recorders had been fitted in both Winnipeg and Ottawa, 3,000 and 5,000 km (1,850 and 3,100 mi) away from the epicenter, showed violent oscillations six and eight minutes after the first shock waves hit Anchorage.

This truly remarkable phenomenon has been studied in far more detail than any previous earthquake. Anchorage suffered $200 million in damage, including destruction of 215 residences and severe damage to 157 commercial properties. Most of the nine deaths and the worst damage resulted from landslides at Turnagain Heights, L Street, and 14th Avenue. Each of these areas is underlain by a gray silty clay with some lenses of sand, well known locally as the "Bootlegger Cove Clay." The U.S. Geological Survey had released a preliminary evaluation of Anchorage geology in 1950 and as a published Geological Survey Bulletin in 1959. This valuable report described the Bootlegger Cove Clay and delimited its extent, presented its physical properties, and discussed conditions under which landslides could be activated. Local geotechnical engineers used this report in their soil and site investigations in the Anchorage area, as did engineers at the Alaska Road Commission and the Alaska Department of Highways. Regrettably, the report was not used as a basis for planning in Anchorage.

> The planning department was relatively new and its early problems concerned more pressing matters. The report is a general treatment and did not zone or classify the ground except by geologic map units. The map is not a document the planners can use directly without interpretation by a geologist. There were no geologists on the planning staff.

It is idle to speculate what this publicly available report could have added to sensible planning of this rapidly developing city.

Seismic-area structures should be founded on solid bedrock wherever possible. Unconsolidated material will fracture and be displaced much more easily than will solid rock. When a rock foundation is not available, foundation design must consider the additional earthquake forces that may be exerted upon the structure. A raft type of foundation, so designed that it will withstand upward forces in addition to the usual downward loads, is one type that has proved satisfactory, as have also continuous foundations carried on fixed-head piles driven to bear on underlying rock. In urban areas, earthquake shaking is liable to fracture water mains; old springs may stop flowing and new ones may be formed. Unconsolidated deposits and fill may consolidate to an extent not possible by any normal means. In the New Zealand earthquake of 1929, all filled ground sank to some extent. One fill which had been in place for six years sank almost 1 m (3 ft), although it was only 6.6 m (22 ft) high.

Steep river banks or artificial cuts may be forced to rapid instability. Tunnel portals are especially susceptible to damage by dynamically induced slides. The 1952 Arvin-Tehachapi earthquake in California affected 15 tunnels along the steeply graded main line of the Southern Pacific Railroad. Four of these were on a loop south of Caliente, adjacent to the White Wolf fault along which the movement took place. Three of the damaged tunnels had to be transformed into open cuts, as repair was uneconomic. Seed and Idriss have correlated damage to buildings with ground conditions. They conclude that "short period structures develop their highest damage potentials when they are located on relatively shallow or stiff soil deposits and long period structure when they are constructed on deep or soft soil deposits."

During the 1970s the phenomenon of *liquefaction* was identified as an earthquake threat in fine-grained soils. The term may be terminologically incorrect, but it is now in common usage and is best defined as

> the sudden large decrease of shearing resistance of a cohesionless soil, caused by a collapse of the structure by shock or strain, and associated with a sudden but temporary increase of the pore fluid pressure. It involves a temporary transformation of the material into a fluid mass.

Potential damage to structures founded on liquefiable soils during an earthquake can well be imagined; all too many examples can be interpreted from historic records of earthquake damage. During the 1906 San Francisco earthquake an

> earth-flow in sandy soil [was observed] at the base of the San Bruno Mountains. ...The sand and water forming this slide came out of a hole several hundred feet long and 46 m wide, flowed down the hill... carried away a pile of lumber and knocked the powerhouse from its foundations.

In urban seismic regions, local building codes usually govern the allowance to be made in structural design to withstand a horizontal force caused by a specified horizontal acceleration, generally one-tenth that of gravity. In other areas, however, regulations will give no guide to the engineer, who will then have to work with engineering geologists to evaluate past seismic history and to locate candidate faults as possible generators of a "design earthquake." It is advisable to select a rock foundation, level ground in preference to undulating or hilly ground; soft ground and uncontrolled fill material should be particularly avoided, as well as sites close to natural slopes such as riverbeds, canals, and coasts.

Many of the factors of interrelation between earthquakes and engineering works were dramatically demonstrated by the effects of the 1959 Hebgen Lake (Montana) earthquake. Hebgen Dam is an earth-fill structure on the Madison River in southwestern Montana; the structure has a concrete core wall, having been built in 1915 by a power company. The epicenter was close to the dam, causing major displacement along a known fault paralleling the reservoir and passing within 210 m (700 ft) of the right abutment. A vertical displacement of about 1.5 m (15 ft) took place on the fault. The dam was heavily fractured and settled as much as 1.2 m (4 ft), but it did not fail, even though overtopped by at least four seismically induced seiche waves. Far more serious was the massive slide which took place 11 km (7 mi) downstream, burying an unknown number of campers under at least 33 million m^3 (43 million yd^3) of rock. Slide debris buried 1.6 km (1 mi) of the stream bed to 120 m (400 ft). The U.S. Army Corps of Engineers quickly excavated a relief spillway but the slide and the lake it created remain as grim reminders of earthquake damage.

Seismic Design Requirements

Earthquakes will continue to take place throughout the world; they cannot be stopped or incrementally released, and needs for engineered structures will continue, in spite of seismic risk. What can be done? Fortunately, seismic design of today is in most cases quite adequate. The mass of theoretical and empirical information collected about earthquakes throughout this century, together with a considerable number of instrumental accelographs obtained at seismically-stressed structures, provide an excellent basis for design.

A description of the effects of any major earthquakes, however brief, must mention some of the engineering consequences as well as their geological implications or significance. It is the task of the geologist to delimit areas of relative seismic risk, and of the engineer to develop design and construction methods which will ensure the maximum degree of performance safety. The two disciplines meet in the joint effort required to delimit and classify areas of seismic risk and then to develop factors to be used in structural design to ensure maximum safety. These specific requirements for structural design and construction are today prominent features of the building codes now in force throughout the developed parts of the world.

As an example, the National Building Code of Canada is a nationally recognized set of design requirements prepared by a national committee. Accompanying each revised edition of this code is a seismic risk map of Canada. The entire country is divided into four numbered zones. Zone 0 indicates those parts of the country in which no earthquakes are to be expected. Zones 1 to 3 are areas with an increasing probability of earthquakes occurring, with corresponding increasing probability of damage being experienced if an earthquake should occur.

There is another factor that must be used in building design, this being what is called the *foundation factor*. Experience throughout the world has shown that much more damage is caused to structures by earthquake shocks when they are founded on soil than when they are directly in contact with firm bedrock. An expert Japanese committee has suggested that the effect of differing ground conditions can be expressed by the following factors.

Site geologic condition	Relative effect
Marshy land	1.5 Amplification
Alluvial ground	1.0 Non-affected
Dilluvial (older and more consolidated ground)	0.7 Reduction
Tertiary-aged rock	0.4 Reduction

There is still much uncertainty about the actual values that should be used for different site geologic conditions, even though the major difference between soil and rock is well attested. The Canadian Code noted above uses the factor of 1.5 for "soils [of] low dynamic shear modulus such as highly compressible soils" and a value of 1.0 for all other soils.

Man-Made Earthquakes

On the basis of the truly magnificent natural proportions of earthquakes, one might imagine that human activities could not possibly bring about earth tremors. Man-made earthquakes are indeed possible. Hoover Dam, on the Colorado River, was completed in 1935. The enormous loading caused by the water in its reservoir (estimated at 36 thousand million tons) caused as much as 17.5 cm (7 in) of land subsidence by 1955. Not surprisingly, the increase in the state of stress in the supporting rocks created micro-earthquakes, the first felt in 1936, the first recording made in 1938. Since then, a very large number of events as large as magnitude 4 on the Richter scale have been recorded. The same experience has been found around some other major reservoirs, such as those formed by Kariba Dam in southern Africa and by Kremasta Dam in Greece. These events serve to emphasize the absolute necessity for detailed geological study of the area to be inundated covered by all major reservoirs to ensure avoidance of highly stressed active faults that could lead to unanticipated tremors. For this reason, as well as to provide an analytic basis for ongoing stability evaluations, it is now common practice to install a small network of strong motion accelerometers around dams and major reservoirs.

Even more remarkable was the release of earthquake energy at the U.S. Army's Rocky Flats Arsenal (Denver) beginning in 1962. Liquid chemical wastes had seeped downward from receiving ponds into the uppermost aquifer. Groundwater contamination apparently caused damage to crops irrigated by the water, and a lined reservoir 43.3 ha (107 acres) in extent was constructed, together with necessary groundwater monitoring wells. Its capacity was exceeded and an injection well (believed to be the deepest of its kind) was sunk to 3,420 m (11,200 ft). Waste liquids were injected into the 300-m (1,000-ft)-thick Fountain Formation of porous sandstone, overlain by an impermeable stratum of shale 900 m (3,000 ft) thick. The well terminated at the underlying Precambrian granite and started with a diameter of 60 cm (24 in) at the surface. Together with a treatment plant to remove colloidal solids from the liquid, the well went into operation in March 1962; within a month, a series of minor earthquakes began in the Denver area, reaching 700 in number by November 1965. These were eventually attributed to the effect of increased cleft water pressure produced by the vast amounts of liquid being injected in quantities as much as 11.4 million L (3 million gal) per month. There was no doubt about the positive correlation between injection and earthquake incidence when disposal was stopped in February 1966. As is often the case, this unfortunate experience brought beneficial results of worldwide significance as a potential means of controlled seismic stress release and earthquake prediction.

Tsunami

Worldwide seismic distribution shows that earthquakes also occur beneath the sea. Underwater fault displacements create severe seawater upheavals which form anomalously large sea waves known by the Japanese name *tsunami*. Quite unlike the ordinary waves on the ocean surface, tsunami may extend for 160 km (100 mi) in width and be as much as 1,000 km (600 mi) from crest to crest. In the deep waters of the Pacific Ocean they have been calculated to travel at the amazing speed of 800 kmh (500 mph). Since they may be from 6 to 18 m (20 to 60 ft) in height, their devastating impact on shorelines is not difficult to imagine.

The Good Friday Alaskan tsunami struck all the adjoining coastline of Alaska with destructive force, and were recorded on tide gauges as far away as Japan, Hawaii, and California, where damage was caused. The tsunami created by the eruption of Krakatoa in Indonesia in 1883 raced across the Pacific and created heavy seas as far away as San Francisco. These seismic sea waves are much more common in the Pacific Ocean than elsewhere, seldom occurring in the Atlantic Ocean. Japan has been subjected to 15 destructive tsunami since 1596, eight of them being disastrous. The Hawaiian Islands are impacted about once every 25 years; a particularly bad tsunami struck in 1946. An ingenious instrumented warning system was developed by the U.S. Coast and Geodetic Survey and installed at critical Pacific locations. The system has chronicled 15 tsunami with not a life lost.

As with earthquakes, tsunami cannot be stopped, though they may be detected in advance. Lives will be spared, but nothing can save fixed property from the inexorable force of the sea. Shoreline development must be considered in relation to possible tsunami damage. Buildings must be firmly anchored to sound foundation units, raised, or relocated to higher ground.

There was a serious tsunami impact at Burin Peninsula, on the south shore of Newfoundland, as a result of the November 18, 1929 earthquake. This 7.3-magnitude event took place in the Grand Banks, about 270 km (170 mi) south of the peninsula, and was felt throughout eastern Canada and in the New England states. Tsunami impact reached the coastline almost exactly at high tide, striking the southern end of the Burin Peninsula and the west side of Placentia Bay with great force. With no warning at all, 27 lives were lost and almost 100 buildings were destroyed. So, "it can happen here" is an old tag not to be disregarded by any seacoast residents.

LANDSLIDES

The term *mass wasting* is used to denote one of the most widespread and active geologic processes—the gravitational displacement of rock and soil from their normal positions. In popular language the term *landslide* serves the same purpose, and so will be used here, even though strictly speaking it should be restricted to movements of soil when the structural stability of a slope is disturbed.

Landslides are a subject touching both geology and engineering and which has attracted attention ever since the start of modern civil engineering. An 1846 French volume dealing only with canal landslides was the start of a steady stream of publications which continue to appear. A notable late nineteenth-century volume was called *Slopes and Subsidences on Public Works*.

All slope movements are caused fundamentally by gravity, dependent completely on the nature of the materials involved and on their relative arrangement; in other words, on site geology. All types of slope failure may occur naturally; all may develop during the course of civil engineering work. If they occur naturally, they must be regarded as an inevitable part of the general geologic cycle, since they contribute to the erosion of parts of the earth's surface and thus are important factors in the processes that are developing new topographic features. If they occur during construction, they represent some interference with natural stability. In all cases, although the exact cause may be difficult to determine, they will be due to either one or a combination of several natural factors, which can usually be determined in the course of an engineering-geologic investigation.

Although in the minds of some, naturally occurring slides are classed as "acts of God," the cause of a slide is no mystery except for earthquake-induced slides. The essential instability of a slope develops slowly until a critical point has been passed; it is actually possible therefore to anticipate many landslides. Based on preliminary investigations, in which site geology is studied with unusual care, many of these events may be predicted with a good deal of certainty. Landslides are a regular and normal feature of landscape development, and obey the basic laws of slope stability.

Slope Stability

Typical earth slopes which present potential stability problems are cuts, piles, and bodies of fill material, and natural hillsides that have to be incorporated into civil construction. Whenever such slopes are composed of material that does not occur in bedded strata, the determination of stability becomes a matter of investigating the mechanical stability of a mass of unconsolidated material conforming to the outline of the slope under consideration. Once the physical properties of the material are estimated, the laws of mechanics can be applied to determine whether or not a specified slope will be stable. If an earth slope consists of coarse-grained soils—sand or gravel—then the maximum stable slope will correspond with the angle of internal friction of the material, readily determined by simple tests. There are necessary precautions to be taken to protect such slopes, but since they will be self-draining, there is rarely any question about their stability. Slopes of fine-grained soils, and especially clay, however, represent an entirely different challenge.

As early as 1846 the fundamentals of the stability of clay slopes were recognized by a gifted French engineer, Alexandre Collin. His field studies, primarily on canals under construction, showed that slides on clay slopes were all deep, rotational slips, generally on a cycloidal surface, starting with a tension crack at the top of the slope. He also carried out the first shear tests on clay, in order to determine soil properties. Reading the English translation of

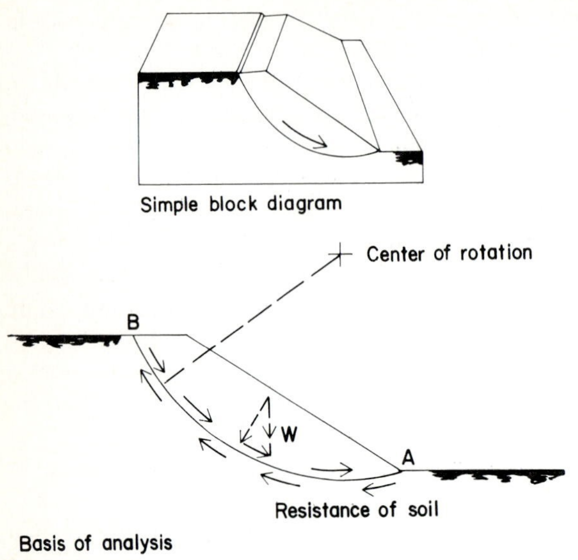

FIGURE 13.4
Elements of the stability of clay slopes.

Collin's book of 1846 makes clear that Collin recognized the importance of quick shear tests, the importance of "undisturbed sampling," and the dependence of all theoretical stability analyses upon local geologic structure. This dependence is an essential counterpart of the most meticulous theoretical stability calculations; assumptions made in calculations must always include geologic peculiarities of the site. Stability studies can often be aided by observing stable natural slopes in the project area that are under the same general conditions as those being considered. Correlation of natural and artificial slopes has been outlined by Ward, and applied by Skempton and DeLory in the London Clay (Fig. 13.4).

Natural Landslides

Natural landslides affect the work of the civil engineer in many ways; some are ancient and may be responsible for present-day topographic features, such as the deflection of the Columbia River at the present site of Bonneville Dam. Dams have, however, been successfully founded at these otherwise undesirable sites. One such example, the Farmers' Union Reservoir on the Rio Grande River in Colorado, blocks off one-quarter of the width of the true valley in the local lava flows; the remaining three-quarters is dammed effectively, with little leakage, by a mass of landslide material.

Recognition of ancient landslides is an important part of geological site characterization. Among the distinguishing features are rock fragments that are angular but neither polished (like fragments in glacial deposits) nor rounded (like stream gravel). Earth material will also be the same as that found in adjacent strata but the disposition of the particles or beds of materials in the mass will be irregular or chaotically jumbled. Landslide motion, both of natural slides and those artificially caused, varies from an almost imperceptible rate of failure to what may be appalling suddenness. The great Gros Ventre slide in Wyoming filled 2.4 km (1.5 mi) of a river valley 300 m (1,000 ft) wide, to an average depth of 54 m (180 ft). Eyewitnesses stated that the 38-million-m^3 (50-million-yd^3) bedding-plane slide occurred in less than five minutes. Similarly, the flow of the Thompson River in British Columbia was completely blocked in October 1881 by a 60-million-m^3 (79-million-yd^3) slide caused by collapse of a small, amateur-built irrigation dam on a terrace above the incised river valley.

Landslides in Sensitive Clays

Some of the largest recorded landslides are in regions characterized by sensitive clays—those that appear to be stiff but which may easily lose their shear strength. Slides in sensitive clays can develop from small beginnings to massive flows as the excess pore water is released to form a slurry viscous enough to carry large blocks of still solid soil. These flow slides are common in Scandinavia and in the St. Lawrence and Ottawa valleys of eastern Canada. A small Quebec village with the significant name of Les Eboulements (French for "landslides") is located 104 km (65 mi) downstream from Quebec City, on the consolidated fan caused by an ancient flow slide.

One of the most serious of all slides in the Leda Clay took place at St. Jean-Vianney, in the Lake St. John district of Quebec, shortly after 10 P.M. on May 4, 1971. Failure of the bank of the Rivière aux Vases swept away 40 homes and 31 persons (Fig. 13.5) along with 6.9 million m^3 (9 million yd^3) of soil. All of this took place within the crater of a much larger slide of about 300 years ago. Many flow slides in sensitive clay are initiated by purely natural causes with no human intervention, often including natural erosion of river banks. Ground loading and vibrations associated with local construction activity are often thought to contribute to the failure.[1]

Causes of Landslides

There are so many ways in which a landslide may be started that it is almost impossible to classify their causes. An almost invariable feature is water, once again a reminder of the importance of climate to engineering construction. Many natural slopes are in a marginally potentially stable condition so that their failure can readily be triggered. Among common causes are earthquake shocks. Differential erosion of a softer stratum may eventually undermine harder beds and cause a slide, and toe-slope erosion of unconsolidated material, all of which

FIGURE 13.5
View from a helicopter of the disastrous slide at St. Jean-Vianney in the Lake St. John area of Quebec, in which 31 lives were lost; the "flow" nature of the slide can clearly be seen. (*Courtesy Department of National Defense, Ottawa.*)

may remove essential support from the material above, which will start to move downward until stability is restored again. This happens even more easily on bedding planes which slope toward or into cuts, and are always a source of possible weakness.

Faults constitute another frequent cause of slides, whenever such planes are oriented and spaced so as to isolate blocks of material which are thus left free to move. The equilibrium of material in its natural position may sometimes be affected if heavy loads are placed upon it. This is often the case when excavation spoil is dumped too near the crowns of cuts. Many of the slides on the Panama Canal were induced by these, among other causes. Erection of buildings on unconsolidated slope material will sometimes have the same effect.

Probably the most important stability control factor is a change in groundwater conditions. This may be caused by interference with natural drainage conditions, excessive evaporation from normally damp ground, or an increase in groundwater level due to excessive rainfall, which may also erode the toe of a slope and so intensify failure tendencies. Rarely will surface erosion in itself lead to trouble; the landslides that form under intense rainfall are generally triggered by a corresponding change in groundwater conditions. When surface erosion strips off vegetation and bares the ground to rainfall infiltration, this may lead very quickly to an increase in groundwater and its three negative effects: (1) an increase of the effective weight of the material that it saturates; (2) buildup of pore pressure; and (3) weakening of many materials, including weak rock and unconsolidated materials with any clay content. One or more of these effects combined with the consequences of heavy rainfall will explain why many slides occur in wet weather and why internal and external drainage is so often an effective remedy for sliding.

Czechoslovakia is burdened with more than its fair share of landslides, especially in the Carpathians of Slovakia. A survey registered no less than 9,100 as part of extensive studies published in English in one of the most notable books on landslides. Handlová, an important coal-mining town in central Slovakia, supports a large brown coal-fired power station. Prevailing winds carried fly ash to the south, depositing the acid-generating dust over grazing land. Some grazing land was then plowed, allowing rain to percolate and disturb a very delicate groundwater condition. Handlová was then subjected to unusually high rainfall in 1960 (1,045 mm or 41 in), compared with the long-term average of 689 mm (27 in). Combined circumstances brought a rise in the groundwater surface and led to a start of slope failure early in December 1960. Accurate measurements initiated by December 22 recorded a maximum slope displacement of almost 150 m (460 ft) per month. A total volume of about 20 million m^3 (26 million yd^3) of clay, clayey silt, and rock debris was moving downhill, threatening to engulf the town.

The slide was quickly controlled by the end of January 1961. Most of the movement had been greatly reduced, but not before 150 houses had been destroyed. Here again, drainage was the key to the control measures. Today the great hill is again stable; to sit on its grassy slopes and to look at the town below, which could have been so seriously damaged, and all because of fly ash interfering with sheep grazing, is to be very forcibly reminded of the geologic complexities of slope drainage. Flow from the drainage tunnels and pipes that were so hastily installed is now so limited that one has to have a real appreciation of soil mechanics to appreciate the meaning of such a small quantity of water, before it was thus extracted from the slope mass (Fig. 13.6).

Preventive and Remedial Works

Landslide control is best undertaken on the basis of expert advice and only after full study and determination of the real cause of the earth movement. Sometimes "unloading" the upper portions of the slope will be called for. In other cases of small slides, retaining structures may hold the slopes so that the

FIGURE 13.6
The head of the 1960 slide at Handlová , Czechoslovakia, this being the first surface indication of the great slide. (*Photo: Q. Zaruba, Prague.*)

area below can safely be used. In extreme cases of very valuable land, attempts may be made to stabilize the soil by chemical means if local conditions are suitable. On construction works, temporary freezing of a large mass of moving soil has been utilized until structural concrete is in place and has cured. The engineering solution must be chosen to fit the particular case, after all the local factors are fully known.

Landslide investigation is a complex and difficult task. In the majority of cases, water will be found to be the culprit, and measures to drain the water out from the critical locations should be given first consideration. This unspectacular work can be misunderstood unless the drainage conditions are appreciated. In some cases, full-sized tunnels or large drainage pipes have been driven into soft and wet slide material to relieve internal water pressure. Another simpler and widely utilized technique is the use of drilled or jetted pipes with near horizontal gradients, perforated at frequent intervals. Water usually collects easily and eliminates the danger of work in the slide material. Effectiveness depends on accurate location of the drainage pipes in relation to site geology and especially to the surface of sliding. In all cases there is the necessity of controlling the surface water inflow to the slide area. Although surface drainage is essential, it is ameliorative rather than remedial.

Each remedial measure must be evaluated for its effect on the overall stability of the entire mass. Such remedial measures as removal of toe material and construction of either pile, crib, or reinforced earth retaining walls at the toe of a slide should not be used, except possibly as temporary emergency measures. These measures may be effective in very small slides, but the general result of toe removal will be to lessen the resistance to movement and to increase the extent and rate of movement instead of stopping it. The use of piles, in particular, is usually clear evidence that those responsible for the remedial works do not appreciate the basic character of earth movement in landslips; the idea of retaining a deep-seated movement of a large mass of soil by means of slender piles is on a par with King Canute's efforts to stem the tide. Also, the vibration necessary to drive piles may further decrease stability in waterlogged or fabric-sensitive materials. Retaining structures are sometimes considered to the neglect of the fundamentals of engineered design. Retaining structures should be designed to transmit a predetermined load to a foundation bed of known capacity, and certainly not the generally unpredictable load induced by movement of an entire slide. There are exceptions to this general rule of unsuitability, but generally only in situations where walls are small and made to serve not so much as retainers but as drainage structures for the lower parts of otherwise stable slopes.

Slope readjustment is another important possibility as a remedial measure, taken to initiate limited controlled failure. Now that stabililty calculations for slopes are so well established, landslides caused by excessive slopes or inadequate drainage should steadily decrease in number. Slope drainage is always a prime candidate for remediation. Drainage efforts begin with runoff control, but interior drainage is usually necessary to reduce pore pressure along the sliding plane. To be effective in preventing or remedying slides, drainage must intercept even small amounts of groundwater which can cause extensive slides. Features such as thin beds of sand, only 2.5 to 5.0 cm (1 to 2 in) thick and only slightly water-bearing, and even a 2.5-cm (1-in) seam of decomposed coal, have led to extensive slides. Drainage can be applied effectively, therefore, only on the basis of a full understanding of the cause of the slides and the source and movement of the groundwater that is contributing to movement. Slopes must be characterized through a judicious program of geologic mapping and exploratory drilling; needless to say, such work is best carried out before, rather than after, any movement has taken place.

Drainage is also an essential ingredient in the maintenance of stable slopes. Many ingenious drainage methods have been developed for use with unstable slopes. Highway construction in California now relies on routing drainage installations where the cost of landslide-stabilizing works can exceed one-third of the total annual State expenditure on highway maintenance. Both drilled and jack-in-place horizontal drains make use of small-diameter perforated metal pipes, and the newer, larger-diameter PVC varieties.

Probably one of the worst individual landslide situations ever to be faced and solved by civil engineers was at the Folkestone Warren, a 3.2-km (2-mi) stretch of Britains' famed White Cliffs of Dover. This main line segment of British Railways, connecting the ports of Folkestone and Dover between the Martello Tunnel near Folkestone and the Abbotsford Tunnel nearer Dover, has suffered extensive landslips from the earliest historical times, possibly since the breaching of the Straits of Dover in Neolithic times. The double-track railway line runs in a shallow cut at the base of the main cliffs, which rise about 150 m (500 ft) above sea level.

Geotechnical conditions at the Warren in 1951 are shown in Figure 13.7. Since its opening in 1844, maintenance has been plagued by earth movements; major remedial work was begun in 1948 following one of the most intensive programs of subsurface exploration on record. Site characterization started in 1939 but had to be stopped because of World War II. Between 1940 and 1948 the position of the shear surface became visible in two of the timbered drainage adits that had been constructed before the war. The movement continued slowly throughout the war, without a major slide. Engineers in charge completed the site characterization in close cooperation with the Geological Survey of Great Britain; their joint efforts produced a solution to the century-old problem. Previous slips had displaced the original geological stratigraphy, making it necessary to rely on a detailed study of the fossils in the Gault Formation obtained from boreholes. It would be hard to imagine a better example of the complete interdependence of such engineering and geologic studies. The Folkestone Warren landslips are unique, but the published records are invaluable guides even for remedial work on a much smaller scale. The subsurface was adequately explored; the cause of the slide was found; and the main contribution was dealt with by a carefully planned drainage system. These steps characterize most landslide preventive and remedial works.[2]

Stability of one large California hillside was maintained because of an extensive hot-air pumping system which circulates warm air through tunnels excavated in the host clay in order to evaporate groundwater seepage. This unusual scheme worked so well that it was possible to fill the tunnels with gravel and discontinue drying, keeping drying as an option that can be resumed if necessary. The drying operation was carried on from 1933 to 1939, and it has not proved necessary to repeat the process. This extreme example is included to show what can be done to remediate relatively small but economically serious slides.

As landslides are so numerous, so varied in type and size, and always so dependent upon special local circumstances, it is hoped that this general review will at least provide a useful background for more detailed consideration of the "landslide situation" in specific localities. Many urban areas have more than their fair share of slope instability. The Los Angeles area in particular has experienced so many small landslides associated with hillside development that a special "grading ordinance" was passed in 1952, since amended and strengthened. Geologists here have shared their hard-won experience in a number of useful publications; public convenience has been enhanced and recognition for the importance of engineering geology ensured.

There is no part of civil engineering that calls more clearly for the joint efforts of engineering geologists and geotechnical engineers than does the struggle with landslides—but always, to the maximum extent possible, *before* they occur. If a landslide *does* occur the following course of action might well be prescribed:

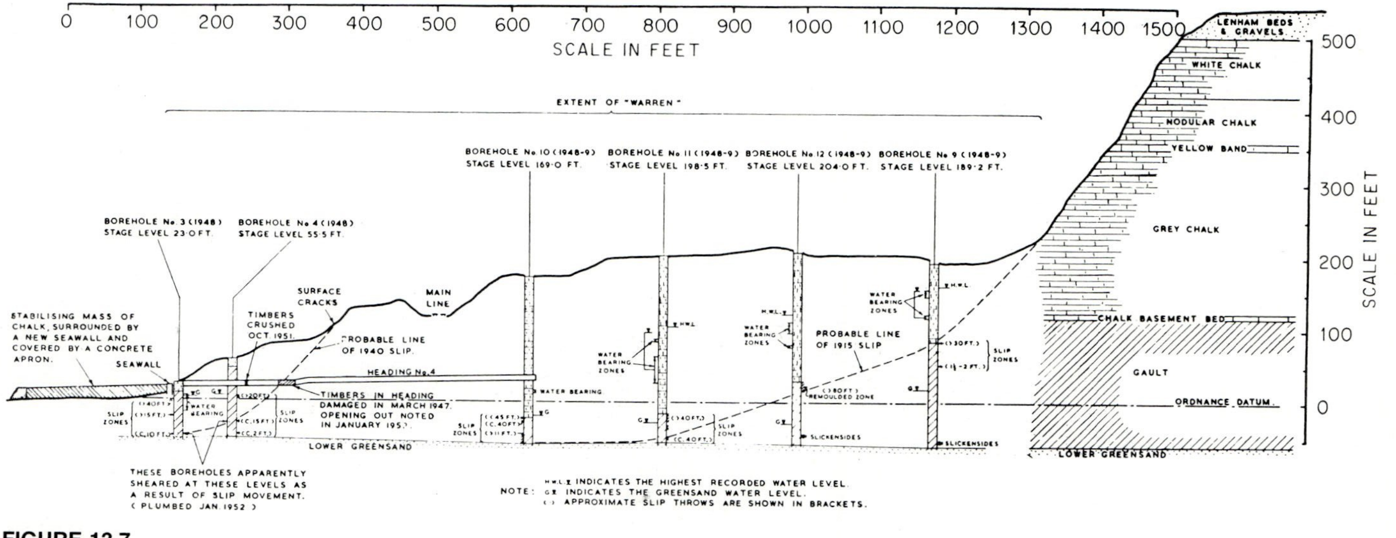

FIGURE 13.7
General cross section of the ground conditions at the Folkestone landslide, Kent, England, showing something of the test borings and remedial works carried out. (*Courtesy Municipal Engineering Publications Ltd., London.*)

1 Don't panic.
2 Don't rush out and "drive piles" or grade the toe.
3 Visit the site; measure, photograph, geologically map surface features; develop key topographic profiles.
4 Study all water exhibited at site; remove ponded water.
5 Review local geology; determine site geology.
6 View site from air; procure and interpret pre- and postslide aerial photos.
7 Study any nearby older landslides that can be located, if any.
8 If a slip-slide has occurred, try to locate evidence of failure surface.
9 Ask "Why did the slide happen?" before seeking solution.
10 Study local climate records, long-term and recent.

And only then:

11 Put down any necessary exploratory borings.
12 Design candidate solutions, using slope stability calculations.
13 Install necessary internal and surficial drainage facilities and/or stabilizers.
14 Prepare best possible geotechnical site record and ensure its deposition where it will be regularly used in inspection and maintenance.

ROCKFALLS

Rockfalls, a major *mass wasting* subgroup, are sometimes difficult to categorize in a hard-and-fast division between soil and rock, and so the suggested categories of ground movement have some overlap. The term *rockfall* does connote collapses of all types of rock and some kinds of weak rock. Rockfalls are generally more determinate and understandable than other types of earth movement. They occur when a mass of unsupported rock becomes detached from surrounding host rock and thus becomes free to move downward, under the influence of gravity. Loosening of masses of rock is almost always associated with such features as bedding planes, joints, foliation, or a local shear or fault zone or plane. The immediate cause of loosening will be a change in internal stress distribution, which will usually be the result of weathering, cleft water accumulation, or natural erosion of the toe of the slope mass.

Natural rockfalls are evident on any steep mountain face; talus slopes are their untimate result. Many large rockfalls take place in uninhabited mountain regions and are of interest mainly to the geologist. One great rockfall was not so located. Turtle Mountain, Alberta, is composed of Paleozoic limestones overthrust upon a syncline of Mesozoic shales, sandstone, and coal beds. In the early morning of April 20, 1903, the crest of this great mountain broke off and slid to the valley below, taking at least 70 lives, damaging much property, and burying nearly 2,200 m (7,000 ft) of the Crow's Nest Railway. The catastrophe was closely investigated by Canadian geologists and engineers, who concluded that the major cause was failure of the rock along the prominent limestone joint

FIGURE 13.8
Sign at the lookout that has been constructed adjacent to the new road, replacing that buried 60 m (200 ft) beneath this spot by the Hope Rockfall, British Columbia. (*Photo: R. F. Legget.*)

planes, although operations in a coal mine located at the foot of the mountain may have contributed to the strain that initiated the rockfall. Sixteen miners miraculously dug themselves an escape way through the soft coal seam where it outcropped outside of the slide mass.

Western Canada was the scene of the very great recent rockfall at Hope, British Columbia. This rockfall combined with a rockslide took place just before dawn on January 9, 1965 in a narrow valley some 16 km (10 mi) to the east of the town of Hope, 160 km (100 mi) to the east of Vancouver, and along the relatively new transmountain highway to Princeton. The original road is now buried 78 m (240 ft) under the 130 million tonnes of rock that took four lives; miraculously, passengers in a bus and other automobiles escaped. The rock involved is massive to slightly schistose metavolcanic with intrusive sheets of felsite that dipped nearly parallel to the 30° slope of the mountainside. Rainfall did not appear to have been a contributing factor, but two small earthquakes appear to have been the triggering mechanism. Today, a new road has been constructed over the top of the great pile of fallen rock, as well as parking lot and observation area (Fig. 13.8). To stand here and see the bare face of the rock from which this vast quantity fell so quickly is a remarkable experience for any traveler who has any appreciation of geology at all.[3]

FIGURE 13.9
Reinforced concrete protective works at Vriog Cliff, near Barmouth, Wales, on the western region of British Railways. (*Courtesy British Railways, Western Region.*)

Some Smaller Rockfalls

Rockfalls encountered in ordinary civil engineering practice are, of course, much smaller than these great natural phenomena. Rarely will vast failures be encountered, even in the construction of transportation routes in mountainous areas. Small rockfalls are common along those routes that are adjacent to natural rock cliffs and in deep rock cuts. An example is illustrated in Figure 13.9, which shows the reinforced-concrete protective "tunnel" constructed by the former Great Western Railway Company near Barmouth on the west coast of Wales. Vriog Cliff rises 450 m (1,500 ft); on March 4, 1933, a train was swept off the tracks at this location by a slide that happened after a severe blizzard and a period of thaw. The slip took place along a thrust fault in Ordovician strata, consisting here of dark flags and slates, one or two beds of limestone, and abundant volcanic tuff. The protective works were constructed after a complete site survey, to avoid any possibility of trouble from slides that might develop in the future because of continued weathering of the exposed weak rocks. As if the great landslides at Folkestone Warren were not enough of a plague upon British Railways, this same line has been the scene of major rockfalls in the Dover Chalk, particularly after heavy rains. Between the Abbotscliff and Shakespeare tunnels, the line is protected by a 3.2-km (2-mi) seawall. Steep chalk cliffs abut the line. A 19,000-m^3 (25,000-yd^3) rockfall took place in November 1939. Since trouble was anticipated, trains were stopped just before the fall occurred.

Hydroelectric powerhouses, commonly located in rocky gorges, are liable to rockfall damage; this is always an important feature to be accommodated in design. Where probable rockfalls can be detected, it will prove economical to remove the loose rock under control with the aid of skilled "rockmen." If this cannot be done, then some form of wire netting or concrete deflection works may be constructed so that falling rocks will not cause serious damage.

On June 7, 1956, a catastrophic rockfall dislodged a sequence of limestones and shales in the Niagara Gorge. Niagara Formation rocks are well exposed in the upper part of the gorge, above the old 300,000-kw Schoellkopf hydroelectric station of the Niagara Mohawk Power Company. On the day of the failure, serious seepage developed in the cliff and water flowed into the powerhouse. Pumping was in progress when, at 5:10 P.M., 38,000 m^3(50,000 yd) of rock demolished two-thirds of the powerhouse, killing one worker and causing damage of about $100 million.

Exposed rock faces of doubtful stability can often be made safe by *rock bolting*. The basic idea is simple—drilling a hole into rock suspected of instability and inserting a long steel bolt, suitably wedged at its end so that, through driving, it can be securely fastened or cemented in the hole, thus anchoring the outer layer of rock to that in which the bolt is keyed (Fig.13.10). Extensive rock bolting was used to anchor the exposed faces of pink sandstone at the Glen Canyon Dam on the Colorado River in Arizona. After thorough exploration, the U.S. Bureau of Reclamation employed 1,000 mine-roof bolts (Fig. 13.11), 1.8 and 2.4 m (6 and 8 ft) long and 1.9 cm (3/4 in) in diameter, to "pin" potentially loose slabs of rock. A similar precautionary measure was taken in 1953 at the Nevada valvehouse of Hoover Dam. Although no movement of the rock had been noticed, it was decided to "pin back" a slab 54 by 36 m (180 by 120 ft) by means of 349 5-cm (2-in) anchor bolts grouted into place in 75-cm (3-in) holes. The work was carried out from a spectacular piece of steel scaffolding.

Minor Rockfalls

One hesitates to describe any rockfall as "minor," since even cobbles falling from a great height can wreak much damage. The term is useful, however, to indicate the fall of individual pieces of rock, made harmless if they do not strike human activity. During regular inspection and maintenance of civil engineering works one commonly encounters the evidence, especially along main highways bearing "falling rocks" signs. Following the railway example, some road authorities have had to install safety fences to protect road users from this hazard. On the Nuuanu Pali Highway in Hawaii, running between Honolulu and Windward on Oahu, a double row of strong steel fencing serves at one critical location. In Quebec City, even the historic Cape Diamond rock mass, familiar to all who have sailed up the St. Lawrence, occasionally drops rocks on the narrow street that winds along the shore below it. City authorities have installed wire rock baskets to intercept falling boulders.

At the other extreme of protective measures are the electrically operated protection fences that are used for particularly dangerous stretches of road or

FIGURE 13.10
Edinburgh Castle, Scotland, and its famous rock, some reinforcement of which has been necessary. (*Courtesy Scottish Tourist Board, Edinburgh.*)

rail. When the source is identified, another alternative is to cover the site with strong steel-wire mesh, adequately secured. An outstanding example of thispractice was carried out when Interstate 70 was opened through Clear Creek Canyon, west of Denver. In the Floyd Hill–Idaho Springs section of the road, twin one-way, two-lane tunnels were constructed through a promontory dipping at about 45° across the line of the road. Expecting weathering to loosen rock on exposed bedding planes, the Colorado Department of Highways included protective chain-link fencing in the construction contract. No. 9 galvanized wire net was secured into place by 252 rock bolts, 1.8 to 2.4 m (6 to 8 ft) long, on 2.7-m (9-ft) centers. After nearly two decades of service, this "rockfall guard" has proved its usefulness.

Rockslides

Rockfalls are obvious hazards to civil engineering works. Rockslides are not, but their usually larger masses make them of equal importance. The term applies to masses of loose rock which, either alone or in association with soil, move under the force of gravity along an inclined surface of competent bedrock. There are cases (as in the Carpathian Mountains) where immense blocks of rock, separated by jointing but having the appearance of

FIGURE 13.11
Installing rock bolts in the sandstone cliffs at the site of Glen Canyon Dam on the Colorado River, Arizona. (*Courtesy U.S. Bureau of Reclamation.*)

bedrock, are actually slowly moving along an underlying stratum of clay. More common, however, are masses of closely bedded and jointed rock on sloping hillsides which, in the course of geologic time, have become so consolidated that they appear to be solid rock but which are nevertheless moving slowly.

Visitors to Switzerland who travel the narrow-gauge Rhaetian Railway to the mountains about Chur pass through the pleasant town of Klosters. Here the railway crosses a small river on a slender reinforced-concrete bridge. Closer inspection of the bridge will disclose a reinforced-concrete beam between the two abutments, designed to serve as a convenient footbridge between the two parts of the village. After the bridge was placed in service, regular inspection detected a slight rise at the center of the bridge. Under full investigation, it was found that the southern abutment was founded on slowly moving rock talus. The only solution was to brace one abutment against the other, by means of the "footbridge." Embedded in it was the necessary strain-measuring instrumentation, which has now confirmed the validity of this design. The vital importance of regular inspection during routine maintenance will be clear. Once again it can be seen that inspection must not only examine structures, but the geologic surroundings on which their stability ultimately depends.

LAND SUBSIDENCE

Extraction of water, oil, and gas in some areas carries with it the penalty of land subsidence. Unlike subsidence caused by the mining of ores and coal, subsidence caused by the withdrawal of fluids will rarely be sudden but will usually be so slow as to be imperceptible under ordinary observation. It will usually cover so wide an area that even when extensive subsidence has been caused it will grade so gradually toward undisturbed ground as to be difficult to appreciate unless there is some man-made structure to act as a telltale. The well-known case of such subsidence at Long Beach, California, has reached a total of 7.8 m (25.5 ft) at the center of an area of 5,200 ha (12,300 acres).

In general terms, the basic mechanism of subsidence is simple and yet individual details are generally complex. When subsurface fluids are extracted, the surrounding ground will tend to readjust itself to changes of pressure just as it will to any other change in applied stress. Where underground pressure has been operative in holding apart individual particles of silt or clay, its release may permit such a redistribution of underground stress that the clay will begin to consolidate with a consequent decrease in volume. In these and other ways, the very slow movements associated with consolidation eventually reach the ground surface as subsidence.

Pumping of groundwater in the London basin has led to slow consolidation of the great stratum of London Clay beneath the city. The result has been a general ground subsidence, though few Londoners are aware of it. The latest estimate of maximum settlement since 1865 is 2.14 cm (0.85 in), occurring near Hyde Park. There has been a corresponding drop in the groundwater level from the 1820 elevation of 9 m (30 ft) above sea level to today's level of as low as 90 m (300 ft) below the same datum. This pattern is repeated around the world: A natural supply of groundwater exists beneath the site of a city, pumping is initiated, and overpumping leads to serious lowering of the original water surface. Finally, destruction of artesian conditions (if they ever existed), occurs along with increased pumping costs, possible seawater contamination, and subsidence of the overlying land surface. Geologic factors conducive to these worst-case developments are, fortunately, not always present.

In North America, pioneer hydrologist Dr. O. E. Meinzer discovered pumping-induced ground subsidence in connection with his groundwater investigations in the Dakotas and at the Goose Creek oil field, Harris Co., Texas. Groundwater withdrawal is responsible for 520 km^2 (200 mi^2) of settlement in which the maximum slightly exceeds 1.5 m (5 ft) in the Santa Clara Valley, about 32 km (20 mi) south of San Francisco, California. The maximum occurred in the business district of the city of San Jose, California, which has a population of 80,000. This subsidence was first detected by survey in 1920, as part of the San Francisco Bay shoreline was already protected by dikes. Settlement was under control by 1938 through restrictions on wellpumping which raised the groundwater level by 15 m (50 ft). Subsidence once initiated remains as a permanent form of deformation.

More serious subsidence has brought the city of Long Beach, situated over the largest oil field in California, an anticipated ultimate subsidence of 7.5 m (25 ft). Settlement was first noted shortly after installation of wells in December 1936. Combined withdrawal of oil, gas, and groundwater produced an elliptical area measuring 8 by 5 km (5 by 3 mi) by 1950, and much of the harbor was seriously affected. Maximum settlement was 3.3 m (11 ft), increasing at the rate of about 0.5 m (1.5 ft) per year. Horizontal displacements of 1 part in 1,200 were also induced. Remedial works were begun about this time. A major bridge was jacked up 22.5 m (7.5 ′ft), and protection walls and other works were constructed at the naval yard. By 1959, 52 km^2 (20 mi^2) had settled as much as 7.8 m (26 ft) with 0.6 m (2 ft) at 14.4 km (9 mi) from the center of the area. Approximately two-thirds of the oil reserves had been withdrawn. Since a $175 million U.S. Naval Station was at the center of the area, the Federal government authorities brought an injunction to close the field. A crash program was initiated by the Long Beach Harbor Commission. Eight pumping plants were built, with an installed capacity of 142 mld (37.5 mgd) (capable of increase to 167 mld, or 44 mgd), to pump saltwater back into the oil-bearing strata under pressure ranging from 60 to 87 kg/cm^2 (850 to 1,250 psi). By the end of 1959, almost 75 mld (20 mgd) were being pumped underground and subsidence was brought to a halt.[4]

Mexico City lies in a closed basin about 2,500 m (8,200 ft) above the sea. Beneath the surficial alluvial deposits is a complex geological structure dating from the end of the Cretaceous Period, and greatly affected by volcanism as recently as 2,000 years ago. Subsoil units include soft, fine-grained volcanic ash and water-transported sediments, as well as sand and gravel with interbedded clayey silt. The latter are a fine aquifer, naturally inviting pumping. The overlying beds of silts and clays are highly compressible. The heavy Aztec monuments and later Spanish governmental buildings have experienced early settlement (despite the deep-pillar foundations of the Aztecs). General ground subsidence, however, seems to have started at the end of the nineteenth century. Benchmarks show a general subsidence of about 1.5 m (5 ft) between 1900 and 1940, but then a rapid increase happened, with total settlement of the general ground level of as much as 7 m (23 ft) by 1960. Visitors can see evidence of the subsidence in older buildings, and more dramatically in the protrusion of old well casings originally flush with the surface but now protruding as much as 6 m (20 ft) above ground (Fig. 13.12). Water is now brought into the city in order to reduce the pumping, which is now strictly controlled. Recharge wells have been installed. Subsidence is irreversible, but further development of the city need not induce further settlement.

It is predictable that Texas should suffer from pumping-induced subsidence,

FIGURE 13.12
Mexico City, Mexico: a well casing, the top of which was once at ground level, illustrating the 6 m (20 ft) of ground subsidence that has taken place here. (*Photo: C. B. Crawford.*)

especially in the underconsolidated strata of the Gulf coastal plain. Withdrawal of oil, gas (initial pressure of 70.3 kg/cm^2, or 5 psi), and water from GooseCreek oil field led to the first recorded case of this subsidence. In 1925 an area of 4 km (2.5 mi) long by 3 km (1.9 mi) wide had subsided by more than 0.9 m (3 ft). The area affected conformed with the area tapped by the producing wells. Some of this land became permanently flooded. Texas subsidence is worse around Houston and its suburbs of Pasadena, Baytown, and Bellaire. Pumping of oil, gas, and water had affected (by 1962) an area of almost 180,000 ha (445,000 acres), with subsidence of at least 30 cm (12 in) and maximum settlements exceeding 0.9 m (3 ft). The area is underlain by the consolidation-prone Beaumont Clay. Subsidence-related "growth faults" frequently break, and abrupt changes in elevation of as much as 60 cm (24 in) are evident along roadway surfaces. Typical of the associated problems was the rupture of a natural gas supply line to an industrial plant. This incident, which occurred on a Sunday afternoon without loss of life, does demonstrate that urban problems follow subsidence. Houston-area problems are mitigated by placement of sewers in twin pipes, within a zone of 150m (500 ft), as a precautionary measure to resist vertical deformation.

Overpumping of groundwater does not necessarily cause ground subsidence. Geologic conditions leading to subsidence are usually relatively thick sequences of soils (especially clays), with intervening aquifers. Precise leveling is used to detect the onset of subsidence. Within Savannah, Georgia, subsidence of as much as 10 cm (4 in) can be directly attributed to a substantial increase in groundwater withdrawal from the Ocala Limestone and associated Eocene limestones. Memphis, Tennessee, on the other hand, has experienced a decline in the artesian head of more than 30 m (99 ft) within confined Eocene aquifers yet does not have appreciable subsidence. This is not to defend overpumping; it merely emphasizes that even excessive drawdown does not necessarily mean damaging subsidence. The possibility, however, always exists. Its probability can be determined only by geologic study, including regular observations of ground levels to verify preliminary geological predictions. Limitations on pumping can have beneficial results: limiting the magnitude of subsidence, conserving groundwater, resisting saline contamination in coastal locations, and possibly maintaining satisfactory foundation conditions for large structures.

Mining-Induced Subsidence

Subsidence due to mining operations is more universally expected than that due to pumping. The records of mining operations prior to about 1960 provide many examples to confirm this suggestion. A great deal of "hard rock" mining has always taken place in igneous rocks. If the underground excavations are not too large, such rock is competent enough to be self-supporting. A good deal of modern mining is carried out at depths of 1 to 4 km and without influence at the ground surface. For more shallow operations, especially for coal and in less competent rocks, most of the developed countries have regulations designed to minimize surface damage. A typical requirement is for backfilling with inert material, which unfortunately sometimes consumes large quantities of sand and gravel. Better yet are those operations making use of mine spoil, ore processing wastes, and utility-generation fly ash and boiler slag, thus solving two environmental problems simultaneously.

Throughout the Old World, unexpected subsidence occurs when old mine workings are encountered during construction excavation. J.E. Richey cites a case in which an excavating machine disappeared into shallow workings covered only by a thin roof of rock and lodgement till. The site had to be opened up and the old workings backfilled before construction could proceed. Unsuspected coal workings were encountered beneath the site for a new science building at the University of Glasgow, Scotland, and had to be fully investigated; necessary roof reinforcement had to be provided to support the new building. An extreme case of mining subsidence affected the "ironstone" region around Barrow-in-Furness, immediately southwest of the English Lake District. Its

FIGURE 13.13
The cavity formed on the (former) Furness Railway at Lindale, near Barrow, Lancashire, England, by coal workings; a locomotive disappeared into the hole in 1892. (*Photo: E. H. Scholes.*)

underground iron mines had been worked for many decades along unsurveyed locations. On September 22, 1892, a Furness Railway locomotive was shunting iron-ore wagons at the railyard at Lindale when the ground beneath failed, taking rails, locomotive, and tender (Fig. 13.13). Only the tender was recovered by a crane; the locomotive was never seen again.

North America also has its share of land subsidence due to mining operations even though its mines are relatively new. As recently as 1968, serious property damage took place in the borough of Ashley, 6.4 km (4 mi) south of Wilkes-Barre, Pennsylvania, damaging streets, bridges, water and gas lines, homes, and churches. Wherever old mine workings exist at relatively shallow depths beneath the ground, the possibility of damage exists. Subsidence may take place suddenly as a result of long-term groundwater erosion of support pillar material in the workings. Ground-surface activities such as blasting or water ponding also affect the stability of these underground openings. Construction of foundations for new structures over mined areas always requires unusual care in site investigation.

A score of urban areas could be used to illustrate these problems. Pittsburgh, Pennsylvania, is a typical example. Coal was first mined here in 1759 by British soldiers on Mount Washington overlooking Fort Pitt. This busy city sprawls across hills and floodplains, with a topographic relief of about 160 m (530 ft). The main Pittsburgh coal seam, the bottom unit of the local Monongahela Formation, has been eroded over one-third of the area. Other seams have been mined locally to depths of as much as 180 m (590 ft) below the surface. The Pittsburgh Seam outcrops on hilltops in the city area but then dips to depths of over 190 m (620 ft) to the south of the city. It has been extensively mined, the right to mine it being originally purchased with waivers against all surface damage. Starting in 1957, one company recognized the social implications of its operations and guaranteed surface safety from subsidence if approximately 50 percent of the coal under a private property was purchased by the home owner and left in place. This practice has been extended. In one 10-year period, one guaranteeing company mined under 635 homes and had to repair only 2 percent of the total. The 1966 enactment of the Bituminous Mine Subsidence and Land Conservation Act of the Commonwealth of Pennsylvania was designed to prevent undermining that would damage public buildings or noncommercial property. Such a wealth of Pittsburgh-area experience is now applied by consultants to a wide variety of subsidence problems. Much of this knowledge is available for the guidance of others facing similar problems.

Successful use of these previously mined sites requires accurate knowledge of all subsurface conditions, especially the extent, layout, and age of old workings. Ingenious methods are used for those workings that cannot actually be entered and surveyed safely. One method has been to drill 15-cm- (6-in)-diameter holes from the surface into the mine workings, then to lower into them a rotating borehole video camera to assess mine workings visually. With this fortuitous method and other means, it is possible today to prepare dimensioned maps of old workings and then to position the overlying structure for optimum foundation conditions. For a proposed school it was determined that an alternative site about 240 m (800 ft) distant had 19.8 m (65 ft) of rock between the foundation and the mined-out Pittsburgh coal seam, 70 percent of which remains in place as a safe foundation.

Methods of Avoiding Mining Subsidence

When sites or layouts cannot be changed, other expensive methods to avoid subsidence must be used. Concepts of providing coal mine roof and pillar support start with construction of solid concrete piers or walls directly beneath the foundation locations for the structure to be erected above. Where the old workings cannot be entered, alternatives include grouting, in some cases after sand and gravel have been poured into the old workings. A significant amount of engineering judgment is required, and this can be exercised only on the basis of complete geologic information about subsurface conditions. In some areas grouting or backfilling is needed for worked-out coal seams as deep as 45 m (148 ft). An alternative procedure, if subsidence is considered likely, is to include provisions for jacking the building back into place after postconstruction settlement.

Modern gold mining can produce notable ground subsidence. Johannesburg was founded at the site of near-surface gold strikes. The old workings now severely restrict construction within an old reef zone running east to west and immediately south of the central business district. Ground movements still take place, and so this area has been zoned for limited development. Freeways now skirt most of the subsiding area, yet the main east-west route lies above it. Design engineers were able to specify upper limits of movements to be compensated by design, and to hold strict control over any future mining.

Structural support piers, weighing as much as 600 tonnes, are equipped with bearings that can be jacked either vertically or horizontally to compensate for any adjacent ground movement.

If mining-induced subsidence is a possibility in any area of proposed urban development, study must be made to establish its extent and magnitude. Special design requirements can then be prepared to achieve satisfactory performance, based on adequate maintenance and inspection. Any suggestion of the existence of old mine workings must be followed up to determine if such workings do exist. In new areas and for new cities, the possible existence of economic minerals, coal, or sand and gravel must be determined and included in detailed planning. Deposits close to the surface may be extracted during site grading. If such minerals are at shallow depths but not available by surface working, then regulations can be drawn up in advance governing their future extraction, so that surface subsidence will be uniform and under strict control. Many fine examples of such regulations are available for guidance, especially those of the Dortmund Board of Mines (West Germany). If future deep mining is contemplated, it will be possible to develop advance subsidence estimates with reasonable certainty.

SINKHOLES

Sinkhole is a self-explanatory geologic term denoting a localized ground surface depression that has been caused by removal of underlying rock or soil. Sinkholes are usually roughly circular to eliptical in plan and as large in diameter as 100 m (330 ft) or more. They usually result from dissolution of underlying limestone, gypsum, or salt. Crater Lake, Saskatchewan, is thought to be due to dissolution of underlying salt beds. This phenomenon is also found in the Canadian west. Culpepper's Dish, England, has a 180-m (600-ft) circumference and a depth of about 45 m (150 ft). A district in the "Cockpit Country" of Jamaica is a terrane of sinkholes in white limestone; these pits are as much as 150 m (500 ft) deep. Sewage of the city of Norwich, England, was conveyed to and discharged on a farm at Whitlingham, soon producing 0.9- to 1.5-m (3- to 5-ft) pits of varying depth. Sewage had washed away sand pockets in the underlying chalk beds. Usually, however, it is the slow dissolution of limestone by acidic water that forms cavities near the surface or hidden at depth by an arch of soil or rock. Most of the dissolution is by mildly acidic rainfall and humic acids in soil. Rainfall on the Kentucky karst fields will dissolve a layer of limestone 1 cm thick in about 66 years. Sinkholes are common throughout the world, in both limestone and dolomite terrane. Spanish karst forms in gypsum strata. Those who viewed either of the 1970 films *Patton* or *Cromwell* may be surprised to know that both were filmed on the karstic Urabasa Plain in Spain. Occasionally, armored vehicles *(Patton)* and horses *(Cromwell)* fell into thinly bridged dissolution cavities.

Many parts of the United States, especially the mid-south, harbor great areas in which sinkholes are significant hazards to development, particularly in the carbonate rock region served by the Tennessee Valley Authority. Sites in all limestone and dolomitic areas should be explored on the assumption that sinkholes may be present. Sinkhole geometry is naturally irregular. Carbonate dissolution channels start out along joint planes but develop quite heterogeneously. There can be no certainty about the absence of sinkholes on building sites in "sinkhole country." The only practicable solution after preliminary site studies is to sink probe borings at the site of every important load-bearing foundation member. This is costly at first sight but a very desirable "insurance factor." One has only to think of the 1969 collapse of part of the roof of a six-year-old department store in Akron, Ohio, which killed one and injured ten, all attributable to settlement of one column that had been founded over an undetected sinkhole. Such an incident shows the need for detailed subsurface investigations at such a building site.

An extensive network of dissolution channels in limestone underlain by Knox Dolomite groundwater was detected during careful exploration for a new paper mill at Calhoun, Tennessee. As much as 10 percent of rock volume had been dissolved. Cavities extended to more than 60 m (200 ft) below the waters of the adjacent Hiwassee River. There is little doubt that structures can be safely and economically founded on karstic terrane, if adequate subsurface investigations are carried out with respect to local geology.

Most sinkholes occur in limestone country; a small percentage are due to dissolution of dolomite and gypsum, with perhaps the least number found in rock salt. Most sinkholes form beneath residual soil matter 12 to 30 m (40 to 100 ft) in thickness. In Missouri, 46 out of a total of 97 catastrophic surface failures recorded since the 1930s were caused in some way by human activity. Williams and Vineyard cite a notable case of highway failure in Pulaski County, Missouri, in February 1966. The sinkhole appeared instantly due to traffic vibrations. Failure is believed to have been brought about by surface drainage discharge from the highway culvert.

More remarkable was the collapse at the site investigation for a bridge over the Piscola Creek in Georgia in 1976. A 15-cm (6-in) boring had been advanced to 27 m (90 ft) without encountering drillrod drops indicative of dissolution cavities, although the site is underlain by dolomitic limestone. The drill crew left the site in perfect shape on a Friday afternoon, but the next day a local farmer noticed that a hole almost a meter in diameter had opened up, and by Sunday the damage was so serious that the existing bridge had to be closed and was partially swallowed by the sinkhole, which expanded as a result of inflow from the creek.

Sinkholes Due to Pumping

Groundwater withdrawal is often an unanticipated cause of not only gradual subsidence of large areas but also the catastrophically sudden appearance of sinkholes at the surface. Naturally, this occurs only in susceptible karst terrane. A well-recorded case happened at Hershey, Pennsylvania, 14.4 km

(9 mi) east of Harrisburg, where a nearby limestone quarry was kept dry by continuous pumping. The rate of pumping was suddenly doubled in May 1949 and over 100 sinkholes appeared in an area of about 2,600 ha (6,400 acres). Groundwater-recharge was then used to restore groundwater levelsto some extent, but this also increased the pumping requirements at the quarry. Legal action followed, but the recharging was not stopped. The quarry company had to resort to a program of grouting to restore satisfactory conditions. This case was brought about by the interdependence of groundwater conditions over a considerable area under conditions of uninhibited underground flow.

Far more serious, since human lives were involved, is the dewatering of the famous gold-bearing reefs of Johannesburg that was begun in 1960 some 65 km (40 mi) to the west of the city. A thick bed of Transvaal Dolomite and dolomitic limestone (up to 1,000 m or 3,300 ft thick) overlies the gold-bearing Witwatersrand strata. Thick, vertical syenite dikes cut across the dolomite and divide the great bed into compartments such as the Oberholzer, which lies at the center of the new mining area. The pumping program dried up traditional springs, then started to lower the entire groundwater surface within the compartment. Between 1962 and 1966 eight sinkholes larger than 50 m (165 ft) in diameter and deeper than 30 m (100 ft) had appeared, together with many smaller ones. In December 1962 a large sinkhole suddenly developed under the crushing plant adjacent to one of the mining shafts; the whole plant disappeared and 29 lives were lost (Fig. 13.14). In August 1964 a similar occurrence took the lives of five people as their home dropped 30 m (100 ft) into another suddenly developed sinkhole. This is the briefest of summaries of a most complex situation, the circumstances of which are such that it is improbable that any similar occurrence will develop.

Karst Country

"Natural" sinkholes have been known from time immemorial, and those due to human underground activity for many decades. Only since World War II have they received widespread attention. Remarkable developments in road building, in dam construction, and in the rapid expansion of urban areas have led to serious sinkhole troubles. Concurrently, a corresponding increase of knowledge in professional circles has brought about an exchange of information and an appreciation that sinkholes do not occur haphazardly but are confined, in general, to certain well-delimited *karst* regions.

The name *karst* derives from the Dinaric Alps of Yugoslavia, adjacent to the Adriatic coast, and long distinguished for its natural sinkholes. *Karst* and *karstic* are now common usage to describe sinkhole regions. Several states, including Alabama, Florida, Georgia, Kentucky, Missouri, and Pennsylvania, have karstic regions. The International Association of Engineering Geologists arranged a special 1975 "symposium-by-correspondence," including contributions from South Africa, West Germany, Belgium, Brazil, Canada, Spain,

FIGURE 13.14
Sinkhole which engulfed the crusher plant at the West Driefontein Mine, South Africa, in December 1962. (*Courtesy Rand Daily Mail, Johannesburg.*)

France, Great Britain, India, Ireland (Eire), Italy, Switzerland, Czechoslovakia, the United States, and Yugoslavia.

Detection and Remedial Works

It will now be clear that site investigations in limestone or dolomite terrane require every precaution to locate and describe sinkholes. Even a quick review of regional geology will usually reveal some indication of whether natural sinkholes are to be expected. Photogeologic interpretation will usually remove most doubt, since existing natural sinkholes will usually be distinguishable. In nonglaciated areas, however, of thick residual soil mantle exploratory drilling and seismic surveying must be supervised with unusual care so as to detect the slightest indication of underground cavities or channels. All four major types of geophysical exploration have been tried with varying degrees of success. South African experience recorded by Dr. A. B. A. Brink has suggested that the *gravimetric method* has the best chance of success. *Electrical resistivity* and *seismic methods* have proved to be of limited use for South African conditions, with *electromagnetic methods* giving some useful results in the detection of dissolution-enlarged joints. Site-specific experimentation is often required to capitalize on combinations of void geometry groundwater, chemistry, and host-rock density.

Ground-penetrating radar (GPR), a proven void-and-low-density pocketdetector used in hazardous waste cleanup, has some utilization for detection of cavities very near the surface. An experimental survey was conducted in an area of hidden caverns in Calaveras Limestone near Cool, California. A soil cover of 3.6 to 18 m (12 to 59 ft) was penetrated for correct detection in almost every case. This approach is a potentially low-cost alternative whenever the need for closely spaced borings can be avoided. Sinkholes will still occur catastrophically, however. The "December Giant," 97 m (325 ft) long, 90 m (300 ft) wide, and 36 m (120 ft) deep, developed without warning in isolated woods of Shelby County, Alabama, on December 2, 1972.

EROSION AND SEDIMENTATION

The complementary phenomena of erosion and sedimentation are geologic processes of great importance in civil engineering work. Such ancient disturbance introduces difficulties into construction work through the presence of buried valleys and by the variation of unconsolidated sediments and glacial deposits. These vital parts of the geological cycle include such widely different processes as today's glaciation in arctic regions and the cutting of solid rock by the wind-driven sand of tropical deserts. The civil engineer is repeatedly faced with problems caused by erosion. On the seacoast, the sea may be destroying valuable developed property; at bends in rivers, the current steadily encroaches upon bankside property or adjacent transportation routes; and in other areas, the combined action of wind and rain erodes fertile topsoil. Similarly, silt deposition behind river control works causing the silting up of slow-floating rivers, and formation of bars in tidal estuaries present special problems for the engineer to solve.

Already discussed in this brief introductory note are such important types of civil engineering as river-training work, coastal protection, and soil conservation. All are constantly exposed to persistent natural forces that tend to alter them in some way. Civil engineering measures to meet erosion or silting must not only face structural problems but must also consider the collateral effects that the resulting structures will have on the future action of these natural forces. In all these considerations, a knowledge of the fundamental geological processes at work is essential; their neglect may result in the failure of structurally sound engineering works.

Erosion by Streamflow

Water flow in rivers and streams is the best recognized of the major causes of erosion. Complementary problems of silt deposition arise whenever the flow of a silt-laden river is interfered with in any way. The correct solution of the many and diverse engineering problems requires an appreciation of the generalgeologic characteristics of riverflow. The life history of rivers is a matter that should be familiar, at least in general outline, to all engineers engaged on river works.

Visualizing a complete profile of a river from source to mouth, one can imagine the river gradually decreasing in slope as it nears its mouth, increasing in volume and cross section on its way. This theoretical profile is not universal, although many of the large rivers of the world, such as the Amazon and the Ganges, conform to it. Rivers rising in mountainous country near the sea may be steeply graded from source to mouth, whereas other rivers, which have their sources in relatively low-lying ground, will have a flat profile throughout and corresponding characteristics. The curved profile is one that a river will constantly tend to produce, and it is therefore a fundamental concept underlying all river studies; changes in gradient are achieved through the erosive power of running water and of the debris that it transports.

The main value of river history is in connection with erosive processes. *Weathering* is the starting point in the train of erosion that eventually finishes as sedimentation in the depths of the sea or of an inland lake. Gravity and wind bring the products of weathering down from the hillsides. After rainstorms, the flows in even the smallest gullies will start the work of transporting this debris. Stream erosion occurs as three separate processes: (1) transportation, (2) corrasion, and (3) dissolution. *Corrasion* denotes the combined mechanical erosive action of water and of the material that it is transporting along the bed of a stream, not unlike the passing of a continually moving file over the stream bottom.

Stream velocity directly influences erosive action, as does the bedload of a stream. Sand and gravel bedload travels on the stream bottom in a constantly irregular movement, bumping against the host rock and intensifying eddies. In this way, erosion gradually takes place, sometimes producing localized potholes containing pebbles or even boulders. As a telling reminder of their mode of formation at Assouan Dam, Egypt, potholes 1.2 to 1.5 m (4 to 5 ft) in diameter and up to 6 m (20 ft) deep were found in the unwatered riverbed area, eroded out of the hardest granite. The plan and profile of a stream affect flow and, in turn, are related to its corrasive action. Increased erosion will take place in the more steeply graded section of a stream bed, especially since the erosive power of a stream depends on some power of the velocity higher than the square. Another feature of a stream profile may be the existence of geologically influenced waterfalls.

Transportation by Streamflow

Rivers at the state of old age have the minimum possible profiles that are capable of carrying the discharge of the river into the sea. This hydraulic criterion is related to the critical velocity for silt and scour, since there is one particular condition at which a regulated stream will neither erode the exposed material in its bed nor deposit the silt that it carries in suspension. This silt-and-scour criterion is a condition of delicate hydraulic balance, and many problems of river-development work involve some alteration of natural features. These works may be seriously complicated by the continued presence in a riverbed of solid material being transported down the river as a product of erosion. This subject has been the object of engineering research since as early as 1851. Quantitative study of sediment transport is now rightly regarded as an important branch of the science of hydraulics, the basis of which is the fundamental geologic process of erosion. An extreme example of the serious impact of this matter is the sediment load of the San Juan River in Colorado. In flood the river is a moving stream of red mud, 75 percent of its volume being composed of silt and red sand. This extreme case suggests that as much as 250 million tonnes of silt annually pass the lower end of the canyon section of the Colorado River. The Mississippi River carries almost twice this amount of silt into its delta each year. Silt transportation is truly a process of magnitude.

In many parts of the world, the regular floodplain deposition of silt is beneficial to fertile areas; the Nile Valley of Egypt is the best-known example. In other parts of the world, flood deposition of silt can be injurious; New Zealand, South Africa, and the Malay peninsula are seriously affected in this way. The town of Kuala Kubu, Malaysia, was completely abandoned in 1929 and reestablished some kilometers away, because the original site was buried several meters deep in river-deposited silt. Sediment sources are usually confined to the individual catchment areas of particular rivers. Quantity and regularity of movement are dependent on the geological characteristics of the catchment basin.

Preliminary studies of river basins, prior to development or remedial work, should include a rough survey of the geology of the drainage area when analyzed with topographic maps. This will indicate the stage of development of the river and the natural causes of erosion (such as disintegration of steep rock slopes, wind action, surface soil erosion), and, hence, the main sediment source. More detailed study is the sampling of river water and its suspended silt to obtain an estimate of the average quantity of material being transported. Several types of sediment samplers are available. Most consist of a closed container into which river water can be allowed to flow at determinate points in the river cross section.

Results given by sampling present only a partial picture, since transportation also involves two other processes. Simply described, the transport process is a rolling along the riverbed of the heaviest particles, an intermittent bumping up and rolling of the next grade of particles, and finally the transportation in suspension of the finest particles of the river's load. The total load is therefore divisible into two parts: The *bedload* is that transported by the first two types of movement, and the *suspended load* is that moved by the third. The suspended load is capable of accurate measurement by means of samples, but no really satisfactory method of measuring bedload has yet been devised for general application.

Silt is removed at the entrance to the All-American Canal on the Colorado River, about 480 km (300 mi) below Hoover Dam. Although Hoover Dam naturally traps practically all the Colorado River sediment as far down as the dam site, scouring action of the next 19 km (12 mi) immediately downstream from the dam lowered the riverbed an average of about 90 cm (3 ft) in the first five months following the closure at Hoover Dam.

Preliminary design of the canal intake works had assumed a total load of 54.5 million kg (120 million lb) of sediment in the maximum daily diversion of 425 cms (15,000 cfs) of water. A desilting installation at the canal intake consists of six settling basins, slowing the flow velocity to less than 0.07 m per second (0.24 fps); the deposited silt is removed by power-operated rotary scrapers. The U.S. Bureau of Reclamation has found that the efficiency of desilting at normal flow is an admirable 80 percent.

A more usual sediment problem faced by civil engineers is the removal of solid particles before water is allowed to pass through the turbines of water power plants. Only in mountainous areas with high-velocity streams does this become a really serious problem. Marlengo and Tel hydroelectric stations on the river Adige in the southern Alps employ lined transport tunnels to move feed water at high velocity without danger of sediment pickup or transport.

Flood debris is a familiar sight in almost all parts of the world. In the semiarid regions debris will accumulate in tributary channels and then will periodically (and sometimes disastrously) be flushed by low-frequency, high-intensity rainfall. In these regions, engineering works have to be constructed to trap flood debris. The Los Angeles County Flood Control District maintains an extensive series of debris-collecting basins which serve to protect lives and valuable property in the metropolitan district of Los Angeles, so close to the foothills of the San Gabriel Mountains. Some of the basins are relatively shallow and dike-retained; others are dammed in narrow valleys, some of notable size and design. Flood debris is periodically removed from the collecting basins by sluicing into perennial streams or by excavation with earth-moving equipment in the semiarid regions. Such debris are generally placed as compacted fill, above the high-water line, and stabilized with vegetation as not to be subject to continued erosion. An average of 18,000 m^3/km^2 (60,000 yd^3/mi^2) of debris (soil, gravel, and boulders) results from each flood; occasional accumulations measure twice this amount. Actual observations of debris flows have disclosed 85 percent solid material, of the consistency of high-slump concrete.

Erosion on Construction Projects

Erosion and sedimentation are problems encountered by almost every civil engineer engaged in construction work. Construction activities almost always disturb natural ground surface, exposing unprotected soil. The first rainstorm begins erosion and sediment-laden runoff carries this load to the nearest drainage ditch and subsequent watercourse. Spoilation of natural streams was once regarded as an inevitable consequence of construction. Such an attitude is no longer acceptable. Every effort must be made to prevent sediment from reaching natural streams, even to the extent of constructing special sedimentation ponds. During urban and suburban construction in Maryland, sediment loads were found to vary from a few thousand to a maximum of 55,000 tonnes/km^2 (140,000 tons/mi^2) per year, whereas the highest comparative concentration from agricultural lands was 390 tonnes/km^2 (1,000 tons/mi^2 per year). In the same region in Maryland 50 percent of building sites were open for eight months, 60 percent for nine, and 25 percent for more than one year. This serious problem has brought many states and major agencies to promulgate strict regulations governing runoff from construction.

FLOODS

Of all the natural catastrophes that beset humanity, major floods are in some ways the most distressing. One can help to fight a fire; earthquake reconstruction can start quickly; a tornado passes and causes great damage, but quiet is quickly restored. But to see a great river burst its banks and floodwaters creeping ever higher and higher with nothing to be done until the peak is passed tries the endurance of even the most stoical among us. Seacoast storms are a part of living by the sea, as is the rare threat of tsunami damage. Seacoast inundation can also bring flooding as high spring tides coincide with violent winds. The eastern side of the great Ganges delta of Bangladesh was devastated on November 12, 1970 by a combination of typhoon winds and high tides, possibly the most serious disaster of its kind in history. Early estimates placed the number of dead at over a million in this heavily populated, flat-lying area, intersected by seemingly innumerable river channels.

Similar combinations of extremes of natural occurrences in other parts of the world can usually be predicted and necessary precautionary measures taken. North Americans now receive hurricane warnings well in advance and evacuations are accepted as essential under extreme conditions. One hundred fifty thousand people were evacuated from the Louisiana delta in 1965 before the onslaught of Hurricane Betsy. This probably saved 50,000 lives, since the whole area became completely inundated.

These unusual combinations are not limited to tropical areas, even though it is in such areas that hurricanes develop. An unenviable storm reputation attends the North Sea. History records the tragedy of 1421, but this inundation was exceeded on February 1, 1953, particularly along the southeast coast of England and in the coastal area of the Netherlands. One hundred and thirty-three Dutch villages were seriously damaged and 1,800 people died. Dikes were breached and seawalls on the coasts of England badly damaged.

Dutch legislation passed within 20 days initiated the great Delta project of flood protection works.

Catchment Areas

Flood protection must consider the whole watershed, and in a framework of time. This frequently neglected factor may result not only in an underestimation of potential debris but of ways in which the debris can be trapped before reaching areas of potential damage. A catchment underlain by a continuous exposure of impermeable rock would yield gauged runoff estimates equivalent to the rainfall less the amount lost by evaporation. Other losses must also be considered; the most important of these are quantities of water absorbed by vegetation, what is temporarily stored as part of groundwater, and the water that may be deflected into adjacent catchment areas because of geologic peculiarities. The calculation of rainfall losses is a vitally important part of water balance studies. Many empirical formulas have been evolved to suggest a general relation between rainfall and runoff for catchment areas of different sizes and under different climatic conditions. Such general formulas are undoubtedly useful in limited application, but they must be interpreted on the basis of local geologic conditions if they are to be effective.

Loss of rainfall by percolation to pervious underground strata is the first of two critical geologic factors. In all but the most exceptional watersheds (such as at Gibraltar), some portion of the rain will percolate to pervious material and eventually join groundwater. Determinations of the minimum possible dry-weather flow and the probable maximum flood at any gauging station are critical hydrologic extrapolations, since it is usually impossible to measure these quantities by actual gauging. Where records at similar sites of constructed works are available, the possibility of modification by variable geological structure must be considered.

A single temperature-related factor further acted to cause catastrophic floods early in 1936 in the eastern part of North America. Unexpectedly early rains fell on catchment areas that were still frozen hard (and that thus constituted catchment areas almost "ideal" if considered from the standpoint of impermeability). The resulting runoffs caused floods that broke all known records and in some cases exceeded the calculated "1,000-year maximum." The importance of this effect can hardly be overemphasized. The figures quoted illustrate how geologic conditions alone can explain some variation in runoff records. By means of similar studies, not only can records be compared, but the accuracy of recorded flows can be evaluated as possible limiting cases.

Causes of Floods

Every flood is affected by unique aspects of Nature. It is possible, however, to delineate the main causes of floods, those which always result from a combination of several geologic and hydrologic elements. Human activity is not entirely responsible for modern floods. The role of catchment area geology should now be clear. But the surface area of the catchment can be affected by climatic factors; frost in the ground, snow on the surface, and the relative position of the groundwater surface. Precipitation, as snow and rainfall, is the vital factor, not only its amount but also the duration of its fall (in the case of rain) and the possible combination of rainfall and springtime snowmelt. We have, however, introduced another disturbing element by cutting down vegetation, notably trees, and so facilitating runoff from the ground surface.

The relative high cost and great extent of so-called "structural flood-protection measures" are reason enough for the detailed study that flood problems are now receiving, quite apart from the possible danger to human lives. Longtime rainfall records are the basis for statistical estimates of the probable "100-year" rainfall. It is impossible, however, to estimate the rate at which rain will fall. That this can be a serious factor was shown by the floods in early May 1973 in the greater Denver area. Up to 11.13 cm (4.38 in) of rain (a new 100-year record) fell over the area. Starting as a light drizzle, the rain fell steadily for a day and saturated the ground. Extensive damage was done to property by the resulting flood.

Flood Protection

From the earliest times of human settlement, efforts have been made to minimize floodwater damage to urban areas adjacent to watercourses. Straightening stream beds liable to flooding may give some relief to adjacent areas but typically increases storm damage downstream. Erection of protective dikes to confine streams within limits is effective if the dikes are carefully designed but liable to cause upstream damage if not. Small flood control dams placed above developed areas are effective where such dams are well built and efficiently operated, but possibly disastrous if not. Major structural flood-protection measures, such as the great dike systems along the banks of the Mississippi, the concrete floodwalls in urban areas, and the large flood retention dams, are only a part of the full answer to flood protection. There has been much argument down the years between those who favored these measures and those who insisted that "little dams" and soil conversation measures in the watershed generally were the real answer. Both sides were right, to varying degrees in different locations.

The importance of "watershed engineering" cannot now be questioned. One of the major pioneer efforts in this direction was the establishment of the Muskingum Watershed Conservancy District in Ohio in 1933, comprising 18 counties and about one-fifth of the land area of the state. Similar work has been done by the Tennessee Valley Authority, and on a much smaller scale by the provincial government for almost all the small rivers of southern Ontario, where forty such authorities now control conservation measures over a total area of nearly 260,000 km^2 (100,000 mi^2).

Both methods of flood control were used in the 1978 reconstruction of the flood-control dam which formed the Charles River basin between Boston and Cambridge. The U.S. Army Corps of Engineers replaced the old earth-fill dam, originally constructed in 1910, with a new concrete structure incorporating a new ship lock and pumping station. This followed the $5 million flood loss in 1955 after Hurricane Diane. Of unusual importance, however, was the associated controlled use of 8,000 ha (20,000 acres) of marsh and swamp area (''wetlands'') in the upper part of the watershed, designed to ''sponge up'' runoff.[5]

Some very large cities have had to make more drastic steps. One of the first of these was the construction of the Red River Floodway between 1961 and 1965 to protect the major Canadian city of Winnipeg, following its devastation by the 1950 flood on the Red River. Studies suggested that the only solution was to provide an alternative flood route around the city. This was excavated on the east side of the river, being 48 km (30 mi) long from its intake south of the city to its junction with the river well downstream. It has a peak capacity of 1,700 cms (60,000 cfs) and has been proven adequate during several bad floods. Excavation of over 76 million m^3 (100 million yd^3) of gray clay was required, about 40 percent of all the excavation on the Panama Canal. The local groundwater condition returned to new equilibrium within a year after construction (Fig. 13.15).

FIGURE 13.15
The Winnipeg Floodway, Manitoba, in use; looking upstream (toward the city), the Red River being on the right. (*Courtesy National Research Council of Canada and F. W. Render.*)

Chicago has long had a serious flood problem, in part due to paving over much of the 930 km^2 (360 mi^2) of its metropolitan area. An average of 50,000 homes were damaged annually. A final solution to this great problem is now nearing completion, one of the largest of all deep-tunneling projects. Two hundred km (125 mi) of large, deep tunnels will serve to store temporarily polluted storm waters, which will later be pumped to surface treatment plants and then discharged safely into Lake Michigan. The Metropolitan Sanitary District of Greater Chicago will eventually spend $5.6 billion to complete The Aqueduct and Reservoir Plan (TARP). This vast, imaginative project, so complex in engineering detail, has benefited, however, from relatively straightforward geology which was explored in great detail before tunneling started.

Greater London, including the ancient city, is now protected from sea flooding by the Thames Barrage, an imaginative dam built across the Thames River well below the main part of London. Large, movable gates can be raised into position to provide protection from any high tidal surges. The "flood risk area" covers 116 km^2 (45 mi^2) and includes about a quarter of a million dwellings, 50 subway stations, and 35 hospitals (Fig. 13.16). Preparatory geologic studies for this landmark project began with a review of the long history of flooding in London, starting with a flood of 1099. The Woolwich Reach site was selected after a long search for optimal site geology and geotechnical properties of foundation materials. The northern abutment was

FIGURE 13.16
The Thames Barrage during construction (in December 1980); the historic river flows from left to right, the city of London being off left. (*Courtesy Rendell, Palmer and Tritton, London.*)

founded on Thanet Sand and the southern end on the well known Chalk of southern England. Site studies showed that near its upper surface the Thanet Sand had been reworked. Some frost shattering of the upper part of the chalk was found at the southern end of the site, but not where the Thanet Sand was in place over the chalk.

Floodplains

Floodplains are a natural geological feature of most river valleys. They are always flat, and their location close to open water makes them desirable building sites, especially for residential construction. Aside from their sometimes soft and irregular deposits of cohesive soils such lands are excellent building areas apart from the overriding danger of flooding. Despite this clear and obvious danger, almost every city which has a river within its boundaries will be found to have flood-prone buildings on the floodplain.

The experience of Rapid City, a well-settled community of about 50,000 in western South Dakota, should be known to every engineering geologist and civil engineer. On the night of June 9, 1972 a phenomenal meeting of a cold front and warm moist air over the Black Hills resulted in an extreme rainfall, as much as 37.5 cm (15 in) falling in some places in six hours. Rapid Creek, from which the City takes its name, reached a flood flow of 1,410 cms (50,000 cfs). Building had been permitted on the creek's floodplain, for the Pactola Reservoir, 22 km (14 mi) upstream from the city, had given a false sense of security. But the worst of the rain fell in the area between the reservoir and the city. Two hundred and thirty-eight people lost their lives during that dreadful night. The area flooded was almost identical with that covered by a flood in 1907. The U.S. Geological Survey had published 1:24,000 maps of the Rapid City area *before* the flood; the coincidence of the flooded area with the alluvium plotted on the maps was significant.

This was one of the very worst such experiences, but it is far from unique. On October 15, 1954 Hurricane Hazel penetrated as far as the Toronto area in southern Ontario. Again through an unusual combination of meteorological circumstances, extreme rainfall took place over a small area, as much as 27 mm (1.06 in) in one hour at one gauge, a total of 177 mm (7 in) in 24 hours at Brampton. This was, however, only slightly more than a previous maximum of 173 mm (6.90 in) in 1915. The small Humber River, with adjacent streams such as Etobicoke Creek, rose to catastrophic heights. Eighty lives were lost; floodplain residences in the borough of Etobicoke were where most of the fatalities occurred. The local municipality soon passed the necessary ordinances prohibiting all building on this floodplain. Now, with passing years, there has been pressure to relax this lifesaving provision.

Herein lies the problem. Floodplains provide some of the superficially most desirable areas for residential development in all riverside cities, and yet they are the most dangerous areas. A 1971 survey identified 5,200 cities and towns of significant size in the United States, part, or even all, of which were located in floodplains. Fifteen states had then adopted regulations regarding the use of floodplains, and 360 municipalities had already adopted floodplain regulations. When it was surveyed some years ago, Prairie du Chien a pleasant town of6,000 on the Mississippi River in Wisconsin, had 48 buildings in the lower part of the town that needed removal, 157 that should have been relocated, and 33 that should have been raised, if they were to be safe from the 100-year flood for that part of the river. Its valley, 2,400 m (1.5 mi) wide, is here confined between high cliffs, with the river normally in two channels, each 300 m (1,000 ft) wide.

The problem is at last receiving some attention in larger centers, if only because of the emphasis now being given to urban planning. The problem is just as serious, in character if not in degree, along smaller streams and in undeveloped areas. Powers have been delegated by the government of Ontario to the small river conservation authorities of the southern part of the province and the authorities have been encouraged and financially assisted to survey their floodplains and to make such maps publicly available. Sample regulations are available as the basis for restriction or prohibition of fill placement in the floodway, prohibition of construction that impinges on the floodway, and control of any proposed alteration to the natural course of their river or its tributary streams. Wherever floodplain geology is studied and appreciated, risk to life and property will be reduced. Much, however, remains to be done. It will always be necessary to monitor floodplains, even those that have been cleared of buildings and are under strict control. They provide ideal parkland, flooding of which can be permitted. Other uses not endangered by flooding have even included small golf courses. But residential construction should most desirably be completely eliminated from all floodplains in North America. As a recent mayor of Rapid City, South Dakota, has said: "It is stupid to sleep in the floodplain."

CONCLUSION

This chapter, dealing as it does with so vast a subject, has been necessarily general in nature. Brevity inevitably renders discussion of the flood problem, for example, somewhat simplistic at first glance. Some might think that the matters discussed have little to do either with civil engineering or geology—but not those readers who have seen a great river in flood or who have been called upon to assist with flood-rehabilitation work. The flow of water down streams and rivers is a vital part of the dynamics of geology and an important aspect of that area of the science often called *geomorphology*. The relevance of geology to runoff has been explained. The geological implications of floodplains, so easily recognized by all familiar with physical geology, are clear-cut. The significance of flood flows to all civil engineering works either on or near rivers is so obvious, once thought is given to it, that it requires no elaboration.

Beyond all such direct interconnections, however, is the fact that civil engineers and geologists are singularly well-informed about the physical aspects of town and country planning and so have a dual interest, not only as engineers but also as citizens, in the impact of geology on all aspects of planning and on the environment. Accordingly, this chapter provides at least an introduction to aspects of the relation of geology to civil engineering that arerapidly assuming increasing importance, but which go some distance beyond the strictly professional subjects normally thought of as civil engineering.

Prediction of the occurrence of geologically influenced natural hazards will remain a high priority in all of the developed nations and will become increasingly so in the developing countries of the world. Preliminary warnings will enable preparations to be made for necessary evacuations. Lives will still be lost and physical damage will still be wrought but means will be increasingly available for reducing such suffering and loss. All civil engineers must be ever aware of the possibility of natural hazards and, with their increasing appreciation of their role as "stewards of the land," give increasing attention to environmental protection to minimize the troubles caused by controllable hazards.

Earthquakes and volcanoes in our own time make it easier to appreciate what some of the great eruptions of the past must have been like. Human imagination, however, is still insufficient to enable one to grasp the specter of the prehistoric eruption of Santorin. The remains of this small island, lying due north of Crete, are still beautiful but its wide internal bay (with a 29-km or 18-m coastline) is probably the crater of the Bronze Age eruption which "blew the head off" this lovely island. The rain of volcanic ash on Crete—100 km (60 m) away—may have been the cause of the disappearance of the Minoan civilization! And the tsunami which resulted must have caused devastation all around the lovely isles of Greece.

REFERENCES

1 F. Tavenas, J. Y. Chagnon, and P. LaRochelle, "The Saint Jean-Vianney Landslide: Observations and Eyewitness Accounts," *Canadian Geotechnical Journal, 8,* p. 463 (1971).

2 J. N. Hutchinson, E. N. Bromhead, and J. F. Lupini, "Additional Observations on the Folkestone Warren Landslides," *Quarterly Journal of Engineering Geology, 13,* p. 1 (1980).

3 W. H. Mathews and K. C. McTaggart, "The Hope Slide, British Columbia," *Proceedings of the Geological Association of Canada, 20,* p. 65 (1969).

4 "Water Buoys Land That Sank as Oil Was Removed," *Engineering News-Record, 163,* p. 26 (November 12, 1959).

5 F. Notardonato and A. F. Doyle, "Corps Takes New Approach to Flood Control," *Civil Engineering, 49,* p. 65 (June 1979).

CHAPTER **14**

GEOLOGY AND THE CIVIL ENGINEER

The main branches of civil engineering construction have been described in the foregoing chapters and the vital importance of geology on every construction site has been demonstrated. There are many other activities in which civil engineers are engaged that also depend upon a proper appreciation of geology for their success and efficiency. For convenience, the principal examples of these related activities have been grouped in this chapter, each of necessity treated quite briefly. Even such introductory statements will illustrate the all-pervading usefulness of geology in the practice of the civil engineer and will correspondingly suggest how, on occasion, civil engineers can assist geologists by allowing them to examine excavations for study of parts of the earth's crust that otherwise would never be seen. This happy reciprocity will be illustrated by some unusual examples in the last section of the chapter which, it is hoped, will leave the reader with a continuing appreciation of geology, not only as a vital aid in his or her professional work but also as a subject of general cultural interest which can add so much, for example, to appreciation of the delights of scenery.

Planning

Wherever the civil engineer is asked to plan, design, and construct, there will be a host of geologic considerations to be either accommodated or mastered. Approached correctly, this process will not only produce an engineered work of stability and continued functionality, but will make suitable use of earth materials "borrowed" from other locations and brought to the site. Environmental impacts within and around each construction site will have been minimized, and judicious layout and design will avoid creation of additional geological problems. All in all, this is indeed a stout order for any engineer to accomplish, especially in consideration of the usual constraints on time and funding.

Such a string of design-related considerations may be termed as "being in harmony with Nature," a concept which has far more significance than the platitude indicates. Forethought given to development is now generally thought of as *planning,* a process of considering the physical and sociological implications of both urban and rural development. Every civil engineer who has appeared before planning and zoning commissions will appreciate that these bodies have extensive control over the outcome of nearly every project that calls for more than reconstruction on small parcels of land. In one form or another, geologic conditions will take a dominant position, along with air quality, noise, esthetics, and traffic.

ELEMENTS OF PLANNING

Most planning is carried out piecemeal, making the best attempt to fit development needs into societal frameworks of hills, valleys, streams, and streets. As a practical concept, planning is the analysis of human needs in the context of existing physical features, to achieve the long-term goals of the area's residents. Planning by definition must be comprehensive. As such, the results of planning are used mainly by civil engineers, architects, and landscape architects. Those who bear the responsible title of "planner" not only should have specialized training in one or more of the related disciplines, but it is clear that they also must have a lively appreciation of all of the related disciplines, especially geology.

Today, many cities and towns have official plans which are constantly under review and revision, mainly as the result of continued testing from the demands of development and redevelopment. Civil engineers naturally encounter this process and would be well advised to take inspiration from the words of pioneer planning advocate Lewis Mumford:

> The task of regional planning...is to make the region ready to sustain the richest types of human culture and the fullest span of human life, offering a home to every type of character and disposition and human mood; creating and preserving objective fields of response to man's deeper subjective needs.[1]

Planning begins with consideration of the wise use of land. *Land* is the summary expression of geologic processes working through the ages. *Land development* must be planned so as to recognize the dynamic geomorphic forces that created the current landforms. Much of the soundness, beauty, and productive power of the landscape is in far more delicate equilibrium with Nature than most citizens image.

"DESIGN WITH NATURE"

An important focal concept for planning arrived in 1969 with the publication of Ian McHarg's widely acclaimed *Design with Nature*. A number of vivid examples were utilized to show how geology can and must be employed in the planning process. McHarg's principles apply mainly to the first-time development of "new towns" and other open spaces, but his technique is applicable to all smaller areas. Transparent basemaps, each made to detail a significant planning factor, are overlaid to indicate where geologic constraints (and other important planning factors) may impact negatively on a specific project.

In commenting on development conditions on Staten Island, New York, McHarg noted:

> It is a special place—its geological history made it so. Silurian schists form the spine of the island, but the great Wisconsin glacier...left its mark, for there lies the evidence of the terminal moraine. There are glacial lakes, ocean beaches, rivers, marshes, forests, old sand dunes, and even satellite islands....The serpentine ridge and the diabase dyke...can only be comprehended in terms of historical geology. The superficial expression of the island is a consequence of Pleistocene glaciation. The climatic processes over time have modified the geological formations, which account for the current physiography, drainage, and distribution of soils.[2]

These are the words of a planner who employs geologic analysis in exactly the way the authors have suggested. Better still would be the words of the great thinker Francis Bacon (1561–1626), who said: *"Nature, to be commanded, must be obeyed."*

PLANNING AND GEOLOGY

The principles of geologic application to planning do not vary, although the particular features of different sites will. Planning begins with identification of the main geologic features that will influence the environmental impact of design, construction, and operation. Some features, such as the nature and distribution of soils, will apply to all sites; others may be encountered only occasionally.

For all sites, a consideration of the *general geologic setting* is the first essential. Type, structure of bedrock, depth of rock, and the nature of overlying soils, all must be considered. There are now geologic maps for all parts of North America. Study of the geologic cross sections that accompany many such maps will identify many potential subsurface conditions that will affect most aspects of planning.

All geologic data will include consideration of *climate* as the second basic requirement. Access to the longest available weather records from the nearest recording station will influence how geologic features may be used or indicate how they will respond to construction and to operation of the facility. Rainfall records will often control *hydrogeology*, which is not normally as visible as soil and rock types, but which can be disastrous if not recognized. The usual requirement for detection of seasonal water level variations is at least one annual period. Annual records may indicate troublesome areas of near-surface groundwater that could affect construction or operation.

The whole of *site hydrogeology* and its interrelationship with soil and rock must be integrated into planning. Evaluation of natural drainage forms the basis for accommodating hydrologic impacts. Running streams must be assessed in terms of extreme high and low flows. Floodplains, in particular, must be delimited and reserved for nonresidential purposes. *Natural drainage* will be found to affect road and street layout. In turn, paving of all such areas will cause 100 percent runoff and a discharge volume far in excess of any natural drainage guides.

Water supply must be considered at a very early stage in planning. If water can be obtained from a river or lake, the location of the necessary intake and treatment plant will be a key element in planning. Groundwater withdrawal will require a detailed study of the entire groundwater regime; the possibility of land subsidence resulting from even moderate pumping must not be overlooked.

Soils must be accurately mapped, sampled, and tested, so that their basic geotechnical character can be judged for use in preliminary zoning or site layout. Any unusual soils such as expansive or sensitive clays, or soil problems which may lead to landslides, can be anticipated.

Such nonrenewable resources as sand and gravel must be located, tested, and prepared for extraction. All-too-many deposits of valuable sand and gravel have now been covered up by urban buildings. Such resources should have been exploited before the buildings were erected in the application of *sequential land use*.

If sand and gravel are not readily available, then *geologic reconnaissance* must be employed to search for sources, or for bedrock suitable for quarrying. Transport costs are now so high as to make worthwhile every planning effort to locate suitable supplies. *Depth to bedrock* should be determined in a general way, accurate to perhaps ±3m. Scattered exploratory borings and groundwater well data will be useful supplements, along with confirmatory seismic refraction surveys. For till plains or deep alluvial valleys where bedrock may be hundreds of meters below the surface, usually only soil studies will be needed. Near-surface bedrock will influence the type of structure that can be used, as well as the installation of underground services. Consider the case of the urban housing development that was "planned" without this factor being considered; all the water mains and sewers had to be subsequently installed in carefully excavated rock trenches, drilled and blasted at an outrageous cost overrun.

Some services may require *tunneling,* especially wastewater and stormwater sewers, both requiring treatment before the disposal of their contents. Bedrock studies should seek to identify major or capable *faults*. Their presence gives warning of the need for specific subsurface investigations of all installations adjacent to the fault trace. *Natural caverns* and presence of any *underground space* will have to be considered in planning, and possibly put to use if appropriate. Only in rural or selected suburban cases will it be possible to create new underground space, but this option should be kept in mind for planning purposes. The advantages for energy conservation by the use of underground space will be a steadily increasing feature of urban development.

Finally, capable faults present an element of *seismic risk*. Many nations have now produced countrywide assessments of seismic risks based on detailed study of records of historic earthquakes. Zones of different seismic-risk intensity are now often defined by responsible authorities. Such initial guides must be supplemented by study of local records of earthquakes throughout recorded history.

Once listed, geological features must be considered individually as the basis for planning. Land cannot be regarded as a great "blank space" on which the planner can exercise creative imagination without regard for physical factors. As soon as specific structures are considered, detailed investigations must be made of every site. And finally, site geology must be assessed as it is revealed by excavation, to ensure that design assumptions are correct and that no unusual or unsuspected features are present.

At the very least, sound geological assessments can assist in minimizing flood damage, in preventing trouble from landslides or rockfalls, in ensuring that valuable sand and gravel deposits are not lost, and in design to withstand seismic forces. In other words, geology can provide significant insurance against damage from natural hazards, some of which will happen sooner or later at almost all sites. Geologic assistance must be clearly presented to planners to be fully effective. Simple, straightforward terms (such as *soil* and *rock*) should be used whenever appropriate. Maps should be prepared in detail; geologic sections are often helpful and appropriate for use in planning. This vital matter of the transfer of information was recognized in a helpful 1979 publication by the Geological Society of America, and in the National Academy of Sciences book, *The Earth and Human Affairs:*

> Knowledge of the earth ought to be available at all governmental levels where environmental concerns are likely to come up for consideration. We must find ways for reliable, complete, and understandable information to be efficiently and promptly provided for persons in government and the public at large.[3]

NATURAL HAZARDS AND PLANNING

Natural hazards are part of the order of Nature and cannot be totally avoided. If recognized in planning, however, much can be done to mitigate their effects geologically, mainly with regard to human lives. This damage is extensive; for 1974 the annual natural hazards property damage in the United States—of which earthquakes and floods are the most serious—was on the order of $10 billion. Engineers are undertaking widespread efforts to gain public recognition of the serious nature of these potentially great losses, so that they can be considered logically and dispassionately, rather than in the panic which naturally follows serious natural disasters.

In California alone, a careful estimate placed the 1970–2000 damage from natural hazards in the state at $55 billion; 98 percent due to earthquakes, loss of mineral resources, landslides, and floods. An estimated $38 billion could be avoided by the application of geologically significant loss-prevention methods at a cost of $6 million, giving a benefit:cost ration of 6.2:1.

These estimates are to be found in a master plan of urban geology for California, which regrettably does not refer to "natural hazards," but "geological hazards." This latter term is so unfortunate as to reduce public appreciation of the fact that it is the constructive use of geology that can minimize the threat of natural hazards. This lamentable term came into frequent use in the 1960s, especially in California. It has appeared in otherwise notable and constructive legislation in Colorado, and in a governor's conference in Utah. The writers urge all readers to use the more appropriate terms, *natural hazards* or *geologic constraints,* to bring attention to the very positive contributions of geology.

SOME EXAMPLES

Some examples of geologic input to planning are in order, if only to illustrate the state of the art and to indicate some of its different approaches.

Washington, D.C.

In January 1968, the Metropolitan Washington Council of Governments (COG) published a 50-page booklet with the stated purpose

> ...to present and analyze, under one cover, certain basic data essential to an understanding of the natural environment of the Washington Metropolitan Area. Within this urban region, approximately 25 square miles per year—or about one percent of the metropolitan area annually—will be converted from farm, swamp, and hillside to subdivision, street and park between now and the year 2000.

Metropolitan Washington includes 610,000 ha (1.5 million acres) of varied geology, including several geological provinces, and 600 million years of geologic time (Fig. 14.1). A concise text explains how geology influences some of 11 key factors in urban development. Typical are slope stability, foundation conditions, excavation characteristics, groundwater features, and suitability for septic tanks. Other maps show generalized industrial mineral distribution ($25 million a year), ground elevations, slope, soil distribution, streams and drainage basins, floodplains, groundwater distribution, woodlands—all with supporting statistics and explanatory notes—and, finally, a general composite map accompanied by notes on the policies that must be considered in the further development of the area. The information was assembled from many sources, notably N. H. Darton's masterly report of 1950, summarizing his studies over a period of 50 years. His work has been followed by a 1961 series of modern geological maps, at 1:24,000.

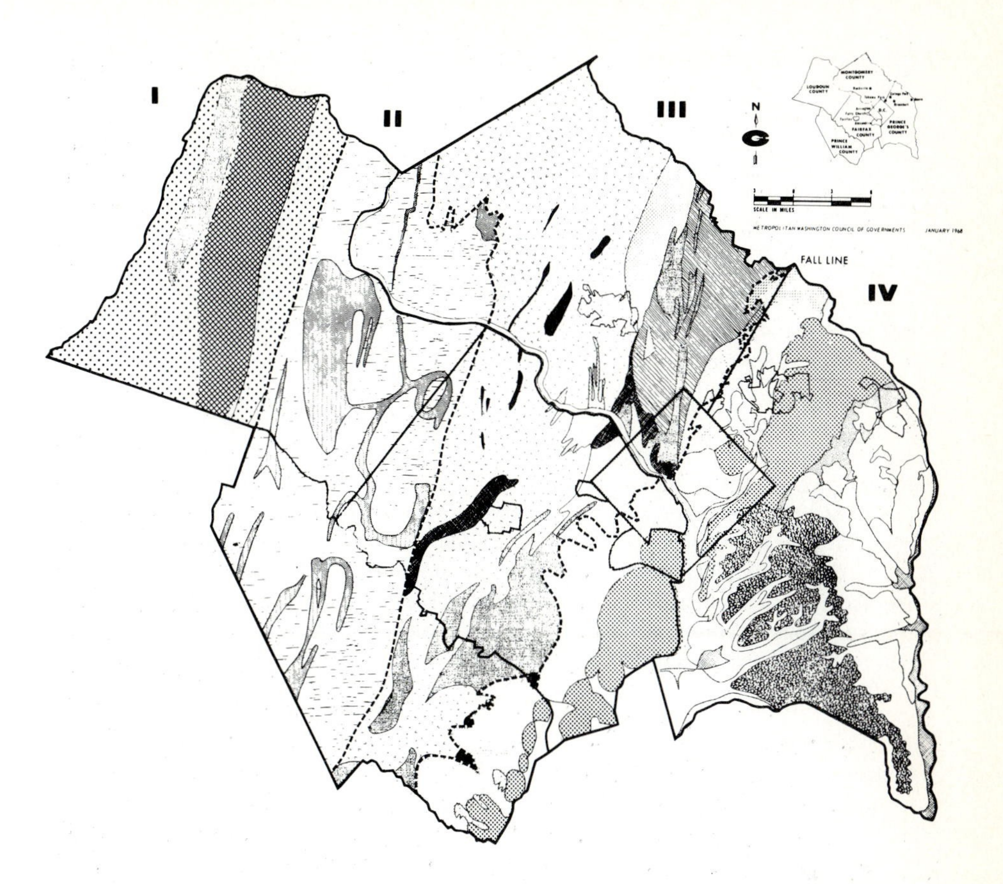

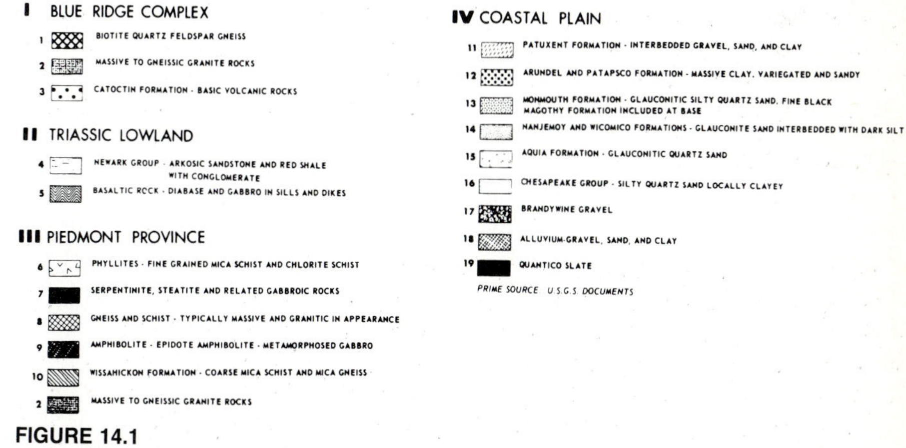

FIGURE 14.1

Generalized geology of the Washington metropolitan area, based on USGS publications and prepared for planning use. (*Courtesy Metropolitan Washington Council of Governments.*)

The San Francisco Bay Area

Two of the pioneer engineering geologic maps of the U.S. Geological Survey were prepared for this area. The 1957 "Areal and Engineering Geology of the Oakland West Quadrangle, California," by Dorothy H. Radbruch, was followed a year later by the "Geology of the San Francisco North Quadrangle, California," by Schlocker, Bonilla, and Radbruch. Both sheets are not only excellent geologic maps, but contain descriptions of the geology and tables of engineering properties of earth materials. Geologic sections complete the remarkable coverage. The manner of mapping is inscribed on the legend of the San Francisco sheet:

> The study in no way pretends to supplant detailed site studies. Rather it tries to supply an accurate background picture of the lithology and of the geologic processes that change or modify the earth materials involved. As much of the quadrangle that does not lie beneath the bay is hidden by streets, buildings and other man-made structures, geologic observations were confined largely to undeveloped lots and to current excavations for utility lines and building foundations and had to be supplemented by data obtained from boreholes and earlier foundation construction. Several thousand logs of boreholes drilled by private firms as well as federal, state, county and municipal agencies provided data that were invaluable in filling out the geologic story of the area.

This excellent summary of what engineering-geologic maps can and should indicate was expanded in the *Program Design,* published in 1971 to describe the operation.

Denver, Colorado

This rapidly growing metropolis is significantly affected by geology, a fact obvious to even the most casual observer. Along with Boulder, 32 km (20 miles) to the northwest, and other satellite communities, Denver faces many common geologic problems. As a result of the 1967 controversy over a proposed hillside subdivision near Boulder, City officials turned to the U.S. Geological Survey to produce a manuscript geologic map of the Boulder quadrangle. Almost coincidentally, the Denver Regional Council of Governments (DRCOG) requested the USGS to conduct a regional planning study of geological conditions as they affect the use of land along the eastern margin of the mountains near Denver. The project was started in 1967, and was the first such project shared by the USGS and a regional planning organization. This cooperative study covered an area of 116,000 ha (287,000 acres), with the first engineering-geologic maps being released in 1971. The objectives of the program are

> ...to explore how geologic information can be made more understandable and useful to more non-geologists, in part by developing methods of representation of map information, text arrangements, and verbal expression.

A summary by Hansen shows that unexpected "by-products" include State (of Colorado) land subdivision regulations requiring "earth testing" and other geotechnical criteria. Lakewood, one of Denver's suburban communities, added a requirement for a site report by an engineering geologist. The City of Boulder employed a full-time geologist on its engineering staff.

Allegheny County, Pennsylvania

This Pittsburgh area county, with an area of 1,886 km^2 (727 mi^2), has no less than 60 percent of its area urbanized. The summary report of extensive USGS investigations in the 1970s for the Appalachian Regional Commission notes that the legacies of two centuries of development include such environmental problems as flooding, landsliding, and mine subsidence. More than 2,000 recent and prehistoric landslide areas were mapped in this one county. Slopes that are susceptible to landsliding underlie about 15 percent of the county. The final results of the studies were published with a series of special maps, three of which relate to landsliding. These have since been supplemented by further maps, which, for adjoining Beaver County, show more than 3,000 landslide areas.

St. Lawrence County, New York

Remarkable rural-area work in St. Lawrence County, New York is an example of what can be done in land-use planning. The county occupies the extreme northwestern corner of central New York State, and is thus bounded by the international border and the St. Lawrence River. It is a large county (7,280 km^2 or 2,800 mi^2) with only four towns, such as Potsdam and Canton, having populations of more than 10,000. The county planning board, with the aid of a small grant from the Federal government, engaged a small support staff. The county also has the good fortune to have an interested and concerned citizen-geologist as a member of the planning board. This resulted in preparation of a general land-use map for the county as a basis for all future planning. Then similar land-use plans with appropriate explanations were prepared for a number of the towns and hamlets in the county. Each was printed in bright colors, sometimes even with a small generalized map of land-use areas, delimiting the major land uses recommended. The town of Fowler is served on one sheet (28 by 41.5 cm or 11 by 16.5 in), with concise and simple explanations for the development of the hamlet printed on the back. Macomb (population 1,000) utilizes a 48-page, pocket-sized, typewritten pamphlet with four pages of photographs, and a land-use map fixed to the back cover.

All cities and towns, along with quiet rural areas, have the same planning needs. A basic appreciation of geology is the starting point for the land-use map. Such a map can provide the framework for all detailed development, with natural hazards being anticipated from geological studies and precautionary measures ensured through good regulations and sound engineering.

GEOLOGIC INFORMATION

Published geologic reports always provide the starting point for all geological contributions to planning. Rarely, however, are geologic maps available at a scale large enough to give a general geological picture of the region for direct use in planning. Few major urban areas in North America do not have some sort of guide to local geology. One remarkable local geologic record is the *Saskatoon Folio,* a fine compilation prepared by a group of volunteer workers. An area of 324,000 ha (800,000 acres) centered on the city is described by 54 maps and diagrams showing bedrock, surficial geology, pedology, climate, and geotechnology, with helpful descriptive notes.

Recognizing the really desperate need for urban geologic information, the Association of Engineering Geologists, through its quarterly *Bulletin,* inaugurated in 1982 "Cities of the World." This is a series of uniformly presented papers describing all geologic, hydrogeologic, geotechnical, seismic, earth resource, underground space, and waste management aspects of major cities. The first cities to be described were Denver, Long Beach, Indianapolis, Albuquerque, Montreal, Dallas, Johannesburg, Boulder and Pretoria.

Earth-Science Information in Land-Use Planning is one of the useful by-products of the San Francisco Bay Region Environment and Resources Planning Study published by the USGS. This 28-page brochure is subtitled *Guidelines for Earth Scientists and Planners,* as it contains helpful suggestions for the best use of geological information in the planning process.

Finally, and most unfortunately, few cities have banks of earth science information which have been revealed through borings and/or excavation at urban sites. This strange void exists despite the fact that building plans must always be registered with civic authorities as a basis for building permits. This is a serious general deficiency in need of urgent correction. The preparation of the first engineering-geologic maps of San Francisco and of Minneapolis-St. Paul, the latter prepared with the aid of about 4,000 test boreholes and 900 water-well logs, are examples that should be widely followed.

THE FUTURE OF PLANNING

The future of urban and regional planning is assured as an essential prerequisite to the best use of land for all human purposes. Planning can only be effectively carried out against a thorough background of geologic knowledge. During construction, a constant watch is necessary to record actual details of geological conditions for feedback into the planning process. It is incumbent upon all geologists to appreciate the uses to which their science can be put in the public service. Special responsibility rests upon civil engineers to make sure that their subsurface information is publicly available for consultation by those interested in local geology and by professionals whose work may be speeded and assisted by such information. In cities where this information has not been prepared, local civil engineering societies should establish a system of assembling information on their own subsurface—before it is too late.

Man-Made Geological Problems

The imprint of our development of the earth is so intense that 10 percent of the land surface of England and Wales is covered with structures such as buildings, railways, roads, and streets. Even in so large a land as the United States, more than 1 percent of the total land area has now been consumed by streets and roads alone. Only when viewed in this broad perspective does the significant change that man has caused in the character of the surface of the earth become clear. Much human activity in changing the face of the earth has, however, been beneficial.

The remarkable achievements of civil engineering must be assessed for their magnitude of effect. Rome's hidden catacombs, originally developed as quarries, have a length of 885 km (550 mi). Those of Paris required the excavation of 10 million m^3 (13 million yd^3), about four times the volume of the Great Pyramid. The great "wet docks" of Liverpool, England, still the greatest continuous area of enclosed harbor space in the world, required the excavation of 4.6 million m^3 (6 million yd^3) of soil, from which 30 million bricks were manufactured. An estimated volume of 30 billion m^3 (40 billion yd^3) of soil and rock was excavated in Great Britain for all purposes prior to the start of World War I. These impressive figures belie the fact that engineered works in many ways affect local geologic conditions. The materials removed are generally inert, but their removal and disposal have created serious problems. Recognition of the generic nature of these problems will assist the civil engineer in their avoidance.

LARGE DAMS

Dam construction inevitably interferes with existing natural conditions. Groundwater conditions will be immediately changed upstream, and possibly downstream; the reservoir area will be flooded and removed, more or less permanently, from human sight and use. Vegetation in the flooded area will be removed before inundation. Animal and insect life will be displaced from the reservoir area. And a new load, the weight of the retained water, will be placed upon the rocks that form the bed of the reservoir. Hoover Dam is 218 m (726.4 ft) high; its impounded Lake Mead is 184 km (115 mi) long, with an area of 66,000 ha (163,000 acres) and an estimated volume of 38 billion m^3 (31 million acre-feet). A surprising discharge of muddy water through the outlet gates began within eight months of infilling. This demonstrated that density currents were active in the lake and pointed the way to advances in sedimentation processes. Silting of the reservoir was allowed for in the design of the dam. Somewhat unexpected was the change in the chemical balance of reservoir water, in which large quantities of salts, particularly gypsum, are going into solution at a rate of more than 1.5 million tonnes per year. As much as 800,000 tonnes of salts, mainly calcium carbonate, are being deposited on the floor of the reservoir annually (Fig. 14.2).

FIGURE 14.2
Lake Mead, created by Hoover Dam on the Colorado River. (*Courtesy U.S. Bureau of Reclamation.*)

DIVERSION OF WATER

Power generation and water supply dams have brought dramatic changes as water is diverted from one watershed to another, usually by means of tunnels. The Big Thompson scheme of the U.S. Bureau of Reclamation diverts water from the Pacific to the Atlantic watersheds in the vicinity of Denver, Colorado. In northern Ontario, the Long Lac and Ogoki projects divert water from the Arctic watershed of the Nelson River into Lake Superior and the St. Lawrence watershed. The water thus gained from Hudson Bay flows through powerhouses at Sault Ste. Marie, Niagara Falls, and Cornwall-Massena before reaching sea level below Montreal. In northern British Columbia, the great Kemano powerhouse takes water from the upper Fraser River and, by means of an impounding dam and long tunnel, brings it to the sea through the Coast Range. And in Australia, the Snowy Mountains project diverts water through the Great Dividing Range so that it may usefully generate power before being made available for irrigation purposes in one of the great fruit-growing districts of the world.

FIGURE 14.3
The Canso Causeway across the Strait of Canso, Nova Scotia, linking Cape Breton Island with the Canadian mainland. (*Courtesy National Film Board, Ottawa.*)

Other engineering projects have resulted in similar diversions. One general example is the causeway embankment that connects the mainland to islands, thereby interfering with tidal currents. Most such projects are small, although some have caused accumulations of moving sand. The rock-fill causeway between Cape Breton Island and the mainland of Nova Scotia in northeastern Canada (Fig. 14.3) was an alternative to a bridge with unusually deep piers. The $23 million embankment was built by direct tipping of rock working outward from land and with a navigation lock to pass shipping through the Strait of Canso, which has typically strong currents. These have now been eliminated with the complete embankment barrier to flow. It will be many years before the possible effects of the causeway upon marine conditions and upon the climate can be evaluated. One totally unexpected, beneficial result has been that the eastern (Atlantic) side remains clear of ice all winter. Depths of 30 m (100 ft) of water are available in this fjordlike strait, providing a major year-round, deepwater port.

The Winnipeg, Manitoba, floodway, designed to carry peak flooding on the Red River, lies below the local groundwater surface. There is usually a normal flow of groundwater into the canal of about 9.100 lpm (2,000 gpm). Major spring floods quickly raise the water level so that water seeps *from* the canal into the surrounding ground. This was anticipated, and possible groundwater contamination has proven to be quite restricted and not of a serious nature.

MORE PROBLEMS WITH WATER

The civil engineer may not always appreciate the degree to which groundwater supplies are now seriously depleted throughout the world. Engineers cannot be blamed for the characteristic overpumping of many water-supply systems; but they do inherit the associated problems and necessary remedial works. All-too-many examples could be cited of the sometimes disastrous results of uncontrolled pumping from natural groundwater reservoirs. Some historic figures are typical of this dilemma:

1 Discharge of 233 wells in the Great Australian Artesian Basin of New South Wales decreased by 39.92 percent between 1914 and 1928, i.e., from about 380 to 230 mld (100 to 60 mgd).

2 Artesian pressure at Woonsocket, South Dakota, in the Great Dakota Artesian Basin in the United States, dropped from 17.5 kg/cm^2 (250 psi) in 1892 to 2.5 kg/cm^2 (35 psi) by 1923.

3 In the Paris syncline, France, the groundwater surface dropped about 75 m (250 ft) in 93 years, prior to 1930.

4 Chicago-area groundwater fell on an average of about 1.2 m (4 ft) per year before 1930.

5 Underground water level in the London Basin, England, has dropped about 45 cm (18 in) per year since about 1878.

Depletion of groundwater is, therefore, not a new threat. One might think that there would have been some amelioration in place by this time. But, as in so many human affairs, things have to get worse before they get better. Today there is a general appreciation of the vital need to protect all groundwater reserves. Regulations limiting withdrawals are in widespread use. And in a growing number of locations, water is now being "recharged" into the subsurface, as workers of today make valiant efforts to undo yesterday's harm. These procedures were in place in 1881 at Northampton in England, where water was fed back into the Lias limestone. Experiments were conducted in the London Basin by the East London Water Company as early as 1890. Today the practice is widespread, often utilizing storm waters or treated sewage effluents for recharging, a combination of solutions to two man-made geological problems.

POLLUTION

Just as water pollution control appears to be making substantial progress, a new source of pollution has appeared in the form of salts used to clear ice from streets in wintertime. When snow and ice along highways melt and run off to drains and so to adjacent waterways, they carry much of this salt as a new pollutant. Much of the salt accumulates on automobiles and is concentrated at car-wash establishments. If this drainage discharges into a small watercourse, serious pollution can result. Every advance in modern convenience seems to bring at least one new problem; all too many of a man-made geologic nature.

Waste management is perhaps *the* environmental concern of our times. It was estimated in 1977 that the United States will soon be handling no less than 3 billion tonnes of solid wastes in one year. Sixty-three percent of this will be from mines (much of it from coal mines), 22 percent from agriculture, 9 percent from industrial sources, 5 percent from municipal sources, and about 1 percent from sewage plants. That 5 percent from municipal sources seems minimal; yet it represents at least 150 million tonnes a year; others estimate the quantity as high as 250 million tonnes. Few North American cities have yet been able to take the logical, but politically unattractive, step of building plants for the conversion of domestic wastes into heat and power; yet by 1985 the tables were turning. Montreal now has such a plant, and other cities are following in an increasing number. Even then, with ash residue, disposal of waste to landfills will have to continue. And here is where geology comes in.

The now-outlawed "garbage dumps" (evidence of communal concern for domestic waste disposal) have led to widespread pollution of groundwater and of surface water. North America's "affluent society" is now served by "sanitary landfills"—generally designed with geologic conditions in mind and so operated that refuse is quickly covered up with soil, giving new reclaimed open land to the municipality. It is not yet universally recognized, however, by landfill operators that snowmelt or rain falling on garbage will seep through it, become contaminated, and, unless geological conditions are very favorable, seep eventually into the groundwater beneath the site. Studies of sites in Iowa suggest that steady-state conditions may be reached in about three years, one of the sites investigated having a contamination plume 2,100 m (7,000 ft) long, 1,350 m (4,500 ft) wide, and in places over 18 m (60 ft) deep.

Correspondingly, regulations are now in operation in most states and provinces strictly limiting the use of land for disposal sites until the necessary *geologic* and topographic surveys have been made, in addition to the usual studies of the social effects of site location. Typical of the new attitude toward this problem is the useful paper, "What Is a Geologist Doing in a Health Department?" There are associated problems. The first is the uses to which completely filled and reclaimed disposal sites shall be put. They are often conveniently located and attractive for development, but if possible they should be reserved for use as parkland or for other recreational purposes so that two further problems do not arise—the settlement of structures upon consolidating wastes and the danger of explosive concentrations of methane gas. These problems will still have to be faced when it is found that old disposal sites underlie areas that have been purchased for active development. If preliminary subsurface investigations are properly conducted and the history of the sites investigated, previous use for garbage disposal can be determined, original ground level can be located, and then foundation design studies can be carried out in the usual way.

Disposal of semisolid wastewater treatment sludge, about 20 percent liquid, is also a significant civic problem. Modern sewage treatment plants produce an effluent that is of such quality that it can be discharged safely into surface waters, but sludge remains in the plant treatment tanks. Burning is an expensive way of disposing of the sludge, but it does contain combustible material. Seaboard cities traditionally dumped sludge in the sea, but this

FIGURE 14.4
Fly ash being used as fill on the Baldock Bypass of the Great North Road (Highway A1) of England; the fill was placed and compacted against bridge abutments, a relative compaction of 90% being achieved. (*Courtesy Central Electricity Generating Board, London.*)

practice has been fought in the courts since the early 1980s. In relatively recent years as much as 150,000 m^3 (174,000 yd^3) of sludge from New York's sewage plants were deposited annually at a specially assigned location at the head of the (underwater) Hudson Canyon in the continental shelf. Possibly the best of all means of disposing of sewage sludge is the method practiced now for many years by Milwaukee in the preparation of its well-known fertilizer, Milorganite. A similar composting facility was placed in operation in 1985 by Philadelphia.

Disposal of industrial wastes is often a difficult and costly matter. Industries now routinely include resource recovery as a part of their overall operations. In some cases, such as the disposal of the sludge obtained from the refining of bauxite for the manufacture of aluminum, there is no solution at present other than to provide storage ponds in which the sludge can settle out from the water with which it is processed. Satisfactory but costly methods of providing this storage have been developed. Not all industrial wastes are hazardous. In the manufacture of steel, for example, large quantities of slag are widely used an engineered fill. As with all natural materials used directly for fill, knowledge of slag properties is essential; for example, certain types of open-hearth slag have un undesirable expansion potential.

Correspondingly, the largest quantities of industrial waste (fly ash from large, coal-fired utility stations) present a massive problem of disposal. This interesting man-made geologic product is being used as stabilizing fill in old mine workings, as a filler for bituminous mixtures, as a constituent of stabilized soil, as an acid-neutralizing medium for water treatment, and for the manufacture of bricks and concrete. Immense quantities of fly ash are produced today (Fig. 14.4). British Railways operates four special trains for transporting fly ash from just three steam power stations in the English Midlands to the Fletton area. Here the ash is being used in a 35-year reclamation program in worked-out clay pits and an old reservoir. Similar remedial works are needed whenever the landscape has been disfigured by massive mining excavation.

California faces unusual problems from hydraulic placer gold mining carried out in years before 1884 when the U.S. Circuit Court of Appeals prohibited uncontrolled deposition of hydraulic-mining debris. These debris piles are still responsible for reduction by sedimentation of the capacity of reservoirs, such as that behind the Combie Dam, by as much as 125 hectare-meters (1,000 acre-feet). Dredged tailings were used in the construction of the great Oroville earth dam. Some success has been achieved in developing ways of manufacturing building materials from this old waste material.

Geology is involved also in the disposal of liquid industrial wastes, since one solution to this problem has been to inject them under pressure through wells into deep-lying strata. There are today possibly as many as 300 deep injection wells in use in North America; some for "untreatable" industrial wastes, although the majority are for the more acceptable disposal of brine from oil fields. Typical injection wells are at least 1,000 m (3,300 ft) deep. The geological implications will at once be obvious. Only after the most searching geological investigation can this method of disposal be followed, the possibility of contamination being one that must be eliminated beyond all doubt.

PROBLEMS WITH METHANE

Methane gas, as encountered in landfills, is particularly insidious since it has no smell; it can form a highly explosive mixture with varying percentages of air (oxygen), while being lethal to human beings at much lower concentrations than would be explosive. Fortunately, there are instrumental means of detecting the gas, and these should always be employed if there is any possibility of methane being present in the subsurface of any site upon which building is proposed. Most reclaimed landfills, however, are rejected as unsuitable for any structures which would enclose space.

The most difficult problems, and occasional tragedies, arise when buildings have been erected over old disposal sites without the possible danger of methane having been recognized. Planners of residential developments must be especially wary, since there is no simple way of getting rid of the methane other than waiting for time to pass. There are all-too-many records of explosions in buildings, usually in basements, because of accumulation of unsuspected methane. In the case of established buildings, there are means of providing vents from the source of the methane, and some details of design can minimize the danger of explosion, but these must be based on thorough exploration of the subsurface, a job that should have been done before construction. Once safety precautions are instituted, they must be vigilantly maintained. One Canadian school, built over an old landfill, had to be vacated as drifting snow had plugged up steel culverts being used as vents from the fill beneath.

A remarkable Florida case of 1969, believed to be a "first," involved 32 explosions which took place in precast, prestressed, reinforced hollow concrete piles that had been driven into place for the new Buckman Bridge near Jacksonville. As was then customary, the cardboard pile forms were left in place. Water got into the piles by wave action before the piles were capped. The court ruled that the plans for the bridge had failed to call for venting of the hollow piles, bacteria in the water had reacted with the cardboard of the inner forms, and the explosions had resulted. And the bacteria? They were found to have come from polluted river water. It may be said that this is not a geologic problem, but if water can be so polluted as to cause such an accident, geologists and civil engineers should feel just as involved as all other citizens, if not more so.

MINE WASTES

Waste from mines has long been deposited in large piles adjacent to mines, often to the disfigurement of the local landscape—a matter at last receiving attention from local authorities and mining companies. The notable instability of these small artificially created mountains has caused disasters such as that of Aberfan, south Wales. This well-known geologic failure came at 9 A.M. on the foggy morning of October 21, 1966. No. 7 tip, above the village of Aberfan, failed and created a flow slide that rushed down the valleyside at a speed of nearly 32 km/h (20 mph), engulfing a farmhouse, destroying the village school, and finally wrecking some homes in the village. In all, 144 people died, of whom 116 were children. A complete public tribunal was appointed by the British government, consisting of a learned judge supported by a prominent geotechnical engineer. Their masterly published report is accompanied by two companion volumes of detailed reports, covering every aspect of the disaster, including the geology of the tip site, which was of critical importance. This tragic disaster has led to the constructive result that spoil dumps are now routinely inspected for threat and remediation in many countries. A 1959 paper entitled "The Stability of Colliery Spoilbanks" was published in *Colliery Engineering* by a mining engineer and geologist of wide experience seven years before the Aberfan disaster, but apparently not taken to heart.

Geologists have been in the forefront of stability investigation of mine tips carried out in other countries since the Aberfan disaster. The U.S. Geological Survey undertook such an inquiry, reported in 1967 in an important paper:

> Following the disaster at Aberfan, Wales, 60 waste banks were studied in the bituminous region of Virginia, West Virginia, and Kentucky. The banks occur as long mounds and ridges both across valleys and on their flanks, as isolated cones and ridges on flat ground, and as flat valley fills. The banks are as much as 800 ft. high and 1 mile long. Some are used as dams to contain settling ponds for wash water from coal processing plants. The banks consist primarily of coal, shale, and sandstone initially of cobble and boulder size, but the material is broken in dumping and quickly slakes to plates and chips as much as inch across. Of an estimated 1500 banks in the 3 states, 600 are on fire or have burned, converting the waste to red dog, a weakly

FIGURE 14.5
Remains of the Buffalo Creek No. 3 waste dam, as seen from the right abutment, near Saunders, West Virginia. (*Courtesy U.S. Geological Survey; photo W. E. Davies.*)

> fused mass of angular blocks and coarse platy ash. Many of the banks are subject to failure from a variety of causes: slippage along shear planes within the bank, saturation of materials by heavy rains, explosion of burning banks, overloading of foundations beneath banks, overtopping and washout of banks that dam valleys, rock and soil slides on valley walls and hillsides breaking through banks, deep gullying, and excavation of toes of banks. In the last 40 years in the 3 states there have been 9 refuse bank failures claiming at least 25 lives; of the 60 banks examined, 38 now show signs of instability.

It is regrettable that one major waste embankment was built and failed after this report was published with the consequent loss of 125 lives. Perhaps the best comment is still that made by one connected with the official inquiry of the Buffalo Creek disaster in West Virginia of February 1972. "We feel that the tragedy was doubly sad because the basic cause was a misdirected and ill-fated attempt to improve the environment by clearing the stream. The dam's failure is an example of the results of hastily conceived legislation—legislation requiring instant conformance without technological assurance that its regulations can be safely met." (Fig. 14.5).

Conserving the Environment

Civil engineers have long held a well-earned appreciation for conservation of the environment. In September 1948, long before the voice of the environmentalist was heard in the land, long before environmental impact statements were a legal requirement, there was held in London a conference entitled *Biology and Civil Engineering*. Its sponsor, the Institution of Civil Engineers, gathered international speakers on such subjects as soil erosion and conservation, the influence of vegetation on floods, and the use of vegetation in stabilizing sand dunes and artificial slopes. All of this was illustrated by what civil engineers had done around the world in promoting the growth of vegetation to enhance the environment of their works. Works described at the meeting were in no way unique, being rather typical of sound civil engineering practice. One of the speakers had this to say:

> If a man cuts his hand, Nature immediately begins to restore the protective layer. Experienced medical aid can help to speed up the process by inducing conditions favourable to Nature. When a man damages the surface of the earth in a comparatively small way, Nature steps in and creates a protective layer, first of grasses, then shrubs, and finally trees. The trouble is that man, in his haste to utilize his handiworks, is unwilling to wait the relatively long time that these unaided natural processes take. It is here that the biologist can be of great assistance to the engineer by creating conditions which remove the element of chance and enable Nature to proceed rapidly from one stage to the next.[4]

This statement is an excellent introduction to the ways in which civil engineers can cooperate with biologists to protect the land on which they work. Engineers and geologists have not been the despoilers of the land, as many shrill voices of today maintain. Sometimes engineers are not able to do all they wish in the way of land restoration when works are complete, but this is almost always due to insufficient funding by owners and financiers rather than designers. And geologists? Geology *is* the study of the environment, and geomorphology the more detailed study of the processes that have given us the landforms of today. In times past, geologists may not always have been sufficiently vocal about the spoiling of the land, but this, too, is changing. Together, geologists and civil engineers, combining their experience and their knowledge of the land, can be in the forefront as "stewards of the land," whose vigilance was never more needed.

RESTORATION OF LAND

Most major civil engineering operations necessitate disturbance on the ground surface, frequently more than removal of topsoil and replacement when the job is done (standard practice on small jobs). Airport construction is a typical case. New York's John F. Kennedy Airport was constructed by pumping 50 million m^3 (65 million yd^3) of sand from Jamaica Bay. The 2,000-ha (4,900-acre) area was vulnerable to almost continual winds. Fortunately, engineers with experience gained at neighboring beaches were able to meet the problem with a mixture of 10 percent poverty grass and 90 percent beach grass.

To speak of biology and civil engineering is to suggest one of the most potent of all scientific combinations in the practice of civil engineering. It is not going too far to say that if Nature can be so utilized as to restore the conditions originally disturbed by engineering works, the best possible remedy will have been achieved and, in some cases, the only practicable solution.

FIGURE 14.6
Bed of Steep Rock Lake, Ontario, being reclaimed after the lake had been drained; seeded vegetation in foreground, naturally seeded vegetation beyond this, and bare soil in background. (*Courtesy National Research Council of Canada; photo W. J. Eden.*)

Steep Rock Iron Mine in western Ontario was operated at a drained lake of 26 km^2 (10 mi^2) area. Exposed glacial-lake clay had to be protected, especially when cutting into slopes and benches for access to the ore body. The completely inert nature of the clay led to failure of the seeding. But within two years Nature had taken a hand and natural seeding took place (Fig. 14.6). Within three years there was a good vegetal cover over almost the entire area of the old lake bed; within five years small bushes were growing.

For most smaller sites, speed of rehabilitation is usually important and topsoil can often be replaced or amended with advice from such agencies as the Soil Conservation Service. Desirable grasses, shrubs, and trees can be incorporated into rehabilitation work. Two years should see tree seedlings well established; within 20 years they will present sturdy windbreaks.

The problem is more serious where dredged material has to be deposited on land. Dredged spoil must be carefully placed and tended to after placement for benefit to result. The U.S. Army Corps of Engineers formed an artificial island in the James River, Virginia, near Windmill Point, 80 km (50 mi) upstream from Norfolk. An area of 6.5 ha (16.25 acres) was enclosed and filled and the soil settled; surplus water was returned to the river. In months, thick wetland vegetation had developed and within two years healthy wildlife was established on the new island.

The U.S. Army Corps of Engineers dredges 249 million m^3 (326 million yd^3) every year, and so has developed a number of successful demonstration projects. Upland habitat has been established on a disposal site at Nott Island

FIGURE 14.7
Sloping face of reconstructed cut reinforced with fabric on British Motorway M4, 20 miles west of Reading. (*Courtesy: Transport and Road Research Laboratory, Department of Transport, Crowthorne.*)

on the Connecticut River. A 9.3-ha (18.25-acre) site has been developed on the waterfront of the harbor of Buffalo, New York, and has already attracted birds (ranging from waterfowl to songbirds) and small mammals, in addition to thick vegetation. Areas such as these compensate in a small way for some of the wetlands that have been lost to earlier uncontrolled development.

SLOPE PROTECTION

Embankments or slopes are essential parts of many finished projects. Even well-designed slopes need protection, especially if the exposed soil is easily erodible and local rainfall is appreciable. Civil engineers have long used vegetation as an immediate supplement to the excavation process. An early method (used with success in California) was the laying of bundles of brush cuttings (for which the old English name of *wattles* is used) at the outer edge of each successive layer of rolled fill, with special compactive effort applied near the brush. The wattles served to give mechanical stability to the slopes until the plantings of western rye grass had time to become well established. Another means of promoting vegetal growth is to cover the slope with one of the recently available "geotechnical fabrics," open-work matting that can be remarkably successful in holding soil in place until grass seeds can germinate and establish root systems, to give the best of all permanent slope surfaces (Fig. 14.7).

STABILIZATION OF SAND DUNES

Wind is also an important agent of erosion; some of the main evidences of its work are drifting sand dunes. Although prominent in desert regions, they often present troublesome local problems on seacoasts with sandy beaches. The travel of coastal sand has wrought great damage in Europe, especially on the shores of the Bay of Biscay. At Liège in France, sand dunes have been moving for two centuries at the rate of 24 m (81 ft) per year, necessitating two successive reconstructions of the local church. Much has been done in the development of plants, grasses, and shrubs that will take root in sand and provide a matting of root material which will effectively bind the surface together.

Large interior areas of dunes—generally of recent origin and sometimes the result of improper land practices—are found in the Great Plains area and especially in New Mexico, Colorado, Texas, Kansas, Wyoming, the Dakotas, and on the south shore of Lake Athabasca, in northern Saskatchewan. Pacific coast dunes are found all the way from Baja California to southern Alaska. Uncontrolled range fires and uncontrolled grazing of cattle are two prime causes of many modern dune areas. Dunes may encroach upon valuable land. "Mechanical" controls, such as fences and brush mats, are only temporary. The use of vegetation is essential for the complete control of drifting sands.

An inland example of such control was applied by the U.S. Army Corps of Engineers at John Martin Dam, across the Arkansas River in southeastern Colorado. Reservoir flooding submerged many miles of the main line of the Santa Fe Railroad. An area of 500 ha (1,200 acres) of sand dunes lay across the most feasible rerouting alternative. The Soil Conservation Service advised the use of stabilizing grasses, to be followed by tree planting after the grass had taken root.

The most extensive sand dunes are found along the low shorelines of the world, bordering the oceans and major lakes. Dutch engineers preserve their fragile coastline with sedge grasses, which are remarkably tolerant of high salinity, open exposure, and lack of humus and groundwater. Notable dune stabilization has reclaimed large areas along the northwestern coast of the United States. Dune stabilization may produce unexpected changes in the dynamic action of natural forces along the coast. Stabilization of dunes along the coast of North Carolina, for example, has been called into question because of accessory changes it appears to have caused. This is a further reminder that one must always be mindful of the great sensitivity of coastal processes.

SAND AND GRAVEL PITS

The sand and gravel business in North America is now so large and so widespread that there are few municipalities that have not felt its impact. Most regulating agencies now have requirements for the complete restoration of worked-out pits as sequential land use. Some Canadian provinces require bonding set aside to

FIGURE 14.8
Golf course with trout ponds in what used to be a sand and gravel pit; a development of Steed and Evans Ltd., Fonthill, Ontario. (*Courtesy Aggregate Producers Association of Ontario.*)

assure final rehabilitation work. The industry has accepted its responsibilities for conservation through orderly operation and careful restoration planning as a regular part of development (Fig. 14.8). Useful guides to the rehabilitation of pits have been published by both government and industry.Typical of thoughtful municipal controls is the 10-year plan developed for the mining of 2.5 million tonnes of high-quality gravel located in the floodplain of Boulder Creek, Colorado. This is close to the city of Boulder and adjacent to the White Rocks area, famous locally as a sensitive nature reserve. Over half the sand and gravel deposits in the vicinity of Boulder have been lost to unplanned development. The operators worked with a number of local groups, including environmentalists, in drawing up the 10-year plan to include seven stages for development and rehabilitation.

Unrestricted gravel-mining operation may have serious effects on the regime of the stream; scour may be increased, with possible danger to bridge structures. Even ephemeral stream channels in areas of low rainfall can be the scene of trouble under unusually heavy floods in excessively mined channels. The 1969 Tujunga Wash flood in California gave vivid evidence of this unusual danger.

MINING AND THE ENVIRONMENT

Desecration of the landscape by strip mining and quarrying has long been the object of protests. Subsurface mining may have more serious effects on surface

FIGURE 14.9
Subsidence pits resulting from the collapse of underground workings at the Monarch Mine near Sheridan, Wyoming; the mine was abandoned in 1914, the photograph taken in 1975. (*Courtesy U.S. Geological Survey; photo F. W. Osterwald.*)

environments than environmentally regulated strip mining. Figure 14.9 is an illustration of the subsidence pits over underground workings at the Monarch mine near Sheridan, Wyoming, abandoned in 1914. Attempts are naturally now being made to limit such harmful results. Almost every state in the Union has regulations requiring some sort of mining restoration. The first attempts to remedy the blight of surface spoil heaps were made voluntarily by industry in the Midwest in the 1920s. The first state regulations were promulgated by West Virginia in 1939, by Illinois in 1943, by Indiana in 1941, and by Pennsylvania in 1945. Thirty-eight states had necessary regulations by 1975. Abandoned mined lands are being slowly rehabilitated, as funded by 20 percent of coal tax monies in the United States.

The problems are many, ranging from control of erosion off spoil banks to the restoration of farmlands uprooted by mining operations. Geology is implicit in the fact that local geologic stratigraphy and structure control the method of mining coal or any ore. Geologic considerations are critical in reclamation planning. A 1976 publication of the U.S. Geological Survey presents a helpful matrix of information on the reclamation requirements of all states. The 24 columns of different requirements and provisions is an indication of the complexity of the reclamation effort now under way. In Great Britain, strip mining of coal is now widespread, all by the National Coal Board. Restoration of land so used has been an integral part of the Board's program from the start, costing about $600 (£230) per acre. This includes such items as cultivation, fertilization, cropping, permanent drainage, and restoration of ditches, fences, and roads. The list is a good indicator of the varied operations necessary to return mined land to its original state.

ENVIRONMENTAL IMPACT STATEMENTS

In a landmark action to conserve the environment, the Congress of the United States of America in 1969 enacted the National Environmental Policy Act. This Act requires that a statement of the probable environmental impact of a project, and alternatives to it, must be prepared for all projects by Federal agencies and for all projects using Federal funds. The Council on Environmental Quality, of the Office of the President, has issued guidelines and oversees preparation of *environmental impact statements,* now numbering in the thousands. The legal requirement has made certain that the environment can never again be neglected in the prosecution of public works. To supplement the guidelines, the U.S. Geological Survey has issued *A Procedure for Evaluating Environmental Impact,* with a "super checklist" matrix for those who prepare statements. Table 14.1 is a mere outline of this large chart; the complete chart contains columns for 98 actions or proposed works which may affect the environment, and almost the same number of items characterizing the existing environment. Naturally, not all categories will apply to any one project, but the idea that almost 8,500 combinations of environment factors exist is clear indication of the complexity of preparing concise and useful documents.

Soil Erosion and Conservation

Sediment in streams, rivers, and estuaries, as retained behind dams, and as found in the great deltas of the world, is not just "silt." Almost all of it was once topsoil, gracing the land served by the river or stream, the soil in which vegetation large and small had its roots. Vegetation cushions the impact of rain reaching the ground. Runoff following heavy rains may flow over surface vegetation with only negligible erosion. It is only when vegetation is lacking or through uncontrolled tree felling, improper cultivation practices, and abuse of the land that erosion can become serious (Fig. 14.10).

Widespread effects of soil erosion have plagued North American and other Western countries in the twentieth century. Soil erosion is no new thing. The deserts of north China, Iran, Iraq, and North Africa were not always deserts; they have become deserts within the span of recorded history because of the gradual but steady depletion of their topsoil. The glory that was Carthage is now a waste of sand, a waste caused by soil exhaustion, crop failures, land abandonment, and the encroachment of the desert, and not merely as a result of warfare. The Adriatic took its name from the ancient port of Adria, located between the Po and the Adige. It was accessible to large vessels in the time of

TABLE 14.1
AN OUTLINE OF AN INFORMATION MATRIX FOR ENVIRONMENTAL IMPACT STATEMENTS

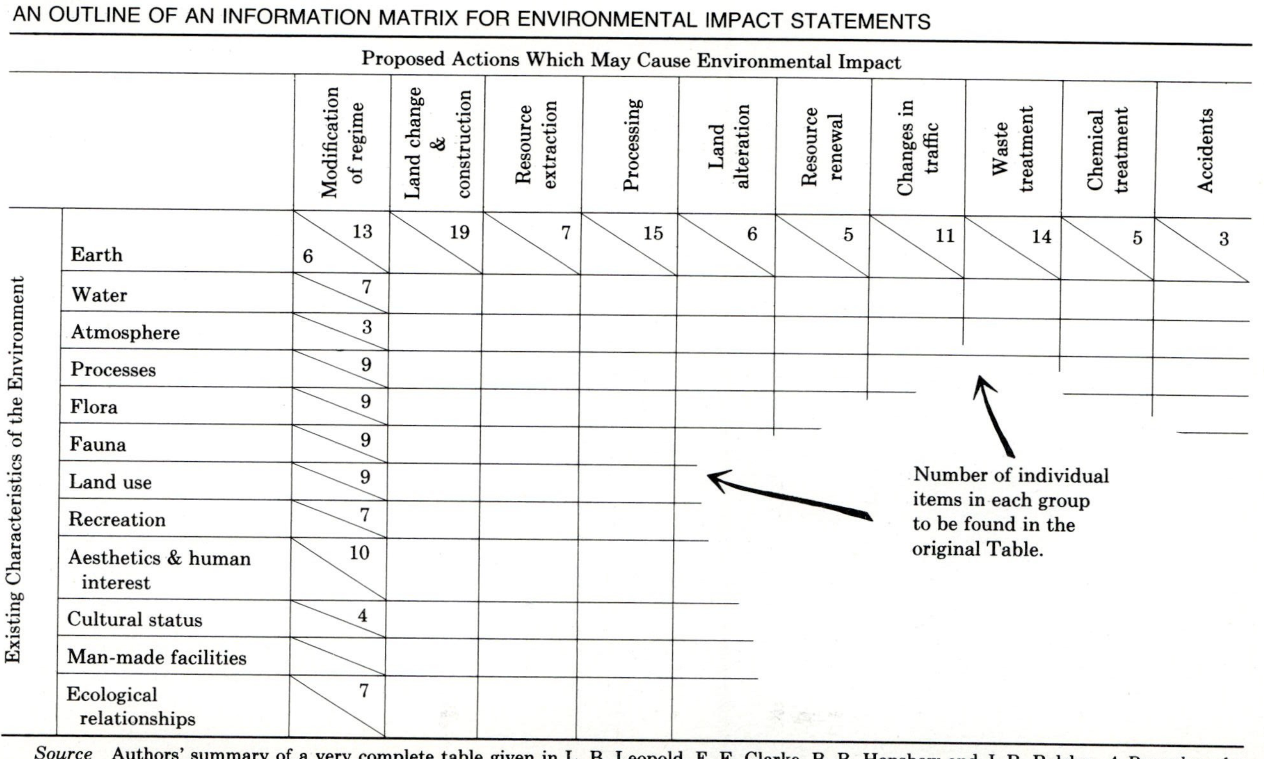

		Proposed Actions Which May Cause Environmental Impact									
		Modification of regime	Land change & construction	Resource extraction	Processing	Land alteration	Resource renewal	Changes in traffic	Waste treatment	Chemical treatment	Accidents
Existing Characteristics of the Environment	Earth	6 13	19	7	15	6	5	11	14	5	3
	Water	7									
	Atmosphere	3									
	Processes	9									
	Flora	9									
	Fauna	9									
	Land use	9									
	Recreation	7									
	Aesthetics & human interest	10									
	Cultural status	4									
	Man-made facilities										
	Ecological relationships	7									

Source Authors' summary of a very complete table given in L. B. Leopold, F. E. Clarke, B. B. Hanshaw and J. R. Balsley, *A Procedure for Evaluating Environmental Impact* (Reference 49.14).

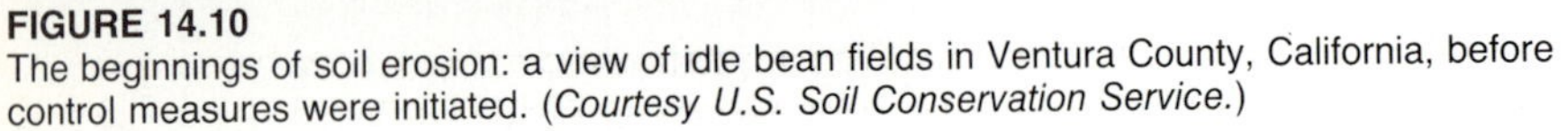

FIGURE 14.10
The beginnings of soil erosion: a view of idle bean fields in Ventura County, California, before control measures were initiated. (*Courtesy U.S. Soil Conservation Service.*)

Caesar Augustus, but is now 22 km (14 mi) inland; the coast has been silted up by soil washed down the two rivers. Vast areas of China's loessal region resemble gigantic battlefields, scarred deeply in the battle with Nature. Nature makes short work of human effort by carving deep channels into any land abandoned in the unequal fight.

All "developed" countries have felt the effects of soil erosion, although these effects have been less profound in older countries with well-established agricultural practices and equitable temperate climates, such as France and Great Britain. In South Africa, soil erosion was described by Field Marshal Smuts as a scourge second only to war. Large areas have been eroded even in so pleasant a land as New Zealand. In South Australia, it has been said that over 2,600 km^2 (1,000 mi^2) of pastoral country have reverted to desert.

In Canada, soil erosion was first noticed at Indian Head, Saskatchewan, as early as 1875. The Magdalen Islands, once a fertile group in the center of the Gulf of St. Lawrence, have had all their topsoil removed; only sand, and not a stick of timber, now remains on the islands. And in the United States, it has been estimated that 50 million acres of good cropland have been ruined almost permanently, with another 50 million acres almost as seriously depleted, and an equivalent total area affected to some degree. Overall figures are so astronomical as to be almost unbelievable; as much as 5.4 billion tonnes of topsoil have been lost in one year in the United States, at a cost in erosion and associated damage to engineering structures of well over $3.8 billion.

SOIL CONSERVATION

Soil erosion can damage engineering structures. This attracts the attention of the civil engineer who constructs or maintains susceptible facilities. Fortunately, most countries maintain active soil conservation agencies whose responsibility is to deal with the problem. Conservation measures necessarily vary widely, but fundamental features are that all designs should aid natural processes, runoff over bare soil must be prevented, and contour plowing is a first step always recommended on farmlands. The same principles can be applied on engineering works; before the end of construction, bare soil has to be temporarily seeded, especially on slopes.

A CONSERVATION ETHIC FOR ENGINEERS

Conserving the environment is one of the great challenges now facing the human race. Much harm has been done in the past, but the problems are now being recognized in time. Civil engineers are not blameless, but they must lay claim, with many others, to being "stewards of the land". They should be in the forefront as supporters of soil-conservation measures, knowing that every bit of soil retained on the ground is soil kept out of reservoirs, rivers, or harbors. They should be the first to encourage sound reforestation projects, knowing that these are one further means of controlling runoff and so of assisting in river and flood-control work. In excavation work, they should see that bare soil is never left to the mercies of the elements but is protected as quickly as possible by the natural cover vegetation so fully provides. In studies of groundwater, they should always remember the natural balance and see to it that their works do not permanently impair this condition. In the construction of dams, they should appreciate that these structures are only a partial answer to river control, an essential part in many cases, but fully effective only in association with what has been called "upstream engineering." In river-training schemes, they must let Nature "have her head" under control, remembering again that there is a natural balance, even of stream regimes, that must not be too seriously disturbed. In bank- and shore-protection work, rather than attempting the impossible, they must fit protection works insofar as is possible into the natural order. In all operations they should be working with Nature. And so, never hesitating to be an imitator of natural processes as they carry out works, they may well recall the crusty words of Thomas Carlyle: "Nature keeps silently a most exact Savings Bank and official register correct to the most evanescent item...and at the end of the account, you will have it all to pay, my friend, there's the rub."

How Civil Engineers Can Aid Geology

Significant contributions to the science of geology have been made by civil engineers. Opportunities for making such contributions abound at almost every civil engineering construction project. Each one will reveal some previously unknown information about the earth through its necessary subsurface investigation and as exposed in its excavation and grading work. It is therefore important that all engineering geologists and civil engineers connected with construction be constantly on the alert for information that will contribute usefully to geologic or archaeologic knowledge. The following few examples are given to encourage a new interest in what may be revealed when excavation work starts on a new project.

This reciprocal link between civil engineering and geology was recognized by William Smith (1769–1839), "Father of British Geology," who practiced and usually signed himself as a civil engineer. He took extensive assignments on canal construction and drainage works. As he traveled on horseback inspecting construction work, he observed the exposed geology. Smith was thus able to prepare the famous "Map of the Strata," first published in 1815. This first real geologic map thus established a link with civil engineering at the start. One of the great pioneers of geology was Charles Lyell of Scotland, who first visited North America in 1841. Landing in Boston, he observed within a few hours:

> Several excavations made for railways,...through mounds of stratified and unstratified gravel and sand, and also through rock, [which] enabled me to recognize the exact resemblance of this part of New England to the less elevated regions of Norway and Sweden....[5]

Throughout travels as far as Pittsburgh, he was on the lookout for excavations. He was guided around Niagara Falls and in the Toronto area by civil engineer Thomas Roy, who had presented a paper to the Geological Society of London in 1837 on the various levels of the Great Lakes; Roy's paper was based on information assembled on railway surveys.

Professor Eaton, of Rensselaer Polytechnic, observed geology along the Erie Canal construction and had published a report in 1824. Lyell met him and discussed his work. During Lyell's second visit (1845–1846), he saw the excavation for the ill-fated Brunswick Canal, between the Altahama and Turtle rivers. This was 14 km (9 mi) long and was even then regarded as a rash undertaking, the only beneficial results being Hamilton Coupar's finds of 45 different species of echinoderm fossils in the excavation! Engineers of today are following a rich tradition when they remember that excavations may yield valuable geologic information or specimens.

EXPLORATORY BORINGS

Every exploratory boring or drill hole provides the civil engineer with what many geologists wish for, a verification of the accuracy of deductions made from surface observations. It is small wonder, then, that boring logs often yield information of value to geologic studies. Typical was the use made by J. T. Hack of records of borings at 14 locations in Chesapeake Bay in connection with projected engineering works. Hack was able to correlate soil descriptions and penetration resistance values with the main types of substrata soil of the bay, varying from gravel and gravelly sand to river mud. The bay is 280 km (175 mi) long and 40 km (25 mi) wide, with Recent and Pleistocene deposits up to 60 m (200 ft) in depth. Hack was able to prepare a profile of the earlier river system, now submerged beneath the bay, thus adding a useful chapter to local geology. In a similar way, study of the bridge foundation borings across Mackinac Straits (between lakes Huron and Michigan) revealed the existence of a deep, preglacial river valley which added significantly to knowledge of the geologic development of the Great Lakes.

Many borings have been taken for erection of offshore drilling structures. Fisk and McClelland analyzed the correlation of many drilling logs in the Gulf of Mexico off the mouth of the Mississippi River and thus added a new chapter to knowledge of the geology of the continental shelf. At a depth of 49 m (164 ft), for example, a sharp increase in soil shear strength was noted in one location, clearly marking the top of the weathered late Pleistocene surface known to underlie the more recent deltaic deposits. Bridge boring samples taken from the bed of the Quinnipiac River, Connecticut (made available by an interested State Highway Department soils engineer), led to carbon-14 dating of organic matter at depths of 9 to 12 m (30 and 40 ft) below sea level. Along with pollen analyses, the results indicated a steady rise of sea level, at New Haven, at an average rate of slightly more than 1.8 mm (0.07 in) per year, a figure that supported estimates. Microfossils separated from 131 soil samples from six borings taken across the Mississippi River provided the basis for significant sedimentary interpretation of the river deposits.

GEOTECHNICAL STUDIES

Geotechnical theory and a well-developed suite of laboratory testing procedures provide an almost untapped potential for geologic utilization. Some useful applications have been made over the years, including the use by Hubbert and Rubey of Karl Terzaghi's concept of pore pressure in their investigation of overthrust faulting. Using simple index parameters of glacial soils from the bed of glacial Lake Agassiz, Rominger and Rutledge established the five basic stratigraphic units of this great inland seabed. Harrison, in a similar way, used the results of consolidation tests to study probable ice pressures exerted upon glacial sediments near the edge of the ice mass. Kenney developed a significant correlation between geological evidence of eustatic sea-level movements and the results of consolidation tests on soils from Boston, Nicolet, Ottawa, and Oslo.

Misiaszek, in an unpublished work, studied natural moisture contents and penetration resistance from about 200 boring logs taken on the campus ofthe University of Illinois to determine the interfaces between separate lodgement till deposits. This significant technique is now widely used to avoid construction difficulties associated with variable till density.

OPEN EXCAVATION

Examination of a relatively small excavation for a railway underpass structure in the Ontario city of Guelph led to the discovery of stratified sand and silt and a weathered diamicton beneath two tills. Exposed also was an underlying lens of organic material which was found to be resting on a paleosol developed on sand and gravel. Pollen content, fossils found in the organic material, and carbon-14 dating showed a date of over 45,000 years, corresponding to the local Port Talbot period. The find was so significant geologically that three continuously cored borings were drilled to supplement the information before the exposure was covered by concrete. As is usually the case with engineering works, such investigations must be carried out on short notice, so geologists must be ready to make their examination of excavations without delay.

Quite different were exposures revealed during the excavation for Units 2 and 3 containment structures at the San Onofre nuclear generating station, north of San Diego County, California—features not detected in careful preliminary borings. Several vertical and near-horizontal conjugate shear planes in three directions were found to be older than 70,000–130,000 years, but were of no consequence to the seismic safety of the power plant.

An accumulation of observations on a number of foundation excavations on Beacon Hill, Boston, enabled Kaye to develop a new explanation for the existence of this notable feature of the city, long described as a drumlin. Kaye suggested that it is really an end moraine formed by a glacial readvance. Excavations revealed a complex of sand, gravel, and clay, instead of the anticipated logement till, with thrusting and other deformations clearly formed within the ice in the terminal zone. Slumping had occurred on deglaciation, forming complex secondary structures on the lower flanks of the ridge. This is a particularly revealing case, based on patient and meticulous observations of relatively small excavations in the center of a great city, leading to a complete revision of a geologic explanation that has long been accepted without question.

Excavation of the Elbe-Seiten Canal in West Germany, which links the river Elbe with the Mittelland Canal, enabled Stephen to examine an almost complete profile through the Pleistocene deposits of lower Saxony. Study of the till fabric disclosed information related to ice movements responsible for formation of lodgement till.

Bulldozer operators can be helpful in this field of joint interest. Finds have

FIGURE 14.11
Dinosaur tracks, first noticed by a bulldozer operator, and then saved for public benefit, now preserved in Dinosaur Park, Hartford, Connecticut. (*Courtesy Connecticut Geological and Natural History Survey.*)

been made in frozen silt in permafrost regions in the Soviet Union; perfectly preserved mammoths have thus been revealed, so well preserved that, in some cases, undigested vegetation has been found in their bodies. In the Kolyma-Berezovka permafrost region of Siberia in 1978 an alert bulldozer operator carefully excavated a perfectly preserved seven- to eight-month-old baby mammoth, a find of the greatest scientific interest. In another instance, a new testing laboratory for the Connecticut State Highway Department was to be erected to the south of Hartford. During clearing of the soil overburden, the bulldozer operator noticed some strange marks when he got down to bedrock and called them to the attention of the engineer and architect on the site, who notified the State Geologist. Within only three weeks after the initial find, the Governor announced that the 3.1-ha (7.5-acre) site would be set aside as Dinosaur State Park. These perfectly preserved dinosaur tracks (Fig. 14.11) caused the highway department to find an alternate laboratory site.

TUNNEL SHAFTS

Large-diameter shafts are necessary as construction access to tunnel workings. As they are sunk, they reveal subsurface conditions previously known only imperfectly from exploratory borings. Close examination of the exposed strata should always engage the attention of geologists, such as in Blank's summary of observations made on the Fordham Gneiss which underlies Long Island,

New York. Observations made during shaft sinking for the Board of Water Supply showed a thick zone of weathered gneiss buried under glacial deposits, in contrast to the relatively fresh surfaces found at outcrops. In addition, a concretionary zone was recognized in Shaft 14A of City Tunnel No. 2 as an ancient laterite. In the sinking of a 6.1-meter- (20-ft)-diameter access shaft for a cable tunnel under the Thames River east of London (between Tilbury and Gravesend), the strata encountered were just as observed in preliminary borings, except that the upper surface of the well-known Chalk (bedrock under London) was found to be badly disintegrated to a depth of 6 m (20 ft). Geological interest centered about the interpretation that the disintegration was the result of a permafrost condition during the last glacial period in the British Isles, as had been previously seen exposed in Czechoslovakia and in Canada. This unusual geologic feature should be regarded as a possibility in glaciated areas.

TUNNELS

Tunnels present such unique opportunities for geologic study that one would think that there would be no need to call for complete geologic mapping of their walls. Unfortunately, many tunnels are excavated and then lined without any record having been made of the revealed geology. Unless there is some geological input to the preparation of official construction records, these opportunities will be forever lost. Later problems associated with tunnel operation or maintenance, however, could well find such information critical to many types of engineering decisions. Boston has benefited from the geologic examination of its major tunnels by Professor Marland Billings, his students and associates, as recorded in a series of valuable papers presented to the Boston Society of Civil Engineers, synthesized in "The Geology of the Boston Basin." There are few cities in which tunnel geology has been so well recorded, except for, perhaps, the outstanding records of the tunnel geology of New York City recorded by Professor Berkey of Columbia University and his student the late Thomas W. Fluhr.

HIGHWAYS

Few civil engineering works have so close a relation with geology as do highways. As it extends over great distances, every major highway project may be expected to encounter a variety of geological conditions. The highway cuts necessary to give uniform superhighway grades provide a multitude of geologic exposures. Such cuts frequently reveal details of local geology that would otherwise remain unknown. Useful guides and even highway geologic maps are now available; some call attention to unique exposures. Construction of the great U.S. interstate highway system has provided significant exposures such as on Interstate 70 some 20 kilometers west of Denver.

FIGURE 14.12
The Hogsback Cut on Interstate Highway 70, outside Denver, Colorado, shown without the varied colors that add so much to this geological exhibit. (*Courtesy Colorado Department of Highways.*)

When the state highway department announced the need for a deep cut through the "Hogsback," there was an immediate outcry from local environmentalists. Progress was almost stalled until a senior member of the U.S.Geological Survey happened to explain to one of the leaders of the opposition what value the cut would have for geology and how it could be developed as an educational feature of great public interest and value. Opposition was thus turned into support. The cut was completed and the highway department finished it off in convenient benches, with good parking facilities and clear, useful signs (Fig. 14.12). Hogsback Cut is now one of the most interesting of scenic viewpoints around Denver—evidence of what engineers and geologists of Denver have been able to do in helping the public appreciate the interesting geology of their area. A special geologic map has been also prepared and published by the Colorado School of Mines.

Another excellent geologically significant road cut example lies on the Red Mountain Expressway at the entrance to Birmingham, Alabama. Yet another is on Interstate 71 at the entrance to Cleveland, Ohio, at Brookside Park. A large excavation was made during highway construction for the purpose of uncovering one of the most prolific deposits of Devonian fossil fishes known in North America, the existence of which was first noticed by then-city engineer of Cleveland, G. Sowers. Salvage of excavation was made possible by a far-sighted provision in the U.S. Federal Highway Act of 1957. This extra excavation gave the Cleveland Museum of Natural Science one of the largest collections of such fossils to be found anywhere today.

A GEOLOGIC CHALLENGE FOR CIVIL ENGINEERS

In this last chapter of a book dealing with the application of geologic principles to enhance the utility and effectiveness of engineered works, it is clear that those who are involved are people who have come to appreciate geology and its importance in civil engineering construction. A growing number are discovering the pleasures which some knowledge of geology can give to the joy of living, in that it makes the beauty of scenery more meaningful, adds interest to so many everyday materials, and opens the mind to great questions of the ages, as Teilhard de Chardin and others have shown so well. It was with sound reason that Charles Kingsley a century ago called geology "the people's science." It should be part of the scientific background of every civil engineer and a guide to all that is seen when sites are opened up.

One of the most moving of all geologic exhibits is to be found near Graniteville, Missouri, at a former 50-ha (124-acre) granite quarry, source of dimension stone for the famous Eads Bridge at St. Louis. Elephant Rocks State Park was given to the state by a geologist in March 1970. A "Braille Trail" was arranged as a series of knotted rope guides for blind visitors, along a mile-long trail through an oak and hickory forest.

In the words of George Perkins Marsh, "Sight is a faculty; seeing is an art," one of many percipient sayings found in his century-old *The Earth as Modified by Human Action*. It is the art of seeing that the authors hope will be further developed by their readers: the art of seeing the significance of every detail of the surface geology of each site they are called upon to investigate and of the subsurface geology, as their borings supplement geologic deductions and gradually reveal a three-dimensional picture of rock, soil, and water. It is the art of seeing the exact interrelationships of the geology of the site and the requirements of the structure under design and of seeing the correlation of what excavation actually reveals with what was expected. It is also seeing the geological significance of all that is displayed by excavation, never forgetting the possible scientific importance of what has been disclosed, and always, without question, accepting as a challenge to their capabilities as civil engineers all the complexities presented by the geology of the site. Then they will be able to accept gladly and adopt as their own those words which John Milton in *Paradise Lost* put into the mouth of the Angel (with the "contracted brow"):

> Accuse not nature, she hath done her part;
> Do thou but thine....

REFERENCES

1 Lewis Mumford, *The Culture of Cities,* Harcourt Brace, New York, p. 336 (1938).

2 I. McHarg, "Ecological Values and Regional Planning," *Civil Engineering, 40,* p. 40 (August 1970).

3 *The Earth and Human Affairs* (Report of Committee on Geological Sciences of the National Academy of Sciences, Ian Campbell, Chairman), Canfield, San Francisco (1972).

4 *Biology and Civil Engineering* (Proceedings of the Conference held at the Institution of Civil Engineers, September 1948), Institution of Civil Engineers, London (1949).

5 C. Lyell, *Travels in North America in the Years 1841–2, with Geological Observations in the United States, Canada, and Nova Scotia,* 2 vols., Wiley and Putnam, New York (1845).

GLOSSARY OF GEOLOGICAL TERMS

This glossary is merely an introduction to geological terminology. Compound names such as **gabbro-syenite** are not generally included, and no mention is made of the names of minerals—whether used to describe minerals only or, as in such cases as **serpentine,** rocks as well. Names in the geological "timetable" are not included. Many words in normal use have been adopted literally for geological use, such as **analogy,** and there are naturally many commonly used words, such as **avalanche,** that would appear in a complete glossary of geology but which are not included here. The glossary is intended to provide the engineer with a guide to the unfamiliar terms that may be encountered in geological reading.

accessory A term applied to minerals occurring in small quantities in a rock, and whose presence or absence does not affect its diagnosis.

acidic A descriptive term applied to those igneous rocks that contain more than 66 percent SiO_2; a term contrasted with **intermediate** and **basic.**

adobe A name applied to clayey and silty deposits found in the desert basins of southwestern North America and in Mexico, where the material is extensively used for making sun-dried brick.

aeolian A term applied to deposits whose constituents have been carried by, and laid down from, the wind.

agglomerate Contemporaneous pyroclastic rocks containing a predominance of rounded or subangular fragments greater than 32 mm in diameter, lying in an ash or tuff matrix and usually localized in volcanic necks. The form of the fragments is in no way determined by the action of running water.

alluvium The deposits made by streams in their channels and over their floodplains and deltas. The materials are usually uncemented and are of many kinds and dimensions.

anticline A fold, generally convex upward, the core of which contains the stratigraphically older rocks.

argillaceous An adjective meaning "clayey" and applied to rocks containing clay.

ash (volcanic) Uncemented pyroclastic debris consisting of fragments mostly under 4 mm in diameter.

basalt A microlithic or porphyritic igneous rock of a lava flow or minor intrusion, having an aphanitic texture as a whole or in the groundmass and composed essentially of plagioclase and pyroxene, with or without interstitial glass.

basic A term applied to igneous rocks having a relatively low percentage of silica; the limit below which they are regarded as basic is about 52 percent.

batholith A huge intrusive body of igneous rock, without a known floor, in contrast to a laccolith, which rests on a floor. Like other intrusive bodies, batholiths become accessible to human observation only as a result of their exposure by erosion.

bedding plane (of rock) Refers to the plane of junction between different beds or layers of sedimentary rock.

bentonite The plastic residue from the devitrification and attendant chemical alteration of glassy igneous material, usually volcanic tuff or ash. It swells greatly with the addition of water and forms a milky suspension.

boulder A detached rock mass, somewhat rounded or otherwise modified by abrasion in transport; 256 mm has been suggested as a convenient lower limit for diameter.

breccia A clastic rock made of coarse angular or subangular fragments of varied or uniform composition (and of either exogenetic or endogenetic origin).

calcareous An adjective applied to rocks containing calcium carbonate.

caliche Irregular and discontinuous layers of hardened, carbonate-cemented soil common to arid regions, and which may create substantial difficulties in site grading and excavation.

chalk A fine-grained, somewhat friable foraminiferal limestone of Cretaceous age occurring widely in Britain and in northwestern Europe.

chert A more or less pure siliceous rock, occurring as independent formations and also as nodules and irregular concretions in formations (generally calcareous) other than the chalk. The fracture is generally conchoidal. When worked, it is called **flint.**

clay A fine-grained, unconsolidated material which has the characteristic property of being plastic when wet and which loses its plasticity and retains its shape upon drying or when heated.

clay mineral One of a group of layered hydrous aluminosilicates commonly formed by weathering. Characterized by small particle size and the ability to absorb large quantities of water and of exchangeable cations.

cleavage (1) The property of minerals whereby they can be readily separated along planes parallel to certain possible crystal faces; (2) the property of rocks, such as slates, that have been subjected to orogenic pressure, whereby they can be split into thin sheets.

cobble A rock fragment between 64 and 256 mm in diameter, rounded or otherwise abraded in the course of aqueous or glacial transport.

conglomerate A cemented clastic rock containing rounded fragments corresponding in their grade sizes to gravel.

current- or cross-bedding A structure of sedimentary rocks, generally arenaceous, in which the planes of deposition lie obliquely to the planes separating the larger units of stratification.

diabase A rock of basaltic composition, consisting essentially of labradorite and pyroxene, and characterized by ophitic texture. Rocks containing significant amounts of olivine are olivine diabases. In Great Britain basaltic rocks with ophitic texture are called **dolerites,** and the term **diabase** is restricted to altered dolerites.

dike An injected tabular intrusion cutting across the bedding or other parallel structures of the invaded formation.

diorite A phanerocrystalline igneous rock composed of plagioclase and mafic minerals such as hornblende, biotite, and augite; hornblende is especially characteristic.

dip The angle of inclination of the plane of stratification with the horizontal plane.

discontinuity Blanket name for all macroscopic rock fractures; includes bedding planes, joints, shear planes, and faults.

dolerite American authors use the term **diabase** in a sense synonymous with dolerite. In Great Britain, dolerite is the name used for an igneous rock occurring as minor intrusions, consisting essentially of plagioclase and pyroxene, and distinguished from basalt by its coarser grain and the absence of glass.

dolomite A carbonate rock consisting predominantly of the mineral dolomite (such as dolomitic limestone and magnesian limestone), in which dolomite and calcite are present, the latter predominating; it is also called **dolostone,** which is the currently preferred term.

exfoliation The "peeling" of a rock surface in sheets due to changes of temperature or to other causes.

fault A fracture of the earth's crust accompanied by displacement of one side of the fracture (in a direction parallel to the fracture), resulting in an abrupt break in the continuity of beds or strata.

ferruginous An adjective applied to rocks with a prominent iron content.

flint A more or less pure siliceous rock composed mainly of granular chalcedony together with a small proportion of opaline silica and occurring in nodules, irregular concretions, layers, and veinlike masses. The fracture is conchoidal. Flint is **chert** worked by man.

foliation A structure represented more characteristically in schists and due to the parallel disposition of layers or lines of one or more of the conspicuous minerals of the rock; the parallelism is not a direct consequence of stratification.

foot wall The wall rock beneath an inclined vein or fault.

formation A term applied stratigraphically to a set of strata possessing a common suite of lithological characteristics.

gneiss A foliated or banded phanerocrystalline rock (generally, but not necessarily, felspathic and of granitic or dioritic composition) in which granular minerals, or lenticles and bands in which they predominate, alternate with schistose minerals. The foliation of gneiss is more "open," irregular, or discontinuous than that of schist.

granite A phanerocrystalline rock consisting essentially of quartz and alkali feldspars with any of the following: biotite, muscovite, and amphiboles and pyroxenes.

gravel An unconsolidated accumulation of rounded rock fragments predominantly larger than sand grains. Depending on the coarseness of the fragments, the material may be called **pebble gravel, cobble gravel,** or **boulder gravel.**

groundwater A term that should be applied only to the water in the zone of rock fracture below the earth's surface, in which interstitial water occurs, i.e., in the zone of saturation.

hanging wall The wall rock above an inclined vein or fault.

hardpan A term to be avoided, in view of its wide and essentially popular local use for a wide range of materials.

igneous (rocks) A general term for those rocks which have been formed or crystallized from molten magmas, the source of which has been the earth's interior.

inclusion A general term for foreign bodies (gas, liquid, glass, or mineral) enclosed by minerals; also extended in its English usage to connote **enclosures** of rocks and minerals within igneous rocks.

inlier An area or group of rocks surrounded by rocks of younger age.

intermediate A term applied to rocks intermediate in silica content between the basic and the acidic groups.

intrusive A term applied to igneous rocks that have cooled and solidified from magma under the cover of the rock masses of the outer shell of the earth; they therefore become accessible to view only after they have been exposed by erosion of the overlying rock.

joints (rock) Rock fractures on which there has been no appreciable displacement parallel with the walls. They may sometimes be characterized, according to their relation with the rock bedding, as **strike joints** and **dip joints.**

laccolith A dome-shaped intrusion with both floor and roof concordant with the bedding planes of the invaded formation; the roof is arched upward as a result of the intrusion.

laterite A residual deposit, often concretionary, formed as a result of the decomposition of rocks by tropical weathering and groundwaters and consisting essentially of hydrated oxides of aluminum and iron.

limestone A general term for bedded rocks of exogenetic origin, consisting predominantly of calcium carbonate.

löess A widespread deposit of aeolian silt, a buff-colored, porous, but coherent deposit traversed by a network of generations of grass roots.

magma A comprehensive term for the molten fluids generated within the earth from which igneous rocks are considered to have been derived by crystallization or other processes of consolidation.

marble A general term for any calcareous or other rock of similar hardness that can be polished for decorative purposes (petrologically restricted to granular crystalline limestones); it is derived from the metamorphism of limestone.

metamorphic A term used for rocks derived from preexisting rocks by mineralogical, chemical and structural alterations due to endogenic processes; the alteration is sufficiently complete throughout the body of the rock to produce a well-defined new type.

mica schist A schist composed essentially of micas and quartz; the foliation is mainly due to the parallel disposition of the mica flakes.

mineral A naturally occurring inorganic element or compound having an orderly internal structure and characteristic chemical and physical properties.

muskeg A Canadian term of Cree Indian origin, meaning a moss-covered muck or peat bog. In the Far North all swamps are called muskegs.

oölite A rock made up of spheroidal or ellipsoidal grains formed by the deposition of successive coats of calcium carbonate around a nucleus.

outcrops (rock) Those places where the underlying bedrock comes to the surface of the ground and is exposed to view.

outlier An isolated mass or detached remnant of younger rocks, or of rocks overthrust upon others, separated by erosion from the main mass to which they belong and now surrounded, areally, by older or at least underlying rocks; opposite of **inlier.**

pebble A rock fragment between 4 and 64 mm in diameter, which has been rounded or otherwise abraded by the action of water, wind, or glacial ice.

plutonic A general term applied to major intrusions and to the rocks of which they are composed, suggestive of the depths at which they were formed, in contradistinction to most minor intrusions and all volcanic rocks.

quartzite A granulose metamorphic rock, representing a recrystallized sandstone, consisting predominantly of quartz. The name is also used for sandstones and grits cemented by silica that has grown in optical continuity around each fragment.

rock flour A general term for finely comminuted rock material corresponding in grade to mud but formed by the grinding action of glaciers and ice sheets and therefore composed largely of unweathered mineral particles.

sand Fine, granular material derived from either the natural weathering or the artificial crushing of rocks, generally limited in diameter to between 2 and 0.05 mm.

sandstone A cemented or otherwise compacted detrital sediment composed predominantly of quartz grains, the grades of the latter being those of sand. Varieties such as **argillaceous, calcareous,** and **ferruginous** are recognized according to the nature of the cementing material. Corresponding rocks of coarser grades are sometimes called **grits.**

schist A general term for foliated metamorphic rocks, the structure of which is controlled by the prevalence of lamellar minerals such as micas—which have normally a flaky habit—or of stressed minerals crystallized in elongated forms rather than in the granular forms generally assumed.

sedimentary A general term for loose and cemented sediments of detrital origin, generally extended to include all exogenetic rocks.

shale A laminated sediment, in which the constituent particles are predominantly of the clay type. The characteristic cleavage is that of bedding and such other secondary cleavage or fissibility as is approximately parallel to bedding.

shingle Coarse beach gravel, relatively free from fine material and commonly of loose texture.

siliceous Of or pertaining to silica; containing silica, or partaking of its nature. Containing abundant quartz.

sill A tabular sheet of igneous rock injected along the bedding planes of sedimentary or volcanic formations.

silt An unconsolidated clastic sediment, most of the particles of which are between 0.05 and 0.002 mm in diameter.

slate A general term for compact aphanitic rocks, formed from shales, mudstones, or volcanic ashes and having the property of easy fissility along planes independent of the original bedding.

slickensides Polished and grooved surfaces on rock faces that have been subjected to faulting.

stratum A layer that is separable along bedding planes from layers above and below; the separation arises from a break in deposition or a change in the character of the material deposited.

strike (of rocks) The direction of the line of intersection of the plane of stratification with the horizontal plane.

structure (rock) A term reserved for the larger features of rocks; e.g., a layered or laminated structure generally indicates sedimentary origin.

syncline A fold, generally concave upward, the core of which contains stratigraphically younger rocks.

talus A term used for the accumulation of fine, coarse, or mixed fragments and particles, fallen at or near the base of cliffs.

texture (rock) The appearance, megascopic or microscopic, seen on a smooth surface of a homogeneous rock or mineral aggregate due to the degree of crystallization, the size of the crystals, and the shapes and interrelations of the crystals or other constituents.

till The unstratified or little-stratified deposits of glaciers. This term should always be modified by **lodgement (basal)** for hard, cohesive material, or by **ablation,** for loose, cohesionless soil, as appropriate.

tuff A rock formed of compacted pyroclastic fragments, some of which can generally be distinguished as such by the naked eye. The indurated equivalent of volcanic ash or dust (cf. **agglomerate**).

unconformity A break in the geologic sequence where there is a surface of erosion or nondeposition separating two groups of strata.

vein An irregular sinuous igneous injection or a tabular body of rock formed by deposition from solutions rich in water or other volatile substances.

weathering The destructive alteration and decay of rocks by exogenetic processes acting at the surface and down to the depth to which atmospheric oxygen can penetrate.

There are now a considerable number of useful **dictionaries of geology** to which reference may be made to supplement this short glossary based on the reading and work of the authors. The varied uses of the words **soil, compaction,** and **consolidation** are discussed in the appropriate sections of the text. The American Geological Institute has placed all geologists and engineers in its debt by publishing the 788-page **Glossary of Geology,** now in its third edition (1987). This may be rightly regarded as the final authority for all geological definitions.

SOME USEFUL JOURNALS

This is a personal list of journals known to and used by the authors. There are doubtless other equally useful journals serving the field of geology and engineering, but it is hoped that this short list will prove of assistance at least as an introduction to current literature in the field. To this end, the following signs are provided to give some indication of the journals' special coverage:

- ● Case histories are a regular feature.
- ○ Case histories are occasionally featured.
- ■ Current news items of relevant interest are a regular feature.

GEOLOGICAL JOURNALS

○ **American Journal of Science,** monthly, from Kline Geological Laboratory, Yale University, New Haven, Conn., 06520.

● **Bulletin of the Association of Engineering Geologists,** quarterly, from Allen Press, 1401 New Hampshire Street, Lawrence, Ks., 66044. Note that the first few issues of this excellent publication were called **Engineering Geology,** which causes confusion in referencing; see below.

○ **Bulletin of The Geological Society of America,** monthly, from G.S.A., 3300 Penrose Place, Boulder, Colo., 80302.

● **Bulletin of the International Association of Engineering Geology,** from I.A.E.G., c/o Laboratoire Central des Ponts et Chaussées, 58 Blvd. Lefebvre, 74732 Pais, Cedex 15, France.

○ **Canadian Journal of Earth Sciences,** monthly, from the National Research Council of Canada, Ottawa, K1A OR6.

- • **Case History Series,** of Engineering Geology Division of The Geological Society of America, 3300 Penrose Place, Boulder, Colo., 80301.
- • **Engineering Geology,** published quarterly by Elsevier Publishing Co., P.O. Box 211, Amsterdam, the Netherlands.
- ○ **Geology,** monthly, from The Geological Society of America, 3300 Penrose Place, Boulder, Colo., 80301.
- ○ **Geological Magazine,** eight times a year, from Cambridge University Press, 32 East 57th Street, New York, N.Y., 10022.
- ■ **Geotimes,** monthly, available from American Geological Institute, 5205 Leesburg Pike, Falls Church, Va., 22041.
- ○ **Journal of Geology,** eight times a year, University of Chicago Press, 5801 Ellis Avenue, Chicago, Ill., 60637.
- • **Quarterly Journal of Engineering Geology** (of the Geological Society of London), obtainable from Blackwell Scientific Publications, Ltd., Osney Mead, Oxford OX7 OEC, England.

GEOTECHNICAL JOURNALS

- • **Canadian Geotechnical Journal,** quarterly, from the National Research Council of Canada, Ottawa, KIA OR6, Canada.
- ○ **Geotechnique,** quarterly, from the British Geotechnical Society, c/o the Institution of Civil Engineers, Great George Street, London XWIP 3AA, England.
- ■ **Ground Engineering,** bimonthly, from Foundation Publications, Ltd., 33 Short Croft, Doddinghurst, Essex, England.
- ○ **International Journal of Rock Mechanics and Mining Science,** (including Geomechanics Abstract), monthly, from Pergamon Press, Head Hill Hall, Oxford OX3 OBW, England; or Maxwell House, Fairview Park, Elmsford, N.Y., 10523.
- ○ **Journal of the Geotechnical Engineering Division,** monthly, from the American Society of Civil Engineers, 345 East 47th Street, New York City, N.Y., 10017.
- ○ **Rock Mechanics,** quarterly, from Springer-Verlag, P.O.B. 367, A-10011 Wien, Austria.
- ■ **Tunnels and Tunnelling,** bimonthly, from Morgan-Grampian, Ltd., 30 Calderwood Street, London SE18 6QH, England.

CIVIL ENGINEERING JOURNALS

- ○ **Civil Engineering,** monthly, from the American Society of Civil Engineers, 345 East 47th Street, New York City, N.Y., 10017.
- ○ **Civil Engineering** (formerly **C.E. and Public Works Review),** monthly, from Calderwood Street, London, SE18 6QH, England.
- ■ **Engineering News-Record,** weekly, from McGraw-Hill, 1221 Avenue of the Americas, New York, N.Y., 10020.
- ○ **New Civil Engineer,** weekly, and **New Civil Engineer International,** monthly, both from Institution of Civil Engineers, Great George Street, London SWIP 3AA, England.

OTHER SOURCES

There are many other sources of useful information in this field, published either regularly, irregularly, or only occasionally with other material. Typical of the second category are the renowned publications of the U.S. Geological Survey, the respective state geological survey, and other national and local surveys (such as those of Canada and the Canadian provinces). Typical of the third are the publications of the Transportation Research Board (formerly the Highway Research Board), 2102 Constitution Avenue, Washington, D.C., which sometimes contain useful information on engineering geology.

ABSTRACT JOURNALS

Engineering Geology Abstracts, quarterly, from American Geological Institue, 4220 King Street, Alexandria, Va., 22302.

Geotechnical Abstracts, (of the International Society of Soil Mechanics and Foundation Engineering), published by the German Society of S.M. and F.E. (Deutsche Gesellschaft für Erd-und Grundbau), 35a Kronprinzenstr., 43 Essen, W. Germany.

Soil Mechanics Information Service, from Geodex International, Inc., P.O. Box 385, Glen Ellen, Calif., 95442.

GUIDE TO GEOLOGICAL LITERATURE

Sources of Information for the Literature of Geology, by J. W. Mackay, published by the Geological Society of London, Burlington House, London WIV OJU, England, 2nd ed., 1974; an invaluable guide to the vast literature of geology, it is in effect a "bibliography of bibliographies of geology."

AND WHEN ALL ELSE FAILS...

Ulrich's International Periodicals Directory, 15th ed., 1973–1974, from R. R. Bowker Co. (a Xerox company), New York and London, will be found in most libraries; it will guide the inquirer to full details of publication of any periodical the title of which is known.

NAME INDEX

SUBJECT INDEX